Networks and Devices
Using Planar
Transmission Lines

Networks and Devices Using Planar Transmission Lines

Franco Di Paolo, Ph.D.

CRC Press
Taylor & Francis Group
Boca Raton London New York

CRC Press is an imprint of the
Taylor & Francis Group, an **informa** business

CRC Press
Taylor & Francis Group
6000 Broken Sound Parkway NW, Suite 300
Boca Raton, FL 33487-2742

First issued in paperback 2019

ISBN-13: 978-0-8493-1835-1 (hbk)
ISBN-13: 978-0-367-39841-5 (pbk)

Library of Congress Card Number 00-008424

Library of Congress Cataloging-in-Publication Data

Di Paolo, Franco.
    Networks and devices using planar transmission lines / Franco di Paolo.
        p. cm.
    Includes bibliographical references and index.
    ISBN 0-8493-1835-1 (alk. paper)
      1. Strip transmission lines.   2. Electric lines—Carrier transmission—Mathematics.
3. Telecommunication—Mathematics.   4. Electronic apparatus and
appliances. I. Title.
    TK7872.T74 P36 2000
    621.381′32—dc21
                                           00-008424
                                            CIP

Visit the Taylor & Francis Web site at
http://www.taylorandfrancis.com

and the CRC Press Web site at
http://www.crcpress.com

ABSTRACT

This book has one objective: to join in one text all the practical information and physical principles that permit a planar transmission line device to work properly. The eight appendices have been written with the aim of helping the reader review the theoretical concepts in the 11 chapters.

This book is intended for microwave engineers studying the design of microwave and radio frequency planar transmission line passive devices in industry, as well as for students in microwave and RF disciplines. More than 500 up-to-date references make this book a collection of the most recent studies on planar transmission line devices, a characteristic that also makes this book attractive to researchers.

Chapters are dedicated to the analysis of planar transmission lines and their related devices, i.e., directional couplers, directional filters, phase shifters, circulators, and isolators.

A special feature is a complete discussion of ferrimagnetic devices, such as phase shifters, isolators, and circulators, with three appendices completely dedicated to the theoretical aspect of ferrimagnetism. Also provided are more than 490 figures to simply and illustrate the input–output transfer functions of a particular device, information that is otherwise difficult to find.

This book is highly recommended for graduate students in RF and microwave engineering, as well as professional designers.

The Author

Franco Di Paolo was born in Rome, Italy, in 1958. He received a doctorate in Electronic Engineering in 1984 from the Università degli studi di Roma, "La Sapienza."

His first job was with Ericsson-Rome, designing wide band RF and microwave circuits for RX and TX optical networks. He has been a senior research engineer at Elettronica-Rome Microwave Labs. Currently he is chief research engineer at Telit, Microwave Satellite Communication Division, in Rome.

Dr. Di Paolo is author of other technical publications and is an IEEE member. He is an associate of the Microwave Theory and Techniques Society, the Ultrasonics, Ferroelectrics and Frequency Control Society, and the Circuit and Systems Society.

CONTENTS

PREFACE

By "planar transmission line" we mean a transmission line whose conductors are on planes. Examples are microstrips and slotlines. By "device" we mean a component that is capable of having some electrical property in addition to the obvious "RF" connecting characteristic. Examples are directional couplers and phase shifters. All the devices we will study are made of planar transmission lines. By "network" we mean a set of complicated "RF" transmission lines without any additional performance beyond interconnecting capability.

While the author has made an effort to explain in a simple way all the theoretical concepts involved in this text, a graduate-level knowledge of electromagnetism and related scientific areas, such as mathematical analysis and physics, is required.

Chapter 1 introduces all the concepts of the general theory of transmission lines. Chapter 2 is dedicated to microstrip networks that are widely diffused in planar devices. Chapter 3 is dedicated to the stripline, perhaps the first planar transmission line developed. Chapter 4 introduces the main problems that can be encountered in planar transmission line networks and devices like discontinuities and higher order modes. Chapter 5 is dedicated to a very important microstrip network, i.e., the coupled microstrip structure, while Chapter 6 is the stripline counterpart, i.e., the coupled stripline structure. Chapter 7 is the largest chapter of this text. It introduces the most used microstrip devices, like directional couplers, phase shifters, and more. Chapter 8 is the stripline counterpart of Chapter 7, and stripline devices are studied. Chapter 9 introduces the slotline, a full planar transmission line, i.e., a transmission line with both conductors on the same plane. This chapter also studies the most important devices that can be built with slotlines. Chapter 10 is dedicated to the coplanar waveguide, another full planar transmission line. Also in this chapter, the most typical devices employing coplanar waveguides are studied. Finally, Chapter 11 introduces the coplanar strips transmission line, which is mainly suited for transmitting balanced signals, requiring a small "PCB" area.

Appendix A1 reviews the theory of the solution methods for simple electrostatic problems. Appendix A2 introduces the most important concepts of wave theory. Appendix A3 is dedicated to the external properties of networks, like the "[s]" parameter matrix. Appendix A4 reviews the main concepts regarding resonant circuits. A common note holds for Appendices A5 and A6. These introduce only the main formulas and concepts for a proper understanding of Appendix A7, and must not be evaluated as an alternative to dedicated texts on physics. Appendix A5 is dedicated to physical relationships among charges, currents, and magnetic fields. Appendix A6 introduces the magnetic properties of materials. Appendix A7 is dedicated to the most important aspects of the electromagnetic field inside ferrimagnetic materials. Finally, Appendix A8 reports all the symbols and some useful relationships used throughout in this text.

To further help the reader, at the end of each chapter and appendix are additional references where some particular issue is analyzed in more detail. If the reference is difficult to find, when possible we have reported alternate texts where the topic under study can be found.

Filters, other than planar transmission line devices, are not the goal of this text and are not included here.

The author hopes this text will help the reader understand the world of planar transmission line networks and devices and will aid in deciding how to choose the proper device. The author also hopes this text will stimulate the reader to study and research other new devices.

Franco Di Paolo
January 2000

CHAPTER 1

Fundamental Theory of Transmission Lines

1.1 GENERALITIES

In telecommunication theory, "transmission line" means a region of the space where "RF" signals can propagate with the best compromise between minimum attenuation and available region of the space. The particular shape of the transmission line can suggest the frequency range where the best compromise exists. In fact, depending on the transmission line shape, it will be best suited to transmit some frequencies and not others.

We can divide transmission lines* into four types:

1. Coupled wires
2. Parallel plates
3. Coaxial
4. Waveguide**

The first three types belong to a family usually called "two conductor" transmission lines (t.l.), while waveguides belong to "one conductor" transmission lines. Other types of t.l. can be included in one of the previous four types. For example, twisted wires lines, parallel wire lines, and slotlines belong to type 1 above (slotlines will be studied in Chapter 9).

A two conductor transmission line can be "balanced" or "unbalanced." An unbalanced transmission line is characterized as having one conductor fixed to a potential, usually the ground one, while the potential of the other conductor moves. A balanced transmission line is characterized as having both conductors as moving potentials with respect to ground potential. In general, the choice of which t.l. to use depends on the type of the generator or load we have to connect to our line. However, physical dimensions of the t.l. greatly influence the natural propagation mode of the line, i.e., whether it is best suited for a balanced or unbalanced propagation.

Every transmission line permits only a fundamental particular polarization*** of the "RF" fields and only a fundamental mode**** of propagation, and these characteristics can also be used to distinguish among lines. Of course, polarization and mode of propagation are strongly a frequency-dependent phenomena, and at some frequencies other modes than the fundamental one can propagate.*****

---

* In this text transmission lines will be called "lines" or abbreviated with "t.l."
** Waveguide transmission lines also are not strongly pertinent to the arguments of this text and will be discussed in Appendix A2.
*** Polarization will be studied in Appendix A2.
**** Modes of propagation will be studied in Appendix A4.
***** This multimode propagation will be discussed for any transmission line we will study in this text.

Two sets of equations exist that can be applied to every transmission line, which relate the voltage "v" and current "i" along the t.l. with its series impedance "Z_s" for unit length (u.l.) and its parallel admittance "Y_p" for u.l. These equations are called "telegraphist's equations" and "transmission line equations" and will now be described.

1.2 "TELEGRAPHIST" AND "TRANSMISSION LINE" EQUATIONS

Let us examine Figure 1.2.1. In part (a) of this figure we have indicated a general representation of a transmission line. The two long rectangular bars represent two conductors, one of which is called "hot conductor" (or simply "hot") and the other "cold conductor" (or simply "cold"). The reader who is familiar with microstrip or stripline* circuits should not confuse the representation in Figure 1.2.1 with two coupled lines.** Similarly, the reader who knows the waveguide mechanics can be dubious about this representation, but we know that modes in waveguides can also be represented with an equivalent transmission line.*** So, Figure 1.2.1 can be used to generically represent any transmission line.

Let us define a positive direction "x" and take into consideration an infinitesimal piece "dx" of this coordinate. Let us consider the t.l. to be lossless, so that the line will only have a series inductance "L" for u.l. and a shunt, or parallel, capacitance for u.l.

With these assumptions, a variation "di" in the time "dt" of the series current "i" will produce a voltage drop "dv" given by:

$$dv = -Ldx\frac{di}{dt} \tag{1.2.1}$$

where the minus sign is a consequence of the coordinate system of Figure 1.2.1. This signal also means that a positive variation "di" of current produces a variation "dv" that contrasts such "di."

Similarly, we can note that a variation "dv" in the time "dt" of the parallel voltage "v" will produce a current variation "di" given by:

$$di = -Cdx\frac{dv}{dt} \tag{1.2.2}$$

where the minus sign means that a positive variation "dv" of voltage produces a variation "di", which is in a direction opposite to the positive one. From the previous two equations we can recognize how "v" and "i" can be set as functions of coordinates and time, and so they can be written more appropriately as:

$$\frac{\partial v}{\partial x} = -L\frac{\partial i}{\partial t} \tag{1.2.3}$$

$$\frac{\partial i}{\partial x} = -C\frac{\partial v}{\partial t} \tag{1.2.4}$$

These last two equations are called "telegraphist's equations," and relate time variation of voltage and current along a t.l. with its physical characteristics as inductance "L" and capacitance "C" per

---

* Microstrip and stripline transmission lines will be studied in Chapter 2 and Chapter 3.
** Generic coupled line theory will be studied later in this Chapter.
*** See Appendix A2 for transmission line equivalents to propagation modes in waveguide.

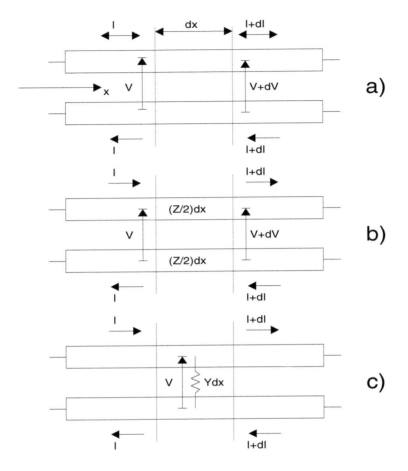

Figure 1.2.1

u.l. From Equations 1.2.3 and 1.2.4 it is possible to obtain two equations where only voltage and current exist. Deriving 1.2.3 with respect to coordinate "x" we have:

$$\frac{\partial^2 v}{\partial x^2} = -L\frac{\partial i}{\partial t}\frac{\partial i}{\partial x} \tag{1.2.5}$$

and inserting 1.2.4 it becomes:

$$\frac{\partial^2 v}{\partial x^2} = -LC\frac{\partial^2 v}{\partial t^2} \tag{1.2.6}$$

Similarly, we can obtain an equation where only current appears:

$$\frac{\partial^2 i}{\partial x^2} = -LC\frac{\partial^2 i}{\partial t^2} \tag{1.2.7}$$

So, voltage and current must satisfy the same equation. Whichever equation, 1.2.6 or 1.2.7, that we take into consideration is called a "monodimensional generalized wave equation." Since:*

$$[LC] \equiv (m/sec)^{-2} \qquad (1.2.8)$$

it is common practice to set:**

$$LC \overset{\perp}{=} 1/v^2 \qquad (1.2.9)$$

and 1.2.7, for example, becomes:

$$\frac{\partial^2 i}{\partial x^2} = \frac{1}{v^2} \frac{\partial^2 i}{\partial t^2} \qquad (1.2.10)$$

where "v" is called "propagation velocity."

A general solution for the monodimensional wave equations does not exist, and it must be found case by case. The only general consequence that can occur is that the general solution "F(t,x)" must satisfy the condition:

$$F(t, x) \equiv F(t - x/v) \qquad (1.2.11)$$

A very familiar aspect assumes the "monodimensional generalized wave equation" when a sinusoidal time variation exists. In this case the time dependence can be written with a multiplication by "$e^{j\omega\tau}$," where "ω" is the angular frequency of voltage or current. With this assumption, the Equation 1.2.6, for example, becomes:

$$v(x) = v^+ e^{(-kx)} - v^- e^{(kx)} \qquad (1.2.12)$$

which is called the "monodimensional wave equation," a particular case of the general "wave equation" studied in Appendix A2. Of course, a similar equation holds for current, and can be obtained substituting "v" with the current "i," and in this case it is called the "monodimensional wave equation."

To introduce the "transmission line equations" let us evaluate part b of Figure 1.2.1. Now, suppose that the t.l. also possesses a series resistance "R" and a parallel conductance "G_p" so that we can write:

$$Z_s = R + j\omega L \qquad (1.2.13)$$

$$Y_p = G_p + j\omega C \qquad (1.2.14)$$

Applying the "Kirchhoff*** voltage loop law" at the network in Figure 1.2.1b, we can write:

---

* Throughout this text, symbols inside square brackets are used to show dimensions. We think that confusion is avoided if square brackets are used in equations. Unless otherwise stated, MKSA unit system will be used.
** With the symbol "$\perp$" we will indicate an equality set by definition.
*** Gustav Robert Kirchhoff, German physicist, born in Koenigsberg in 1824 and died in Berlin in 1887.

$$v(x) = i(x)(Z_s/2)dx + \left[v(x) + dv(x)\right] + i(x)(Z_s/2)dx \qquad (1.2.15)$$

that is:

$$\frac{dv(x)}{dx} = -Z_s i(x) \qquad (1.2.16)$$

Applying the "Kirchhoff current law" at the network in Figure 1.2.1c, we can write:

$$i(x) = v(x)Y_p dx + \left[i(x) + di(x)\right] \qquad (1.2.17)$$

that is:

$$\frac{di(x)}{dx} = -Y_p v(x) \qquad (1.2.18)$$

Equations 1.2.16 and 1.2.18 are called "transmission line equations" and, together with the "telegraphist's equations," form a set of equations widely used in all transmission line problems, and in coupled line cases as will be shown in the next section. Of course, at high frequency, voltages and currents along the lines are not determined in the same way,* i.e., these quantities are not obtained from the general relationships:

$$\Delta v = \int_a^b \underline{E} \cdot d\underline{\ell} \qquad (1.2.19)$$

$$i = \int_s \underline{J} \cdot \underline{n} dS \qquad (1.2.20)$$

where "$\underline{E}$" is the electric field vector, "$d\underline{\ell}$" is an increment vector, "$\underline{J}$" is the surface current density vector, and "$\underline{n}$" is a versor orthogonal to surface "S." These equations are very important and will be of great help for many arguments in this text.

1.3 SOLUTIONS OF TRANSMISSION LINE EQUATIONS

From transmission line equations it is possible to obtain two equations where only voltage and current are present. Deriving with respect to "x" in Equation 1.2.16, we have:

$$\frac{d^2 v(x)}{dx^2} = -Z_s \frac{di(x)}{dx} \qquad (1.3.1)$$

and inserting Equation 1.2.18 it becomes:

$$\frac{d^2 v}{dx^2} = Z_s Y_p v \qquad (1.3.2)$$

* See Appendix A2 to see how voltages and currents are defined along high frequency transmission lines.

Of course, a similar equation can be obtained for current, i.e.:

$$\frac{d^2i}{dx^2} = Z_s Y_p i \qquad (1.3.3)$$

The two previous equations are mathematically equivalent since they are examples of second order linear differential equations. In mechanics theory, equations of this type are called "harmonic motion equations." The solution of this equation is simple, and with reference to 1.3.3, can be found by setting $i(x) \triangleq ie^{kx}$. With this substitution in 1.3.3 we have:

$$Z_s Y_p = k^2 \rightarrow !k = \left(Z_s Y_p\right)^{0.5} \qquad (1.3.4)$$

and the general solution is a linear combination of exponentials:

$$i(x) = i^+ e^{(-kx)} + i^- e^{(kx)} \qquad (1.3.5)$$

Equation 1.3.5 is not the only representation for the solution. Since hyperbolic sinus and cosinus are defined as

$$\cosh(x) \triangleq \frac{e^x + e^{-x}}{2} \qquad \text{senh}(x) \triangleq \frac{e^x - e^{-x}}{2}$$

the solution of 1.3.3 can also be set as a linear combination of hyperbolic sinus and cosinus, i.e.:

$$i(x) = A\cosh(kx) + B\,\text{senh}(kx) \qquad (1.3.6)$$

All the quantities "i^+," "i^-," "A," and "B" are constants, in this case with the unit "Ampere." The quantity "k" obtained from Equation 1.3.4 is called the "propagation constant," and its units are "1/m" in MKSA. Note that with the insertion of 1.3.4 in 1.3.2 or 1.3.3, these equations are mathematically the same as those in the previous section, i.e., Equation 1.2.12.

Of interest is the case where the quantity "k" is imaginary, that is when "Z_s" and "Y_p" are only imaginary, as a consequence of 1.3.4. In this case the solution of 1.3.3 is a linear combination of sinus and cosinus, i.e.:

$$i(x) = A'\cos\left(k_j x\right) + B'\,\text{sen}\left(k_j x\right) \qquad \text{with } k \triangleq jk_j. \qquad (1.3.7)$$

Choosing the best solution between 1.3.5 and 1.3.7 depends on the known boundary conditions of the electromagnetic problem. Exponential solution 1.3.5 is useful when one extreme of the t.l. goes theoretically to infinity, while hyperbolic solutions are useful when considering limited length transmission lines. The term that contains the negative exponential is called "progressive,"* since it decreases in amplitude in the positive direction of "x," while the other is called "regressive," which decreases in amplitude when "x" decreases in amplitude in the negative direction of "x." Note that this procedure can also be applied to obtain the solution 1.3.2.

Once the solution of 1.3.2 or 1.3.3 is extracted, it is possible to obtain the other electrical variable easily, i.e., current or voltage. If we employ the exponential solution of 1.3.5 for current, we can obtain the voltage "v," from 1.2.18 which is given by:

---

* Note that the progressive term decrease in amplitude when "x" increases.

$$v(x) = \left[i^+ e^{(-kx)} - i^- e^{(kx)} \right] k/Y_p \tag{1.3.8}$$

If we had used the hyperbolic solution 1.3.6 for current, from Equation 1.2.18 we would have:

$$v(x) = -\left[A\operatorname{senh}(kx) + B\cosh(kx) \right] \left(k/Y_p \right) \tag{1.3.9}$$

The quantity:

$$\zeta \overset{\perp}{=} k/Y_p \tag{1.3.10}$$

is called "characteristic impedance" of the t.l.* Remembering 1.3.4, the previous equation can be transformed in a very well-known aspect, i.e.:

$$\zeta \overset{\perp}{=} \left(Z_s/Y_p \right)^{0.5} \tag{1.3.11}$$

It is important not to confuse the characteristic impedance with the series impedance of the line; in fact, both series impedance and shunt admittance of the t.l. compose its characteristic impedance. The reciprocal of this quantity is called "characteristic admittance" and is identified as "σ."

Note that with the introduction of "ζ," the Equation 1.3.8 can be written as:

$$v(x) = v^+ e^{(-kx)} - v^- e^{(kx)} \tag{1.3.12}$$

where

$$v^+ \overset{\perp}{=} \zeta i^+ \quad \text{and} \quad v^- \overset{\perp}{=} \zeta i^- \tag{1.3.13}$$

It is interesting at this point to note that from the solution of the monodimensional wave equation for current 1.3.3, we obtained the exponential expression of current with a "+" sign between terms and the exponential expression for voltage, i.e., 1.3.12, with a minus sign. If we had started our study by resolving the monodimensional wave equation for voltage, we would have obtained the exponential expression of voltage with a "+" sign and the exponential expression for current with a minus sign. This sign diversity for the same equation for current or voltage is only analytical and has no influence in the physical problem. This is because the constants that appear in the expressions for current or voltage are generic, and the sign only depends on the effective physical problem. What is always true in the general case is that if in one exponential solution there is the sign "+" between terms, there will be the sign "−" in the other exponential solution. In any case, the true sign will depend on the contour conditions of the particular electromagnetic problem.

For the case where losses can be neglected, useful relationships can be obtained from Equation 1.3.11. In fact, for the lossless case, 1.3.11 becomes:

$$\zeta \overset{\perp}{=} (L/C)^{0.5} \tag{1.3.14}$$

and using 1.2.9 we have:

* Sometimes we will simply named "impedance."

$$\zeta = Lv \equiv 1/Cv \tag{1.3.15}$$

Expressions 1.3.14 and 1.3.15 are used every time a particular transmission line is studied and are assumed to be in the most simple case of no losses.

1.4 PROPAGATION CONSTANT AND CHARACTERISTIC IMPEDANCE

The propagation constant "k" defined by Equation 1.3.4 is in general a complex number. Inserting in that definition the general expression 1.2.13 for "Z_s" 1.2.14 and "Y_p" we have:

$$k = \left[(R + j\omega L)\left(G_p + j\omega C\right)\right]^{0.5} \pm k_r + jk_j \tag{1.4.1}$$

where "k_r" and "k_j" are real numbers. Remember that in the previous equation the quantities "R," "L," "G_p," and "C" are defined for t.l. so that the dimension of "k" is [1/m]* more theoretically exact [Neper/m]. The word "Neper" reminds us that "k" appears in an exponential form "$e^{\pm kx}$." Consequently, to extract "k" we have to perform an operation of natural logarithm "ln." In other words, "k" is proportional to the natural logarithm of the signal amplitude along the t.l. From a value of "α" in [Neper/m], it is simple to calculate the value of "α_{dB}" in [dB/m] using the obvious relationship:

$$\alpha_{dB} = 20 * \log\left(e^{\alpha}\right)$$

For example, 1 [Neper/m] = 8.686 [dB/m].

If now we square Equation 1.4.1 and equate real with real and imaginary with imaginary terms, we have:

$$k_r = \left\{RG_p - \omega^2 LC + \left[\left(RG_p - \omega^2 LC\right)^2 + \omega^2\left(RC + LG_p\right)^2\right]^{0.5}\right\}^{0.5}\bigg/\sqrt{2} \tag{1.4.2}$$

$$k_j = \left\{\omega^2 LC - RG_p + \left[\left(RG_p - \omega^2 LC\right)^2 + \omega^2\left(RC + LG_p\right)^2\right]^{0.5}\right\}^{0.5}\bigg/\sqrt{2} \tag{1.4.3}$$

The ideal lossless lines are those where $R = 0 = G_p$, and in this case from 1.4.2 and 1.4.3:

$$k_r = 0 \quad \text{and} \quad k_j = j\omega\left(LC\right)^{0.5} \tag{1.4.4}$$

In practice, lines are never without losses. So, the practical approximation to the lossless case is when the length "ℓ" of the t.l. is so that:

$$\ell \ll \zeta/R \quad \text{and} \quad \ell \ll \sigma/G_p \tag{1.4.5}$$

---

* Remember that unless otherwise stated we will use the MKSA system unit.

When losses cannot be neglected it is possible to simply obtain an expression for "k_r." For this purpose, let us evaluate the case of a very long t.l., so that we can only use progressive terms for current and voltage and write:

$$i(x) = i^+ e_r^{(-k\,x)}\, e_j^{(-jk\,x)} \quad \text{and} \quad v(x) = v^+ e_r^{(-k\,x)}\, e_j^{(-jk\,x)} \tag{1.4.6}$$

The mean power "W_t" transmitted along the line will be:

$$W_t \overset{\pm}{=} \mathrm{Re}\left[v(x)i^*(x)\right]/2 \tag{1.4.7}$$

and 1.4.6 becomes:

$$W_t = v^+ i^+ e_r^{(-2k\,x)}/2$$

or, using 1.3.13:

$$W_t = v^{+2} e_r^{(-2k\,x)}/2\zeta \tag{1.4.8}$$

The mean power "W_r" dissipated in "R" and "W_g" dissipated in "G_p" are given by:

$$W_r = R\left(i^+\right)^2/2 \quad \text{and} \quad W_g = G_p\left(v^+\right)^2/2 \tag{1.4.9}$$

and, remembering Equation 1.3.13, the total mean power "W_{dt}" dissipated will be:

$$W_{dt} = \left(G_p + R/\zeta^2\right)v^{+2}/2 \tag{1.4.10}$$

The decrease along "x" of "W_t" will be equal to "W_{dt}," so we can write:

$$-dW_t/dx = W_{dt}$$

whioh with 1 4 8 gives:

$$k_r = W_{dt}/2W_t$$

or, using 1.4.8 valuated simply for $x = 0$ and 1.4.10:

$$k_r = \left(G_p\zeta + R/\zeta\right)/2 \tag{1.4.11}$$

In the most general case, "k_r" is given by the sum of two quantities, one dependent on the conductor loss and one dependent on the dielectric loss, i.e., the medium that surrounds the conductor that contains the e.m. field. These two quantities are indicated with "α_c" and "α_d," and so:

$$k_r = \alpha_c + \alpha_d \tag{1.4.12}$$

where:

$$\alpha_c = W_c/2W_t \quad \text{and} \quad \alpha_d = W_d/2W_t \qquad (1.4.13)$$

where "W_c" is the mean power dissipated in the conductors and "W_d" is the mean power dissipated in the dielectric. Appendix A2 shows that for any "TEM," t.l. dielectric losses are governed by the same expression, while conductor losses are in general different.

A more general definition of the propagation constant can be obtained when the signal propagates inside a medium with the following characteristics:

1. μ_r = relative permeability
2. ε_r = relative permittivity
3. g = conductivity

In this case, the propagation constant is given by:*

$$k = j\omega\left(\mu\varepsilon_c\right)^{0.5} \qquad (1.4.14)$$

where:

$$\mu \doteq \mu_0\mu_r \quad \varepsilon_c \doteq \varepsilon - jg/\omega \quad \varepsilon \doteq \varepsilon_0\varepsilon_r \doteq^{**} \varepsilon_{ar} - j\varepsilon_{aj} \quad \varepsilon_r \doteq \varepsilon_{rr} - j\varepsilon_{rj} \quad (1.4.15)$$

Note that from the two previous definitions it follows:

$$\varepsilon_{ar} \equiv \varepsilon_0\varepsilon_{rr} \quad \text{and} \quad \varepsilon_{aj} \equiv \varepsilon_0\varepsilon_{rj} \qquad (1.4.16)$$

and the second definition of 1.4.15 becomes:

$$\varepsilon_c \doteq \varepsilon_{ar} - j\left(\varepsilon_{aj} + g/\omega\right) \qquad (1.4.17)$$

For some simple transmission lines, for example, the coaxial cable, the equivalent inductance "L_s," and capacitance "C" can be simply related to "μ" and "ε."*** The reader interested in the relationships between general transmission line theory and wave propagation can read Appendix A? From 1.4.15 and 1.4.14 it is simple to recognize that if the medium is lossless, i.e., $\varepsilon_{rj} = g = 0$, then "k" is purely imaginary, as in the case of 1.4.4. Other coincidences between waves and transmission lines can be obtained remembering the wave theory, as given in Appendix A2, where it is shown that for any mode of propagation it is possible to associate an equivalent transmission line. Not considering gyromagnetic dielectrics,**** from Equation 1.4.14 we note that "k" is imaginary until "ε_c" is a real quantity. Note that the dielectric constant "ε_r" is in general a complex quantity, independent of the presence of a dielectric conductivity "g," since "ε_{rj}" is due to a damping phenomena associated with the dielectric polarizability.[1,2,3]***** Using this concept, a dielectric is often characterized by a "tangent delta" "$\tan\delta$," (also called a loss tangent) defined as:

---

* See Appendix A2 for other expressions of propagation constant.
** The subscript "a" recalls the significance "absolute."
*** The relative relationships among "L_s," "C," "μ," and "ε" for coaxial cable are given in Appendix A2.
**** Gyromagnetic materials will be studied in Appendix A7, while devices working with gyromagnetic materials are studied in the following chapters.
***** Dielectric polarizability is assumed to be known to the reader. Fundamentals about this argument can be found in the references at the end of this chapter.

$$\tan \delta \overset{\perp}{=} \left| \omega \, \mathrm{Im}\left(\varepsilon_c\right)\right| \Big/ \left| \omega \, \mathrm{Re}\left(\varepsilon_c\right)\right| \equiv \left(\omega\varepsilon_{aj} + g\right)\Big/\omega\varepsilon_{ar} \tag{1.4.18}$$

At μwave frequencies, usually $\omega\varepsilon_{aj} \gg g$ and "tanδ" assumes the well-known expression:

$$\tan \delta \equiv \varepsilon_{rj}/\varepsilon_{rr} \tag{1.4.19}$$

Sometimes the so-called "power factor" is used, indicated with "senδ."

We want to conclude this section noting that impedance "ζ" can also be decomposed into real and imaginary parts. This means that inserting 1.2.13 and 1.2.14 into 1.3.11, in general we have:

$$\zeta = \zeta_r + j\zeta_j \tag{1.4.20}$$

while for a lossless transmission line from 1.3.14 we have $\zeta \equiv (L/C)^{0.5}$, i.e., it is a real quantity.

While all used transmission lines can be practically considered to have real impedances, in the following chapters we will study other transmission* lines where the impedance can be imaginary and the propagation cannot take place.

1.5 TRANSMISSION LINES WITH TYPICAL TERMINATIONS

Quite often t.l. are terminated with short or open circuits. In both cases if this line is in shunt to another line, then the short or open terminated t.l. is called a "stub." It is important to study such cases of simple terminations since stubs are frequently employed in planar transmission line devices, especially for tuning purposes.

We will study cases where these terminations are at the beginning of the t.l., and when they are at the end. Since we are evaluating limited length transmission lines, we will use the hyperbolic form for current and voltage.

a. Terminations at the INPUT of the Line

Our environment is a transmission line of length "ℓ" with a longitudinal axis "x" with origin x = 0 at the beginning of the line.

a1. OPEN circuit at the INPUT

The current at x = 0 will be zero, while the voltage is known. From 1.3.6 we have:

$$i\left(0\right) \equiv 0 \rightarrow A = !0 \quad \text{and} \quad B = \text{any finite value} \tag{1.5.1}$$

The value of "B" cannot be defined with only the condition $i(0) = 0$. We need to have a further condition. If we introduce the condition A = !0 in the hyperbolic voltage expression 1.3.9 evaluated for x = 0 we have:

$$B = ! \quad -v(0)/\zeta \tag{1.5.2}$$

Inserting 1.5.1 and 1.5.2 in 1.3.6 and 1.3.9 we have the expression of voltage and current along the t.l.:

---

* See Appendix A7 and, among others, chapters 7 and 8 where ferrimagnetic devices are studied.

$$i(x) = -v(0)\operatorname{senh}(kx)/\zeta \tag{1.5.3}$$

$$v(x) = v(0)\cosh(kx) \tag{1.5.4}$$

a2. SHORT circuit at the INPUT

The voltage at x = 0 will be zero, while the current is known. From 1.3.9 we have:

$$v(0) \equiv 0 \rightarrow B = !0 \quad \text{and} \quad A = \text{any finite value} \tag{1.5.5}$$

To define the value of "A" we introduce the condition B = !0 in the hyperbolic current expression 1.3.6 evaluated for x = 0 we have:

$$A = ! \quad i(0) \tag{1.5.6}$$

Inserting 1.5.5 and 1.5.6 in 1.3.6 and 1.3.9 we have the expression of voltage and current along the t.l.:

$$i(x) = i(0)\cosh(kx) \tag{1.5.7}$$

$$v(x) = -\zeta i(0)\operatorname{senh}(kx) \tag{1.5.8}$$

a3. GENERAL termination at the INPUT

In this case, a voltage "v(0)" and a current "i(0)" are present at the input. To have the expression of voltage and current along the t.l., as a function of the general termination at the input, we can use the superposition effect principle and apply the solutions of the previous points a1 and a2. So, for this case we have:

$$i(x) = i(0)\cosh(kx) - v(0)\operatorname{senh}(kx)/\zeta \tag{1.5.9}$$

$$v(x) = -\zeta i(0)\operatorname{senh}(kx) + v(0)\cosh(kx) \tag{1.5.10}$$

b. Terminations at the OUTPUT of the Line

To have a simple expression for the constant "A" and "B" we will apply the transformation variable:

$$x' \doteq \ell - x \tag{1.5.11}$$

to the hyperbolic expression of current, and write:

$$i(x') = A\cosh(kx') + B\operatorname{senh}(kx') \equiv i(\ell - x) \tag{1.5.12}$$

This transformation corresponds to having the new axis origin at the end of the t.l. and the positive direction of "x" in the opposite direction with respect to the previous case a. Deriving 1.5.12, i.e., applying the 1.2.18, we have:

$$v(x') = \zeta\left[A\operatorname{senh}(kx') + B\cosh(kx')\right] \equiv v(\ell - x) \tag{1.5.13}$$

With 1.5.12 and 1.5.13 we can repeat the previous points a1 through a3.

b1. OPEN circuit at the OUTPUT
The current at $x' = 0$ will be zero, while the voltage is known. From 1.5.12 we have:

$$i(x' = 0) \equiv 0 = i(x = \ell) \rightarrow A = !0 \text{ and } B = \text{any finite value} \tag{1.5.14}$$

The value of "B" cannot be defined with only the condition $i(0) = 0$. We need to have a further condition. If we introduce the condition $A = !0$ in the hyperbolic voltage expression 1.5.13 evaluated for $x' = 0$ we have:

$$B = !\quad v(x' = 0)/\zeta \equiv v(\ell)/\zeta \tag{1.5.15}$$

Inserting 1.5.14 and 1.5.15 in 1.5.12 and 1.5.13 we have the expression of voltage and current along the t.l.:

$$i(x) = \text{senh}\left[k(\ell - x)\right] v(\ell)/\zeta \tag{1.5.16}$$

$$v(x) = v(\ell) \cosh\left[k(\ell - x)\right] \tag{1.5.17}$$

b2. SHORT circuit at the OUTPUT
The voltage at $x' = 0$ will be zero while the current is known. From 1.5.13 we have:

$$v(x' = 0) \equiv 0 = v(x = \ell) \rightarrow B = !0 \quad \text{and} \quad A = \text{any finite value} \tag{1.5.18}$$

To define the value of "A" we introduce the condition $B = !0$ in the hyperbolic current expression 1.5.12 evaluated for $x' = 0$ we have:

$$A = !i(x' = 0) \equiv i(x = \ell) \tag{1.5.19}$$

Inserting 1.5.18 and 1.5.19 in 1.5.12 and 1.5.13 we have the expression of voltage and current along the t.l.:

$$i(x) = i(\ell) \cosh\left[k(\ell - x)\right] \tag{1.5.20}$$

$$v(x) = \zeta i(\ell) \text{senh}\left[k(\ell - x)\right] \tag{1.5.21}$$

b3. GENERAL termination at the OUTPUT
In this case, a voltage "v(0)" and a current "i(0)" are present at the output. To have the expression of voltage and current along the t.l., as a function of the general termination at the input, we can use the superposition effect principle and apply the solutions of the previous points b1 and b2. So, for this case we have:

$$i(x) = i(\ell) \cosh\left[k(\ell - x)\right] + \text{senh}\left[k(\ell - x)\right] v(\ell)/\zeta \tag{1.5.22}$$

$$v(x) = \zeta i(\ell) \text{senh}\left[k(\ell - x)\right] + v(\ell) \cosh\left[k(\ell - x)\right] \tag{1.5.23}$$

b4. Input impedance with known termination at the OUTPUT
It is useful to have the input impedance "Z(0)" when the load impedance "Z(ℓ)" is known and of finite value, i.e., when:

$$v(\ell)/i(\ell) \equiv Z(\ell) \tag{1.5.24}$$

Calculating the ratio of 1.5.23 with 1.5.22, both evaluated for x = 0, and using 1.5.24 we have:

$$Z(0) = \frac{\zeta^2 \operatorname{senh}(k\ell) + \zeta Z(\ell)\cosh(k\ell)}{\zeta \cosh(k\ell) + Z(\ell)\operatorname{senh}(k\ell)} \tag{1.5.25}$$

Of course, if we do the reciprocal of 1.5.24, we can also calculate the input admittance of the t.l. "Y(0)" given by:

$$Y(0) = \frac{\sigma^2 \operatorname{senh}(k\ell) + \sigma Y(\ell)\cosh(k\ell)}{\sigma \cosh(k\ell) + Y(\ell)\operatorname{senh}(k\ell)} \tag{1.5.26}$$

Particular cases of 1.5.25, or 1.5.26, are when:
b4a. Input impedance with open circuited line
In this case $Z(\ell) = \infty$ and from 1.5.25:

$$Z_{oc}(0) \equiv \zeta \cotgh(k\ell) \tag{1.5.27}$$

b4b. Input impedance with short circuited line
In this case $Z(\ell) = 0$ and from 1.5.25:

$$Z_{sc}(0) \equiv \zeta \tgh(k\ell) \tag{1.5.28}$$

b4c. Input impedance with matched terminated line
A t.l. is said to be matched if the load impedance* "Z_ℓ" is so that:

$$Z_\ell \equiv \zeta \tag{1.5.29}$$

Then, in this case $Z(\ell) = \zeta$ and from 1.5.25:

$$Z(0) \equiv \zeta \tag{1.5.30}$$

When the lines can be approximated with the ideal case of zero losses we know from 1.4.4 that $k \equiv jk_j$, and:

$$\operatorname{senh}(jk_j x) \equiv j\operatorname{sen}(k_j x) \tag{1.5.31}$$

$$\cosh(jk_j x) \equiv \cos(k_j x) \tag{1.5.32}$$

In this case the expressions 1.5.25, 1.5.27, and 1.5.28 assume a very simple aspect given by:

---

* A completely matched line has both source "Z_g" and load "Z_ℓ" impedance equal to "ζ."

$$Z(0) = \frac{j\zeta^2 \, tg\left(k_j \ell\right) + \zeta Z(\ell)}{\zeta + jZ(\ell) \, tg\left(k_j \ell\right)} \tag{1.5.33}$$

$$Z_{oc}(0) \equiv j\zeta \, cotg\left(k_j \ell\right) \tag{1.5.34}$$

$$Z_{sc}(0) \equiv j\zeta \, tg\left(k_j \ell\right) \tag{1.5.35}$$

The last three expressions are widely used in many transmission line networks such as filters and matching networks.

1.6 "TRANSMISSION" AND "IMPEDANCE" MATRICES

With "transmission matrices" we have a representation of the transmission line that simply relates input and output line excitations.

Let us examine Figure 1.6.1 a, where a t.l. of length "ℓ" and characteristic impedance "ζ" is excited at one extreme with voltage "v_i" and current "i_i." The excitation extreme is set as the origin of the "x" axis coordinate. We want to evaluate the voltage "v(x)" and current "i(x)" at a distance "x" from the origin. We can write:

$$v_i \overset{\perp}{=} v^+ - v^- \tag{1.6.1}$$

$$i_i \overset{\perp}{=} i^+ + i^- \tag{1.6.2}$$

from which (using 1.3.13) we have:

$$v^- = \frac{\zeta i_i - v_i}{2} \qquad v^+ = \frac{\zeta i_i + v_i}{2} \tag{1.6.3}$$

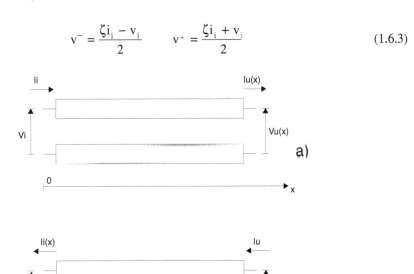

Figure 1.6.1

From 1.3.5 and 1.3.13 and using the previous relationships we have:

$$i_u(x) = \frac{\zeta i_i + v_i}{2\zeta} e^{(-kx)} + \frac{\zeta i_i - v_i}{2\zeta} i^- e^{(kx)} \tag{1.6.4}$$

Collecting together terms with "i_i" and "v_i" and remembering that:

$$e^{(kx)} + e^{(-kx)} = 2\cosh(kx) \tag{1.6.5}$$

$$e^{(kx)} - e^{(-kx)} = 2\operatorname{senh}(kx) \tag{1.6.6}$$

Eq. 1.6.4 becomes:

$$i_u(x) = i_i \cosh(kx) - (v_i/\zeta)\operatorname{senh}(kx) \tag{1.6.7}$$

Inserting 1.6.3 into 1.3.12, collecting together terms with "i_i" and "v_i," and applying 1.6.5 and 1.6.6 we have:

$$v_u(x) = v_i \cosh(kx) - i_i \zeta \operatorname{senh}(kx) \tag{1.6.8}$$

Equations 1.6.7 and 1.6.8 can be written as:

$$\begin{pmatrix} v_u(x) \\ i_u(x) \end{pmatrix} = \begin{pmatrix} \cosh(kx) & -\zeta\operatorname{senh}(kx) \\ -\sigma\operatorname{senh}(kx) & \cosh(kx) \end{pmatrix} \begin{pmatrix} v_i \\ i_i \end{pmatrix} \tag{1.6.9}$$

where the square matrix is indicated with "T_f" and called the "forward transmission matrix," i.e.:

$$T_f = \begin{pmatrix} \cosh(kx) & -\zeta\operatorname{senh}(kx) \\ -\sigma\operatorname{senh}(kx) & \cosh(kx) \end{pmatrix} \tag{1.6.10}$$

Let us consider part b of Figure 1.6.1 and attempt to evaluate the input voltage "$v_i(x)$" and current "$i_i(x)$," with "v_u" and "i_u" known. This can be simply done by looking at Equation 1.6.9. We have.

$$\begin{pmatrix} v_u(x) \\ i_u(x) \end{pmatrix} = T_f \begin{pmatrix} v_i \\ i_i \end{pmatrix} \rightarrow T_f^{-1} \begin{pmatrix} v_u(x) \\ i_u(x) \end{pmatrix} = \begin{pmatrix} v_i \\ i_i \end{pmatrix} \tag{1.6.11}$$

So, inverting* the Equation 1.6.10 we have:

$$\begin{pmatrix} v_i(x) \\ i_i(x) \end{pmatrix} = \begin{pmatrix} \cosh(kx) & \zeta\operatorname{senh}(kx) \\ \sigma\operatorname{senh}(kx) & \cosh(kx) \end{pmatrix} \begin{pmatrix} v_u \\ i_u \end{pmatrix} \tag{1.6.12}$$

where the square matrix is indicated with "T_r" and called the "reverse transmission matrix," i.e.:

---

* Matrix inversion operation can be found in many mathematical books.

$$T_r = \begin{pmatrix} \cosh(kx) & \zeta\,\mathrm{senh}(kx) \\ \sigma\,\mathrm{senh}(kx) & \cosh(kx) \end{pmatrix} \qquad (1.6.13)$$

It is very important to say that transmission matrices are always relative to a well-defined orientation along the t.l. and a well-defined orientation of currents. Note that to define the "T_r," we have not changed the orientation in Figure 1.6.1b with respect to that in Figure 1.6.1a. In fact, there is no input or output of a line until we do not define a positive direction for it. In other words, if we do not define a positive direction for a transmission line, there is no reason to speak about forward or reverse transmission matrix. With these concepts clear, we can say that "T_f" and "T_r" both give the voltage and current at the opposite extreme of the t.l., i.e., the extreme where there is no excitation, but "T_f" is relative to the same direction defined as positive along the t.l. while "T_r" is relative to the opposite direction. Note that "T_r" can be obtained with the same procedure used to obtain "T_f," i.e., using transmission line theory, instead of using the inversion matrix operation.

The reader who knows the "chain" or "ABCD" matrix representation of a two-port network* can recognize how "T_r" is the ABCD matrix of our transmission line. Voltage and current direction used to define the chain matrix are the same as those we used to define "T_r."

An interesting application of the transmission matrix, which is useful when analyzing coupled lines,** is the case of a lossless t.l. open terminated and excited by a current generator of value "i_i." The input voltage "v_i" will be:

$$v_i = -j\zeta\,i_i\,\cotg\!\left(k_j\ell\right) \qquad (1.6.14)$$

and from the forward transmission matrix the output voltage "v_u" will be:

$$v_u = -j\zeta\,i_i\,\cotg\!\left(k_j\ell\right)\cos\!\left(k_j\ell\right) - j\zeta\,i_i\,\mathrm{sen}\!\left(k_j\ell\right) \equiv -j\zeta\,i_i\big/\mathrm{sen}\!\left(k_j\ell\right) \qquad (1.6.15)$$

Another matrix that is sometimes used in transmission line problems is the impedance matrix "[Z]" defined as:

$$(Z) = \begin{pmatrix} Z_{11} & Z_{12} \\ Z_{21} & Z_{22} \end{pmatrix} \qquad (1.6.16)$$

where:

$$Z_{11} \doteq V_1/I_1 \quad \text{for } I_2 = 0$$

$$Z_{22} \doteq V_2/I_2 \quad \text{for } I_1 = 0$$

$$Z_{12} \doteq V_1/I_2 \quad \text{for } I_1 = 0$$

$$Z_{21} \doteq V_2/I_1 \quad \text{for } I_2 = 0$$

* See Appendix A3 for ACBD matrix definition.
** This example will be used in Chapter 6

Subscripts "1" and "2" indicate ports. Note that since the transmission line is a linear reciprocal device, it is only necessary to evaluate one term between "Z_{11}" and "Z_{22}" and one term between "Z_{12}" and "Z_{21}."

Since to evaluate "Z_{11}" or "Z_{22}" we have an open circuit at one extreme, these values can simply be obtained from 1.5.27 as follows:

$$Z_{11} \overset{\perp}{=} V_1/I_1 \equiv \zeta \cot gh(k\ell) = Z_{22} \overset{\perp}{=} V_2/I_2 \qquad (1.6.17)$$

To evaluate the other parameters, we extract "I_1" from the previous equation:

$$I_1 \equiv V_i \, tgh(k\ell)/\zeta \qquad (1.6.18)$$

To obtain "V_2" we use "T_f" and write:

$$V_2 = V_1 \cosh(k\ell) - I_1 \, \zeta senh(k\ell) \qquad (1.6.19)$$

which, with insertion of 1.6.18 becomes:

$$V_2 = V_1 \left[\cosh(k\ell) - senh(k\ell) \, tgh(k\ell) \right] \qquad (1.6.20)$$

Performing the ratio of 1.6.20 with 1.6.18 we have:

$$Z_{21} = \zeta cosech(k\ell) = Z_{12} \qquad (1.6.21)$$

and all the matrix "[Z]" parameters are now defined.

1.7 CONSIDERATIONS ABOUT MATCHING TRANSMISSION LINES

As we said before, transmission lines are quite often used for impedance matching. This section will discuss operating "bandwidth" of such matching. First of all, the term "bandwidth" will be defined. The operating bandwidth of a device is the frequency interval where some, or all of its frequency characteristics are evaluated as acceptable for the device purpose. For instance, the operating bandwidth of a band pass filter is the frequency interval where the value of its attenuation is included between the attenuation at center frequency and a number of dB, typically 1 or 3, below this value. Bandwidth is usually indicated in three manners:

a. *Ratio of Bandwidth Limits* — If we indicate with "f_h" and "f_l," respectively, the maximum and minimum frequency of the operating bandwidth, then:

$$n \overset{\perp}{=} f_h/f_l$$

where with the symbol "$\overset{\perp}{=}$" we indicate an equality by definition.

b. Fractional — It is defined as:

$$B \doteq 2\frac{f_h - f_l}{f_h + f_l}$$

or:

$$B\% \doteq 2\frac{f_h - f_l}{f_h + f_l}100$$

if "in percent." The relationship between the definition in point a and b is:

$$B \doteq 2\frac{n - 1}{n + 1}$$

or:

$$B\% \doteq 2\frac{n - 1}{n + 1}100$$

if "in percent."

c. Octave — Each octave is a multiplication by "2." For instance, if from a minimum frequency of 6 GHz the operating bandwidth extends for one octave, it means that the upper frequency of the bandwidth is 12 GHz; if the operating bandwidth extends for two octaves, it means that the upper frequency of the bandwidth is 24 GHz, and so on. If one half an octave is added to a number "o" of octave, the resulting multiplication number "m_{oh}" depends on "o" according to the following relationship:

$$m_{oh} = 2^o + \left[2^{(o+1)} - 2^o\right]/2 \qquad \text{for } o = 1, 2, 3, \ldots$$

For instance, the multiplication factor "m_{1h}" for one octave and a half is "3" while $m_{2h} = 6$, $m_{3h} = 12$ and so on.

Of course, if "B" is known, the "n" may be obtained very simply by:

$$n = \frac{2 + B}{2 - B}$$

After these important definitions, let us rewrite Equation 1.4.12 evaluated for zero losses. We have:

$$k \equiv k_j = j\omega(\mu\varepsilon)^{0.5} = j2\pi/\lambda \qquad (1.7.1)$$

where "λ" is the signal wavelength along the line.* With 1.7.1, Equations 1.5.34 and 1.5.35 become:

$$Z_{oc}(0) \equiv -j\zeta\,\mathrm{cotg}(2\pi\ell/\lambda) \tag{1.7.2}$$

$$Z_{sc}(0) \equiv j\zeta\,\mathrm{tg}(2\pi\ell/\lambda) \tag{1.7.3}$$

In Figure 1.7.1 the shape of $|Z_{oc}(0)/\zeta|$ and $|Z_{sc}(0)/\zeta|$ is represented as a function of $\rho \triangleq \ell/\lambda$. We see how any value of input impedance can be obtained, i.e., positive, negative, zero, or infinite. From this point of view, stubs work like transformers. If we assume that the wavelength of the signal is fixed, this figure represents the possible impedance values that we can report at the input of the stubs when varying the t.l. length "ℓ." If we want to operate in a region where tolerances on the exact required value of "ℓ" are permitted, we have to work in a region with a small slope vs. "ρ." This means that $\ell \approx (2n+1)\lambda/4$ for open circuited stubs and $\ell \approx n\lambda/2$ for short circuited stubs, where "n" is an integer number. The same conclusions hold if we assume that the length of the stub is fixed while the wavelength of the signal is varied. In other words, the higher the transformation ratio that is needed, the lower the resulting operating bandwidth is, since small variations in frequency cause a large change in the reported impedance, which results in a mismatching.

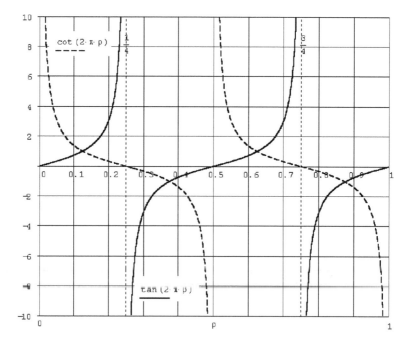

Figure 1.7.1

Other important matching characteristics can be obtained from the general input impedance formula, i.e., Equation 1.5.25, which, with 1.7.1 becomes:

$$Z(0) = \frac{j\zeta^2 \mathrm{sen}(2\pi\ell/\lambda) + \zeta Z(\ell)\cos(2\pi\ell/\lambda)}{\zeta\cos(2\pi\ell/\lambda) + jZ(\ell)\mathrm{sen}(2\pi\ell/\lambda)} \tag{1.7.4}$$

---

* Unless otherwise stated, in this text signal wavelength will always be relative to a guided case, i.e., the signal travels inside a transmission line.

Let us evaluate the following cases:

a. Line Length Equal to Integer Number of Half Wavelength — In this case we have:

$$\ell = n\lambda/2 \quad \text{with n} = 1, 2, 3... \tag{1.7.5}$$

which, inserted in 1.7.4 gives:

$$Z(0) \equiv Z(\ell) \tag{1.7.6}$$

This means that whatever the characteristic impedance "ζ" of a t.l. is, if Equation 1.7.5 holds, the termination impedance "$Z(\ell)$" is reported at the input of the transmission line.

b. Line Length Equal to an Odd Number of Quarter Wavelength — In this case we have:

$$\ell = (2n + 1)\lambda/4 \quad \text{with n} = 1, 2, 3... \tag{1.7.7}$$

which, inserted in 1.7.4 gives:

$$Z(0) \equiv \zeta^2/Z(\ell) \tag{1.7.8}$$

This relationship represents the most useful effect of a stub that can be realized according to 1.7.8. This characteristic is also used to do simple filters. Suppose we need to remove a tone of wavelength "λ" from a signal passing in a t.l. If we insert a stub open terminated, i.e., with $Z(\ell) = \infty$, then from the previous equation we have $Z(0) = 0$, which means that the desired signal is shorted to ground. Of course, problems arise if the signal to be shorted possesses a bandwidth since this type of filtering is narrowband, as we said previously. Filter theory is not the topic of this text, although in the next sections we will quite often study networks, which have characteristics that are near to filtering properties.

c. Line Terminated With Matched Load — In this case we have:

$$Z(\ell) \equiv \zeta \tag{1.7.9}$$

which, inserted in 1.7.4 gives:

$$Z(0) \equiv \zeta \tag{1.7.10}$$

This equation means that whichever is the length of the t.l. when the termination is equal to the characteristic impedance of the line, the input impedance is always equal to this characteristic impedance. Note that since 1.7.10 is independent of frequency, the matching condition is the broadest possible bandwidth relationship for a transmission line.

1.8 REFLECTION COEFFICIENTS AND STANDING WAVE RATIO

In the previous paragraphs we have shown how a matched transmission line has the widest operating bandwidth. In addition, matching condition is also helpful, which will now be discussed.

Suppose that a transmission line of characteristic impedance "ζ" is fed at one extreme by a generator of impedance $R_g = \zeta$ and at the other extreme is terminated by a load of impedance $Z_\ell = \zeta$. In this case the t.l. is completely matched, i.e., matched at both ends. It is useful to define an impedance "$Z(x)$" function of "x" given by:

$$Z(x) \overset{\perp}{=} v(x)/i(x) \qquad (1.8.1)$$

Taking into account expressions 1.3.5 for "$i(x)$" and 1.3.12 for "$v(x)$," remembering 1.3.13, and assuming that only progressive terms exist, then:

$$Z(x) \equiv \zeta \qquad (1.8.2)$$

while if only regressive terms exist, then:

$$Z(x) \equiv -\zeta. \qquad (1.8.3)$$

So, Equation 1.8.1 gives the characteristic impedance "ζ" as a result, only if a monodirectional wave exists. The negative sign in 1.8.3 has no physical effect, since it comes out from the conventional sign for "v^+," "v^-," "i^+," and "i^-." It is common practice to explain the different signs in 1.8.2 and 1.8.3 as showing that voltage is a "parallel" quantity and doesn't change sign with "x," while current is a "longitudinal" quantity and does change sign with direction of "x." However, if this explanation works to explain the signs in 1.8.2 and 1.8.3, sometimes the assumptions on "parallel" and "longitudinal" quantities can lead to error. To avoid such a possibility, it is always convenient to refer to the general expressions of "$v(x)$" and "$i(x)$." Equation 1.8.1 is also true at the termination point, where the load is connected. At this coordinate, 1.8.1 means that all the voltage "$v(x)$" must be at the load terminals, and all the current "$i(x)$" must pass inside it. If this doesn't happen, it means that at the termination there is not the proper impedance, i.e., $Z_\ell \neq \zeta$. So, if the t.l. is matched, no reflection exists, and consequently, in a matched transmission line, only a monodirectional wave exists. These results could lead one to think it is possible to connect a matched load in any section of the line without affecting the matching. This is not true. In fact, if we insert a matched load $Z_\ell = \zeta$, for instance, at the middle coordinate "$x = x_h$" of a matched transmission line, we have the half line on the right report "ζ" in parallel to $Z_\ell = \zeta$. This situation is comparable to a transmission line with impedance "ζ" and length "x_h" terminated with a load $Z_\ell = \zeta/2$, which is not the matching condition for the t.l. The case when reflections exist inside a transmission line is said to be a "standing wave" phenomena.

After such introduction, let us suppose that the line of length "x" is terminated by a generic load "Z_ℓ." We can define a current "i^t" passing inside the load and a voltage "v^t" between its terminals and the following parameters:

voltage reflection coefficient: $\Gamma_v \overset{\perp}{=} v^r/v^p$ $\qquad (1.8.4)$

voltage transmission coefficient: $T_v \overset{\perp}{=} v^t/v^p$ $\qquad (1.8.5)$

current reflection coefficient: $\Gamma_i \overset{\perp}{=} i^-/i^+$ $\qquad (1.8.6)$

current transmission coefficient: $T_i \overset{\perp}{=} i^t/i^+$ $\qquad (1.8.7)$

All these parameters are, in general, complex quantities.

At the termination coordinate, current "i^t" passing across the load and voltage "v^t" at its terminals must satisfy the following relationships:

$$v^t = v^p + v^r \tag{1.8.8}$$

$$i^t = i^+ + i^- \tag{1.8.9}$$

where:

$$v^p \doteq \zeta i^+ \quad \text{and} \quad v^r \doteq -\zeta i^- \tag{1.8.10}$$

The quantities "v^p" and "v^r" are respectively the amplitudes of the progressive and regressive, or reflected waves, i.e., traveling in the positive and negative direction of "x." Since $Z_\ell \equiv v_t / i_t$, Equation 1.8.8 with 1.8.10 becomes:

$$Z_\ell i^t = \zeta i^+ - \zeta i^- \tag{1.8.11}$$

Indicating with "$Z_{\ell n}$" the value of "Z_ℓ" normalized to "ζ," the previous equation becomes:

$$Z_{\ell n} i^t = i^+ - i^- \tag{1.8.12}$$

Summing and subtracting the previous equation to 1.8.9 we have, respectively:

$$\left(Z_{\ell n} + 1\right) i_t = 2 i^+ \quad \text{and} \quad \left(Z_{\ell n} - 1\right) i_t = -2 i^- \tag{1.8.13}$$

Using 1.8.6, the ratio of the two equations in 1.8.12 becomes:

$$\Gamma_i = \frac{1 - Z_{\ell n}}{1 + Z_{\ell n}} \tag{1.8.14}$$

Dividing 1.8.11 by "i^+" and using 1.8.7 and 1.8.14, we have:

$$T_i = \frac{1 - \Gamma_i}{Z_{\ell n}} = \frac{2}{1 + Z_{\ell n}} \tag{1.8.15}$$

To extract "Γ_v" and "T_v" we can begin to rewrite 1.8.9 as:

$$v^t / Z_\ell = v^p / \zeta - v^r / \zeta \tag{1.8.16}$$

or

$$v^t / Z_{\ell n} = v^p - v^r \tag{1.8.17}$$

Summing and subtracting the previous equation to 1.8.8, we can proceed in a manner similar to that used for 1.8.14 and 1.8.15, obtaining:

$$\Gamma_v = \frac{Z_{\ell n} - 1}{1 + Z_{\ell n}} \equiv -\Gamma_i \qquad\qquad (1.8.18)$$

$$T_v = 1 + \Gamma_v = \frac{2Z_{\ell n}}{1 + Z_{\ell n}} = Z_{\ell n} T_i \qquad\qquad (1.8.19)$$

It is simple to evaluate the limits value for all transmission and reflection coefficients we have defined. Indicating with a subscript "M" or "m," respectively, the maximum and minimum parameter value and considering positive values for the normalized impedance,* we have:

$$\left|\Gamma_v\right|_M = 1 \qquad \text{for } Z_{\ell n} = 0 \text{ or } Z_{\ell n} = \infty$$

$$\left|\Gamma_v\right|_m = 0 \qquad \text{for } Z_{\ell n} = 1$$

$$\left|T_v\right|_M = 2 \qquad \text{for } Z_{\ell n} = \infty$$

$$\left|T_v\right|_m = 0 \qquad \text{for } Z_{\ell n} = 0$$

$$\left|\Gamma_i\right|_M = 1 \qquad \text{for } Z_{\ell n} = 0 \text{ or } Z_{\ell n} = \infty$$

$$\left|\Gamma_i\right|_m = 0 \qquad \text{for } Z_{\ell n} = 1$$

$$\left|T_i\right|_M = 2 \qquad \text{for } Z_{\ell n} = 0$$

$$\left|T_v\right|_m = 0 \qquad \text{for } Z_{\ell n} = \infty$$

The situation is different when the terminating impedances assume negative values for their resistive parts. Passive components always have positive values of resistance, but active devices can possess negative resistances under particular conditions of bias and loading network. In this case, let us suppose that an active device possesses an input impedance "Z_i" purely resistive and of negative value, i.e., $Z_i \equiv -R_i$, and that the source impedance "Z_g" is also purely resistive and positive, i.e., $Z_g \equiv R_g$. Normalizing these impedances to "R_g," from 1.8.18 we have:

$$\left|\Gamma_v\right| = \frac{Z_{in} + 1}{\left|1 - Z_{in}\right|} \qquad \text{with } Z_{in} \doteq R_i/R_g$$

from which we see that if:

$$\left|1 - Z_{in}\right| < 1 + Z_{in}$$

then $|\Gamma_v| > 1$. Theoretically, if $Z_{in} = 1$, then $|\Gamma_v| = \infty$.

The use of negative resistance presented by an active device is one of the main foundations of oscillator circuits. Oscillators are one of the most attractive devices of all electronic circuits. This topic is not treated in this text, but the interested reader can refer to the articles and books indicated in references. [4,5,6,7,8,9]

---

* We will soon return to this assumption regarding positive values for terminating impedance.

It is possible to define transmission and reflection coefficients using power. In this case we define:

$$\text{power reflection coefficient:} \qquad \Gamma_w \overset{\perp}{=} w^r / w^p \qquad\qquad (1.8.20)$$

$$\text{power transmission coefficient:} \qquad T_w \overset{\perp}{=} w^t / w^p \qquad\qquad (1.8.21)$$

where transmitted "w^t," progressive "w^p," and reflected "w^r" powers must satisfy:

$$w^p \overset{\perp}{=} w^t + w^r \qquad\qquad (1.8.22)$$

Note that inserting the previous equation in 1.8.20 we have:

$$\Gamma_w \overset{\perp}{=} w^r / w^p \equiv \left(w^p - w^t\right)/w^p = 1 - T_w \qquad\qquad (1.8.23)$$

These coefficients can easily be obtained from the previous ones. In fact, for "T_w" we have:

$$T_w \equiv \mathrm{Re}\left(T_v\, T_i^*\right) \qquad\qquad (1.8.24)$$

where "T_i^*" is the complex conjugate of "T_i." From 1.8.15 and 1.8.19 we have:

$$T_w = 4\,\mathrm{Re}\left[\frac{Z_{\ell n}}{\left(1 + Z_{\ell n}\right)\left(1 + Z_{\ell n}^*\right)}\right] = \frac{4\,\mathrm{Re}\left(Z_{\ell n}\right)}{\left(1 + Z_{\ell n}\right)\left(1 + Z_{\ell n}^*\right)} = \frac{2\left(Z_{\ell n} + Z_{\ell n}^*\right)}{\left(1 + Z_{\ell n}\right)\left(1 + Z_{\ell n}^*\right)} \qquad (1.8.25)$$

Inserting this last equality in Equation 1.8.23 we have:

$$\Gamma_w \equiv \left|\Gamma_i\right|^2 \equiv \left|\Gamma_v\right|^2 \qquad\qquad (1.8.26)$$

and from 1.8.23:

$$T_w = 1 - \left|\Gamma_v\right|^2 \qquad\qquad (1.8.27)$$

Another parameter often used, especially in filter network theory, is the power attenuation factor "A_w" defined by:

$$A_w \overset{\perp}{=} w^p / w^t \equiv 1/T_w = 1/\left(1 - \left|\Gamma_v\right|^2\right) \qquad\qquad (1.8.28)$$

Reflections and transmission coefficients can also be obtained using admittances. To do that let us rewrite Equation 1.8.8 as:

$$i^t / Y_\ell = i^+ / \sigma - i^- / \sigma$$

which, with the definition of normalized load admittance $Y_{\ell n} \overset{\perp}{=} Y_\ell / \sigma$ it becomes:

$$i^t / Y_{\ell n} = i^+ - i^-$$
(1.8.29)

Summing and subtracting this expression to 1.8.9 we have:

$$i^t \left(1 + 1/Y_{\ell n}\right) = 2i^+$$
(1.8.30)

$$i^t \left(1/Y_{\ell n} - 1\right) = -2i^-$$
(1.8.31)

Calculating the ratio between 1.8.31 with 1.8.30 we have:

$$\Gamma_i \triangleq \frac{i^-}{i^+} = \frac{Y_{\ell n} - 1}{1 + Y_{\ell n}}$$
(1.8.32)

and directly from 1.8.30:

$$T_i \triangleq \frac{i^t}{i^+} = \frac{2Y_{\ell n}}{1 + Y_{\ell n}}$$
(1.8.33)

Rewriting 1.8.9 as:

$$v^t Y_{\ell n} = v^p - v^r$$

and summing and subtracting this expression to 1.8.8 we can proceed as we did before, obtaining:

$$\Gamma_v \triangleq \frac{v^r}{v^p} = \frac{1 - Y_{\ell n}}{1 + Y_{\ell n}}$$
(1.8.34)

$$T_v \triangleq \frac{v^t}{v^p} = \frac{2}{1 + Y_{\ell n}}$$
(1.8.35)

It is interesting to observe that voltage, current reflection, and transmission coefficients are complex numbers if the impedances or admittances are complex. Power reflection and transmission coefficients are always real numbers, since they are related to the modulus of reflection coefficient, as indicated by 1.8.26 and 1.8.27.

After defining these parameters, it is very interesting to show that the reflection coefficient "$\Gamma_v(x)$" along "x" has a simpler expression than "$Z(x)$" as we have seen in Section 1.7. Let us take as the origin of axes "x," the point where the load "Z_ℓ" is connected, and as negative direction we choose the left side. This situation is indicated in Figure 1.8.1.

If $\Gamma_v \triangleq \Gamma_v(0)$ is the reflection coefficient of the load, we can write:

$$v^r(x) \equiv \Gamma_v(0) v^p(x) = \Gamma_v(0) v^p e^{kx}$$
(1.8.36)

The positive sign of the exponential is due to the fact that the negative sign in the exponential of $v^p(x) \triangleq v^p e^{-kx}$ has to be changed by the negative direction of propagation along "x." Generalizing the definition in 1.8.4 of "Γ_v" and inserting in it the dependence with "x," we have:

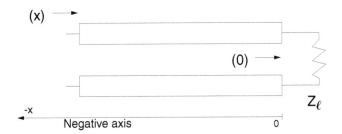

Figure 1.8.1

$$\Gamma_v(x) \overset{\perp}{=} \frac{v^r(x)}{v^p(x)} \equiv \frac{\Gamma_v(0)v^p e^{kx}}{v^p e^{-kx}} = \Gamma_v(0)\, e^{2kx} \tag{1.8.37}$$

which is clearly a simpler expression than "$\zeta(x)$." From 1.8.37 we can recognize how "$\Gamma_v(x)$" moves along "x" with two times the dependence of "$v(x)$." This result will be very useful when we use the Smith* chart to study matching problems.

The value of 1.8.37 also lies in the fact that from "$\Gamma_v(x)$," it is possible to have the normalized impedance "$\zeta_n(x)$" presented by the t.l. at the coordinate "x," still obtaining a simpler expression. In fact, from the expression 1.8.18 of "Γ_v" we have:

$$\zeta_n(x) = \frac{1 + \Gamma_v(x)}{1 - \Gamma_v(x)} \tag{1.8.38}$$

The impedance and voltage reflection coefficient along "x" are strictly related from 1.8.18. Of course, similar to "$\zeta(x)$" and "Γ_v," "$\Gamma_v(x)$" is, in general, a complex number.

Another important parameter is the voltage standing wave ratio, abbreviated with "VSWR," also indicated briefly with "SWR." This is defined by:

$$\text{VSWR} \overset{\perp}{=} |v|_M / |v|_m \tag{1.8.39}$$

where "$|v|_M$" and "$|v|_m$" indicate respectively the maximum and minimum of voltage modulus. From 1.8.39 we can recognize that the VSWR is always a real number. Also this parameter can be obtained from "Γ_v," and can be set as function of coordinate "x." Let us start to rewrite 1.3.12 using 1.8.10, i.e.:

$$v(x) = v^p e^{(-kx)} + v^r e^{(kx)} \tag{1.8.40}$$

which, with the definition of "Γ_v" becomes:

$$v(x) = v^p e^{(-kx)} + \Gamma_v v^p e^{(kx)} \tag{1.8.41}$$

Since "Γ_v" is a complex number, we can write:

$$\Gamma_v \overset{\perp}{=} |\Gamma_v| e^{j\varphi}{}_v \tag{1.8.42}$$

* Definition and use of the Smith chart will be studied in Section 1.12.

Assuming a lossless t.l., (i.e., $k \equiv jk_j$), inserting 1.8.42 in 1.8.41 and separating real and imaginary parts, 1.8.41 becomes:

$$v(x) = v^p\left[\cos\left(k_j x\right) + \left|\Gamma_v\right|\cos\left(k_j x + \varphi_v\right)\right] + jv^p\left[\left|\Gamma_v\right|\mathrm{sen}\left(k_j x + \varphi_v\right) - \mathrm{sen}\left(k_j x\right)\right] \quad (1.8.43)$$

whose modulus is:

$$\left|v(x)\right| = \left|v^p\right|\left[1 + \left|\Gamma_v\right|^2 + 2\left|\Gamma_v\right|\cos\left(2k_j x + \varphi_v\right)\right]^{0.5} \quad (1.8.44)$$

Using expression 1.7.1 for "k_j," Equation 1.8.44 can be written as:

$$\left|v(x)\right| = \left|v^p\right|\left\{1 + \left|\Gamma_v\right|^2 + 2\left|\Gamma_v\right|\cos\left[2\pi/(\lambda/2)x + \varphi_v\right]\right\}^{0.5} \quad (1.8.45)$$

where it is simple to recognize how the modulus of "$v(x)$" moves along "x" with a period "$\lambda/2$." In other words, every integer of half wavelength "$|v(x)|$" assumes the same value. Maximum and minimum of 1.8.45 correspond to the value of "1" or "-1" for the cosinus. Consequently, we have:

$$\left|v(x)\right|_M = \left|v^p\right|\left(1 + \left|\Gamma_v\right|\right) \quad (1.8.46)$$

$$\left|v(x)\right|_m = \left|v^p\right|\left(1 - \left|\Gamma_v\right|\right) \quad (1.8.47)$$

So, from the definition of VSWR we have:

$$\mathrm{VSWR} \overset{\perp}{=} \frac{1 + \left|\Gamma_v\right|}{1 - \left|\Gamma_v\right|} \quad (1.8.48)$$

Maximum VSWR_M and minimum VSWR_m value of VSWR can simply be evaluated since $|\Gamma_v|_M$ and $|\Gamma_v|_m$ have been given before. Therefore, we have:

$$\mathrm{VSWR}_m = 1 \quad \text{when} \quad \left|\Gamma_v\right| \equiv \left|\Gamma_v\right|_m = 0 \quad (1.8.49)$$

$$\mathrm{VSWR}_M = \infty \quad \text{when} \quad \left|\Gamma_v\right| \equiv \left|\Gamma_v\right|_M = 1 \quad (1.8.50)$$

The measure of VSWR is sometimes used to determine the value of a load impedance when it is known if its value is higher or lower than the reference impedance. If it is not known that the value of the load impedance is higher or lower than the reference impedance, then the measure is ambiguous, due to presence in 1.8.48 of the modulus of "Γ_v." Given a normalized load "$Z_{\ell n}$" it is simple to show that the associated "$|\Gamma_v|$" is also obtained for another load $Z_{\ell n}' = 1/Z_{\ell n}$. So, denormalizing expression 1.8.18 for "Γ_v" and inserting it into 1.8.48, we have that if $\Gamma_v > 0$, i.e., if $Z_\ell > \zeta$, then:

$$Z_\ell = \zeta\mathrm{VSWR} \quad (1.8.51)$$

while if $\Gamma_v < 0$, i.e., if $Z_\ell < \zeta$, then:

$$Z_\ell = \zeta/\text{VSWR} \tag{1.8.52}$$

We can conclude this section noting that the measure of the reflection coefficient is very useful, since impedances and VSWR are related to this parameter. We will show later how the reflection coefficient has an important role in the Smith chart.

1.9 NONUNIFORM TRANSMISSION LINES

A nonuniform transmission line is a line where characteristic impedance is a function of its longitudinal coordinate. If we want to use the concepts of line series impedance "Z_s" and line parallel admittance "Y_p," as we did in Section 1.2, in this case these quantities are also a function of coordinates. Nonuniform transmission lines are usually generically represented with a tapered profile, as indicated in Figure 1.9.1, just to remind one that characteristic impedance is a function of "x." The reader who is familiar with the representation with the technology of planar transmission lines did not make the error of thinking of Figure 1.9.1 as the case of two coupled microstrip or striplines. The two "conductors" indicated in Figure 1.9.1 just means that if one conductor is the hot one, then the other is the cold one as we used in Section 1.2 to describe the uniform transmission line. In the most general case, i.e., without restrictions about the shape of the t.l., the general transmission line theory cannot be used in this case, since for this theory the characteristic impedance must be constant along the line. Only the applications of Maxwell's* equations are correct in our case. In this text, Maxwell's equations are assumed to be known, but we have summarized them in Appendix A2. Complete explanations of these fundamental equations are found in a lot of texts on electromagnetism.[10,11,12] To explain how these equations are important, we say that without Maxwell's equations, the whole of electromagnetism would still be an obscure physics argument.

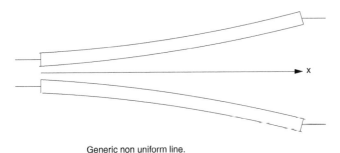

Generic non uniform line.

Figure 1.9.1

If we impose the restriction that for a coordinate increment "dx," the new impedance "$Z(x+dx)$" is only a small percent different than "$Z(x)$," then we can apply the theory used in Section 1.2 to the study of nonuniform transmission lines. So, in our case we can write:

$$\frac{dv(x)}{dx} = -Z_s(x)\, i(x) \tag{1.9.1}$$

---

* James Clark Maxwell, English physicist, born in Edinburgh in 1831, died in Cambridge in 1879.

$$\frac{di(x)}{dx} = -Y_p(x)v(x) \qquad (1.9.2)$$

Indicating the derivative operation with a prime " ′ " sign, and again deriving 1.9.1 it becomes:

$$v''(x) = -\left[Z_s'(x)i(x) + Z_s(x)i'(x)\right] \qquad (1.9.3)$$

Inserting in this equation the value of "i(x)" dervied from 1.9.1 and the value of "i′(x)" derived from 1.9.2, we have:

$$v''(x) - \frac{Z_s'(x)}{Z_s(x)}v'(x) - Z_s(x)Y_p(x)v(x) = 0 \qquad (1.9.4)$$

Of course, a similar equation can be obtained for current, resulting in:

$$i''(x) - \frac{Y_p'(x)}{Y_p(x)}i'(x) - Z_s(x)Y_p(x)i(x) = 0 \qquad (1.9.5)$$

In contrast with the case of uniform transmission lines, now there does not exist a general simple solution "v(x)" or "i(x)" for the second order nonlinear differential Equations 1.9.4 or 1.9.5, contrary to the simple exponential or hyperbolic solutions 1.3.5 or 1.3.6 for the uniform transmission lines. Simple solutions can only be found for particular expressions for "$Z_s(x)$" and "$Y_p(x)$." An example is the "exponential" lossless t.l., i.e., a line where its series impedance and parallel admittance can be written as:

$$Z_s(x) \equiv j\omega L e^{sx} \quad \text{and} \quad Y_p(x) \equiv j\omega C e^{-sx} \qquad (1.9.6)$$

where the constant "s" is related to the geometrical shape of the t.l. With 1.9.6, Equations 1.9.4 and 1.9.5 become:

$$v''(x) - sv'(x) + \omega^2 LCv(x) = 0 \qquad (1.9.7)$$

$$i''(x) + si'(x) + \omega^2 LCi(x) = 0 \qquad (1.9.8)$$

which are simple second order linear differential equations. Setting

$$v(x) \doteq v e^{k_v x} \qquad i(x) \doteq i e^{k_i x} \qquad (1.9.9)$$

respectively in 1.9.7 and 1.9.8 we have:

$$k_v = \frac{s \pm \left(s^2 - 4\omega^2 LC\right)^{0.5}}{2} \qquad (1.9.10)$$

$$k_i = \frac{-s \pm \left(s^2 - 4\omega^2 LC\right)^{0.5}}{2} \qquad (1.9.11)$$

Note that "k_v" and "k_i" are never completely imaginary, which means that propagation is affected by losses. In particular, for:

$$\omega < s/2\,(LC)^{0.5}$$

these quantities are purely real, which means that there will be a complete attenuated propagation. The frequency:

$$f_c = s/4\pi\,(LC)^{0.5} \tag{1.9.12}$$

is called the "cutoff frequency," and represents the minimum frequency that must be overcome to have propagation, also if attenuated.

It is possible to have an equation for a nonuniform t.l. that doesn't require any limitation about the variation of "$\zeta(x)$." This equation relates the variation of reflection coefficient along the line with the variation of its characteristic impedance. To do that, let us define with "Z_i" the input impedance of the t.l. at the coordinate "x," and "$Z_i + dZ_i$" the impedance at the coordinate "$x + dx$." The impedance "$Z_i + dZ_i$" can be regarded as the load impedance for the line length "dx," whose input impedance "Z_i" we want to evaluate. For simplicity we will suppose the t.l. to be lossless. From Equation 1.5.33 we have:

$$Z_i(x) = \frac{j\zeta_m^{\,2}\,\mathrm{tg}\left(k_j dx\right) + \zeta_m\left(Z_i + dZ_i\right)}{\zeta_m + j\left(Z_i + dZ_i\right)\mathrm{tg}\left(k_j dx\right)} \tag{1.9.13}$$

with "ζ_m" the mean value of the characteristic impedance in the element "dx." We now suppose that "dx" is so small that $\mathrm{tg}(k_j dx) \approx k_j dx$ and that the product of infinitesimal terms can be neglected. So, the previous equation becomes:

$$Z_i(x) = \left[j\zeta_m^{\,2}k_j dx + \zeta_m\left(Z_i + dZ_i\right)\right]\left[\zeta_m + jZ_i k_j dx\right]^{-1} \tag{1.9.14}$$

Now, expand the last term in the McLaurin series, stopping at the second term due to the small value of "dx." We have:

$$\left[\zeta_m + jZ_i(x)k_j dx\right]^{-1} \approx \zeta_m^{-1}\left[1 - jZ_i(x)k_j dx/\zeta_m\right]$$

which, inserted in 1.9.14 results in:

$$\frac{dZ_i(x)}{dx} = jk_j\left[Z_i^{\,2}(x)/\zeta_m - \zeta_m\right] \tag{1.9.15}$$

From 1.8.38, denormalized and applied to our case, we have:

$$Z_i(x) = \frac{1 + \Gamma_v(x)}{1 - \Gamma_v(x)}\,\zeta_m \tag{1.9.16}$$

which, inserted in 1.9.15 results in:

$$\frac{dZ_i(x)}{dx} = jk_j\left(\frac{1 + \Gamma_v(x)}{1 - \Gamma_v(x)} - \zeta_m\right) \tag{1.9.17}$$

Deriving the Equation 1.9.16 with respect to the coordinate "x" we have:

$$\frac{dZ_i(x)}{dx} = \frac{1 + \Gamma_v(x)}{1 - \Gamma_v(x)}\frac{d\zeta_m}{dx} + \frac{2\zeta_m}{\left[1 - \Gamma_v(x)\right]^2}\frac{d\Gamma_v(x)}{dx} \tag{1.9.18}$$

Inserting 1.9.17 into 1.9.18 we have:

$$\frac{d\Gamma_v(x)}{dx} = 2jk_j\Gamma_v(x) - \frac{1 - \Gamma_v^2(x)}{2}\frac{d\left[\ln\left(\zeta_m\right)\right]}{dx} \tag{1.9.19}$$

Similar to 1.9.4 or 1.9.5, a general simple solution of 1.9.19 doesn't exist. The previous equation is also said to be a Riccati* equation.

If we still apply the same restriction used above for variation of "ζ" with coordinate, we can simply evaluate the voltage reflection coefficient at the input of the nonuniform transmission line. Let us examine Figure 1.9.2. Part a represents the original nonuniform t.l., while part b indicates a possible approximation, made with steps of uniform lines. We can observe how an impedance variation "$d\zeta$" moving from "x" at "$x + dx$" will generate a variation "$d\Gamma_v$" given by:

$$d\Gamma_v \equiv \frac{(\zeta + d\zeta) - \zeta}{(\zeta + d\zeta) + \zeta} \tag{1.9.20}$$

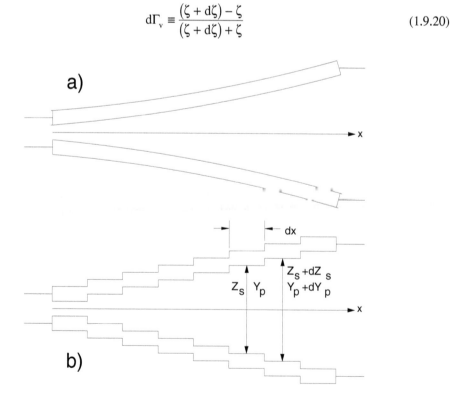

Figure 1.9.2

* I.F. Riccati, Italian mathematician, born in Venezia in 1676 and died in Treviso in 1754.

Assuming "dζ" is negligible with respect to "2ζ," the previous equation becomes:

$$d\Gamma_v \approx \frac{d\zeta}{2\zeta} \equiv \frac{d\left[\ln\left(\zeta\right)\right]}{2} \qquad (1.9.21)$$

Now, suppose the t.l. is lossless. According to 1.8.37, at the input of the section "dx," the reflection coefficient "$d\Gamma_{vi}$," will be:

$$d\Gamma_{vi}(x) = d\Gamma_v e^{-2jk_j x} \qquad (1.9.22)$$

With respect to Equation 1.8.37, now the exponential changes sign, because in this theory the distance "x" is considered an absolute sign while for 1.8.37 the distance "x" has a negative sign, as shown in Figure 1.8.1.

If we assume the multiple reflections along the t.l. are negligible, which still means there is a slow variation of "ζ" with the coordinate, we can evaluate the input reflection coefficient "Γ_v" as the continuous sum of the previous terms, i.e.:

$$\Gamma_v = \int_0^\ell e^{-2j\beta x} d\Gamma_{vi}(x) \equiv \frac{1}{2} \int_0^\ell e^{-2j\beta x} \frac{d\left[\ln\left(\zeta\right)\right]}{dx} dx \qquad (1.9.23)$$

where "ℓ" is the length of the nonuniform t.l. When we study tapered directional couplers* we will use the previous formula to define the coupling for such important devices.

1.10 QUARTER WAVE TRANSFORMERS

The argument of the "quarter wave" transmission line transformers is of great importance in all RF and microwave devices. For this reason, we think it is necessary to go more deeply into the theory of this topic. We will only consider the ideal case of "TEM," lossless transmission lines, since the specialization of the general theory is carried out in the following chapters for every planar transmission line used as a transformer.

As we said in Section 1.7, the matching between a load and a source using a quarter wave transmission line is ideally perfect only at the design frequency, i.e., at that frequency where the electrical length of the line is a quarter wavelength of the signal guided by the t.l. The operating situation is indicated in Figure 1.10.1. The line is characterized by a characteristic impedance "ζ" and an electrical length "θ," and will be evaluated as lossless.

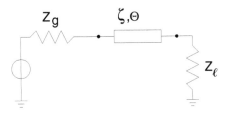

Figure 1.10.1

* Directional coupler will be studied in Chapter 7 and Chapter 8.

In most practical cases it is not required that the matching be exactly perfect, i.e., with a VSWR = 1, but usually it is required that the VSWR be lower than a fixed value. The consequence of this less stringent requirement is that we no longer have a sole frequency for matching, but instead we have an operating bandwidth where the matching can be accepted. So, it is useful to determine the frequency characteristics of the single quarter wave transformer. This can be performed very easily, remembering expression 1.5.33, which gave us the input impedance "Z_i" of a transmission line of impedance "ζ" and electrical length "θ" terminated in a load "Z_ℓ." We have:

$$Z_i = \frac{Z_\ell \zeta + j\zeta^2 \tan(\theta)}{\zeta + jZ_\ell \tan(\theta)}$$

(1.10.1)

and the voltage reflection coefficient "Γ" and VSWR at its input, for the situation in Figure 1.10.1 are:

$$\Gamma = \frac{Z_i - Z_g}{Z_i + Z_g}$$

(1.10.2)

$$\text{VSWR} = \frac{1 + |\Gamma|}{1 - |\Gamma|}$$

(1.10.3)

where "Z_g" is the impedance of the generator which feeds the transformer. In Figure 1.10.2 we have represented the two previous equations vs. the normalized frequency "f_n" for two values of "R," where $R \triangleq Z_\ell/Z_g$. "f_n" is given by the ratio of the general variable frequency "f" and that frequency where the transmission line is $\lambda/4$ long. We see how the matching is exact for frequencies where the transformer is an odd multiple of a quarter wavelength, regardless of the transformer ratio "R." For frequencies outside the designed one, the "VSWR" value depends on the transformer ratio. This is simple to understand. Remember that if the load impedance is equal to the source one from 1.7.8, it follows that the impedance of the transformer transmission line is equal to this value, and no mismatching occurs.

At this point, remembering what we have said in the previous sections, the reader should realize that no theoretical reason exists not to realize the matching using more than one section since each section can transform the impedance in a value that can still be transformed to the desired value by the other quarter wavelength section. This situation is represented in Figure 1.10.3, where a number of "s" sections of the quarter wave transformer are connected in series. In the most general case, the only requirements between the impedances of the transmission line are given by the following relationships:

$$\hat{\zeta}_1 = \sqrt{R} \qquad \text{for } s = 1$$

(1.10.4)

$$\frac{\hat{\zeta}_1 \hat{\zeta}_3 \hat{\zeta}_5}{\hat{\zeta}_2 \hat{\zeta}_4 \hat{\zeta}_6} \cdots \frac{\hat{\zeta}_{s-2}}{\hat{\zeta}_{s-1}} \hat{\zeta}_s = \sqrt{R} \qquad \text{for } s = 3, 5, 7..$$

(1.10.5)

i.e., an odd number "s" of sections, or

$$\frac{\hat{\zeta}_1 \hat{\zeta}_3 \hat{\zeta}_5}{\hat{\zeta}_2 \hat{\zeta}_4 \hat{\zeta}_6} \cdots \frac{\hat{\zeta}_{s-1}}{\hat{\zeta}_s} = \frac{1}{\sqrt{R}} \qquad \text{for } s = 2, 4, 6..$$

(1.10.6)

i.e., an even number "s" of sections. The symbol "∧" over the name of the impedances means that their values are normalized to the system measurement impedance, usually 50 Ohm. It is clear that

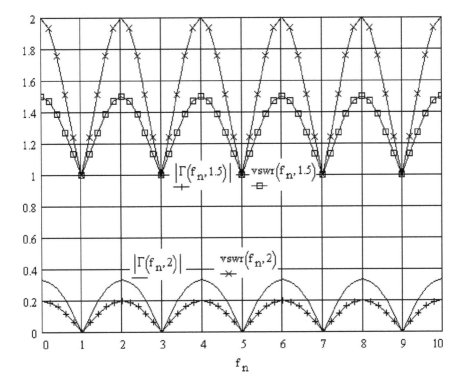

Figure 1.10.2

in the case of a multiple quarter wavelength transmission line transformer, the design is not unique since it is theoretically possible to choose the impedances in an infinite number of ways, with the only requirements stated by Equations 1.10.5 and 1.10.6.

The first study on the problem to design the network indicated in Figure 1.10.3 was made by R.E. Collin.[13,14,15,16,17] In his work, Collin studied the case of synthesizing the network with a VSWR in the matching bandwidth of Chebyshev shape. In Appendix A4 we have reported the expressions of Chebyshev polynomials, together with their shape. Here we report the first three Chebyshev polynomials together with the recursion formula, as follows:

$$\begin{cases} T_1(x) = x \\ T_2(x) = 2x^2 - 1 \\ T_3(x) = 4x^3 - 3x \\ T_4(x) = 2xT_{n-1}(x) - T_{n-2}(x) \end{cases} \qquad (1.10.7)$$

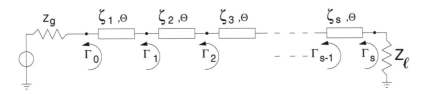

Figure 1.10.3

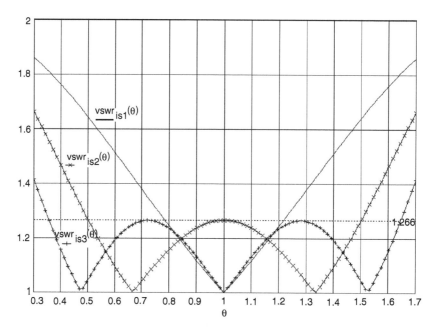

Figure 1.10.4

It is known that Chebyshev polynomials* can be used in the design of signal handling networks such as filters, directional couplers, transmission line transformers. The resulting network has the greatest possible useful bandwidth, after fixing an acceptable ripple in the desired bandwidth. Conversely, after fixing an operating bandwidth, the design with Chebyshev polynomials results in a network with the lowest possible ripple in the bandwidth. It is also known that accepting a higher value of ripple, the bandwidth increases. Applying these concepts to our case, we find that when using a transformer as indicated in Figure 1.10.3, with s ≥ 2, it is possible to have a much wider useful bandwidth with respect to the single quarter wave transformer, when a desired value of VSWR has been defined. In Figure 1.10.4 we have indicated the input VSWR of one, two, and three sections of quarter wave transmission line transformer, with an impedance ratio R = 2. From this figure it is evident that we have a large increase in the operating bandwidth when the number of sections is increased.

To see how the synthesis of the network in Figure 1.10.3 can be performed using Chebyshev polynomials, we will follow the original simplified theory of Collin. In this theory, it is assumed that the reflection coefficient value between each transition is so small that multiple reflections due to different transitions may be neglected. In other words, the reflection coefficient "Γ" at the input of the network in Figure 1.10.3 can be written as:

$$\Gamma = \rho_0 + \rho_1 e^{-j2\theta} + \rho_2 e^{-j4\theta} + \rho_3 e^{-j6\theta} + \ldots + \rho_{s-1} e^{-j2(s-1)\theta} + \rho_s e^{-j2s\theta} \qquad (1.10.8)$$

Then, it is assumed that the reflection coefficients are symmetrical with respect to the middle of the transformer, so that we may write:

$$\Gamma_0 \equiv \Gamma_s, \; \Gamma_1 \equiv \Gamma_{s-1}, \; \Gamma_2 \equiv \Gamma_{s-2}, \; \ldots \qquad (1.10.9)$$

---

* See Appendix A4 for Chebyshev polynomials and their characteristics.

It is important to observe that the previous relationship does not assume that the impedance value of the transmission lines used in the transformer is symmetrical with respect to its middle. In fact, from 1.8.18, which defines the reflection coefficient, it is evident that two reflection coefficients may also be equal if the impedance values are different. With the use of 1.10.8, the Equation 1.10.7 may be written as:

$$\Gamma = e^{-js\theta}\left[\rho_0\left(e^{js\theta}+e^{-js\theta}\right)+\rho_1\left(e^{j(S-2)\theta}+e^{-j(S-2)\theta}\right)+\rho_2\left(e^{j(S-4)\theta}+e^{-j(S-4)\theta}\right)+...+L(\theta)\right] \qquad (1.10.10)$$

where

$$L(\theta) = \rho_{(S-1)/2}\left(e^{j\theta}+e^{-j\theta}\right)$$

for an odd number of sections, or

$$L(\theta) = \rho_{S/2}$$

for an even number of sections. If now we remember the trigonometric relationship:

$$2\cos x = e^{jx}+e^{-jx}$$

the Equation 1.10.10 may be rewritten as:

$$\Gamma = 2e^{-js\theta}\left[\rho_0\cos(s\theta)+\rho_1\cos(s-2)\theta+\rho_2\cos(s-4)\theta+..+\rho_i\cos(s-2i)\theta+..+L'(\theta)\right] \qquad (1.10.11)$$

where

$$L'(\theta) = \rho_{(S-1)/2}\cos\theta$$

for an odd number of sections, or

$$L'(\theta) = 0.5\rho_{S/2}$$

for an even number of sections. In both of these situations, Equation 1.10.11 may also be regarded as a polynomial in "$\cos\theta$" of degree "s" with "π" as periodicity, and in the most general case with only even or odd powers of "$\cos\theta$." The reader who is familiar with network synthesis can remember how a transfer function that has such a polynomial representation may be synthesized according to Chebyshev polynomials. So, in this case, we can also synthesize a multisection quarter wavelength transformer with a VSWR with a Chebyshev shape, as indicated in Figure 1.10.4. The first thing to do is to transform the interval in "x" of extension "± 1" in the interval "$\pm\theta_w$" where "Γ" has an equal-ripple shape. To do that, we substitute the variable "x" in $T_n(x)$ so that:

$$x \overset{\perp}{=} \cos\theta/\cos\theta_m$$

where "θ" is the electrical length and "θ_m" is the absolute bandwidth limit with $\theta = \pi/2$ as reference. So, the operating bandwidth is "$\pi/2 \pm \theta_m$." Another definition of bandwidth used is called "fractional bandwidth" and is defined as

$$BW = 4\left(\pi/2 - \theta_m\right)\big/\pi \tag{1.10.12}$$

We can recognize that the Chebyshev function $T_n(\sec\theta_m\cos\theta)$ has an equal ripple shape with unity as the maximum value just in the operating bandwidth "$\pi/2 \pm \theta_m$." We may now write:

$$\Gamma \doteq \rho_m e^{-js\theta} T_n\left(\sec\theta_m\cos\theta\right) \tag{1.10.13}$$

where "ρ_m" is the maximum desired value of "Γ" in the bandwidth. The value of "ρ_m" can be easily related to the impedance ratio "R" and Chebyshev functions. In fact, for $\theta = 0$ we have:

$$\Gamma \equiv \frac{Z_\ell - Z_g}{Z_\ell + Z_g}$$

Now, from the series expansion:[18]

$$\ln(Z) = 2\left[\frac{Z-1}{Z+1} + \frac{(Z-1)^3}{3(Z+1)^3} + \frac{(Z-1)^5}{5(Z+1)^5} + \ldots\right]$$

and remembering the hypothesis of small reflections, we may write:

$$\ln(Z) \approx 2\frac{Z-1}{Z+1} \tag{1.10.14}$$

Thus, from 1.10.13 for $\theta = 0$ and 1.10.14 we have:

$$\rho_m = \frac{\ln(R)}{2T_n\left(\sec\theta_m\right)} \tag{1.10.15}$$

Note that since "θ_m" is known, it is clear from 1.10.7 that $T_n(\sec\theta_m\cos\theta)$ are polynomials of degree "n" in $\cos\theta$. Since we have the relationship:

$$(\cos\theta)^n = 2^{1-n}\left[c_0\cos n\theta + c_1\cos(n-2)\theta + c_2\cos(n-4)\theta + .. + c_i\cos(n-2i)\theta + .. + U(\theta)\right] \tag{1.10.16}$$

where "c_i" is the binomial coefficient, given by:

$$c_i = \frac{n!}{(n-i)!i!} \tag{1.10.17}$$

and

$$U(\theta) = c_{(n-1)/2}\cos\theta$$

for an odd number of sections, or

$$U(\theta) = 0.5 c_{n/2}$$

for an even number of sections, $T_n(\sec\theta_m\cos\theta)$ may also be expressed as a polynomial in $\cos(n\theta)$ with "n" as an integer number. So, from 1.10.7 and 1.10.16 for the first four Chebyshev polynomials we may write:

$$\begin{cases} T_1\left(\sec\theta_m \cos\theta\right)=\sec\theta_m \cos\theta \\[2mm] T_2\left(\sec\theta_m \cos\theta\right)=\sec^2\theta_m \cos 2\theta + \sec^2\theta_m - 1 \\[2mm] T_3\left(\sec\theta_m \cos\theta\right)=\sec^3\theta_m \cos 3\theta + \left(3\sec^3\theta_m - 3\sec\theta_m\right)\cos\theta \\[2mm] T_4\left(\sec\theta_m \cos\theta\right)=\sec^4\theta_m \cos 4\theta + \left(4\sec^4\theta_m - 4\sec^2\theta_m\right)\cos 2\theta + 3\sec^4\theta_m - 4\sec^2\theta_m \end{cases} \qquad (1.10.18)$$

At this point, the synthesis of a Chebyshev transformer can be made equating the second member of 1.10.11 with the second member of 1.10.13. As an example, let us assume a three section transformer with an impedance ratio of "2" and an operating bandwidth "BW" of 1. From 1.10.12 and 1.10.15 we have:

$$\theta_m = 0.785 \quad \text{and} \quad \rho_m = 0.022$$

Then, using 1.10.11, 1.10.13, and 1.10.18 we have:

$$\rho_0 = 0.5\rho_m \sec^3\left(\theta_m\right) = 0.032$$

$$\rho_1 = 1.5\rho_m \left[\sec^3\left(\theta_m\right) - \sec\left(\theta_m\right)\right]$$

For a two section transformer with the same "BW" and "R" we have:

$$\theta_m = 0.785 \quad \rho_m = 0.116$$

$$\rho_0 = 0.5\rho_m \sec^2\left(\theta_m\right) = 0.116$$

$$\rho_1 = \rho_m \left[\sec^2\left(\theta_m\right) - 1\right] = 0.116$$

The dependence vs. "θ" of the reflection coefficient "Γ_2" and "Γ_3" for a two or three element transformer is given in Figure 1.10.5. We can see how, for a fixed bandwidth, the three section transformer gives a smaller ripple when compared to a two section transformer. We may also decide to fix "ρ_m" and "R," instead of "BW" and "R." In this case, a typical situation is indicated in Figure 1.10.6. We see how for a fixed ripple, the three section transformer gives a wider operating bandwidth when compared to the two element transformer.

We could suppose that it should be desirable to use a higher number of sections for a good transformer. It is rare that a number of sections higher than three is used due to the increasing attenuation and discontinuities that are encountered in this type of transformer. In the coming chapters we will see how every change of impedance in planar transmission line technology is always a source of discontinuity, and for this reason it is preferable to contain the number of sections

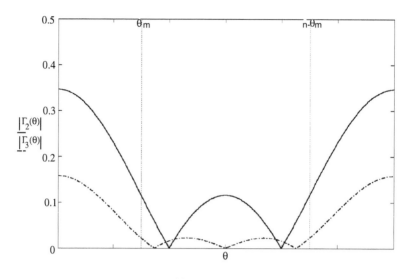

Figure 1.10.5

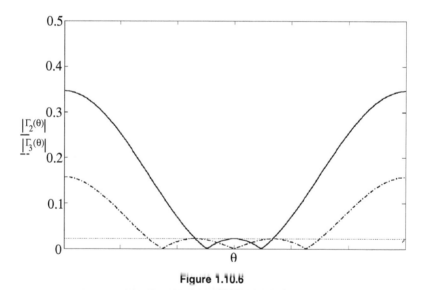

Figure 1.10.6

to a small value, in practice, lower than three. With the theory studied in this section the reader can easily synthesize a transformer for any desired "R" and "BW," with up to four sections.

1.11 COUPLED TRANSMISSION LINES

Two lines are said to be coupled when the electromagnetic* field supported by a t.l. can induce an e.m. field on another t.l., whose intensity is defined as negligible. Theoretically, when an electromagnetic field is emitted from a source, its intensity is zero only at an infinite distance from the source. So, any other structure that is able to support an e.m. field when placed in a region where an e.m. field exists will receive such a field, unless it is set at an infinite distance. In practice,

---

* The word "electromagnetic" will often be simply indicated with "e.m."

when the two structures, one of which generates an e.m. field, are at a distance for which the induced e.m. to the other structure cannot be used due the low intensity, they are said to be "practically electromagnetically isolated," or simply "isolated." So, the definition of e.m. coupling is always practical, and sometimes also subjective. Vice versa, as we said, if the induced field is not negligible, the two structures are said to be "coupled."

In general, two t.l.s can be set in close proximity for two reasons:

a. Coupling is desired
b. It is not possible to separate more than a value these lines

For case a, the concept of coupling is used, while for case b, since the proximity is not desired, the concept of "crosstalk"* is used. This argument will be studied at the end of this section.

Examples where coupled lines are required are directional couplers,** filters, and to transmit differential signals in "ECL"*** circuits.

Coupled lines are simply indicated as shown in Figure 1.11.1. In general, since transmission lines are built to have minimum losses, they have low e.m. radiation. So, coupling between lines is obtained in mechanically different ways. For example, coaxial lines are not very suitable to be simply coupled, and to do that requires opening the outer conductor to set the hot conductors of the two cables close together. In microstrip technology, coupling is much more simple, and it is enough to set the two hot conductors close together.

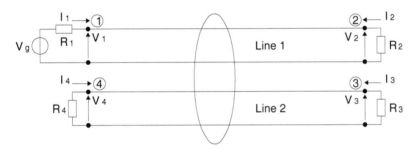

Figure 1.11.1

Coupled lines can be studied using the general transmission line theory we described in Section 1.2. In this case, in addition to the series impedance "Z_s" and parallel admittance "Y_p" of each t.l., we also have mutual series admittance "Z_{sm}" and mutual parallel admittance "Y_{pm}." These last two new quantities are respectively representative of magnetic and electric coupling. In addition, if "Z_q" and "Y_p" are relative to each t.l. they are evaluated when the other line is present.

From the theory developed in Section 1.2 we can write:

$$\frac{dv_1(x)}{dx} = -Z_{s1}\, i_1(x) - Z_{sm}\, i_2(x) \tag{1.11.1}$$

$$\frac{dv_2(x)}{dx} = -Z_{s2}\, i_2(x) - Z_{sm}\, i_1(x) \tag{1.11.2}$$

$$\frac{di_1(x)}{dx} = -Y_{p1}\, v_1(x) - Y_{pm}\, v_2(x) \tag{1.11.3}$$

* The word "crosstalk" is often abbreviated with "xtalk."
** Directional couplers will be studied in next chapters.
*** "ECL" is the abbreviation of "Emitter Coupled Logic," a logic circuitry used in fast digital devices.

$$\frac{di_2(x)}{dx} = -Y_{p2}\, v_2(x) - Y_{pm}\, v_1(x) \tag{1.11.4}$$

Note that in analogy with Section 1.2, all the "Z" and "Y" are defined per u.l.

These equations can be simply extracted using Figure 1.11.2 as reference. In part a we have assumed a single t.l., for example line "1," with a voltage controlled generator of value "$Z_{sm}i_2$," which represents the effect of the magnetic coupling. So, Equation 1.11.1 results from the application of the "Kirchoff voltage loop law" at the network in Figure 1.11.2a. To obtain Equation 1.11.2 it is sufficient to apply the same law to Figure 1.11.2a, substituting every "1" and "2" with "2" and "1." In part b we have assumed a single t.l., for example line "1," with a current controlled generator of value "$Y_{pm}v_2$," which represents the effect of the voltage coupling. So, Equation 1.11.3 results from the application of the "Kirchoff current law" at the network in Figure 1.11.2b. To obtain Equation 1.11.4 it is sufficient to apply the same law to Figure 1.11.2b, substituting every "1" and "2" with "2" and "1."

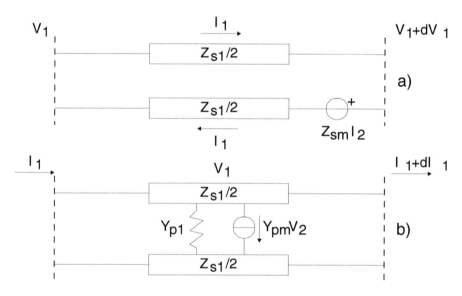

Figure 1.11.2

To obtain something similar to the monodimensional wave equation we defined in Section 1.2, let us start to derive Equations 1.11.1 and 1.11.2. Assuming uniform coupled lines, i.e., impedances and admittances do not vary with coordinate, we have:

$$\frac{d^2 v_1(x)}{dx^2} = -Z_{s1}\frac{di_1(x)}{dx} - Z_{sm}\frac{di_2(x)}{dx} \tag{1.11.5}$$

$$\frac{d^2 v_2(x)}{dx^2} = -Z_{s2}\frac{di_2(x)}{dx} - Z_{sm}\frac{di_1(x)}{dx} \tag{1.11.6}$$

Inserting Equations 1.11.3 and 1.11.4 in the previous two equations we have:

$$\frac{d^2 v_1(x)}{dx^2} - a_1 v_1(x) - b_1 v_2(x) = 0 \tag{1.11.7}$$

$$\frac{d^2 v_2(x)}{dx^2} - a_2 v_2(x) - b_2 v_1(x) = 0 \qquad (1.11.8)$$

where:

$$a_1 \overset{\perp}{=} Y_{p1} Z_{s1} + Y_{pm} Z_{sm} \qquad (1.11.9)$$

$$a_2 \overset{\perp}{=} Y_{p2} Z_{s2} + Y_{pm} Z_{sm} \qquad (1.11.10)$$

$$b_1 \overset{\perp}{=} Y_{pm} Z_{s1} + Y_{p2} Z_{sm} \qquad (1.11.11)$$

$$b_2 \overset{\perp}{=} Y_{pm} Z_{s2} + Y_{p1} Z_{sm} \qquad (1.11.12)$$

From 1.10.7 we have:

$$\frac{d^2 v_2(x)}{dx^2} = \left(\frac{d^4 v_1(x)}{dx^4} - a_1 \frac{d^2 v_1(x)}{dx^2} \right) \bigg/ b_1 = 0 \qquad (1.11.13)$$

Extracting "$v_2(x)$" from 1.11.8 we have:

$$\frac{d^2 v_1(x)}{dx^2} - a_1 v_1(x) - b_1 \frac{d^2 v_2(x)}{a_1 dx^2} + \frac{b_1 b_2}{a_2} v_1(x) = 0 \qquad (1.11.14)$$

and inserting 1.11.13 in this equation we have:

$$\frac{d^4 v_1(x)}{dx^4} - (a_1 + a_2) \frac{d^2 v_1(x)}{dx^2} - (a_1 a_2 - b_1 b_2) v_1(x) = 0 \qquad (1.11.15)$$

This equation is an example of a fourth order linear differential equation. The solution of this equation is simple and can be found by setting $v_1(x) \overset{\perp}{=} v_1 e^{kx}$. With this substitution in 1.11.15, we have four solutions for "k," given by:

$$k_1 = k_c \quad k_2 = -k_c \quad k_3 = k_p \quad k_4 = -k_p \qquad (1.11.16)$$

with:

$$k_c = \left\{ 0.5(a_1 + a_2) + 0.5 \left[(a_1 - a_2)^2 + 4 b_1 b_2 \right]^{0.5} \right\}^{0.5} \qquad (1.11.17)$$

$$k_p = \left\{ 0.5(a_1 + a_2) - 0.5 \left[(a_1 - a_2)^2 + 4 b_1 b_2 \right]^{0.5} \right\}^{0.5} \qquad (1.11.18)$$

So, the general solution is a linear combination of four exponentials, i.e.:

$$v_1(x) = v_{1c}{}^+ e_c^{(-k_c x)} + v_{1c}{}^- e_c^{(k_c x)} + v_{1p}{}^+ e_p^{(-k_p x)} + v_{1p}{}^- e_p^{(k_p x)} \qquad (1.11.19)$$

It should be clear that proceeding in a similar way, it is possible to obtain equations similar to 1.11.19 for voltage "$v_2(x)$" and currents "$i_1(x)$" and "$i_2(x)$." Analogous to the case of a single transmission line, the exponentials with negative signs are called "progressive" terms, while the exponentials with positive signs are called "regressive" terms. So, we can say that effectively we have to evaluate only two propagation constants, "k_c" and "k_p." This circumstance leads us to say that coupled transmission lines, in general, support two "modes of propagation," that is, only two spatial dependences exist.

It is interesting to evaluate the voltage ratio "v_2/v_1," since it will introduce one important study method for coupled lines. Inserting the expression $v_2(x) \perp v_2 e^{kx}$ in 1.11.8 we have:

$$\frac{v_2(x)}{v_1(x)} \perp R = \frac{b_2}{k^2 - a_2} \tag{1.11.20}$$

Evaluating this expression for the two cases of $k = \pm k_c$ and $k = \pm k_p$, we have:

$$R_c \perp v_2/v_1 \text{ for } k = \pm k_c \equiv \left\{ a_2 - a_1 + \left[(a_1 - a_2)^2 + 4 b_1 b_2 \right]^{0.5} \right\} \bigg/ 2b_1 \tag{1.11.21}$$

$$R_p \perp v_2/v_1 \text{ for } k = \pm k_p \equiv \left\{ a_2 - a_1 - \left[(a_1 - a_2)^2 + 4 b_1 b_2 \right]^{0.5} \right\} \bigg/ 2b_1 \tag{1.11.22}$$

We can now obtain voltage "$v_2(x)$" using the same parameters of 1.11.19, and write:

$$v_2(x) = R_c v_{1c}{}^+ e^{(-k_c x)} + R_c v_{1c}{}^- e^{(k_c x)} + R_p v_{1p}{}^+ e^{(-k_p x)} + R_p v_{1p}{}^- e^{(k_p x)} \tag{1.11.23}$$

Of course, currents can be obtained inserting 1.11.19 and 1.11.23 into 1.11.7 and 1.11.8, resulting in:

$$i_1(x) = Y_{c1} v_{1c}{}^+ e^{(-k_c x)} - Y_{c1} v_{1c}{}^- e^{(k_c x)} + Y_{p1} v_{1p}{}^+ e^{(-k_p x)} - Y_{p1} v_{1p}{}^- e^{(k_p x)} \tag{1.11.24}$$

$$i_2(x) = Y_{c2} R_c v_{1c}{}^+ e^{(-k_c x)} - Y_{c2} R_c v_{1c}{}^- e^{(k_c x)} + Y_{p1} R_p v_{1p}{}^+ e^{(-k_p x)} + Y_{p1} R_p v_{1p}{}^- e^{(k_p x)} \tag{1.11.25}$$

where:

$$Y_{c1} \perp k_c \frac{Z_{s2} - Z_{sm} R_c}{Z_{s1} Z_{s2} - Z_{sm}^2} \tag{1.11.26}$$

$$Y_{c2} \perp \frac{k_c}{R_c} \frac{Z_{s1} R_c - Z_{sm}}{Z_{s1} Z_{s2} - Z_{sm}^2} \tag{1.11.27}$$

$$Y_{p1} \perp k_p \frac{Z_{s2} - Z_{sm} R_p}{Z_{s1} Z_{s2} - Z_{sm}^2} \tag{1.11.28}$$

$$Y_{c1} \perp \frac{k_p}{R_p} \frac{Z_{s1} R_p - Z_{sm}}{Z_{s1} Z_{s2} - Z_{sm}^2} \tag{1.11.29}$$

If the t.l. have the characteristic that:

$$Z_{s2} \equiv Z_{s1} \overset{\perp}{=} Z_s \quad \text{and} \quad Y_{p2} \equiv Y_{p1} \overset{\perp}{=} Y_p \qquad (1.11.30)$$

then $a_2 \equiv a_1$ and $b_2 \equiv b_1$, and in this case:

$$R_2 = 1 \quad \text{and} \quad R_p = -1 \qquad (1.11.31)$$

When the conditions in 1.11.30 hold, the coupled lines are said to be "symmetrical." In this case, the mode corresponding to $R_c = 1$ is said to be "even" since voltages on the two lines are of the same amplitude and phase. Conversely, the mode corresponding to $R_p = -1$ is said to be "odd" since voltages on the two lines are of the same amplitude, but 180° out of phase. For symmetrical lines, Equations 1.11.26 ÷ 1.11.29 become:

$$Y_{c1} = \frac{k_c}{Z_s + Z_{sm}} \equiv Y_{c2} \overset{\perp}{=} Y_e \qquad (1.11.32)$$

$$Y_{p1} = \frac{k_p}{Z_s + Z_{sm}} \equiv Y_{p2} \overset{\perp}{=} Y_o \qquad (1.11.33)$$

and the propagation constants become:

$$k_c = \pm \left[\left(Z_s + Z_{sm} \right) \left(Y_p + Y_{pm} \right) \right]^{0.5} \overset{\perp}{=} k_e \qquad (1.11.34)$$

$$k_p = \pm \left[\left(Z_s - Z_{sm} \right) \left(Y_p - Y_{pm} \right) \right]^{0.5} \overset{\perp}{=} k_p \qquad (1.11.35)$$

Another important case is:

$$Y_{p1} Z_{s1} \equiv Y_{p2} Z_{s2} \quad \text{and} \quad Y_{pm}/Z_{sm} \equiv -\left(Y_{p1} Y_{p2}/Z_{s1} Z_{s1} \right) \qquad (1.11.36)$$

from which it follows that:

$$k_c \equiv k_p \qquad (1.11.37)$$

and the t.l. is said to be in a homogeneous media.

Using the concepts of "even" and "odd" modes for symmetrical lines, the results of $R_c = 1$ and $R_p = -1$ have been generalized and, using the principle of superposition effects, any linear coupled line structure, symmetrical or not, is studied using the "even-odd" excitation method. In this method, "ζ_e" is defined as the impedance of a t.l. to ground, considering the other line present, with an even excitation; similarly, "ζ_o" is defined as the impedance of a t.l. to ground, considering the other line present, with an odd excitation. A representation of these impedances for the case of "side coupled striplines"* is depicted in Figure 1.11.3, respectively, in parts a and b. Here, for simplicity, only the electrical field is indicated.

---

* Side coupled striplines will be studied in Chapter 6.

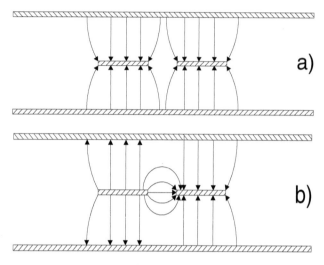

Figure 1.11.3

In the "even-odd" excitation method, which we will use extensively in our text, any excitation to a coupled line can be decomposed as the sum of an even and odd excitation, and the lines are characterized by their "even" and "odd" admittances or impedances, together with other "even" and "odd" line characteristics like propagation constants and phase constants.

A generalized typical decomposition in even-odd excitations is depicted in Figure 1.11.4. In part a an operative situation is indicated, where a generator is connected to a port with the other three ports correctly terminated. In part b, corresponding even excitation is indicated, while the odd excitation is shown in part c. Note that the sum of even and odd excitations still gives the real excitation indicated in Figure 1.11.3a. It is important to observe that for every even and odd excitation the lines are considered decoupled and respectively of impedance "ζ_e"and "ζ_o." The even and odd method will be used in the next chapters where coupled lines devices will be studied.

Another very important topic of coupled line theory is the research of the conditions between "ζ_e," "ζ_o," and termination impedance "Z_ℓ" to have a t.l. input impedance still equal to "Z_ℓ." To begin, let us assume the line to be lossless and to be in homogeneous media. Then we have $k_c \equiv k_p \doteq jk_j$, and the "[Z]" matrix of the line can be written as:

$$(Z_i) = \begin{pmatrix} \zeta_m/\tan\left(k_j\ell\right) & \zeta_m/j\,\mathrm{sen}\left(k_j\ell\right) \\ \zeta_m/\mathrm{sen}\left(k_j\ell\right) & \zeta_m/j\,\tan\left(k_j\ell\right) \end{pmatrix} \qquad (1.11.38)$$

where "ζ_m" can be "ζ_e" or "ζ_o" depending on how we apply 1.11.38. Applying the even-odd method, the input impedance "Z_i" of a single t.l., as indicated in Figure 1.11.4 can be written as:

$$Z_i = \frac{v_{1e} + v_{1o}}{i_{1e} + i_{1o}} \qquad (1.11.39)$$

Applying 1.11.38 at port "1" for the "even" excitation, we can write:

$$v_{1e} = -j\zeta_e\left[\cot g\left(k_j\ell\right)i_{1e} + \mathrm{cosec}\left(k_j\ell\right)i_{2e}\right] \qquad (1.11.40)$$

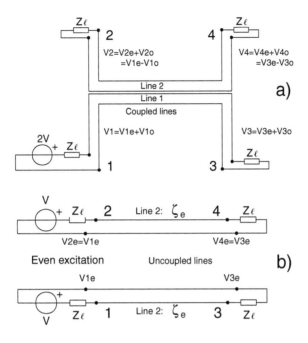

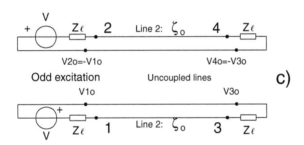

Figure 1.11.4

while to have "v_{1o}" we only need to change in the previous equation the subscript "e" with "o," i.e.:

$$v_{1o} = -j\zeta_o \left[\cotg(k_j\ell)i_{1o} + \cosec(k_j\ell)i_{2o} \right] \qquad (1.11.41)$$

For port "1" we can write:

$$i_{1e} = \left(v - v_{1e}\right)/Z_\ell \qquad (1.11.42)$$

while for port "2" we have:

$$v_{2e} = -j\zeta_e \left[\cosec(k_j\ell)i_{1e} + \cotg(k_j\ell)i_{2e} \right] \qquad (1.11.43)$$

$$v_{2e} = -Z_\ell i_{2e} \qquad (1.11.44)$$

Combining 1.11.42) ÷ 1.11.44 we have:

$$i_{2e} = j\zeta_e \left(v - v_{1e}\right) \Big/ \left[Z_\ell^2 \operatorname{sen}\left(k_j\ell\right) - j\zeta_e Z_\ell \cos\left(k_j\ell\right)\right] \qquad (1.11.45)$$

If now we insert 1.11.45 and 1.11.42 into 1.11.40 we have "v_{1e}" as a function of impedances only. As we said, we can repeat the same procedure for "v_{1o}" and also for currents. The procedure is quite tedious, but doing it and setting the condition:

$$Z_i = ! \, Z_\ell \qquad (1.11.46)$$

we have that 1.11.46 is satisfied if:

$$Z_\ell \equiv \left(\zeta_e \, \zeta_o\right)^{0.5} \qquad (1.11.47)$$

which is a condition that is independent of frequency and length of the coupling region. Equation 1.11.47 is one of the most important relationships of coupled line theory and is fundamental for directional coupler devices.

In coupled line theory another parameter is frequently used, called a "voltage coupling factor" and indicated with "c_v," which is related to "ζ_e" or "ζ_o" by the following relationship:

$$c_v \equiv \frac{\zeta_e - \zeta_o}{\zeta_e + \zeta_o} \qquad (1.11.48)$$

If "c_v" is a defined value, for instance a request for some electronic system, then "ζ_e" or "ζ_o" can be obtained by:

$$\zeta_e = Z_\ell \left(\frac{1 + c_v}{1 - c_v}\right)^{0.5} \qquad (1.11.49)$$

$$\zeta_o = Z_\ell \left(\frac{1 - c_v}{1 + c_v}\right)^{0.5} \qquad (1.11.50)$$

As we will study in the directional coupler theory,* "c_v" is quite often the mean value of the frequency variable coupling "$c_v(f)$," defined as the ratio between the coupled electric field "$E_c(f)$" and the input electric field "$E_i(f)$," i.e.:

$$c_v\left(f\right) \stackrel{\perp}{=} E_c\left(f\right)\big/E_i\left(f\right) \qquad (1.11.51)$$

In any case, the modulus of the coupling factors theoretically have unity as the maximum value, also if in practice the unity cannot be reached.

In many problems coupled lines can be evaluated as lossless, so that the coupling quantities reduce to mutual inductance "M" and mutual capacitance "C_m," as indicated in Figure 1.11.5. Here we have also assumed that the two lines of the couple are the same. In such cases, and supposing the structure supports a "TEM" mode, it is simple to relate "C," "C_m," "L," and "M" to "ζ_e" and

---

* See Chapters 7 and 8 for directional coupler theory.

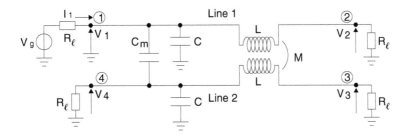

Figure 1.11.5

"ζ_o." To do that, first of all assume that the propagation along a single line has the longitudinal dependence "$e^{-j\beta x}$" and apply the telegraphist's equations. We have:

$$-j\beta i_1 = -j\omega C v_1 - j\omega C_m \left(v_1 - v_4\right) \tag{1.11.52}$$

$$-j\beta v_1 = -j\omega L i_1 - j\omega M i_4 \tag{1.11.53}$$

$$-j\beta i_4 = -j\omega C v_4 - j\omega C_m \left(v_4 - v_1\right) \tag{1.11.54}$$

$$-j\beta v_4 = -j\omega L i_4 - j\omega M i_1 \tag{1.11.55}$$

The previous equations can be rewritten as:

$$\beta i_1 = \omega \left(C + C_m\right) v_1 - \omega C_m v_4 \tag{1.11.56}$$

$$\beta v_1 = \omega L i_1 + \omega M i_4 \tag{1.11.57}$$

$$\beta i_4 = \omega \left(C + C_m\right) v_4 - \omega C_m v_1 \tag{1.11.58}$$

$$\beta v_4 = \omega L i_4 + \omega M i_1 \tag{1.11.59}$$

Now, let us apply the even and odd excitations at ports "1" and "4," so that we write;

$$v_1 = v_{1e} + v_{1o}, \quad v_4 = v_{1e} - v_{1o} \tag{1.11.60}$$

$$i_1 = i_{1e} + i_{1o}, \quad i_4 = i_{1e} - i_{1o} \tag{1.11.61}$$

Inserting equations 1.11.60 and 1.11.61 into 1.11.56 through 1.11.59 we have the following two sets of homogeneous equations, respectively for the even and odd case:

$$\begin{pmatrix} \beta i_e - \omega C v_e = 0 \\ -\omega \left(L + M\right) i_e + \beta v_e = 0 \end{pmatrix} \tag{1.11.62}$$

$$\begin{pmatrix} \beta i_o - \omega \left(C + 2C_m\right) v_o = 0 \\ -\omega \left(L - M\right) i_o + \beta v_o = 0 \end{pmatrix} \tag{1.11.63}$$

The solutions of these two sets are, respectively:

$$\beta \doteq \beta_e = \pm\omega\left[(L+M)C\right]^{0.5} \tag{1.11.64}$$

$$\beta \doteq \beta_o = \pm\omega\left[(L-M)(C+2C_m)\right]^{0.5} \tag{1.11.65}$$

which represent the even and odd phase constants. The even and odd impedances can be easily obtained. In fact, for even excitation "C_m" is between two equipotential points and the magnetic coupling corresponds to having a single inductance of value of "$L + M$." This situation is indicated in Figure 1.11.6. So, applying the definition 1.3.11 for characteristic impedance we have:

$$\zeta_e = \left[(L+M)/C\right]^{0.5} \tag{1.11.66}$$

since $Y_{pe} = \omega C$ and $Z_{se} = \omega(L + M)$.

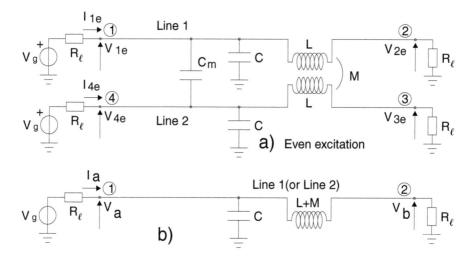

Equiv. network for even excitation of network in a)

Figure 1.11.6

The situation for odd excitation is indicated in Figure 1.11.7, from which we have:

$$\zeta_o = \left[(L-M)/(C+2C_m)\right]^{0.5} \tag{1.11.67}$$

since $Y_{po} = \omega(C + 2C_m)$ and $Z_{so} = \omega(L - M)$.

In the case of a "TEM" mode, "β_e" and "β_o" must satisfy the general condition for isotropic and homogeneous media. In the case of progressive propagation, this constraint is written as:

$$\beta_e = \omega\left[(L+M)C\right]^{0.5} \doteq \omega(\mu\varepsilon)^{0.5} \tag{1.11.68}$$

$$\beta_o = \omega\left[(L-M)(C+2C_m)\right]^{0.5} \doteq \omega(\mu\varepsilon)^{0.5} \tag{1.11.69}$$

from which we have:

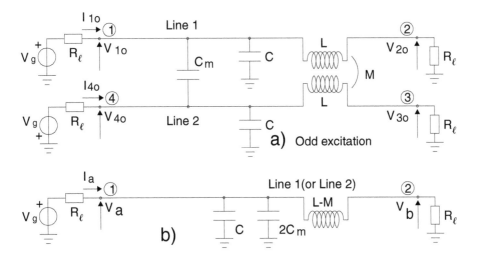

Equiv. network for odd excitation of network in a)

Figure 1.11.7

$$L + M \doteq \mu\varepsilon/C \tag{1.11.70}$$

$$L - M \doteq \mu\varepsilon/(C + 2C_m) \tag{1.11.71}$$

Inserting these last two equations into 1.11.66 and 1.11.67 we have:

$$\zeta_e = (\mu\varepsilon)^{0.5}/C \tag{1.11.72}$$

$$\zeta_o = (\mu\varepsilon)^{0.5}/(C + 2C_m) \tag{1.11.73}$$

From the two previous equations we can recognize that even impedance is always higher than or equal to odd impedance. Once we know "ζ_e" and "ζ_o," the four previous equations permit evaluation of "C," "C_m," "L" and "M." For example, Figures 1.11.8 and 1.11.9 give respectively the values of "C," "C_m" in pf/cm and "L," "M" in nH/cm, as functions of the coupling for a directional coupler.

Xtalk is a phenomenon that represents the effect of a signal traveling in a t.l. toward another t.l. So, crosstalk is an undesired phenomenon present when two or more t.l.s are set close together. The general theory of coupled t.l.s is perfectly applicable for the xtalk case. However, here we want to discuss this phenomenon from a "time domain" and "voltage" point of view, as is customary in hardware digital technique. We will study the case of only two coupled t.l.s where only one is fed at one port, as indicated in Figure 1.11.1. The voltages induced at points "4" and "3" are respectively named "backward," or "near end," and "forward," or "far end" xtalk. In contrast to the case of a directional coupler where even "ζ_e" and odd "ζ_o" satisfy 1.11.47, with "Z_ℓ" the termination impedance, in the present case this relationship in general is not verified, since the coupling is undesired. In the following, unless otherwise stated, we will assume such a case, i.e., $Z_\ell^2 \neq \zeta_e\zeta_o$.

Simple formulas are available to evaluate the xtalk if we suppose:

1. The two t.l.s are identical
2. Each t.l. supports a pure "TEM" mode
3. The termination load is a pure resistance of value "R_ℓ," equal at any port

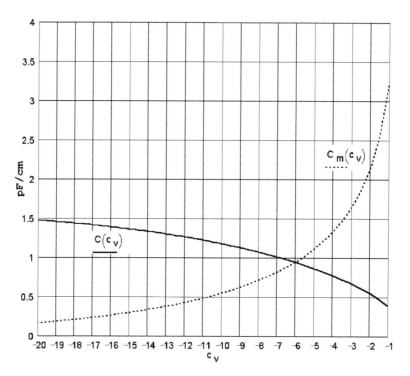

Figure 1.11.8

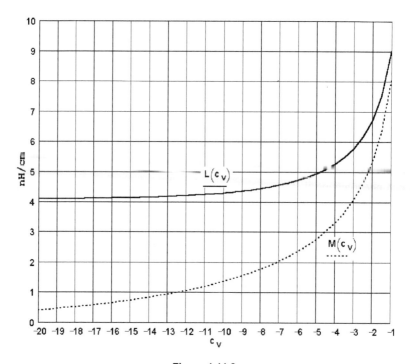

Figure 1.11.9

In such a hypothesis, with reference to the circuit indicated in Figure 1.11.10, in the case where the coupling length "ℓ" is shorter than $\lambda_g/4$, backward "v_b" and forward "v_f" xtalk voltages are approximately given by:

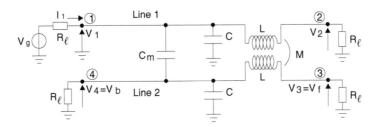

Figure 1.11.10

$$v_b = 0.25 \, v_1 \left(C_m/C + M/L \right) \tag{1.11.74}$$

$$v_f = 0.5 \, v_1 \, \tau_d \ell \left(C_m/C - M/L \right) / \tau_r \tag{1.11.75}$$

In Table 1.11.1 we have indicated some typical values for "C," "C_m," "L," and "M," where inductances are in nH/cm, capacitances in pF/cm, length in cm, "τ_d" in nS/cm, "τ_r" in nS, and voltages in volt:

Table 1.11.1

C	C_m	L	M	τ_d	ℓ	τ_r	V_1	V_b	V_f
0,4	0,17	8	4,12	0,0566	4	1	1	0,235	−0,0102
0,4	0,17	8	4,12	0,0566	10	1	1	0,235	−0,0255
0,4	0,17	8	4,12	0,0566	10	0,5	1	0,235	−0,0509
0,3	0,15	10	4,8	0,0548	4	1	1	0,245	0,0022
0,3	0,15	10	4,8	0,0548	10	1	1	0,245	0,0055
0,3	0,15	10	4,8	0,0548	10	0,5	1	0,245	0,0109

Even and odd impedances are easily related to the elements of the equivalent circuit for the coupled t.l.s, indicated in Figure 1.11.10, through the following equations:

$$C - \left(\mu \varepsilon_0 \, \varepsilon_{re} \right)^{0.5} \Big/ \zeta_e \qquad\qquad \text{F/m} \tag{1.11.76}$$

$$C_m = 0.5 \left(\mu \varepsilon_0 \, \varepsilon_{re} \right)^{0.5} \left(1/\zeta_o - 1/\zeta_e \right) \qquad\qquad \text{F/m} \tag{1.11.77}$$

$$L = 0.5 C \left(\zeta_e^2 + \zeta_o^2 \right) + \zeta_o^2 \, C_m \qquad\qquad \text{H/m} \tag{1.11.78}$$

$$M = 0.5 C \left(\zeta_e^2 - \zeta_o^2 \right) - \zeta_o^2 \, C_m \qquad\qquad \text{H/m} \tag{1.11.79}$$

Of course, if "ℓ" is comparable to "λ_g" then the general equations of coupled t.l.s need to be used, inserting the corresponding "ζ_e" and "ζ_o" for the "w_1," "w_2," "s," and "h" employed for the particular coupled µstrip lines. It is important to say that if the coupled t.l.s have "ζ_e" and "ζ_o" so that $R_\ell \equiv (\zeta_e \zeta_o)^{0.5}$ is satisfied and $\ell = (2n + 1) \, \lambda/4$ with "n" and odd number, then a directional coupler is realized, and the forward xtalk is theoretically always zero.* In Figure 1.11.11 we have reported the case where $\zeta_e = 69.4\Omega$ and $\zeta_o = 36\Omega$, corresponding to a coupling of $|c_m| = |v_4/v_1| = -10$ dB in a 50Ω system, vs. the electrical length "θ." The values above "0" for the vertical axis are in linear scale, while below zero they are in "dB." The direct, coupled, and isolated outputs are respectively indicated with "$d(\theta)$," "$c(\theta)$," and "$i(\theta)$," and correspond to "v_2," "v_4," and "v_3" of Figure 1.11.10. For this figure, Equation 1.11.47 is supposed to be verified, and we can see how the isolation is always infinite since $i(\theta) = 0$ for every "θ." The situation is different if Equation 1.11.47 is not verified, as we have indicated in Figure 1.11.12 where $\zeta_e = 69.4\Omega$ and $\zeta_o = 18\Omega$. We see how "$i(\theta)$" is no longer always zero.

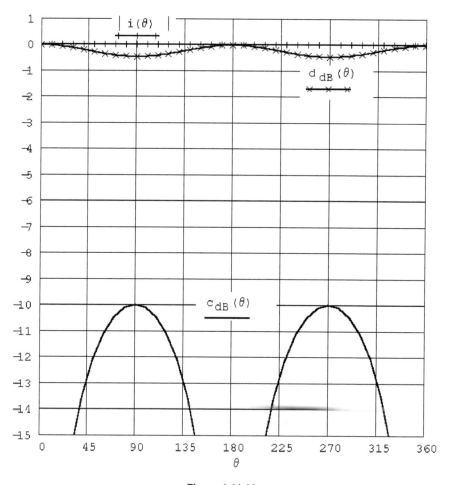

Figure 1.11.11

The Xtalk theory shown here can be applied to any coupled line structure for which every line supports a "TEM" propagation mode. An example is the coupled stripline structures as indicated in Figure 1.11.3 for the "BCS"** case. An example where this theory can be applied with some approximation is the coupled µstrip studied in Chapter 5, since the µstrip does not support a pure "TEM" mode.

---

* This assertion will be verified in Chapter 7, when we will study µstrip directional couplers.
** "BCS" means "Broad Side Coupled Stripline," a coupled line structure studied in Chapter 6.

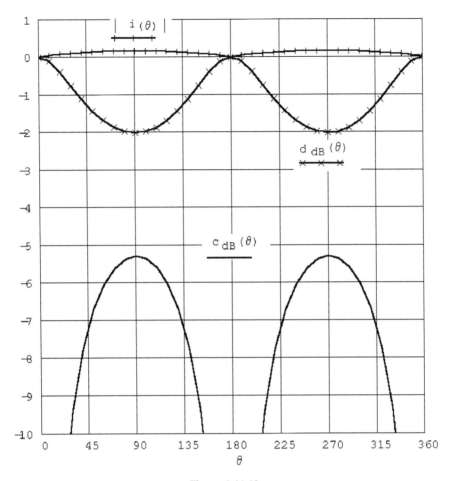

Figure 1.11.12

1.12 THE SMITH CHART

All the formulas given in the previous paragraphs, even if they are not so complicated, still require a calculator. To simplify these calculations, a helpful idea came to researcher P.H. Smith,[*] who introduced a chart on which many transmission line problems can be easily represented. In this section we will study how this chart works and what we can represent on it.

The Smith chart can be introduced starting with the relationship in 1.8.38 between the normalized impedance "$\zeta_n(x)$" presented by the t.l. and voltage reflection coefficient "$\Gamma_v(x)$." We have:

$$\zeta_n(x) = \frac{1 + \Gamma_v(x)}{1 - \Gamma_v(x)} \tag{1.8.38}$$

where:

$$\Gamma_v(x) = \Gamma_v(0)e^{kx} \tag{1.8.37}$$

From 1.8.38 we can write:

---

* P.H.Smith, American researcher, born in 1905 in Lexington and died there in 1987.

$$\Gamma_v(x) = \frac{\zeta_n(x) - 1}{1 + \zeta_n(x)} \tag{1.12.1}$$

Remember that 1.8.37 has been defined when the origin of the "x" axis is at the termination load, and negative values of "x" are on the left. In this section we will use this reference system, unless otherwise stated.

Since "$\Gamma_v(x)$" and "$\zeta_n(x)$" are, in general, complex numbers, we can write:

$$\Gamma_v(x) \overset{\perp}{=} \Gamma_{vr}(x) + j\Gamma_{vj}(x) \tag{1.12.2}$$

$$\zeta_n(x) \overset{\perp}{=} \zeta_{nr}(x) + j\zeta_{nj}(x) \tag{1.12.3}$$

which, inserted into 1.8.38, results in:

$$\zeta_{nr}(x) + j\zeta_{nj}(x) = \frac{1 + \Gamma_{vr}(x) + j\Gamma_{vj}(x)}{1 - \Gamma_{vr}(x) - j\Gamma_{vj}(x)} \tag{1.12.4}$$

Equating imaginary and real parts at both members, we have:

$$\zeta_{nr}(x) = \frac{1 - \Gamma_{vr}^2(x) - \Gamma_{vj}^2(x)}{\left[1 - \Gamma_{vr}(x)\right]^2 + \Gamma_{vj}^2(x)} \tag{1.12.5}$$

$$\zeta_{nj}(x) = \frac{2\Gamma_{vj}(x)}{\left[1 - \Gamma_{vr}(x)\right]^2 + \Gamma_{vj}^2(x)} \tag{1.12.6}$$

Writing the two previous expressions as functions of the variables "$\Gamma_{vr}(x)$" and "$\Gamma_{vj}(x)$" we have:

$$\Gamma_{vj}^2(x) + \left\{\Gamma_{vr}(x) - \zeta_{nr}(x)/\left[1 + \zeta_{nr}(x)\right]\right\}^2 = 1/\left[1 + \zeta_{nr}(x)\right]^2 \tag{1.12.7}$$

$$\left[\Gamma_{vj}(x) - 1/\zeta_{nj}(x)\right]^2 + \left[1 - \Gamma_{vr}(x)\right]^2 = 1/\zeta_{nj}^2(x) \tag{1.12.8}$$

Equation 1.12.7 in the Cartesian plane of abscissae "$\Gamma_{vr}(x)$" and ordinate "$\Gamma_{vj}(x)$" represents a circle with center in:

$$\Gamma_{vr}(x) = \xi_{nr}(x)/\left[1 + \xi_{nr}(x)\right] \text{ and } \Gamma_{vj}(x) = 0 \tag{1.12.9}$$

and radius "r" so that:

$$r = 1/\left[1 + \zeta_{nr}(x)\right] \tag{1.12.10}$$

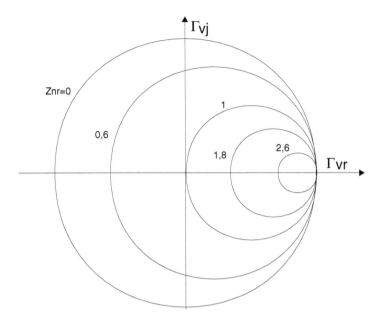

Figure 1.12.1

Figure 1.12.1 represents some circles given by 1.12.7 for some values of "$\zeta_{nr}(x)$." Note that the maximum value of the radius is "1," corresponding to $\zeta_{nr}(x) = \infty$.

In the same coordinate system, Equation 1.12.8 represents a circle with its center in:

$$\Gamma_{vr}(x) = 1 \quad \text{and} \quad \Gamma_{vj}(x) = 1/\zeta_{nj}(x) \tag{1.12.11}$$

with radius "r":

$$r = 1/\zeta_{nj}(x) \tag{1.12.12}$$

Figure 1.12.2 represents some circles given by 1.12.8 for some values of "$\zeta_{nj}(x)$."

Combining Figures 1.12.1 and 1.12.2 in the same graph we obtain the Smith chart, indicated in Figure 1.12.3. So, circles represent normalized resistances while arcs represent normalized reactances.

From the Smith chart, we can obtain the admittances in a very simple manner. From 1.8.38 we have:

$$Y_n(x) = 1/\zeta_n(x) = \frac{1 - \Gamma_v(x)}{1 + \Gamma_v(x)} \tag{1.12.13}$$

So, admittances can be simply obtained by rotating the point that represents the value "$\zeta_n(x)$" clockwise of "π," along a circle. As an example, in Figure 1.12.4 we have indicated a point $Z = 1 + j$. To obtain its normalized admittance "Y," it is enough to rotate clockwise of "π" on a circle, to have the point $Y = 0.5 - j0.5$. Of course, the Smith chart is numbered for every circle drawn on it, and in case we have not numbered the circle for simpler drawing purposes.

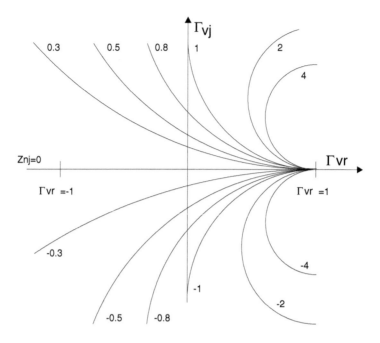

Figure 1.12.2

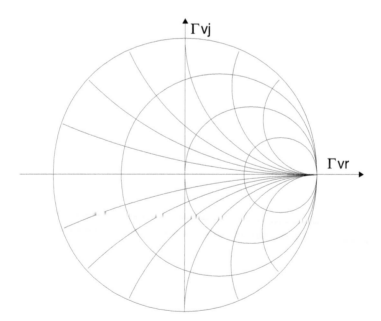

Figure 1.12.3

To again simplify the passage between "Z" and "Y," the Smith chart is represented with another Smith chart on it, but just rotated of "π." This situation is indicated in Figure 1.12.5. For this figure, point "A" has an impedance readable by circles and arcs starting from the right, while its admittance is readable by circles and arcs starting from the left. So, with this chart it is not necessary to make any rotation to pass between "Z" and "Y," since it is the graph that is rotated. Note that left arcs,

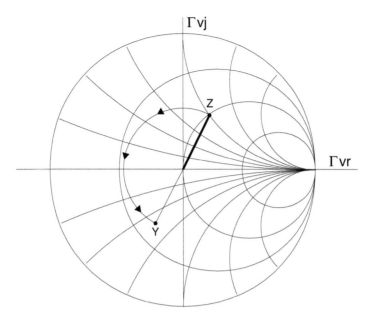

Figure 1.12.4

i.e., susceptances, have negative and positive values respectively above and below the "Γ_{vr}" axis, i.e., in an opposite manner with respect to right arcs, which represent reactances. But this is obvious, since a complex number "$Y \doteq G + jB$," which is the reciprocal of a complex number "$Z \doteq R + jX$," i.e., $Y \doteq 1/Z$, has the imaginary part "B" of opposite sign with respect to the imaginary part "X" of the number "Z."

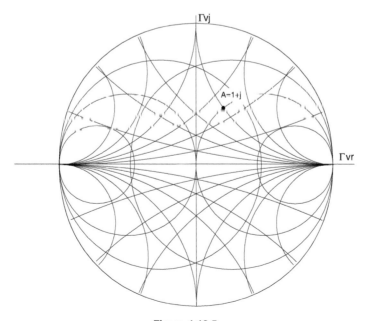

Figure 1.12.5

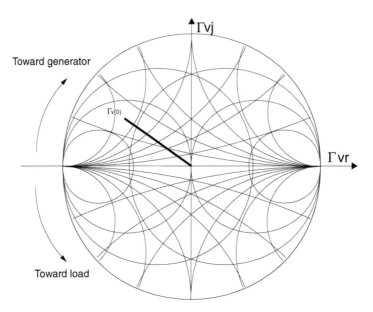

Figure 1.12.6

Another important topic on the Smith chart is the sign of the rotation. From 1.8.37 we see that moving toward the generator, the phase of "$\Gamma_v(x)$" decreases, since the coordinate is negative, while moving toward the load, the phase increases. For this reason, once the point "$\Gamma_v(0)$" is signed on the Smith chart, the distance toward the generator must be measured moving clockwise, while the distance toward the load must be measured moving counterclockwise. This situation is indicated in Figure 1.12.6, where the zero angle is set by definition in $\Gamma_{vr} = -1$ and clockwise movement as positive angle direction.

Inserting the real and imaginary part of "k" in 1.8.37 we have:

$$\Gamma_v(x) = \Gamma_v(0)\,e^{2k_r x}e^{2jk_j x} \tag{1.12.14}$$

from which we can see that if $k_r = 0$, i.e., losses are zero, then the vector "$\Gamma_v(x)$" describes a circle, while if $k_r > 0^*$ then "$\Gamma_v(x)$" describes a decreasing spiral. So, if losses are present and the line is theoretically of infinite length, then the reflection attenuates when returning to the input of the line, reaching the value of zero. Remembering Equation 1.7.1, i.e.,

$$k_j = j\omega(\mu\varepsilon)^{0.5} = j2\pi/\lambda$$

we see how "$\Gamma_v(x)$" gets the same angle for every $x = n\lambda/2$, with "n" an integer number. This means that in the Smith chart, a turn corresponds to a distance "d" in the line equal to half wavelength. For this reason, the outermost circumference is numbered in angles and/or in units of half wavelength.

The Smith chart can also be used to measure "$|\Gamma_v(x)|$." We know from Section 1.8 that $|\Gamma_v(x)| \le 1$, which means that any value of "$|\Gamma_v(x)|$" is surely inside the value of the outermost circle radius "r" given by 1.12.10. To simplify the reading of "$|\Gamma_v(x)|$" quite often another axis parallel to the "Γ_{vr}" axis is drawn with its scale numbered from "0," i.e., the center of the Smith chart, and "1," i.e., the maximum value of "$|\Gamma_v(x)|$," or the maximum value of the outer circle of the Smith chart.

---

* Losses, i.e., "k_r," are represented with a positive number. Remember that 1.8.37, and therefore 1.12.10, have been defined for $x < 0$. So "$e^{k_r x}$" is always negative for $k_r > 0$.

Another parameter that can be measured on the Smith chart is the VSWR, given by Equation 1.8.48, i.e.:

$$\text{VSWR} \triangleq \frac{1 + |\Gamma_v|}{1 - |\Gamma_v|} \tag{1.8.48}$$

Another axis parallel to that of the "Γ_{vr}" or "$|\Gamma_v(x)|$" axis is drawn, with its scale numbered from "1," corresponding to the value $|\Gamma_v(x)| = 0$, and "∞," corresponding to the value $|\Gamma_v(x)| = 1$.

Another parameter that can be measured is the line attenuation constant "k_r." We know that if the line has attenuation, the extreme of "$|\Gamma_v(x)|$" describes a decaying spiral. So, if the line is at least a wavelength long, it surely trespasses twice the "Γ_{vr}" axis, in two different positions whose distance is proportional to "k_r."

From these arguments, we can recognize how the Smith chart is a formidable tool for evaluating a lot of transmission line parameters. Note that the transformation of the Smith chart is theoretically exact, so the unique source of error can be the manual drawing of points and paths along this chart. However, these errors can quite often be neglected.

Another graph that can be drawn in the Smith chart is the so-called "constant-Q" or "transformer" lines. These lines represent the points on the Smith chart where the ratio between the imaginary and real parts is constant. Ideal transformers are devices that move impedances on such lines. "Constant-Q" lines are used for matching using transformers or in filter synthesis. Some of these lines are given in Figure 1.12.7 and are indicated with dashed lines.

The Smith chart can also be generalized to include negative values of resistance. As was said before, some active elements, under particular bias and load conditions, can present negative values for input impedance. The generalized Smith chart is indicated in Figure 1.12.8. To draw this graph it is enough to draw, with the desired negative value of "Z_{nr}," the circles whose coordinates are given by 1.12.9 through 1.12.12. For example, the circle relative to $Z_{nr} = -1$ has a radius equal to infinity, and in Figure 1.12.8 is indicated with the dashed line marked with "$Z_{nr} = -1$." Also note

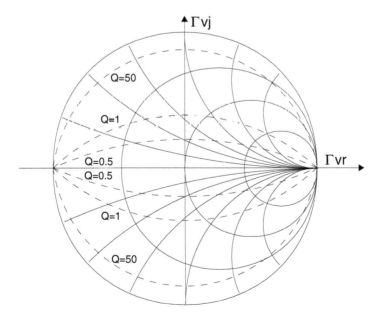

Figure 1.12.7

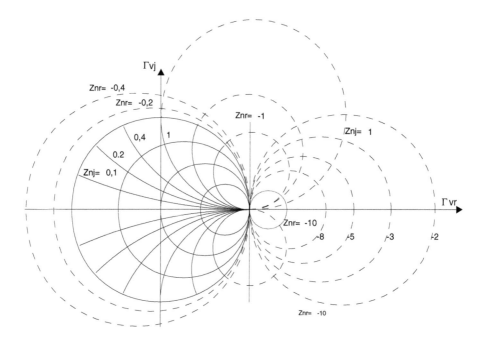

Figure 1.12.8

how any point with a negative real part gives a reflection coefficient greater than "1," since the corresponding modulus of the vector "$|\Gamma_v(x)|$" is greater than the bigger radius of the Smith chart's positive resistance circle, which is just equal to "1."

On the Smith chart, a lot of other parameters can be drawn that are not representative of transmission line problems. These other parameters are "stability circles," "noise figure circles," and "constant gain circles." These parameters are pertinent to amplifier design and are not the topic of this text. The reader interested in such topics can read specific books on these subjects.[19,20]

The reader can recognize how the Smith chart is not only useful in resolving transmission line problems, but can be used in every matching problem. Also, today, when in every problem we can use computer programs to help us, the Smith chart remains a very simple method which is still employed.

1.13 SOME EXAMPLES USING THE SMITH CHART

After having studied the Smith chart in the previous section, it is useful to show some examples of practical problems that can be easily resolved by applying this chart. Of course, we cannot obtain exact result values, which would only be possible using a computer. Many readings in the Smith chart necessarily have to be interpolated. Remember that the procedure for drawing the Smith chart is absolutely mathematically correct, and the only error source is in the reading of the exact position on the curve in the chart. Errors are quite often negligible.

The general way to use the Smith chart is to convert loads to reflection coefficients "Γ_v," to draw this "Γ_v," and to choose a convenient path to realize the matching. Remember that if we move toward the generator we need to rotate clockwise, while if we move toward the load we have to rotate counterclockwise.

Some simple examples will better explain how to use the Smith chart. Unless otherwise stated, by "reflection coefficient" we will mean the "voltage reflection coefficient."

a. Example I

Consider a coaxial cable with $\zeta = 50$ ohm of characteristic impedance. Its length is $\ell = 5$ meters, and its dielectric constant is $\varepsilon_r = 2.3$. In the cable, a signal with frequency $f = 9$ GHz is sent. The cable is terminated with a load of impedance $Z = 75$ ohm. Evaluate the input impedance of the cable.

Result

The load normalized impedance "Z_n" and reflection coefficient "Γ_v" are:

$$Z_n = 1.5 \quad \text{and} \quad \Gamma_v = 0.2 \tag{1.13.1}$$

The point corresponding to "Γ_v" is indicated by "A" in Figure 1.13.1. Now we draw a circle with its center in the Smith chart center and radius equal to "Γ_v." All the impedances presented by the line will be on this circle. The "guided wavelength" "λ_g" is:

$$\lambda_g = \frac{v_0}{f\sqrt{\varepsilon_r}} \approx 21.98 \text{ mm} \tag{1.13.2}$$

where "v_0" is the light speed in the vacuum. The cable length is a number "n" of half wavelength given by:

$$n = 1(\lambda_g / 2) = 454.96 \tag{1.13.3}$$

Since from the previous section we know that a complete turn on the Smith chart corresponds to moving a half wavelength along the line, we need to subtract from "n" its integer part, obtaining $n' \pm 0.96$. Multiplying "n'" for 360° we have the number of degrees "θ" we have to move counterclockwise on the circle. In this case we have $\theta = n' * 360° = 345.6°$. The resulting point is indicated with "B" in Figure 1.13.1, which corresponds to a reading of:

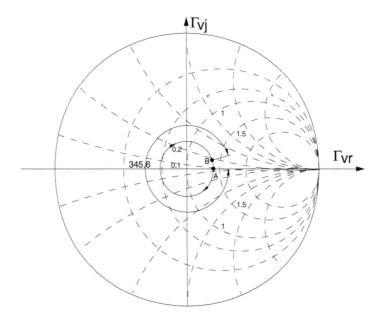

Figure 1.13.1

$$n = 1\left(\lambda_g/2\right) = 454.96 \qquad (1.13.4)$$

which represents the real and imaginary parts of the cable input impedance, normalized to "ζ."

b. Example II

Let us consider a transmission line with $\zeta = 50$ ohm of characteristic impedance. Its length is $\ell = 10$ centimeters and its dielectric constant is $\varepsilon_r = 4.3$. The attenuation factor of this line corresponds to $k_r = 0.3$ m^{-1}. In the line, a signal with frequency $f = 12$ GHz is sent. The line is terminated with a load of impedance $Z = 75$ ohm. Evaluate the input impedance of the line.

Result

The load normalized impedance "Z_n" and reflection coefficient "Γ_v" are:

$$Z_n = 1.5 \quad \text{and} \quad \Gamma_v = 0.2 \qquad (1.13.5)$$

The point corresponding to "Γ_v" is indicated with "A" in Figure 1.13.2. In contrast with the previous Example I, now we cannot move on a circle radius equal to "Γ_v," since now the line is lossy. What we have to do is to draw the circle corresponding to the final value "Γ_f" of "Γ_v," given by equation 1.8.37, i.e.:

$$\Gamma_v(x) = \Gamma_v(0)\, e^{2kx} \qquad (1.8.37)$$

where "x" is a negative number, i.e., the length of the line changed in sign. In this case we have:

$$\Gamma_f \overset{\perp}{=} \left|\Gamma_v(x)\right| \equiv \left|\Gamma_v(0)\right| e^{2k_r x} \overset{\perp}{=} \Gamma_v\, e^{2k_r x} \qquad (1.13.6)$$

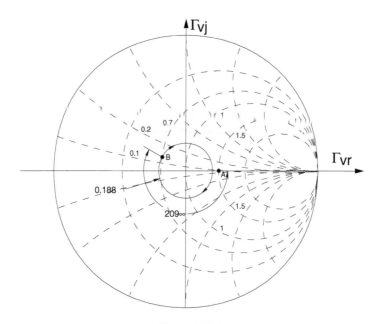

Figure 1.13.2

Since the line is 10 cm long and "k_r" is known, from 1.13.6 we have:

$$\Gamma_f = 0.2\,e^{-2*0.3*0.1} \approx 0.188 \qquad (1.13.7)$$

Now we draw a circle with its center in the Smith chart center and radius equal to "Γ_f." All the impedances presented by the line will be on this circle. The "guided wavelength" "λ_g" is:

$$\lambda_g = \frac{v_0}{f\sqrt{\varepsilon_r}} \approx 12.06 \text{ mm} \qquad (1.13.8)$$

The cable length is a number "n," of half wavelength, given by:

$$n = \ell / \left(\lambda_g / 2\right) = 16.58 \qquad (1.13.9)$$

obtaining $n' \doteq 0.58$ or $\theta = n' * 360° \approx 209°$. The resulting point is indicated by "B" in Figure 1.13.2, which corresponds to a reading of:

$$z_{nr} \approx 0.7 \quad \text{and} \quad Z_{nj} \approx 0.14 \qquad (1.13.10)$$

which represent the normalized real and imaginary parts of the cable input impedance.

c. Example III

Let us consider a generic network with an input normalized impedance equal to $Z = 1.5 - j$ at a frequency of $f = 1$ GHz. Match this network to 50 ohm.

Result

The point corresponding to "Z" is indicated with "A" in Figure 1.13.3. The matching problem can be obtained in a several ways.

The first way is to move along a constant resistance circle, and reach the point "B." Since we move increasing reactance it means we are adding a series inductance "L." The value of "L" is equal to the reactance corresponding to the distance between point "A" and the "Γ_{vr}" axis, evaluated at "f," i.e.:

$$L = 50/2\pi f \approx 7.96 \text{ nH} \qquad (1.13.11)$$

At point "B" only a resistive impedance exists. Then we can move along the "Γ_{vr}" axis until reaching the value "1." Since this movement results in decreasing the resistive part, it is equivalent to inserting a parallel resistor "R." The value of normalized "R," "R_n," is equal to the antiparallel between the resistance "1.5" in "B," and that in the arrival point "C," i.e., "1." So:

$$R_n = 1.5 * 1/(1.5 - 1) = 3 \qquad (1.13.12)$$

or $R = 150$ ohm. In Figure 1.13.3 we have indicated the procedure for matching.

Lossless matching at the desired frequency is quite often desired. The matching indicated in Figure 1.13.3 is quite lossy, since it uses resistors.

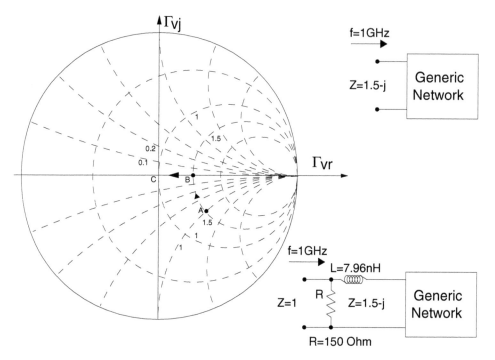

Figure 1.13.3

The second way to match to 50 ohm is indicated in Figure 1.13.4. Using a 50-ohm line we can move from point "A" to point "B," choosing the line length "ℓ" as indicated on the outermost Smith chart circle. In this case $\ell \approx 0.033\lambda$, as the difference between the "λ" corresponding to point "A," and point "B" being "λ" the guided signal wavelength. Then, moving on a constant resistance circle, we can reach point "C." This movement is equivalent to inserting a series inductance "L." The value of "L" is equal to the reactance at point "B," evaluated at "f." At point "B" the denormalized reactance "$|X_B|$" is near 44 ohm, so:

$$L = \left| X_B \right| \big/ 2\pi f \approx 7 \text{ nH} \tag{1.13.13}$$

Of course, other ways exist to do the matching. Let us consider Figure 1.13.5. For instance we can move from "A" to "B" on a constant conductance circle that corresponds to add a shunt capacitor, and then reaches point "C" moving on a constant resistance circle, i.e., adding a series inductor. Alternatively, we can move from point "A" to point "D" adding a shunt inductor, and then from "D" to "C" adding a series capacitor. The corresponding matching sections are indicated in Figure 1.13.5, together with the proper values.

This example can show us that quite often, many matching sections can be found. The most suitable can be chosen evaluating other parameters, like attenuation, physical construction, size, and cost.

In many cases, simple low-pass or high-pass sections work quite well, especially if the matching networks are not always that simple to build. For example, note as the low-pass section indicated in Figure 1.13.5 has a capacitor of only 0.57 pF, which is quite a small value for a concentrated element.

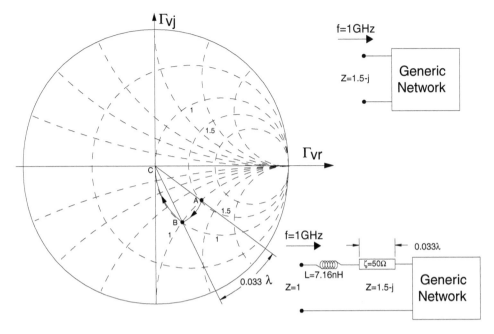

Figure 1.13.4

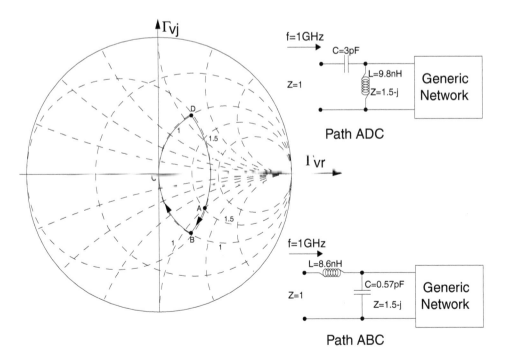

Figure 1.13.5

1.14 NOTES ON PLANAR TRANSMISSION LINE FABRICATION

In this section we want to outline two inevitable issues in planar t.l. fabrication which need to be recognized in order to avoid failures in the designed device. We will concentrate on μwave t.l. fabrication, but the following discussion applies to low frequency "PCB" as well, and of course, is applicable to any planar t.l. studied in this text.

A planar t.l. is built mainly in two ways:

1. Wet chemical etching
2. Selective conductor plating

The first method uses exactly the same technology applied in low frequency "PCB" construction. Briefly, the areas where the conductor is desired are protected against the chemical etching with a resin, as indicated in Figure 1.14.1a. Then the circuit is inserted inside the corrosive liquid,* which removes the conductor from the unprotected area. A consequence of this process is that the cross-section of the planar t.l. hot conductor assumes a trapezoidal shape, as indicated in Figure 1.14.1b. In practice, the conductor is smaller below the photoresist than the desired width "w," while near the substrate, the real width is much closer to "w." This phenomenon is called "undercut," and the corresponding decrease in width has been indicated with "u" in Figure 1.14.1b. Typical values of "u" are very close to thickness "t." In alumina circuits the conductor is gold, with t ≈ 4 ÷ 6 μm, and excluding the case of directional couplers,** where "w" can be near some tens of μm, the undercut can be neglected. In the case where copper is used as a conductor, its thickness is typically 17 μm or 30 μm and quite often the undercut cannot be neglected. In any case, when a t.l. device has to be built, it is convenient to know the exact values for undercut and photolithographic tolerance. Once the undercut value is known, its effects can be compensated for by enlarging the t.l. of the undercut value only, so that once the t.l. is realized it will have a width very near the theoretical designed value.

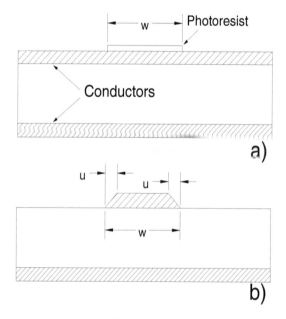

Figure 1.14.1

---

* This liquid is quite often an acid, depending on the metal to be eroded.
** These devices will be studied in Chapter 7.

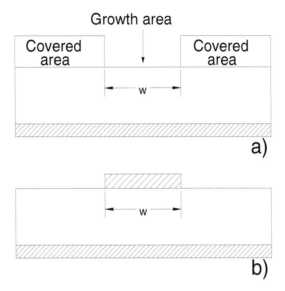

Figure 1.14.2

Another source of error arises in the photolithographic process used to attach the photoresist to the conductor. In microwave technology, this error gives a typical imprecision of ±3 µm in "w."

Selective conductor plating minimizes the undercut. In this case, the area where the conductor is desired is left free of resin, while the rest is protected by some particular resin, as indicated in Figure 1.14.2a. Then the conductor is grown using a galvanic process. Knowing the growth area, the DC current and the metal density inside the galvanic bath, a very precise conductor thickness "t" can be grown. Then the protective resin is removed using a selective erosive liquid, and only the desired conductor remains, as shown in Figure 1.14.2b.

In low frequency "PCB" technology, a very diffused substrate is a glass-resin mixture, called FR4, with a relative dielectric constant ranging from 3.5 to 6. Such boards can have dimensions much wider than RF t.l. devices to be realized. Components on these PCBs are usually soldered using a process called "wave soldering." In practice, the entire board is warmed to 40 or 50 degrees and is set in contact with a wave of soldering material, so that all components are soldered in a few seconds. A large area of conductor is not recommended with this procedure, because it can deform the board due to its heating when it is touched by the soldering wave. This could cause a problem in t.l. devices where the ground plane should be continuous. The problem is solved by building the ground plane as a grid, so that the wider side of it is smaller than λ/10. For low frequency signals the ground plane works as if it were continuous.

REFERENCES

1. D. Sette, *Lezioni di Fisica: Volume III*, Veschi, Ed., Roma, 1967, 139.
2. R. E. Collin, *Foundations for Microwave Engineering*, McGraw Hill, New York, 1992, 23.
3. S. Ramo, J. R. Whinnery, T. Van Duzer, *Fields and Waves in Communication Electronics*, John Wiley & Sons, New York, 1965, 131.
4. F. Di Paolo, Oscillator design achieves fast, wideband modulation, *Microwaves & RF*, 61, June 1994.
5. M. E. Frerking, *Crystal Oscillator Design and Temperature Compensation*, Van Nostrand Reinhold Co., New York, 1978.
6. F. Di Paolo, High frequency VCOs top 100 MHz, *EDN*, 135, Oct. 1991.
7. R. G. Rogers, *Low Phase Noise Microwave Oscillator Design*, Artech House, Norwood, MA, 1991.

8. C. Schiebold, Getting back to the basics of oscillator design, *Microwave J.*, May 1998.
9. S. Alechno, Analysis method characterizes microwave oscillators, *Microwaves and RF*, 82, Nov. 1997.
10. S. Ramo, J. R. Whinnery, T. Van Duzer, *Fields and Waves in Communication Electronics*, John Wiley & Sons, 1965, 131.
11. R. E. Collin, *Foundations for Microwave Engineering*, McGraw Hill, New York,1992, 23.
12. G. Barzilai, Fondamenti di elettromagnetismo, Siderea, Roma, 1975.
13. R. E. Collin, Theory and design of wide band multisection quarter wave transformers, Proc. IRE, February 1955.
14. J. M. Drozd, W. T. Joines, Using parallel resonators to create improved maximally flat quarter wavelength transformer impedance matching networks, *IEEE Trans. on MTT*, 132, Feb. 1999.
15. V. P. Meschanov, I. A. Rasukova, V. D. Tupikin, Stepped transformers on TEM transmission lines, *IEEE Trans. on MTT*, 793, June 1996.
16. H. Oraizi, Design of impedance transformers by the method of least squares, *IEEE Trans. on MTT*, 389, Mar. 1996.
17. J. M. Drozd, W. T. Joines, Using the binomial transformer to approximate the Q distribution of maximally flat quarter wavelength coupled filters, *IEEE Trans. on MTT*, 1495, Oct. 1998.
18. M. Abramowitz, I. A. Stegund, *Handbook of Mathematical Functions*, Dover, New York, 1970.
19. G. Gonzales, *Microwave Transistor Amplifiers*, Prentice Hall, Englewood Cliffs, NJ, 1984.
20. G. D. Vendelin, A. M. Pavio, U. L. Rhode, *Microwave Circuit Design Using Linear and Non Linear Techniques*, Wiley Interscience, New York, 1990.

CHAPTER 2

Microstrips

2.1 GEOMETRICAL CHARACTERISTICS

An illustration of a microstrip t.l.* is provided in Figure 2.1.1 in a cross-sectional view.** As we can see, this t.l. is composed of two parallel plane conductors, separated by a dielectric sheet with dielectric constant "ε_r" and permeability "μ_r." The dielectric sheet is usually called the "substrate." In this chapter unless otherwise stated, we will evaluate the dielectric material as not ferro-ferrimagnetic, i.e., we will suppose $\mu_r = 1$. The case where the dielectric material is also ferrimagnetic will be studied in Chapter 7. One of the two conductors is much wider than the other,*** and it has been indicated as "conductor II" in Figure 2.1.1. The wider conductor is set to the signal ground and for this reason it is also called the "cold conductor" or "ground conductor." Conversely, the shorter conductor, indicated as "conductor I" in Figure 2.1.1, is called the "hot conductor." For its physical construction, microstrips are employed as unbalanced t.l. We will indicate with:

1. "w" the width of the hot conductor
2. "h" the substrate height
3. "t" the conductor's thickness

Microstrips are the most widely used t.l. in all planar circuits, regardless of the frequency range of the applied signals. Especially at lower frequency, let us say until some hundreds of MHz, microstrips are widely used in multilayer printed circuit boards. In these cases, the ground conductor cannot coincide with the board metal housing or 0 Volt signal layer, but it can be a voltage layer, properly filtered. A possible four layers PCB is indicated in Figure 2.1.2. In this case, conductors "M1" and "M2" are two microstrips with layer "2," which is a power supply layer, while layer "3" is available for other purposes. Such PCB configuration is widely used in ECL or TTL boards. Of course, in microwave devices, microstrips are always two layers t.l., as indicated in Figure 2.1.1.

In Chapter 7 we will study some networks and devices employing microstrips, and in particular we will introduce nonreciprocal devices which use ferrimagnetic materials as substrate.

---

* Quite often we will abbreviate the reference to the microstrip transmission line with only the word "microstrip" or "µstrip."
** Whenever no confusion will arise, we will omit the obvious phrase "with a cross-sectional view."
*** In the following chapter we will show that the ground conductor should at least be three times wider than the hot conductor.

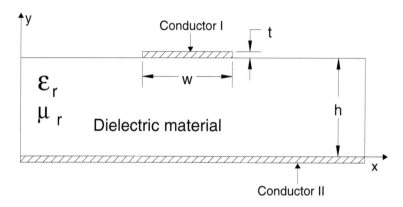

Figure 2.1.1

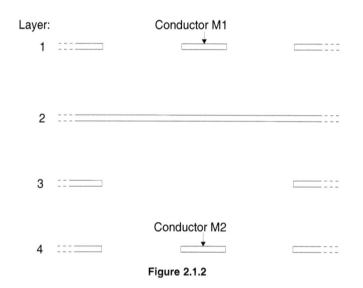

Figure 2.1.2

2.2 ELECTRIC AND MAGNETIC FIELD LINES

Some electric "e" and magnetic "h" field lines for the fundamental "qTEM" mode in microstrips are indicated in Figure 2.2.1 in a defined cross-section and a defined time. In the fundamental mode the hot conductor is equipotential.

The real fields disposition inside a microstrip is frequency dependent. In fact, logitudinal fields components exist due to the substrate discontinuity, so that the resulting propagation mode is called the "hybrid mode." Until near 10 GHz, these longitudinal components can be neglected, and the mode can be evaluated at a first approximation as a pure "TEM." However, due the non- "TEM" propagation, the microstrip is a dispersive* t.l., and this fact is particularly detrimental in wideband circuits operating at center frequencies above 10 GHz. Coupled line directional couplers, studied in Chapter 7, are the most dispersion-sensitive microstrip devices. For this reason, caution needs to be used when employing microstrip devices in wideband precision devices.

However, due to their technological simplicity, microstrip devices are the most widely employed.

---

* Dispersion is studied in Appendix A2.

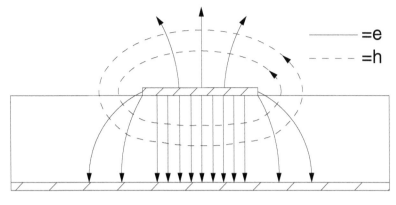

Figure 2.2.1

2.3 SOLUTION TECHNIQUES FOR THE ELECTROMAGNETIC PROBLEM

Three theory groups are mainly employed to study microstrip t.l. They are:

1. *Quasi static group*, where the microstrip is evaluated as a parallel plate transmission line, supporting a pure "TEM" mode. Examples of methods[1] used in this group are the application of the finite difference method for the Laplace equation[2] or conformal transformation method,[3] both studied in Appendix A1.
2. *Dispersion group*, where the microstrip is for example evaluated as a particular coupling between a "TEM" and "TE" t.l.[4,5] or with other dispersion models.[6,7,8]
3. *Full wave group*, where no simplification is made and a full Maxwell's equations solution is found.[9,10,11,12]

The first two groups are quite simple to apply while the third requires more analytical applications.

A common quantity for all three groups is "effective relative permittivity."* The introduction of this quantity can be done with reference to Figure 2.3.1. In part a, the microstrip has been enclosed in a box, with dimensions such that its effect on the field distribution can be neglected. In part b, the microstrip is surrounded by an homogeneous, isotropic dielectric medium with permittivity "ε_{re}," so that the wave phase velocity is the same for cases in Figure 2.3.1 a and b. "ε_{re}" stands for "effective relative permittivity." This quantity is evaluated as a constant in the quasi static group, while for the other groups it is a frequency-dependent function, as it is in reality. So, for points 2 and 3 above, the phase velocity coincidence has to be considered as satisfied at each frequency.

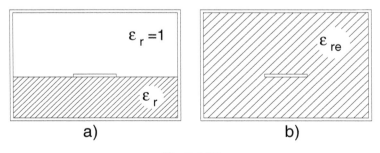

Figure 2.3.1

---

* It is common practice to refer to "effective permittivity as ε_{re}." However, since we think that the omission of the word "relative" can cause confusion, we will add this word.

In general, the determination of "$\varepsilon_{re}(f)$" is a very important point of any group of study. In fact, once "$\varepsilon_{re}(f)$" is known, the microstrip characteristic impedance can be evaluated inserting this function in any of the quasi static impedance expressions that we will give in the next section. The result is that the µstrip impedance also will be a function of frequency.

In the next sections we will discuss each one of the three groups.

2.4 QUASI STATIC ANALYSIS METHODS

These methods consider the microstrip as a static problem, so that it can be transformed in a structure that resembles a parallel plate capacitor. These methods evaluate, in different manners, the capacitance "C" of the µstrip, from which the characteristic impedance "ζ" can be evaluated as $\zeta = 1/Cv$ where "v" is the light speed in the media with effective dielectric constant "ε_{re}."

It is assumed that this parallel plate capacitor represents a "TEM" lossless t.l., for which the characteristic impedance "ζ" and phase constant "β" are evaluated as:

$$\zeta = \zeta_0 \left(C_0/C\right)^{0.5} \stackrel{\perp}{=} \zeta_0 / \left(\varepsilon_{re}\right)^{0.5} \tag{2.4.1}$$

$$\beta = \beta_0 \left(C/C_0\right)^{0.5} \stackrel{\perp}{=} \beta_0 \left(\varepsilon_{re}\right)^{0.5} \tag{2.4.2}$$

In these equations, the subscript "0" individuates quantities with the substrate replaced by the vacuum. The effective relative dielectric constant "ε_{re}" is defined as:

$$\varepsilon_{re} \stackrel{\perp}{=} C/C_0 \tag{2.4.3}$$

In the quasi static analysis, simple relations occur among the previous quantities. In this case, the µstrip can also be considered ideally as a generic "TEM" lossless transmission line whose equivalent circuit is indicated in Figure 2.4.1. First of all, note that if we assume the substrate to be not ferrimagnetic,* the inductance "L" is independent of the substrate "ε_r," i.e., $L_0 \equiv L$. Phase velocities "v_p" and v_{p0}," according to Chapter 1, in this case are:

$$v_{p0} = \left(LC_0\right)^{-0.5} \quad \text{and} \quad v_p = \left(LC\right)^{-0.5} \tag{2.4.4}$$

from which, also using 2.4.3:

$$v_p = v_{p0}\left(C_0/C\right)^{-0.5} \equiv v_{p0}/\sqrt{\varepsilon_{re}} \tag{2.4.5}$$

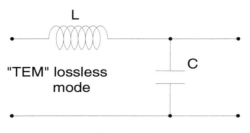

L

"TEM" lossless
mode

C

Figure 2.4.1

* Ferrimagnetism is studied in Appendix A7.

Since $v_{p0} = (\mu_0 \varepsilon_0)^{-0.5}$, from 2.4.4 it also follows:

$$LC_0 \equiv \mu_0 \varepsilon_0 \qquad (2.4.6)$$

The µstrip characteristic impedance "ζ_0" for the case $\varepsilon_r = 1$, referring to the equivalent circuit represented in Figure 2.4.1 and according to Chapter 1, is given by:

$$\zeta_0 = \left(L/C_0\right)^{0.5} \equiv \left(\mu_0 \varepsilon_0\right)^{0.5}/C_0 = \zeta_v \varepsilon_0 / C_0 \qquad (2.4.7)$$

where $\zeta_v \doteq (\mu_0 / \varepsilon_0)^{0.5} = 120\pi$ is the vacuum impedance. So, inserting 2.4.7 into 2.4.1, we have:

$$\zeta = \zeta_v \varepsilon_0 \left(CC_0\right)^{-0.5} \qquad (2.4.8)$$

The quasi static method most suited to be computer executed is the finite difference method, abbreviated as "FDM," applied to the solution of Laplace's equation.* With this method, the microstrip is enclosed in a box, as indicated in Figure 2.3.1a and the potential "V" on the hot conductor is found according to the "FDM." Once the potential "V" has been determined, the electrical field is found using the well-known equation:

$$\underline{E} = -\underline{\nabla}V \qquad (2.4.9)$$

and the charge "q" on the hot conductor is:**

$$q = \varepsilon_0 \varepsilon_r \oint_S \underline{E} \cdot \underline{n} dS \qquad (2.4.10)$$

and consequently the capacity "C" per u.l. is:

$$C = q/V \qquad (2.4.11)$$

In 2.4.10, "S" is the surface that contains a volume "Q"; "$\underline{n}$" is the "normal" to this surface and directed outside the region under study.

Another method used to study the microstrip is the conformal transformation method,*** by which the microstrip is transformed in a real parallel plate t.l. The researcher H.A. Wheeler[13] was the first to study the microstrip with this method, giving useful formulas for the evaluation of the effective relative dielectric constant once known as "ε_r" and the geometrical microstrip dimensions. Successively, Wheeler's formulas have been improved by E.O. Hammerstad,[14] resulting in the following expression for "ε_{re}":

$$\varepsilon_{re} = 0.5\left[\varepsilon_r + 1 + \left(\varepsilon_r - 1\right)F\right] \qquad (2.4.12)$$

where for w/h ≤ 1:

---

* See Appendix A1 for solution of Laplace's equation through "FDM."
** See Appendix A1 for the Gauss's law given by Equation 2.4.5.
*** See Appendix A1 for conformal transformation method.

$$F \overset{\perp}{=} \left(1 + 12h/w\right)^{-0.5} + 0.04\left(1 - w/h\right)^2 \qquad (2.4.13)$$

while for $w/h \geq 1$:

$$F \overset{\perp}{=} \left(1 + 12h/w\right)^{-0.5} \qquad (2.4.14)$$

Expressions 2.4.12 through 2.4.14 have been proven to be accurate inside 1% for $\varepsilon_r \leq 16$ and $0.05 \leq w/h \leq 20$.

From the above expressions we note that the conductor thickness "t" is not taken into account. Wheeler[15] suggests considering the effect of "t" as an extra width "Δw" added to "w," so that the resulting "w_e" can still be evaluated with zero thickness. The quantity "w_e" is called the "effective width" of the hot µstrip conductor, and is given by:

$$w_e = w + 0.5\left(1 + 1/\varepsilon_r\right)\Delta w \qquad (2.4.15)$$

where "Δw" for any value of w/h but for $t/w < 1$ and $t/h < 1$ is given by:

$$\Delta w = \left(t/\pi\right)\ln\left[\frac{4e}{\left(t/h\right)^2 + r^2}\right] \qquad (2.4.16)$$

where "r" is given by:

$$r = 1/\left(\pi w/t + 1.1\pi\right) \qquad (2.4.17)$$

and "e" is the natural number. Other researchers have studied the effect of the strip thickness as we will discuss later. Note that with quasi static methods there are no t.l. characteristics that are frequency dependent. In the next sections we will study other methods where frequency appears as a variable.

It is important at this point to make a distinction if the substrate is ferro-ferrimagnetic. In such cases, an effective permeability "μ_{re}" can be defined, according to:

$$\mu_{re} \overset{\perp}{=} L/L_0 \qquad (2.4.18)$$

where "L_0" individuates the µstrip equivalent inductance when the substrate is evaluated as $\mu_r = 1$, i.e., the substrate is replaced by the vacuum. However, not a unique expression for "μ_{re}" exists since this quantity is strongly dependent on the direction and intensity of the applied static magnetic field and on its direction with respect to the "RF" magnetic field. These expressions are given in Chapter 7 and Appendix A7. In particular, expressions 2.4.1 and 2.4.5 are replaced with:

$$\zeta = \zeta_0 \left(\mu_{re}/\varepsilon_{re}\right)^{0.5} \qquad (2.4.19)$$

$$v_p = v_{p0}/\left(\mu_{re}\varepsilon_{re}\right)^{0.5} \qquad (2.4.20)$$

2.5 COUPLED MODES ANALYSIS METHOD

Among the dispersion group analysis methods, the coupled mode method is the most simple to be studied. With this method, the effective relative permittivity is a function of frequency, and consequently, impedance.

This method supposes that the real propagation mode in a microstrip can be obtained through a coupling between a "TEM" and "TE" mode, as indicated respectively in Figure 2.5.1 parts a and b with their equivalent t.l.* networks. This explanation of the microstrip propagation was first suggested by the researcher H. J. Carlin.[16] The choice of a "TEM" and "TE" line comes from the fact that the fundamental mode can be approximated to a "TEM" while the first higher order mode has been evaluated to be a "TE." The elements' values which appear in Figure 2.5.1 can be found in Appendix A2. According to Carlin, the value of "k_t^2" is:

$$k_t^2 = \frac{\left(\varepsilon_r - \varepsilon_{re0}\right)\zeta^2}{4.143\left(60h\right)^2\left(0.5 + 0.001\zeta^{1.5}\right)} \qquad (2.5.1)$$

where "ε_{re0}" is the effective relative dielectric constant evaluated at DC, given for example by equation 2.4.7. So, the expression for "$\varepsilon_{re}(f)$" is:

$$\varepsilon_{re}(f) = \varepsilon_{re0} - \frac{\left(v_0 k_t\right)^2}{2\omega^2} + \left[\left(\varepsilon_r - \varepsilon_{re0}\right)^2 + \frac{\left(v_0 k_t\right)^4}{4\omega^4}\right]^{0.5} \qquad (2.5.2)$$

where "v_0" is the light speed in the vacuum.

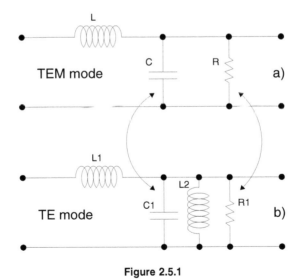

Figure 2.5.1

* See Appendix A2 for equivalence between modes and t.l.

2.6 FULL WAVE ANALYSIS METHOD

With this method, a solution of the Maxwell equations is found for a microstrip enclosed in a box, applying the boundary conditions that the fields must satisfy. The enclosed microstrip, as indicated in Figure 2.6.1 is supposed to have cylindrical symmetry, and a Cartesian coordinate system is applied. Following the classical methods of study for structures with such a symmetry,* the electric "A" and magnetic "F" vector potentials are written as:

$$A(x,y,z) \triangleq L(z) A_t(x,y) \tag{2.6.1}$$

$$F(x,y,z) \triangleq L(z) F_t(x,y) \tag{2.6.2}$$

where "L(z)" is assumed to be in a lossless case and in a reflectionless propagation in the "$\underline{z}_0$" direction, is given by:

$$L(z) = Ce^{-j\beta z} \tag{2.6.3}$$

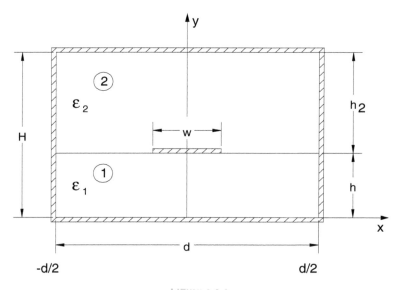

Figure 2.6.1

Then, for every region i = 1,2 indicated in Figure 2.6.1 we suppose the propagation mode to be composed of the sum of infinite "TM" and "TE" modes, obtained using equations A2.6.1 and A2.6.2. So, we can write:**

$$\underline{e}_{zi} = \frac{-k_{ti}^2}{j\omega\varepsilon_i} L(z) A_{ti}(x,y) \underline{z}_0 \quad \text{due to "TM" mode} \tag{2.6.4}$$

$$\underline{h}_{zi} = \frac{-k_{ti}^2}{j\omega\mu_0} L(z) F_{ti}(x,y) \underline{z}_0 \quad \text{due to "TE" mode} \tag{2.6.5}$$

---

* See Appendix A2 for a general treatment on guided propagation.
** See Appendix A2 for general expressions of "TE" and "TM" modes.

$$\underline{h}_{ti} = \frac{1}{j\omega\mu_0}\frac{dL(z)}{dz}\nabla_t F_{ti}(x,y) + L(z)\nabla_t A_{ti}(x,y) \otimes \underline{z}_0 \quad \text{due to "TE + TM" modes} \quad (2.6.6)$$

$$\underline{e}_{ti} = -L(z)\nabla_t F_{ti}(x,y) \otimes \underline{z}_0 \frac{1}{j\omega\varepsilon_i}\frac{dL(z)}{dz}\nabla_t A_{ti}(x,y) \quad \text{due to "TE + TM" modes} \quad (2.6.7)$$

where

$$k_{ti}^2 = -\beta^2 - \omega^2\mu_0\varepsilon_i \quad (2.6.8)$$

and "∇_t" is the transverse gradient operator in the Cartesian system, as defined in Appendix A8.

To consider in the expression of the field all the possible "TE" and "TM" modes, we can express the transverse potential vectors with a series of infinite terms. Of course, the resulting expressions for "A_t" and "F_t" must verify the two-dimensional wave equation and the boundary conditions for the structure in Figure 2.6.1. To this purpose, with reference to Figure 2.6.1, the e.m. field components have to verify:

1. for $y = 0$ and $y = H$, with $|x| \leq d/2$, i.e., top and bottom shields: $e_x = !0$ and $e_y = !0$;
2. for $x = \pm d/2$, with $0 \leq y \leq H$, i.e., the lateral shields: $e_z = !0$ and $e_y = !0$
3. for $y = h$, with $|x| \leq w/2$, i.e., the hot conductor: $e_x = !0$ and $e_z = !0$
4. for $y = h$, with $w/2 \leq |x| \leq d/2$, i.e., the dielectrics interface:
   a. $e_{x1} = !e_{x2}$ and $e_{z1} = !e_{z2}$, i.e., continuous tangential electric components
   b. $\varepsilon_1 e_{y1} = !\varepsilon_2 e_{y2} - q_s$, i.e., variation of "$d$"* normal component is equal to surface charge density "q_s"
   c. $h_{y1} = !h_{y2}$, i.e., continuous normal magnetic components
   d. $h_{x1} = !h_{x2} + i_z(x)$ and $h_{z1} = !h_{z2} - i_x(x)$, i.e., variation of "$h$" tangential component is equal to linear current "i."

Until this point, the relations are very simple and are a direct application of the boundary conditions to a cylindrical guiding structure that supports an e.m. field. Analytic difficulties arise when we apply the boundary conditions 1 through 4 above to the field expressions 2.6.4 through 2.6.7. The most common procedure is to transform the equations resulting from the application of the boundary conditions to the field components,[17,18,19] into a set of homogeneous equations, with "β" as the unknown, whose solution is found zeroing the determinantal equation. The result of this procedure is that the guided wavelength** "λ_g" is not linearly dependent to free space wavelength "λ_0" and frequency "f," as indicated in Figure 2.6.2.

Figure 2.6.2

Concerning the t.l. characteristic impedance evaluation, it is common to all "full wave methods" to define this quantity through the equation regarding the traveling power "W," i.e.:

* $d \overset{\perp}{=} \varepsilon e$ is the electric displacement vector.
** See Appendix A2 for guided wavelength definition.

$$\zeta \doteq 2W/i_z^2 \qquad (2.6.9)$$

where "i_z" is the current flowing in the propagation direction "z." This quantity can be obtained integrating "$i_z(x)$" along "x," i.e.:

$$i_z = \int_{-w/2}^{w/2} i_z(x)\,dx \qquad (2.6.10)$$

Similarly, the power "w" can be obtained integrating the Poynting vector* $\underline{P} \doteq \underline{e} \otimes \underline{h}^*$ inside the transverse surface of the structure indicated in Figure 2.6.1, i.e.:

$$w = \int_{-d/2}^{d/2} \int_0^H \underline{P} \cdot \underline{z}_0\,dx\,dy \qquad (2.6.11)$$

The results of this procedure give an impedance increasing with frequency, as indicated in Figure 2.6.3. However, caution needs to be used when comparing µstrip impedance vs. frequency as obtained by some authors. In fact, these results are strongly dependent on the definition[20] employed to extract the impedance, because the current and voltage definitions at µwave frequencies are not the same.**

Figure 2.6.3

2.7 DESIGN EQUATIONS

A lot of researchers have studied the microstrip, resulting in a lot of closed form analysis equations. We will give the design equations produced by the researchers I. J. Bahl and R. Garg,[21] who have modified Hammerstad's expressions given before to include the effect of the conductor thickness.[22,23,24] The effective relative dielectric constant is:

$$\varepsilon_{re} = \frac{\varepsilon_r + 1}{2} + \frac{\varepsilon_r - 1}{2}F - \frac{(\varepsilon_r - 1)t}{4.6(wh)^{0.5}} \qquad (2.7.1)$$

where "F" is given by 2.4.8 and 2.4.9.

To evaluate the impedance, we introduce the quantity "Δw" named "extra width" given by:

$$\Delta w = (1.25t/\pi)\left[1 + \ln(4\pi w/t)\right] \qquad \text{for } w \le h/2\pi \qquad (2.7.2)$$

---

* See Appendix A2 for Poynting vector definition.
** See Appendix A2 for some definition of current and voltage at µwave frequencies.

$$\Delta w = (1.25t/\pi)\left[1 + \ln(2h/t)\right] \qquad \text{for } w \geq h/2\pi \qquad (2.7.3)$$

so that an effective strip width "w_e" can be defined as $w_e \doteq w + \Delta w$. The microstrip characteristic impedance is:

$$\zeta = \left(60/\varepsilon_{re}^{0.5}\right)\ln\left[8h/w_e + 0.25w_e/h\right] \qquad \text{for } w \leq h \qquad (2.7.4)$$

$$\zeta = \left(120\pi/\varepsilon_{re}^{0.5}\right)\left[w_e/h + 1.393 + 0.667\ln(w_e/h + 1.444)\right]^{-1} \qquad \text{for } w \geq h \qquad (2.7.5)$$

The error produced by these equations has been evaluated as being less than 1% of the measured values.

Wheeler[25] has also obtained microstrip synthesis equations, which give the ratio "w/h" as a function of the µstrip impedance "ζ" according to:

$$w/h = \frac{2}{\pi}\left[\frac{377\pi}{2\zeta\sqrt{\varepsilon_r}} - 1 - \ln\left(\frac{377\pi}{\zeta\sqrt{\varepsilon_r}} - 1\right)\right] + \frac{\varepsilon_r - 1}{2\varepsilon_r}\left[\ln\left(\frac{377\pi}{2\zeta\sqrt{\varepsilon_r}} - 1\right) + 0.293 - \frac{0.517}{\varepsilon_r}\right] \quad \text{for } w \geq 2h$$

$$(2.7.6)$$

$$w/h = \left(e^x/8 - e^{-x}/4\right)^{-1} \qquad \text{for } w \leq 2h \qquad (2.7.7)$$

where:

$$x = (\zeta/60)\left[(\varepsilon_r + 1)/2\right]^{0.5} + \left[(\varepsilon_r - 1)/(\varepsilon_r + 1)\right](0.226 + 0.12/\varepsilon_r) \qquad (2.7.8)$$

In the previous synthesis equations the conductor thickness is supposed to be infinitesimal, i.e., it is neglected. To take into account the thickness effect, we can use equations 2.7.2 and 2.7.3 and instead build the "w" defined by 2.7.6 or 2.7.7, as the quantity $w' \doteq w - \Delta w$.

The effects on the µstrip electrical characteristics of a metallic shield over the structure, as indicated in Figure 2.7.1, have been investigated by the researcher I. J. Bahl.[26] The resulting new expression for "ζ" is:

$$\zeta = \left(60/\varepsilon_{re}^{0.5}\right)\ln(8h/w_e + 0.25w_e/h) - p \qquad \text{for } w \leq h \qquad (2.7.9)$$

$$\zeta = \left(120\pi/\varepsilon_{re}^{0.5}\right)\left[w_e/h + 1.393 + 0.667\ln(w_e/h + 1.444)\right]^{-1} - q \qquad \text{for } w \geq h \qquad (2.7.10)$$

where:

$$p = 270\left\{1 - \tanh\left[0.28 + 1.2(h_0/h)\right]^{0.5}\right\} \qquad (2.7.11)$$

$$q = \left\{1 - \tanh\left[1 + \frac{0.48(w_e/h - 1)^{0.5}}{(1 + h_0/h)^2}\right]\right\}p \qquad (2.7.12)$$

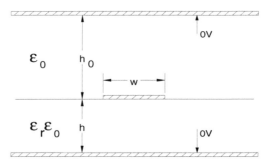

Figure 2.7.1

The resulting new expression for "ε_{re}" is:

$$\varepsilon_{re} = \frac{\varepsilon_r + 1}{2} + (F - C)\frac{\varepsilon_r - 1}{2}\tanh\left(0.18 + 0.235\frac{h_0}{h} - \frac{0.415}{(h_0/h)^2}\right) \tag{2.7.13}$$

where "F" is given by 2.4.8 and 2.4.9, "w_e" as stated above and "C" is given by:

$$C = \frac{\varepsilon_r - 1}{4.6}\frac{t}{(wh)^{0.5}} \tag{2.7.14}$$

The accuracy of the Bahl's equation has been proved to be near 99% compared with the numerical result of full wave analysis. The macroscopic effect of the top cover, as indicated in Figure 2.7.1, is a decrease in the characteristic impedance with respect to the case of the isolated µstrip structure. In any case, for $h_0/h \geq 5$ this effect is negligible. The effect of a complete box enclosing the µstrip has also been investigated.[27] Indicating with "s" the distance between the nearest lateral plane to the µstrip hot conductor edge, it has been proved that if $s \geq 4h$ the effect of the side conductor on the µstrip impedance is negligible.

We can recognize how all the previous formulas do not consider the dispersive nature of the microstrip since no equation contains the frequency as a variable. Dispersion has been evaluated by the researchers M. Kirschning and R. H. Jansen[28] who have given expressions for effective relative dielectric constant and impedance as a function of frequency for t = 0. The resulting expression is:

$$\varepsilon_{re}(f) = \varepsilon_r - \frac{\varepsilon_r - \varepsilon_{re}}{1 + D} \tag{2.7.15}$$

where:

$$D = D_1 D_2\left[(0.1844 + D_3 D_4)10hf\right]^{1.5763} \tag{2.7.16}$$

$$D_1 = 0.27488 + (w/h)\left[0.6315 + \frac{0.525}{(1 + 0.157hf)^{20}}\right] - 0.065683e^{-8.7513w/h} \tag{2.7.17}$$

$$D_2 = 0.33622\left(1 - e^{-0.03442\varepsilon_r}\right) \tag{2.7.18}$$

$$D_3 = 0.0363e^{-4.6w/h}\left[1 - \exp\left(-hf/3.87\right)^{4.97}\right] \tag{2.7.19}$$

$$D_4 = 1 + 2.751\left[1 - \exp\left(-\varepsilon_r/15.916\right)^8\right] \tag{2.7.20}$$

and "ε_{re}" is the static effective relative dielectric constant, evaluated, for example, as indicated in Section 2.4.

In the previous formulas, "exp( )" means the natural number exponential, i.e., "$e^{()}$." The error produced by the previous equations is below 1% from the exact values if $0.1 \leq w/h \leq 100$, $1 \leq \varepsilon_r \leq 20$, and $0 \leq h/\lambda_0 \leq 0.13$, where the frequency "f" is measured in GHz, "h" and "λ_0" in cm, and "λ_0" is the signal wavelength in free space.

2.8 ATTENUATION

Any practical t.l. has three sources of attenuation, due to:

1. Finite conductibility of t.l. conductors
2. Finite resistivity of the substrate and its dumping phenomena
3. Radiation effects

Of course, we are not considering ferrimagnetic materials as substrates that could cause magnetic resonance losses.*

Attenuations defined in points 1 and 2 above are analytically represented with two constants, respectively indicated by "α_c" and "α_d" and called "conductor loss constant" and "dielectric loss constant."** Radiation losses are strongly dependent on the type of t.l. under test. For example, waveguides have no radiation losses, while in our case, since the microstrip is an open t.l., radiation effects are surely present at any discontinuity section. However, for µstrip using high "ε_r" materials and accurate conductor shape and matching, conductor and dielectric losses are predominant in relation to the radiation losses. In the next chapter, radiation and other non- "TEM" effects in µstrip circuits will be studied.

Assuming a pure "TEM" mode in the µstrip, the evaluation of conductor losses[29] can be performed applying Wheeler's[30,31] incremental inductance rule. The foundation of this theory is that the e.m. energy penetrates inside the nonideal conductors.*** A "penetration depth (p)" is introduced, given by:

$$p \overset{\perp}{=} \left(\pi f\mu_c g\right)^{-0.5} \qquad [\text{meters}] \tag{2.8.1}$$

for which the field amplitudes are reduced by "1/e." In Equation 2.8.1, "f" is the signal frequency, "g" and "μ_c" are the conductor conductivity and absolute permeability, respectively. The effect of each penetration can be regarded as an introduction of an additional series inductance**** per u.l., indicated with "L_i" and called "incremental inductance." In this case, the evaluation of all the penetration depths can be done observing that five sides of the conductor are involved in this phenomena as indicated in Figure 2.8.1. Here, with the dashed line we have indicated the penetration depth in each conductor.

---

* Foundations of magnetism applied to t.l. are introduced in Appendix A5, A6 and A7.
** These quantities have been defined in Chapter 1.
*** See Appendix A2 for e.m. energy penetration inside nonperfect conductors.
**** Since we suppose the µstrip only supports a "TEM" mode, we are referring to the simple low pass equivalent network for a line supporting a "TEM" mode. This argument is treated in Chapter 1.

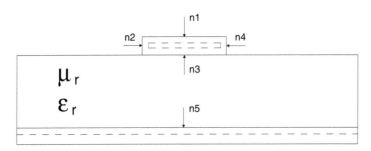

Figure 2.8.1

If "L" is the equivalent series inductance per u.l. of the lumped equivalent t.l. for the µstrip, given by:*

$$L = \zeta/v = \zeta\left(\mu_{re}\,\varepsilon_{re}\right)^{0.5}/v_0 \qquad \left[\text{Henry}/\text{meter}\right] \qquad (2.8.2)$$

each "L_i" is:

$$L_i = \frac{p\mu_{cr}}{2}\frac{\partial L}{\partial n} \equiv \frac{p\mu_c}{2\mu_0}\frac{\partial L}{\partial n} \qquad \left[\text{Henry}/\text{meter}\right] \qquad (2.8.3)$$

where "μ_{cr}" is the conductor relative permeability and "∂n" is an infinitesimal penetration inside the conductor, positive when the vector "$\underline{n}$" is directed into the conductor. The associated reactance** "R_i" of "L_i" is called "incremental resistance" per u.l., and is given by:

$$R_i \overset{\perp}{=} \omega L_i = \mu f p\mu_{cr}\frac{\partial L}{\partial n} \qquad \left[\Omega/\text{meter}\right] \qquad (2.8.4)$$

If we define the quantity "R_s," called "sheet resistance" for the conductor, as:

$$R_s \overset{\perp}{=} \pi f p\mu_c \qquad \left[\Omega/\text{square}\right]*** \qquad (2.8.5)$$

Equation 2.8.4 becomes:

$$R_i = \frac{R_s}{\mu_0}\frac{\partial L}{\partial n} \qquad \left[\Omega/\text{meter}\right] \qquad (2.8.6)$$

To obtain the whole additional inductance "L_a" and resistance "R_a" we must include all the incremental inductances and resistances in our calculations. With reference to Figure 2.8.1 and using 2.8.3 and 2.8.6, we have:

$$L_a = \sum_{j=1}^{j=5}\left(L_i\right)_j = \frac{1}{2\mu_0}\sum_{j=1}^{j=5} p_j\left(\mu_c\right)_j\frac{\partial L}{\partial n_j} \qquad (2.8.7)$$

---

*   See Chapter 1 for relations between t.l. characteristics quantities.
** See Appendix A2 for definition of internal impedance for good conductors.
*** See Appendix A2 for measurement unit of "conductor resistance."

$$R_a = \sum_{j=1}^{j=5}\left(R_i\right)_j = \frac{1}{\mu_0}\sum_{j=1}^{j=5}\left(R_s\right)_j\frac{\partial L}{\partial n_j} \qquad (2.8.8)$$

The conductor attenuation coefficient "α_c"* is defined as:

$$\alpha_c = W_c/2W_t \qquad (2.8.9)$$

where "W_c" and "W_t" are respectively the mean power dissipated in the conductor and the mean transmitted power, given by:

$$W_c = R_a\left|i\right|^2, \quad W_t = \zeta\left|i\right|^2 \qquad (2.8.10)$$

Consequently, the conductor attenuation constant does not depend by the additional inductance "L_a." Using 2.8.8 and 2.8.10, Equation 2.8.9 becomes:

$$\alpha_c = \frac{1}{2\mu_0\zeta}\sum_{j=1}^{j=5}\left(R_s\right)_j\frac{\partial L}{\partial n_j} \qquad (2.8.11)$$

From 2.8.2 we note that:

$$\frac{\partial L}{\partial n} = \frac{1}{v_0}\frac{\partial\left[\left(\mu_{re}\,\varepsilon_{re}\right)^{0.5}\zeta\right]}{\partial n} \perp \frac{1}{v_0}\frac{\partial\zeta_z}{\partial n} \qquad (2.8.12)$$

where $\zeta_z \perp \left(\mu_{re}\varepsilon_{re}\right)^{0.5}\zeta$. Note how "$\varepsilon_{re}$" and "$\mu_{re}$" are also theoretically functions of µstrip dimensions, and for this reason, these quantities need to be derived. This doesn't happen in a real "TEM" t.l., where the dielectric is homogeneous and the concept of effective dielectric constant does not have to be introduced. However, in a lot of cases no appreciable error is made if only "ζ" is derived.

Observing that $\mu_0 v_0 = \zeta_0 \equiv 120\pi$ and using Equation 2.8.12, equation 2.8.11 becomes:

$$\alpha_c = \frac{1}{\zeta 240\pi}\sum_{j=1}^{j=5}\left(R_s\right)_j\frac{\partial\zeta_z}{\partial n_j} \qquad (2.8.13)$$

Since the "ζ" and "ε_{re}" are also functions of "w," "h," and "t," as was shown previously, the derivative "$\partial\zeta_z/\partial n$" is:

$$\frac{\partial\zeta_z}{\partial n} = \frac{1}{dn}\left(\frac{\partial\zeta_z}{\partial w}\,dw + \frac{\partial\zeta_z}{\partial h}\,dh + \frac{\partial\zeta_z}{\partial t}\,dt\right) \qquad (2.8.14)$$

From Figure 2.8.1 we observe that:

$$dw = -2dn, \quad dh = 2dn, \quad dt = -2dn \qquad (2.8.15)$$

---

* We are assuming a longitudinal variation of conductor attenuation with $e^{-\alpha_c z}$. See Chapter 1 for fundamental theory of transmission lines.

and 2.8.13 becomes:*

$$\alpha_c = \frac{1}{\zeta 120\pi} R_s \left(-\frac{\partial \zeta_z}{\partial w} + \frac{\partial \zeta_z}{\partial h} - \frac{\partial \zeta_z}{\partial t} \right) \tag{2.8.16}$$

The incremental inductance rule has been verified to give very accurate results for conductor thickness greater than four times "p." This condition is usually verified for every planar transmission line, since for the typical conductors used, the value of "p" is lower than some micrometer for frequencies greater than 1 GHz.** Simple equations for "α_c" just oriented to computer implementation have been produced by the researchers R. A. Pucel, D. J. Massé, and C. P. Hartwig,[32] simply by applying Equation 2.8.16 to Wheeler's impedance expressions. The resulting equations are:

for $w \leq h/2\pi$:

$$\alpha_c \frac{\zeta h}{R_s} = \frac{8.68}{2\pi} \left[1 - \left(w_e/4h \right)^2 \right] \left[1 + h/w_e + \left(h/\pi w_e \right) \left(\ln \frac{4\pi w}{t} + t/w \right) \right] \tag{2.8.17}$$

for $h/2\pi \leq w \leq 2h$:

$$\alpha_c \frac{\zeta h}{R_s} = \frac{8.68}{2\pi} \left[1 - \left(w_e/4h \right)^2 \right] \left[1 + h/w_e + \left(h/\pi w_e \right) \left(\ln \frac{2h}{t} - t/h \right) \right] \tag{2.8.18}$$

for $w \geq 2h$:

$$\alpha_c \frac{\zeta h}{R_s} = \frac{8.68}{D} \left[w_e/h + \left(w_e/\pi h \right)/\left(w_e/2h + 0.94 \right) \right] \left[1 + h/w_e + \left(h/\pi w_e \right) \left(\ln \frac{2h}{t} - t/h \right) \right] \tag{2.8.19}$$

where:

$$D = \left\{ w_e/h + (2/\pi) \ln \left[2\pi e \left(w_e/2h + 0.94 \right) \right] \right\}^2 \tag{2.8.20}$$

and "e" is the natural number. The "α_c" values obtained from Equations 2.8.17 through 2.8.20 result in "dB/u.l." These researchers suggest using a slightly modified expression for the effective hot strip width "w_e," (with respect to the expressions given previously), which are as follows:

$$w_e = w + (t/\pi) \left[1 + \ln \left(4\pi w/t \right) \right] \quad \text{for } w \leq h/2\pi \tag{2.8.21}$$

$$w_e = w + (t/\pi) \left[1 + \ln \left(2h/t \right) \right] \quad \text{for } w \geq h/2\pi \tag{2.8.22}$$

Concerning the dielectric losses, the researchers M. V. Schneider, B. Glance, and W. F. Bodtmann[33] have given the value of "α_d" in "dB/u.l." for a nongyromagnetic substrate as:

---

* We assume that the bottom conductor has the same surface resistivity as the hot conductor, which it usually does.
** See Appendix A2 for values of penetration depth inside good conductors.

$$\alpha_d = \frac{20\pi}{\ln 10} \frac{1/\varepsilon_{re} - 1}{1/\varepsilon_r - 1} \frac{\tan\delta}{\lambda_0} \sqrt{\varepsilon_{re}} \qquad (2.8.23)$$

where "$\tan\delta$" is the substrate "tangent delta."* Of course, the quantities "ε_r" and "ε_{re}" are all relative to the real part of the substrate dielectric constant.** In general, for the typical substrates employed in MIC circuits, i.e., alumina and quartz, conductor losses are predominant, and "α_c" in dB can also be 10 times the value of "α_d" in dB. Different is the case for semiconductor substrate, employed in MMIC devices, where "α_d" is comparable to "α_c." For magnetic losses, the "α_d" expression can be formally modified multiplying by $\sqrt{\mu_{re}}$. However, this is only a way to remember that the presence of any ferrimagnetic material will need to be evaluated. It is known, in fact, that every ferrimagnetic material possesses an equivalent permeability that is dependent on many parameters such as the reciprocal direction between e.m. energy propagation and direction of the applied magnetic field "H_{dc}," fields' intensity, signal frequency, ferrimagnetic composition, and more.*** So the symbol "μ_{re}" is quite often only a notational simplification. In Chapter 7 we will study some µstrip devices that use the RF interaction with ferrimagnetic materials.

2.9 PRACTICAL CONSIDERATIONS

As we said earlier in this chapter, the µstrip is the most widely diffused "PCB" t.l. In general, this µstrip is not only used in µwave devices but also in low frequency "PCB."

The first thing to consider is the undercut phenomenon, which can cause problems if high impedance t.l. or coupled microstrips have to be built. The reasons for this phenomenon were discussed in Chapter 1 and will not be repeated here.

Another problem in low frequency µstrip devices is that, especially in high density boards, the ground plane is required to be as small as possible. In practice, it has been found[34] that the microstrip impedance "ζ'" with finite ground plane width "w_g," is practically equal to the value "ζ" with infinite width ground plane, until "w_g" is at least greater than 3w as indicated in Figure 2.9.1.

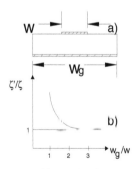

Figure 2.9.1

The progressive reduction of the ground plane, as indicated in Figure 2.9.2a, passes from the unbalanced µstrip structure to the balanced structure, used in mixer devices.[35] The practical situation is indicated in Figure 2.9.2b, and the balanced structure is called the "balanced broadside coupled line." We will simplify this name using the letters "BBCL." Assuming the practical case for which a ≥ 5w and b ≥ 5h, where we can neglect the effect of the enclosure, the characteristic impedance "ζ" between the two conductors of this t.l. can be evaluated using Wheeler's formulas:

* See Chapter 1 for "$\tan\delta$" definition.
** See Chapter 1 for complex permittivity definition.
*** See Appendix A7 for fundamentals of energy exchange between waves and ferrimagnetic materials.

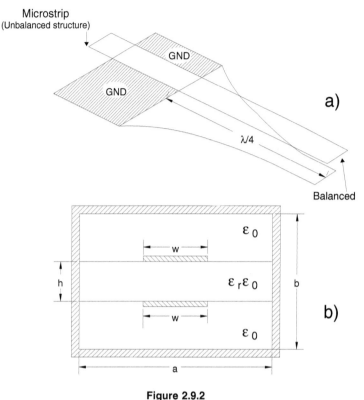

Figure 2.9.2

$$\zeta = \frac{120\pi}{\varepsilon_{re}^{0.5}}\left\{w_e/h + 0.441 + \frac{\varepsilon_r + 1}{2\pi\varepsilon_r}\left[\ln\left(w_e/h + 0.94\right) + 1.451\right] + 0.082\frac{\varepsilon_r - 1}{\varepsilon_r^2}\right\}^{-1} \text{ for } w \ge h \quad (2.9.1)$$

$$\zeta = \frac{120\sqrt{2}}{\left(\varepsilon_{re} + 1\right)^{0.5}}\left\{\ln\frac{4h}{w_e} + (1/8)\left(w_e/h\right)^2 - \frac{\varepsilon_r - 1}{2\left(\varepsilon_r + 1\right)}\left[\ln\left(\pi/2\right) + \frac{\ln\left(4/\pi\right)}{\varepsilon_r}\right]\right\}\text{for } w \le h \quad (2.9.2)$$

The researchers B. Bhattia and P. Pramanick[36] have studied the transmission line characteristics of the boxed structure indicated in Figure 2.9.2 b considering the effect of the enclosure.

REFERENCES

1. Y. J. He, S. F. Li, Analysis of arbitrary cross section microstrip using the method of lines, *IEEE Trans. on MTT*, 162, Jan. 1994.
2. H. A. Stinehelfer, An accurate calculation of uniform microstrip transmission lines, *IEEE Trans. on MTT*, 439, July 1978.
3. H. A. Wheeler, Transmission line properties of parallel strips separated by a dielectric sheet, *IEEE Trans. on MTT*, 172, March 1965.
4. H. J. Carlin, A simplified circuit model for microstrip, *IEEE Trans. on MTT*, 589, Sept. 1973.
5. Haim Cory, Dispersion characteristics of microstrip lines, *IEEE Trans. on MTT*, 59, Jan. 1981.
6. W. J. Getsinger, Microstrip dispersion model, *IEEE Trans. on MTT*, 34, Jan. 1973.

7. A. R. Djordjevic, T. K. Sarkar, Closed form formulas for frequency dependent resistance and inductance per unit length of microstrip and strip transmission lines, *IEEE Trans. on MTT*, 241, Feb. 1994.

8. A. K. Verma, R. Kumar, A new dispersion model for microstrip line, *IEEE Trans. on MTT*, 1183, Aug. 1998.

9. J. B. Knorr, A. Tufekcioglu, Spectral domain calculation of microstrip characteristic impedance, *IEEE Trans. on MTT*, 725, Sept. 1975.

10. R. Mittra, T. Itoh, A new technique for the analysis of the dispersion characteristics of microstrip lines, *IEEE Trans. on MTT*, 47, Jan. 1971.

11. G. Coen, N. Faché, D. De Zutter, Comparison between two sets of basis functions for the current modeling in the Galerkin spectral domain solution for microstrips, *IEEE Trans. on MTT*, 505, Mar. 1994.

12. S. Y. Lin, C. C. Lee, A full wave analysis of microstrips by the boundary elements method, *IEEE Trans. on MTT*, 1977, Nov. 1996.

13. H. A. Wheeler, Transmission line properties of parallel strips separated by a dielectric sheet, *IEEE Trans. on MTT*, 172, March 1965.

14. Erik O. Hammerstad, Equations for microstrip circuit design, Proc. V European Microwave Conf., Cambridge, 1975.

15. H. A. Wheeler, Transmission line properties of a strip on a dielectric sheet on a plane, *IEEE Trans. on MTT*, 631, Aug. 1977.

16. H. J. Carlin, A simplified circuit model for microstrip, *IEEE Trans. on MTT*, 589, Sept. 1973.

17. R. Mittra, T. Itoh, A new technique for the analysis of the dispersion characteristics of microstrip lines, *IEEE Trans. on MTT*, 47, Jan. 1971.

18. M. K. Krage, G. I. Haddad, Frequency dependent characteristics of microstrip transmission lines, *IEEE Trans. on MTT*, 678, Oct. 1972.

19. J. B. Knorr, A. Tufekcioglu, Spectral domain calculation of microstrip characteristic impedance, *IEEE Trans. on MTT*, 725, Sept. 1975.

20. B. Bianco, L. Panini, M. Parodi, S. Ridella, Some considerations about the frequency dependence of the characteristic impedance of uniform microstrips, *IEEE Trans. on MTT*, 182, March 1978.

21. I. J. Bahl, R. Garg, Simple and accurate formulas for a microstrip with finite strip thickness, Proc. IEEE, Nov. 1977.

22. M. Kobayashi, K. Takaishi, Normalized longitudinal current distributions on microstrip lines with finite strip thickness, *IEEE Trans. on MTT*, 866, May 1994.

23. T. S. Horng, A generalized method for evaluating the metallization thickness effects on microstrip structures, Int. Microwave Symp., San Diego, 1994.

24. N. H. Zhu, W. Qiu, E. Y. B. Puns. P. S. Chung, Quasi static analysis of shielded microstrip transmission lines with thick electrodes, *IEEE Trans. on MTT*, 288, Feb. 1977.

25. H. A. Wheeler, Transmission line properties of parallel strips separated by a dielectric sheet, *IEEE Trans. on MTT*, 172, March 1965.

26. I. J. Bahl, Use exact methods for microstrip design, *Microwaves*, 61, Dec. 1978.

27. H. A. Stinchelfer, An accurate calculation of uniform microstrip transmission lines, *IEEE Trans. on MTT*, 439, July 1978.

28. M. Kirschning, R. H. Jansen, Accurate model for effective dielectric constant of microstrip with validity up to millimetre wave frequencies, *Electronics Lett.*, 272, March 1982.

29. G. B. Stracca, A simple evaluation of losses in thin microstrips, *IEEE Trans. on MTT*, 281, Feb. 1997.

30. H. A. Wheeler, Formulas for the skin effect, Proc. of the IRE 30, 1942.

31. R. Sturdivant, Transmission line conductor loss and the incremental inductance rule, *Microwave J.*, 156, Sept. 1995.

32. R. A. Pucel, D. J. Massé, C. P. Hartwig, Losses in microstrip, *IEEE Trans on MTT*, 342, June 1968. See also Corrections to: Losses in microstrip, *IEEE Trans. on MTT*, 1064, Dec. 1968.

33. M. V. Schneider, B. Glance, W. F. Bodtmann, Microwave and millimeter wave hybrid integrated circuits for radio systems, *Bell System Technical J.*, 1703, July-Aug. 1969.

34. C. E. Smith and R. S. Chang, Microstrip transmission line with finite width dielectric and ground plane, *IEEE Trans. on MTT*, 835, Sept. 1985.

35. S. A. Maas, *Microwave Mixers*, Artech House, 1993.

36. B. Bhartia, P. Pramanick, Computer aided design models for broadside coupled striplines and milli-
    meter wave suspended substrate microstrip lines, *IEEE Trans. on MTT*, 36(11), 1476, 1988. See also
    Corrections to: Computer aided design models for broadside coupled striplines and millimeter wave
    suspended substrate microstrip lines, *IEEE Trans. on MTT*, 37(10), 1658, 1989.

CHAPTER 3

Striplines

3.1 GEOMETRICAL CHARACTERISTICS

The geometric structure of a symmetric stripline is shown in Figure 3.1.1 with a cross-sectional view. It requires three layers of conductors, and for this reason it is also called "triplate." The internal conductor is commonly called the "hot conductor," while the other two, always connected at signal ground, are called "cold" or "ground" conductors. The hot conductor is embedded in a homogeneous and isotropic dielectric, of dielectric constant "ε_r." So, unlike the case of μstrip, the word "substrate" is not appropriate since the dielectric completely surrounds the hot conductor. Similarly, there is no need to introduce the concept of "effective relative dielectric constant" which, for striplines, is very important. In the present chapter unless otherwise stated, we will not study the case of the ferrimagnetic dielectric, which is instead studied in Chapter 8.

The structure indicated in Figure 3.1.1 is called "symmetric"* stripline because the hot conductor is at the middle of the distance "b" between the ground planes. However, striplines can also be built in an asymmetric fashion, as indicated in Figure 3.1.2, so that the hot conductor is not at the middle of the distance "b." In this case the stripline is said to be "asymmetric" or "offset."**

In general, we will use the following variables:

1. "w," the width of the hot conductor
2. "b," the separation of the ground planes
3. "t," the conductor's thickness
4. "h," the shorter distance between the hot conductor and one ground plane. If the stripline is symmetric and we neglect the strip thickness, then h = b/2.

Striplines are used in wideband networks and devices because of their low radiation, dispersion, and loss. Unfortunately, they present greater technological difficulties when compared to μstrip counterparts, and for this reason they are less diffused with respect to the same μstrip devices, excluding the case where accurate and precise devices are needed.

A common characteristic of these striplines is the typical lower impedance of μstrip t.l. It is not difficult to reach $\zeta = 20\Omega$, while for a μstrip this is an impractical value that can cause the rise of higher order modes.***

In the following sections we will give design equations for both types of striplines, i.e., symmetric and asymmetric. For simplicity, with the word "stripline" we will refer to the hot conductor of the stripline structure indicated in Figures 3.1.1 or 3.1.2 unless otherwise stated. In addition, the dielectric is assumed to be homogenous, isotropic, and nongyromagnetic, completely surrounding the hot conductor.

* The word "symmetric" is sometimes replaced with "balanced."
** Offset stripline will simply be called "OSL."
*** See Chapter 4 for higher order modes in μstrips and striplines.

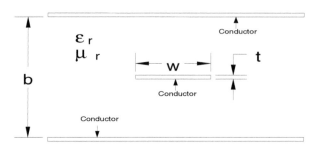

Figure 3.1.1

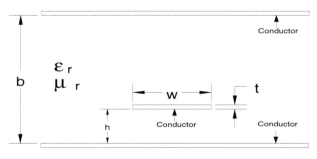

Figure 3.1.2

3.2 ELECTRIC AND MAGNETIC FIELD LINES

Due to the nearly homogenous substrate and a "TEM" propagation mode, the stripline is the less dispersive t.l. among the t.l.s. that we study in our text. Consequently, the stripline has the widest theoretical operative bandwidth. Devices where striplines can be advantageously employed are directional couplers and filters. However, when the device to be constructed needs some elements in shunt configuration or when the t.l. is dispersive, this t.l. is less practical than stripline.

Some electric "e" and magnetic "h" field lines for the fundamental "TEM" mode in stripline are indicated in Figure 3.2.1, in a defined cross-section and a defined time. In the fundamental mode the hot conductor is equipotential. High order modes in striplines will be studied in the next chapter.

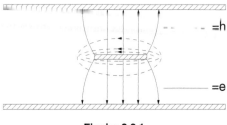

Figuire 3.2.1

3.3 SOLUTION TECHNIQUES FOR THE ELECTROMAGNETIC PROBLEM

Due to the intrinsic ability to support a "TEM" mode, stripline has been studied with quasi static methods. Of course, it is also possible to use full wave methods. The most commonly employed quasi static methods are:

1. The analytical solution of Laplace's equation[1]
2. The finite difference method applied to the Laplace's equation[2]
3. Conformal transformation.[3]

The conformal transformation and finite difference methods are reviewed in Appendix A1. The next section will describe how to obtain the stripline characteristic impedance using the conformal transformation method.

3.4 EXTRACTION OF STRIPLINE IMPEDANCE WITH A CONFORMAL TRANSFORMATION

The conformal transformation method belongs to the quasi static group for the analysis of an e.m. structure. The stripline is regarded as a static problem so that it can be transformed in a structure that resembles a parallel plate capacitor. The capacitance "C" of the stripline is evaluated, and the characteristic impedance "ζ" is obtained as $\zeta = 1/Cv$ where "v" is the light speed in the media with dielectric constant "ε_r."

It is assumed that this parallel plate capacitor represents a "TEM" lossless t.l. for which the characteristic impedance "ζ" is evaluated as:

$$\zeta = Lv \equiv 1/Cv \tag{3.4.1}$$

where "L" and "C" are the inductance and capacitance per unit length of the equivalent low pass network for the "TEM" line,* and "v" is the phase velocity of the light in the medium, i.e.:

$$v = 1/\left(\mu_0 \mu_r \varepsilon_0 \varepsilon_r\right)^{0.5} = v_0/\left(\mu_r \varepsilon_r\right)^{0.5} \tag{3.4.2}$$

In these equations, the subscript "0" individuates quantities with the substrate replaced by the vacuum. Of course, with the q.s. hypothesis, other simple relations occur among the previous quantities. These can be reviewed in Chapter 2 for the µstrip case and replacing "ε_{re}" with "ε_r."

The passages to transform the stripline in a parallel plate capacitor are indicated in Figure 3.4.1. First, the stripline is divided into four regions, and only one is taken as the original structure to be transformed. This situation is represented in part a of Figure 3.4.1. The two dashed lines labeled "mw" represent two ideal magnetic walls, i.e., ideal walls where the RF tangential magnetic component is zero. If "C" is the capacitance to ground of the stripline hot conductor, it is simple to recognize that the capacitance "C_4" to ground for Figure 3.4.1a is $C_4 = C/4$. Then a Cartesian complex reference system is applied, as indicated in Figure 3.4.1b. A first Schwarz-Christoffel transformation** is applied, which transforms the structure in Figure 3.4.1b into that in Figure 3.4.1c. This transformation from "Z" plane into "T" plane is performed by:

$$z = A_1 \int \left[(t+1)t\right]^{-0.5} dt + B_1 \tag{3.4.3}$$

Integrating this equation and using the correspondences $z_0 \leftrightarrow t_0$, $z_2 \leftrightarrow t_2$, and $z_3 \leftrightarrow t_3$ we have:

$$z = \left(-b/\pi\right) \ln\left[t^{0.5} + (t+1)^{0.5}\right] + jb/2 \tag{3.4.4}$$

---

* See Appendix A2 for networks equivalent to propagation modes.
** See Appendix A1 for the theory of Schwarz-Christoffel transformation.

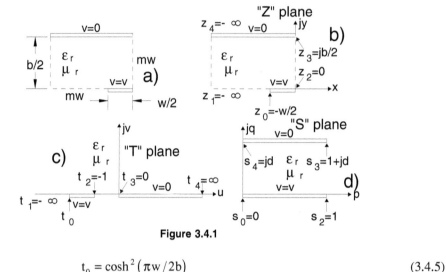

Figure 3.4.1

$$t_0 = \cosh^2\left(\pi w / 2b\right) \tag{3.4.5}$$

A second Schwarz-Christoffel transformation is applied, which transforms the structure in Figure 3.4.1c into that in Figure 3.4.1d. This transformation from "T" plane into "S" plane is performed by:

$$s = A \int \left[\left(t + t_0\right)\left(t + 1\right) t \right]^{-0.5} dt + B_2 \tag{3.4.6}$$

Unfortunately, this equation is not as simple to integrate as that in 3.4.3, but with the substitution $t = -x^2$ we obtain:*

$$s = \frac{2jA}{\sqrt{t_0}} \int \left[\left(1 - x^2\right)\left(1 - \frac{x^2}{t_0}\right) \right]^{-0.5} dx + B_2 \tag{3.4.7}$$

The integral of the previous equation can be transformed in the elliptic integral of first kind** "$F(\xi, p)$," so that equation 3.4.7 becomes:

$$s = A_2 F\left(\sqrt{-t}, \frac{1}{\sqrt{t_0}} \right) + D_2 \tag{3.4.8}$$

where $A_2 = 2jA/\sqrt{t_0}$. Using the correspondences $t_0 \leftrightarrow s_0$, $t_2 \leftrightarrow s_2$, and $t_3 \leftrightarrow s_3$ in the previous equation we have a system of three equations in three unknowns, "A_2," "B_2," and "d," from which the distance "d" of the plates in Figure 3.4.1d is obtained:

$$d = K(p) / K(p') \tag{3.4.9}$$

The quantity "$K(p)$" is the "complete elliptical integral of the first kind," *** defined in Appendix A8, and the parameters "p" and "p'" are:

---

* "x" is, of course, a real number.
** See Appendix A8 for definition of the elliptic integral of first kind "$F(\xi, p)$" and its relation to "$K(p)$."
*** Sometimes in literature "$K(p')$" is indicated with "$K'(p)$." This is only a different symbology since operatively the integral is evaluated for $p' = (1 - p^2)^{0.5}$.

$$p = \sec h\left(\pi w/2b\right) \tag{3.4.10}$$

$$p' \overset{\perp}{=} \left(1 - p^2\right)^{0.5} \equiv \tanh\left(\pi w/2b\right) \tag{3.4.11}$$

Once we have found "d," we can calculate the capacitance per unit length "C_4" of the structure indicated in Figure 3.4.1c, i.e., $C_4 = \varepsilon_0 \varepsilon_r/d$. Consequently, the capacitance "C" per unit length of the stripline indicated in Figure 3.1.1 is:

$$c = 4\varepsilon_0 \varepsilon_r/d \tag{3.4.12}$$

Inserting Equation 3.4.9 into 3.4.12 and using 3.4.1, the stripline characteristic impedance is:

$$\zeta = \frac{30\pi\sqrt{\mu_r}}{\sqrt{\varepsilon_r}} \frac{K(p)}{K(p')} \tag{3.4.13}$$

Note that in this study the thickness "t" of the hot conductor is assumed to be zero. In the next section we will give formulas that take into account the nonzero value of "t."

3.5 DESIGN EQUATIONS

The most simple design equations come directly from the stripline characteristic impedance of the previous section. In fact, the ratio of the elliptic integrals can be approximated in closed form with high accuracy,* typically some percent, according to the parameter range. In this case the parameters "p" and "p'" essentially depend on the ratio "w/b." So, if w/b ≤ 0.5 we have:

$$\frac{K(p)}{K(p')} = \frac{1}{\pi} \ln\left[2\frac{1 + p^{0.5}}{1 - p^{0.5}}\right] \tag{3.5.1}$$

while if w/b ≥ 0.5

$$\frac{K(p)}{K(p')} = \pi\left\{\ln\left[2\frac{1 + (p')^{0.5}}{1 - (p')^{0.5}}\right]\right\}^{-1} \tag{3.5.2}$$

The previous two equations inserted into the stripline characteristic impedance of the previous section permit the evaluation of "ζ" in a very simple way.

In the case of stripline, it is quite simple to have synthesis equations, i.e., equations from which we can derive the ratio "w/b" for a given "ζ" and "ε_r." Using 3.5.1 and 3.4.13 we can write:

$$\zeta = \frac{30\sqrt{\mu_r}}{\sqrt{\varepsilon_r}} \ln\left[2\frac{1 + p^{0.5}}{1 - p^{0.5}}\right] \tag{3.5.3}$$

With simple manipulation, the previous equation gives:

* See Appendix A8 for elliptic integrals and their approximations in closed form.

$$\frac{w}{b} = \frac{2}{\pi} \operatorname{acosh}\left[\left(\frac{e^{\alpha} - 2}{e^{\alpha} + 2}\right)^{-2}\right] \tag{3.5.4}$$

where

$$\alpha \stackrel{\pm}{=} \zeta\sqrt{\varepsilon_r}\big/30\sqrt{\mu_r} \tag{3.5.5}$$

Similarly, using 3.5.1 and 3.4.13 we can write:

$$\zeta = \frac{30\pi^2\sqrt{\mu_r}}{\sqrt{\varepsilon_r}} \ln\left[2\frac{1 + (p')^{0.5}}{1 - (p')^{0.5}}\right] \tag{3.5.6}$$

With simple manipulation, the previous equation gives:

$$\frac{w}{b} = \frac{2}{\pi} \operatorname{atgh}\left[\left(\frac{e^{\beta} - 2}{e^{\beta} + 2}\right)^{2}\right] \tag{3.5.7}$$

where

$$\beta \stackrel{\pm}{=} 30\pi^2\sqrt{\mu_r}\big/\zeta\sqrt{\varepsilon_r} \tag{3.5.8}$$

It has been observed that expression 3.5.7 gives the best mean value of "w/b" with respect to 3.5.4 and should be preferred.

The effect of strip thickness for a nongyromagnetic substrate has been evaluated by S.B. Cohn[4] with a new conformal transformation introducing the fringing capacities at the ends of the strips. With the conditions $t/w \le 0.11$ and $t/b \le 0.25$, the new impedance expressions are:

for $w/(b - t) \le 0.35$

$$\zeta = \frac{60}{\sqrt{\varepsilon_r}} \ln\left(\frac{4b}{\pi\phi}\right) \tag{3.5.9}$$

where "ϕ" is given by:

$$\phi = \frac{w}{2}\left\{1 + \frac{t}{w}\left[1 + \ln\left(\frac{4\pi w}{t}\right) + 0.51\left(\frac{t}{w}\right)^2\right]\right\} \tag{3.5.10}$$

for $w/(b - t) \ge 0.35$

$$\zeta = 94.15\left\{\sqrt{\varepsilon_r}\left[w/(b - t) + C_f/0.0885\varepsilon_r\right]\right\}^{-1} \tag{3.5.11}$$

where "C_f" is the fringing capacitance from one side of the hot conductor to one ground plane, given by:

$$C_f = \frac{0.0885\varepsilon_r}{\pi}\left\{\frac{2b}{b-t}\ln\left(\frac{b}{b-t}+1\right)-\left(\frac{b}{b-t}-1\right)\ln\left[1/(1-t/b)^2-1\right]\right\} \text{ in pf/cm} \qquad (3.5.12)$$

All the previous formulas give an error under 2% when compared to the nonapproximated equations.

3.6 ATTENUATION

Any practical stripline has three sources of attenuation, due to:

1. Finite conductivity of its conductors
2. Finite resistivity and dumping phenomena of the dielectric
3. Magnetic resonances

In contrast to the μstrip case, or other open t.l., in this case radiation losses can be neglected with respect to the other source of attenuation because stripline is a closed t.l. Point 3 above will be studied in Chapter 8 and Appendixes A5 through A7.*

Attenuations defined in points 1 and 2 above are analytically represented with two constants, respectively indicated with "α_c" and "α_d" and called "conductor loss constant" and "dielectric loss constant." **

Assuming a pure "TEM" mode in the stripline, the evaluation of conductor losses can be performed applying Wheeler's[5,6] incremental inductance rule. This way of evaluating the t.l. attenuation has been used throughout this text, and for this reason, we will only give the formulas here that are characteristics of the stripline. For the other concepts and common formulas, see the losses evaluation for the μstrip case in Chapter 2.

The additional inductance "L_a" and resistance "R_a," with reference Figure 3.6.1 are given by:

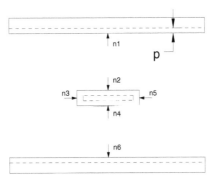

Figure 3.6.1

$$L_a = \sum_{j=1}^{j=6}(L_i)_j = \frac{1}{2\mu_0}\sum_{j=1}^{j=6}p_j(\mu_c)_j\frac{\partial L}{\partial n_j} \qquad (3.6.1)$$

$$R_a = \sum_{j=1}^{j=6}(R_i)_j = \frac{1}{\mu_0}\sum_{j=1}^{j=6}(R_s)_j\frac{\partial L}{\partial n_j} \qquad (3.6.2)$$

---

* Foundations of magnetism applied to t.l. are introduced in Appendix A5, A6, and A7.
** These quantities have been defined in Chapter 1.

where:

a. "L_i" is the "incremental inductance" per u.l.
b. "R_i" is the "incremental resistance" per u.l.
c. "p" is the "penetration depth" [u.l.]
d. "μ_c" is the conductor absolute permeability
e. "R_s" is the conductor "sheet resistance" [Ω/square]*

The conductor attenuation coefficient "α_c"** is defined as:

$$\alpha_c = W_c/2W_t \tag{3.6.3}$$

where "W_c" and "W_t" are respectively the mean power dissipated in the conductor and the mean transmitted power, given by:

$$W_c = R_a|i|^2, \quad W_t = \zeta|i|^2 \tag{3.6.4}$$

Consequently, the conductor attenuation constant does not depend on the additional inductance "L_a." Using 3.6.2 and 3.6.4, Equation 3.6.3 becomes:

$$\alpha_c = \frac{1}{2\mu_0\zeta} \sum_{j=1}^{j=5} (R_s)_j \frac{\partial L}{\partial n_j} \tag{3.6.5}$$

From 3.4.1 and 3.4.2 it follows that:

$$L = \zeta/v = \zeta(\mu_r\varepsilon_r)^{0.5}/v_0 \tag{3.6.6}$$

and so:

$$\frac{\partial L}{\partial n} = \frac{(\mu_r\varepsilon_r)^{0.5}}{v_0} \frac{\partial \zeta}{\partial n} \tag{3.6.7}$$

Observing that $\mu_0 v_0 = \zeta_V \equiv 120\pi$ and using Equation 3.6.7, Equation 3.6.5 becomes:

$$\alpha_c = \frac{\sqrt{\mu_r\varepsilon_r}}{\zeta 240\pi} \sum_{j=1}^{j=5} (R_s)_j \frac{\partial \zeta}{\partial n_j} \tag{3.6.8}$$

Since the "ζ" is a function of "w," "t," and "b," as was shown in the previous section, the derivative "$\partial\zeta/\partial n$" is:

$$\frac{\partial \zeta}{\partial n} = \frac{1}{dn}\left(\frac{\partial \zeta}{\partial w}dw + \frac{\partial \zeta}{\partial b}db + \frac{\partial \zeta}{\partial t}dt\right) \tag{3.6.9}$$

---

* See Appendix A2 for measurement unit of "conductor resistance."
** We are assuming a longitudinal variation of conductor attenuation with $e^{-\alpha_c z}$. See Chapter 1 for fundamental theory of transmission lines.

From Figure 3.6.1 we observe that:

$$dw = -2dn, \quad db = 2dn, \quad dt = -2dn \tag{3.6.10}$$

and 3.6.8 becomes:*

$$\alpha_c = \frac{R_s\sqrt{\mu_r \varepsilon_r}}{\zeta 120\pi}\left(-\frac{\partial\zeta}{\partial w} + \frac{\partial\zeta}{\partial b} + \frac{\partial\zeta}{\partial t}\right) \tag{3.6.11}$$

Of course, the value given by "α_c" is in neper/meter.**

Simple closed form equations for "α_c" for a symmetrical stripline have been produced by the researcher S. B. Cohn,[7] by applying equation 3.6.11 to the impedance expressions. The resulting equations for "α_c" in Neper/u.l. are:

for $w/(b - t) \geq 0.35$:

$$\alpha_c = P\left\{k + (2w/b)k^2 + \left[(1 + t/b)k^2/\pi\right]\ln\left[(k + 1)/(k - 1)\right]\right\} \tag{3.6.12}$$

where:

$$P \overset{\perp}{=} 4R_s\zeta\varepsilon_r\sqrt{\varepsilon_r}/b(120\pi)^2 \quad \text{and} \quad k \overset{\perp}{=} (1 - t/b)^{-1} \tag{3.6.13}$$

for $w/(b - t) \leq 0.35$, $t/b \leq 0.35$, $t/w \leq 0.11$:

$$\alpha_c = R_s/(2\pi b\zeta)\left\{1 + (b/d)\left[0.5 + 0.669t/w - 0.255(t/w)^2 + \ln(4\pi w/t)/2\pi\right]\right\} \tag{3.6.14}$$

where:

$$d \overset{\perp}{=} (w/2)\left\{1 + (t/w)\left[1 + \ln(4\pi w/t) + 0.51(t/w)^2\right]\right\} \tag{3.6.15}$$

where "d" is the radius of a cylindrical conductor equivalent to the hot stripline conductor.[9] Concerning the dielectric losses, we can use the general formula for dielectric losses in a "TEM" t.l.,*** resulting in:

$$\alpha_d = \frac{\pi\sqrt{\varepsilon_r}\tan\delta}{\lambda_0}\,[\text{Neper/u.1}] \quad \text{or} \quad \alpha_d = \frac{27.29\sqrt{\varepsilon_r}\tan\delta}{\lambda_0}\,[\text{dB/u.1}] \tag{3.6.16}$$

where "$\tan\delta$" is the substrate "tangent delta" **** for $g = 0$ in the substrate. Of course, the quantity "ε_r" is relative to the real part of the substrate dielectric constant.***** In general, for the typical substrates employed in MIC stripline circuit conductor losses are predominant.

---

* We assume that top and bottom conductors have the same surface resistivity, as they usually do.
** See Chapter 1 for attenuation constant dimensions.
*** See Appendix A2 for the procedure to have the attenuation constant "α_d."
**** See Chapter 1 for "$\tan\delta$" definition and conversion between Neper and dB.
***** See Chapter 1 for complex permittivity definition.

3.7 OFFSET STRIPLINES

The geometric structure of an offset stripline was depicted in Figure 3.1.2. From a general point of view, there is no reason to use a single offset stripline, but in practice sometimes it cannot be avoided. This happens especially in high density multilayer "PCB," used for low frequency applications, for instance below a given GHz. In such cases more than a layer can be used for DC power distribution, and if striplines are required, use of an offset stripline could be inevitable.

Such t.l. has been studied by the researcher P. Robrish,[9,10,11,12,13] using conformal transformations. For $0.2 \leq h/b \leq 0.8$, $w > t$, and $t/b < 0.2$ the resulting impedance formulas are:

for $w/(b - t) \leq 0.35$

$$\zeta = \left(60/\sqrt{\varepsilon_r}\right)\operatorname{acosh}(A) \tag{3.7.1}$$

$$A \overset{\perp}{=} \operatorname{sen}(\pi h/b)\coth(\pi d/2b) \tag{3.7.2}$$

where "d" is given by 3.6.15);

for $w/(b - t) \geq 0.35$

$$\zeta = \left(120\pi/\sqrt{\varepsilon_r}\right)\left[\rho/\gamma + \rho/(\beta - \gamma) + 2C_f/\varepsilon_0\varepsilon_r\right]^{-1} \tag{3.7.3}$$

where:

$$\rho = w/b + (1 - t/b)^8\left[K(p')/K(p) - (2/\pi)\ln 2 - w/b\right] \tag{3.7.4}$$

"p" and "p'" are given in 3.4.11 and 3.4.12 and the ratio of the elliptical integrals are evaluated as in 3.5.1 and 3.5.2,

$$\gamma \overset{\perp}{=} h/b - t/2b \qquad \beta = 1 - t/b \tag{3.7.5}$$

$$C_f = (\varepsilon_r\varepsilon_0/\pi)\left\{2\ln\left[1/\gamma(\beta - \gamma)\right] + 1/\gamma(\beta - \gamma)\left[f(t/2b) - f(h/b)\right]\right\} \tag{3.7.6}$$

and "$f()$" is a function defined as:

$$f(x) \overset{\perp}{=} (1 - 2x)\left[(1 - x)\ln(1 - x) - x\ln x\right] \tag{3.7.7}$$

For the above specified range of validity, these formulas give a maximum error of 2% when compared to the computer data using a finite difference method.

Explicit closed form formulas for losses are not available for the present case; however, the expressions of the previous section can be applied in the present case as well.

3.8 PRACTICAL CONSIDERATIONS

As with any other planar t.l., stripline is also affected by the undercut* which needs to be compensated for when high impedance lines or coupled striplines are needed. As a result, the hot

---

* See Chapter 1 for undercut phenomenon.

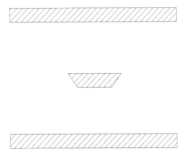

Figure 3.8.1

conductor is in general a trapezoidal one, as indicated in Figure 3.8.1, unless the selective conducting plating is employed.

Stripline is a three-layer configuration, obtained by overlaying two separate "PCBs"; one "PCB" has only the ground layer, while the other "PCB" has one ground layer and one layer of desired tracks. As a result, the dielectric surrounding the hot conductor is never theoretically homogeneous, as indicated in Figure 3.8.2, but the dishomogeneity introduced by the air gap can often be neglected. In fact, typical values of the air gap when copper* conductors are employed are in the range of 20 to 40μm, and 3 to 8μm when gold conductors are used. However, this last value for "t" is relative to sputtered gold on ceramic substrates like alumina, seldom used in stripline devices. So, for μwave devices the air gap tends to decrease departing from the hot conductor, because the typical employed dielectrics are of soft type. In low frequency stripline devices, a very diffused substrate is a glass-resin mixture, called FR4, with a relative dielectric constant ranging from 4 to 6. In such boards the air gap is minimal since the board can be grown in height depositing a resin mixture, called "prepreg," which works as an adhesive for the other layers.

In contrast to other t.l.s studied in this text, striplines are not suitable for use in MMIC. This is another reason why this t.l. has not been studied much in these last years, especially from the point of view of equivalent circuits for discontinuities. In fact, there is no doubt that MMIC devices are the most attractive components in μwave electronics today, where other t.l.s can be advantageously employed.

Stripline is more insensitive than μstrip to lateral ground planes of a metallic enclosure, since the e.m. field is strongly contained near the center conductor and the top–bottom ground planes. This situation is indicated in Figure 3.8.3. For $w' \geq 3w$ the effect of the sidewalls on "ζ" can be neglected.

Concerning transitions between a stripline and other lines, the most common are μstrips and coaxial. Also if transitions with other lines are possible, only μstrips and coaxial are used due to

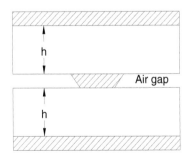

Figure 3.8.2

---

* In practice, at μwave frequencies, copper conductors are gold plated.

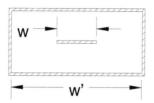

Figure 3.8.3

the bulky aspect of stripline devices. Transition with µstrip is indicated in Figure 3.8.4. The two hot conductors are usually connected together using gold ribbons. The sharing of the same ground plane between the two t.l.s is of course not assured, and at the transition section some step in height for the ground plane could be inevitable. Of course, ground continuity must be assured.

Transition with coaxial cable is indicated in Figure 3.8.5. Coaxial connectors are usually employed in this transition since both t.l.s are bulky.

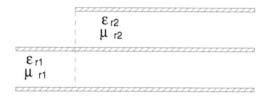

Figure 3.8.4

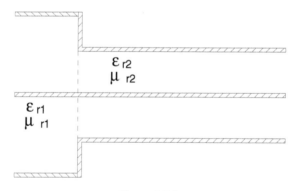

Figure 3.8.5

REFERENCES

1. N. A. Begovich, Capacity and characteristic impedance of strip transmission line with rectangular inner conductors, *IRE Trans. on MTT*, 127, March 1955.
2. M. V. Schneider, Computation of impedance and attenuation of TEM lines by finite difference methods, *IEEE Trans. on MTT*, 793, Nov. 1965.
3. S. B. Cohn, Characteristic impedance of the shielded strip transmission line, *IRE Trans. on MTT*, 52, July 1954.
4. S. B. Cohn, Problems in strip transmission lines, *IRE Trans. on MTT*, 119, March 1955.
5. H. A. Wheeler, Formulas for the skin effect, *Proc. of the IRE*, 30, 412, 1942.
6. R. Sturdivant, Transmission line conductor loss and the incremental inductance rule, *Microwave J.*, 156, Sept. 1995.
7. S. B. Cohn, Problems in strip transmission lines, *IRE Trans. on MTT*, 119, March 1955.
8. N. Marcuvitz, *Waveguide Handbook*, McGraw Hill, New York, 1951, 263.

9. P. Robrish, An analytic algorithm for unbalanced stripline impedance, *IEEE Trans. on MTT*, 1011, Aug. 1990.

10. R. E. Canright, Jr., Comments on "an analytic algorithm for unbalanced stripline impedance," *IEEE Trans. on MTT*, 177, Jan. 1992. The Robrish reply is in the same issue, p. 179.

11. E. Costamagna, A. Fanni, Comments on "an analytic algorithm for unbalanced stripline impedance," *IEEE Trans. on MTT*, 173, Jan. 1993.

12. R. E. Canright, Jr., Comments on "an analytic algorithm for unbalanced stripline impedance," *IEEE Trans. on MTT*, 1718, Sept. 1994. The Robrish reply is in the same issue, p. 1721.

13. E. Costamagna, A. Fanni, Further comments on "an analytic algorithm for unbalanced stripline impedance," *IEEE Trans. on MTT*, 693, Apr. 1994.

CHAPTER 4

Higher Order Modes and Discontinuities in μStrip and Stripline

4.1 RADIATION

Since the μstrip is an open t.l., some of the transmitted energy is not guided by the t.l., but is instead radiated in the surrounding space above the dielectric, as indicated in Figure 4.1.1. This phenomenon is called "radiation." For stripline, since it is a closed t.l., radiation can be neglected. For this reason, unless otherwise stated this section radiation is only pertinent to μstrip devices.

The loss of energy is dependent on the particular μstrip device. For example, μstrip antennas are elements just constructed to radiate, but in the greatest number of devices this effect is not desired. In general, every discontinuity is a radiation source that causes signal attenuation, undesired coupling,* and crosstalk** among other μstrips, if any, that are near the radiation area.[1] Also, commonly used μstrip devices like filters, generate radiation,[2] an effect that sometimes is forgotten. Radiation effects in GaAs MMIC have been studied by T. Rozzi, G. Cerri, and M. Mongiardo.[3,4]

In general, the study of the radiation caused by a discontinuity is made by evaluating the power radiated through the integration of the Poynting vector*** over a surface surrounding the discontinuity.[5,6] According to work by the researchers M. D. Abouzahra and L. Lewin,[7] the normalized**** radiated power "W_r" is given by:

$$W_r = \frac{240(\pi h)^2}{\zeta \lambda_0^2} F_r\left[\varepsilon_{re}\right]$$ (4.1.1)

where:

"h" is the substrate thickness
"ζ" is the μstrip impedance
"λ_0" is the free space wavelength
"$F_r[\varepsilon_{re}]$" is the "radiation function."

The radiation function depends on the discontinuity type and the effective dielectric constant "ε_{re}" of the μstrip. For example,[8] we have:

* Devices working on coupling effects between μstrips will be studied in Chapter 7.
** Crosstalk in μstrips can be studied according to the general theory explained in Chapter 1.
*** See Appendix A2 for Poynting vector definition and application.
**** The normalization is made with respect to the input power.

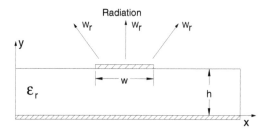

Figure 4.1.1

a. open circuit:

$$F_{ro}\left(\varepsilon_{re}\right) = \frac{\varepsilon_{re}+1}{\varepsilon_{re}} - \frac{\left(\varepsilon_{re}-1\right)^2}{2\varepsilon_{re}^{1.5}} \ln\frac{\sqrt{\varepsilon_{re}+1}}{\sqrt{\varepsilon_{re}-1}}$$

(4.1.2)

b. series resistance:

$$F_{rm}\left(\varepsilon_{re}\right) = F_{ro}\left(\varepsilon_{re}\right)\left|\frac{Z}{Z+2\zeta}\right|^2$$

(4.1.3)

where "Z" is the series impedance and "ζ" the microstrip characteristic impedance.

c. symmetric "T":
   With reference to the network indicated in Figure 4.1.2:

$$F_{rT}\left(\varepsilon_{re}\right) = \frac{\left(3\varepsilon_{re}+1\right)^2}{8\varepsilon_{re}^{1.5}} \ln\frac{\sqrt{\varepsilon_{re}+1}}{\sqrt{\varepsilon_{re}-1}} - \frac{\varepsilon_{re}}{2\varepsilon_{re}-1} \ln\left[\frac{\varepsilon_{re}+\left(2\varepsilon_{re}-1\right)^{0.5}}{\varepsilon_{re}-\left(2\varepsilon_{re}-1\right)^{0.5}}\right] - \frac{\varepsilon_{re}+1}{4\varepsilon_{re}}$$

(4.1.4)

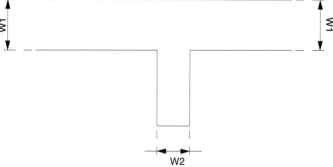

Figure 4.1.2

d. change in width:
   With reference to the network indicated in Figure 4.1.3:

$$F_{rw}\left(\varepsilon_{re}\right) = F_{ro}\left(\varepsilon_{re}\right)\left|\frac{\zeta_2-\zeta_1}{\zeta_2+\zeta_1}\right|^2$$

(4.1.5)

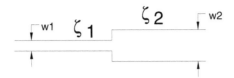

Figure 4.1.3

When the µstrip is enclosed in a box, as it is in the greatest number of cases, radiation is also responsible for resonances, centered on frequencies whose exact value is quite difficult to predict.[9] The study of this phenomenon is made by extrapolating that of some particular rectangular waveguide modes* when they are partially filled with dielectric,[10] as indicated in Figure 4.1.4 from two view points. These modes are called "Longitudinal Section Electric-LSE" and "Longitudinal Section Magnetic-LSM." According to work from the researcher G. H. Robinson,[11] the wavelength "λ_{gn}" of the "n-th" resonance along the greatest width of a box as indicated in Figure 4.1.5 is approximately given by:

$$\lambda_{gn}/\lambda_0 \approx \left\{\left[1-(h/b)\left(1-1/\varepsilon_r\right)\right]^{-1} - \left(n\lambda_0/2w\right)^2\right\}^{-0.5} \tag{4.1.6}$$

where "n" is an integer number. To suppress these higher order modes, the researchers M. V. Schneider and B. S. Glance[12] suggest suspending the microstrip in the middle of the box through two lateral hollows so that they can be approximated to a shorter quarter wavelength transformer. However, if some undesired resonance in the enclosed µstrip device is found, it can be dumped or removed placing RF absorbing material beneath the top metallic cover and/or attaching it to the side walls.

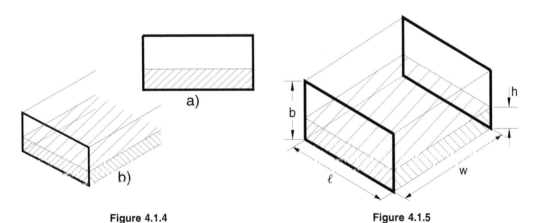

Figure 4.1.4   **Figure 4.1.5**

4.2 SURFACE WAVES

Surface waves are e.m. waves that propagate on the dielectric interface layer of the µstrip. Due to the practical homogeneity of stripline dielectric, this phenomenon can also be neglected in stripline devices. For this reason, this section is pertinent to µstrip t.l.s only. Surface waves are generated at any discontinuity of the microstrip. Once generated, they travel and generate, coupling with other µstrips of the circuit, decreasing isolation between different networks and signal attenuation. A simple

---

* See Appendix A2 for waveguide modes definition.

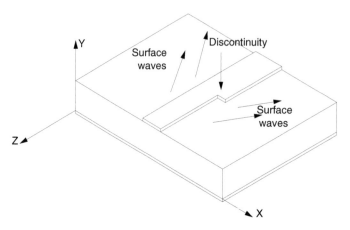

Figure 4.2.1

representation of the propagation of these waves is indicated in Figure 4.2.1. Note that similar to the case of radiation, these waves are not guided by the µstrip. Consequently, surface waves are a cause of crosstalk, coupling, and attenuation in a multiµstrip circuit. For these reasons the surface waves are always an undesired phenomenon. The propagation modes of these surface waves are practically "TE_{m0}" and "TM_{m0}." The second subscript of these modes is assumed to be zero since we evaluate µstrips for which $h \leq 0.1\lambda_g$, i.e., modes have no dependence on "y." Surface wave modes have been investigated by some researchers.[13,14,15] According to work by the researcher Vendelin,[16] the cut-off frequencies for these surface modes are given by:

$$f_{c, TM_m} = m \frac{v_0}{2h(\varepsilon_r - 1)^{0.5}} \qquad\qquad m = 0, 1, 2, \ldots \qquad\qquad (4.2.1)$$

$$f_{c, TE_m} = (1 + 2m) \frac{v_0}{4h(\varepsilon_r - 1)^{0.5}} \qquad\qquad m = 0, 1, 2, \ldots \qquad\qquad (4.2.2)$$

In practice it has been verified that for each one of these surface modes, a frequency is found for which a maximum of coupling exists between the surface mode and the fundamental "qTEM" mode. In such a case, a sharp attenuation of the desired "qTEM" signal is observed due to energy exchange between these modes. Such a frequency is called the "frequency of synchronous coupling" (f_s). The lowest "f_s" is for the "TM_0" mode and indicated with "f_{s,TM_0}" given by:

$$f_{s, TM_0} = \frac{v_0 atg(\varepsilon_r)}{\pi h \left[2(\varepsilon_r - 1)\right]^{0.5}} \qquad\qquad (4.2.3)$$

Note as the "TM_0" surface mode has a theoretical cut-off frequency equal to zero. The "f_s" for the "TE_0" mode is indicated with "f_{s,TE_0}" and given by:

$$f_{s, TE_0} = \frac{3v_0}{4h \left[2(\varepsilon_r - 1)\right]^{0.5}} \qquad\qquad (4.2.4)$$

In Figure 4.2.2 we have shown the curves for the frequencies of synchronous coupling "f_{s,TM_0}" and "f_{s,TE_0}," together with the cut-off frequencies "f_{c,TM_1}" and "f_{c,TE_0}," as functions of "h" in mm

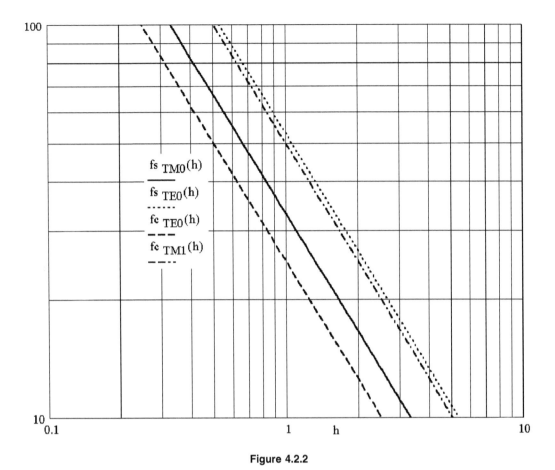

Figure 4.2.2

and with frequency expressed in GHz. The value of "ε_r" is 10. Note the first surface mode to start is the "TE_0," but the first mode for a synchronous coupling is the "TM_0."

4.3 HIGHER ORDER MODES

In contrast to the case of surface waves, higher order modes are waves propagated by the structure, but they do not start from zero frequency as the "qTEM" and "TEM" mode for µstrip and stripline. For both of these t.l.s in general, when the guided signal wavelength is comparable or lower than some transverse dimension, i.e., "w," "h," or "b," modes other than the fundamental one are possible, which are called "higher order modes." In the most general case, these modes are identified by the two subscripts "m" and "n," as in the waveguide case.* The e.m. field disposition of these higher order modes is similar to that of "TE" and "TM" modes, and for this reason they are called "qTE" and "qTM." The integer numbers "m" and "n" give the number of half periods in the signal phase along the coordinates "x" and "y." Higher order modes are always undesired because they cause signal dispersion** and attenuation. We will now concentrate our study on µstrip and stripline.

---

* "TE" and "TM" modes in waveguide are studied in Appendix A2.
** See Appendix A2 for dispersion definition.

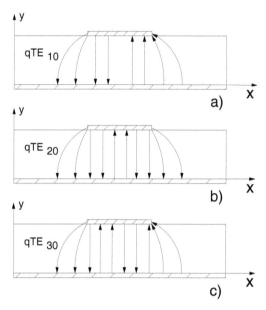

Figure 4.3.1

4.3.1 μStrip Case

Since in actuality the dimension "h" is much smaller than the guided signal wavelength "λ_g," the higher order modes have no dependence on "y," i.e., "n" is always equal to zero. Consequently, these modes can be identified by only one subscript "m," i.e., they are "TE_m" and "TM_m." In addition, since these modes are not perfect "TE" and "TM," they are sometimes noted as "HE_m" and "EH_m," where the first letter indicates the field with the greatest component along the propagation direction "z."

The study of these high order modes has been performed by some researchers,[17,18,19,20,21] mainly applying the full wave analysis method to the μstrip geometry. The result is that the first higher order mode to start is the "qTE_{10}," whose cut-off frequency "$f_{c,TE10}$" can be evaluated setting m = 1 in the following expression:

$$f_{c,TE_{m0}} = m \frac{v_0}{2 w_e \sqrt{\varepsilon_{re}}} \qquad\qquad m = 1,2,\dots \qquad\qquad (4.3.1)$$

where "w_e" is the effective strip width defined in Chapter 2 and "v_0" is the light speed in the vacuum. Equation 4.3.1 gives the cut-off frequency of "qTE_{m0}" modes. The electric field for these modes for the cases of m = 1,2,3 are represented in Figure 4.3.1, in a cross-sectional view. The graph of Equation 4.3.1 vs. "w_e" in mm, with ε_{re} = 5.7, is reported in Figure 4.3.2, where frequency is in GHz.

For the case of a μstrip line the previous equation is approximated, but gives a sufficiently accurate result when w ≥ 10h. This condition is seldom verified in a typical μstrip 50Ω t.l. on alumina substrate. In any case, Equation 4.2.1 can be used to determine a rough estimate of the maximum "qTEM" signal propagation frequency and to avoid propagation with higher order modes. From a comparison among Equations 4.3.1 and 4.2.1 through 4.2.4, we recognize how the dimension "h" strongly affects the surface modes while the dimension "w" affects the transverse resonance of the "qTE" modes.

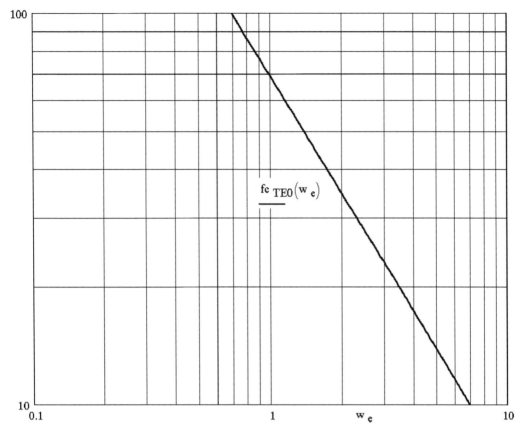

Figure 4.3.2

4.3.2 Stripline Case

A general study of stripline higher order modes has been done by some authors,[22,23,24] special-izing the wave equation to the stripline structure. With reference to Figure 4.3.3, "TM" modes are generated when the signal frequency is in resonance with dimension "b." Assuming the dimension "w" is negligible, for $w/b \leq 10$, the lower order "TM" mode has a cut-off frequency "$f_{c,TM}$" approximately given by:

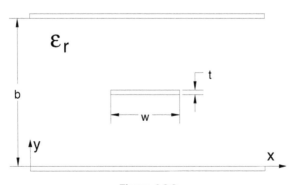

Figure 4.3.3

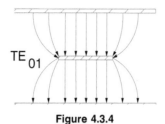

Figure 4.3.4

$$f_{c,TM} = v_0 / 2b\sqrt{\varepsilon_r} \qquad (4.3.2)$$

"TE" modes are generated when the signal frequency is in resonance with dimension "w." In contrast to "TM" modes, the cut-off frequency is now dependent on "w" and "b." In fact, the lower order "TE" mode cut-off frequency is given by:[25]

$$f_{c,TE} = v_0 / (2w + 0.5\pi b)\sqrt{\varepsilon_r} \qquad (4.3.3)$$

From the previous two equations we have:

$$f_{c,TM} / f_{c,TE} = w/b + \pi/4 \qquad (4.3.4)$$

Since in practice "w/b" is quite often higher than 0.215 in such cases, $f_{c,TE} < f_{c,TM}$. The electric field for the fundamental "TE_{01}" mode is represented in Figure 4.3.4, in a cross-sectional view. The graph of Equation 4.3.3 vs. "b" in mm, with $\varepsilon_r = 1$ and w = 0.2, 0.5, and 1, is reported in Figure 4.3.5, where frequency is in GHz.

4.4 TYPICAL DISCONTINUITIES

Every time the center conductor encounters a dimensional variation and/or dielectric constant variation, we say that a "discontinuity" is present. It is important to observe that the discontinuity can be generated not only if the hot conductor is varied in some manner, but also if the ground plane is varied from its theoretically infinite, continuous extent.[26] Of course, the discontinuity effect on the traveling signal is directly proportional to its entity, for example the entity of hot conductor width variation. Here, and in the following sections, we will assume that the discontinuity is only generated by the hot conductor. We will study the equivalent circuits for discontinuities in µstrip and stripline, giving for each t.l. the element's value for the most used equivalent circuit to the discontinuity under test.

In Figure 4.4.1 we have represented the most frequent discontinuities in any circuit. For any one of the networks there is a corresponding equivalent lumped network, composed of inductors and capacitors. We will describe each discontinuity separately in the following sections.

In the next sections we will use the letters "w," "ε_r," "h," and "b" to indicate respectively the strip width, the dielectric constant, and the substrate thickness, for the µstrip and stripline cases. The reader can observe Figures 4.3.1 and 4.3.3 for a closer insight into the geometry of these t.l.s.

While stripline discontinuities were studied earlier than their µstrip counterparts, the latter have been more fully investigated.[27,28,29] This is because µstrips are more practical to use, and they require a smaller PCB volume to be constructed. Unless otherwise stated, stripline discontinuities design formulas used are those obtained by the researchers A. A. Oliner[30] and H. M. Altschuler.[31] All the stripline discontinuities are relative to the symmetrical stripline, (see Figure 4.3.3).

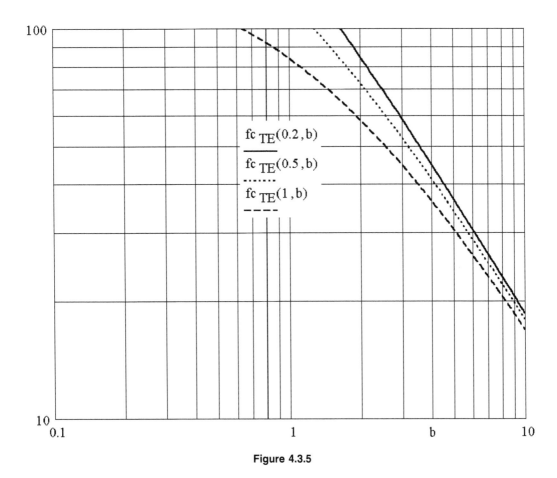

Figure 4.3.5

In contrast to the stripline case, µstrip discontinuities design formulas have been obtained by a number of researchers, applying the equation procedures that best fit the measured data.

The analysis methods of the references reported in this chapter belong to the quasi static group, already discussed in Chapters 2 and 3. As a consequence of this method, dispersion is not evaluated and all the element values associated with the network have a constant value with respect to frequency.

4.5 BENDS

Bends are the most frequently encountered discontinuities, and for the µstrip case they have been studied and measured by many authors.[32,33,34,35]

The most simple bend is the 90° bend, as indicated in Figure 4.5.1. This bend doesn't work well above some GHz, due to a high VSWR. The same holds true for bends with angles "α" greater than 90°. Experiments on various bends have proven that a decrease in the input reflection coefficients for network in Figure 4.5.1 can be achieved if the corner is chamfered, as indicated in Figure 4.5.2. In this case a "chamfer percentage" or "miter percentage" is defined as:

$$m \stackrel{\perp}{=} (S/d)100 \qquad (4.5.1)$$

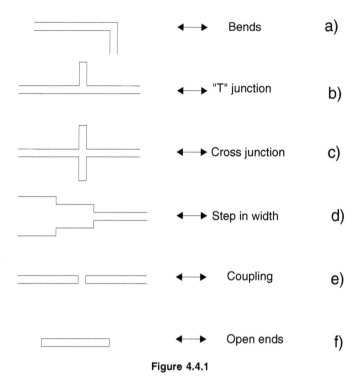

Figure 4.4.1

Measurements of the bends shown in Figures 4.5.1 and 4.5.2 yields the approximate equivalent lumped network is that shown in Figure 4.5.3. Of course, the element values are different if the lumped network has to represent the bend in Figure 4.5.1 or 4.5.2.

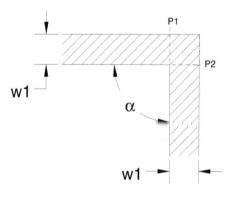

Figure 4.5.1

4.5.1 μStrip Case

The researchers M. Kirschning, R. H. Jansen, and N. H. L. Koster[36] have obtained the following equations for the elements indicated in Figure 4.5.3 applied to Figure 4.5.1, i.e., a noncompensated 90° bend:

$$C = 10^{-3} h \left[(10.35 \varepsilon_r + 2.5) (w/h)^2 + (2.6 \varepsilon_r + 5.64) (w/h) \right] \qquad (4.5.2)$$

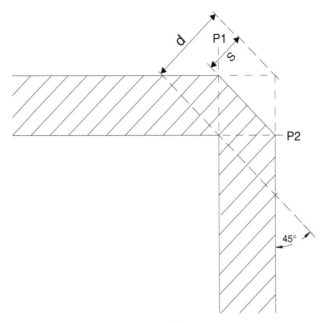

Figure 4.5.2

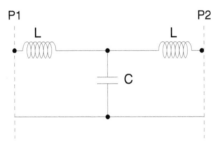

Figure 4.5.3

$$L = 0.22h\left[1 - 1.35\exp\left(-0.18\left(w/h\right)^{1.39}\right)\right] \qquad (4.5.3)$$

while if the network indicated in Figure 4.5.3 is applied to Figure 4.5.2 we have:

$$C = 10^{-3}h\left[\left(3.93\varepsilon_r + 0.62\right)\left(w/h\right)^2 + \left(7.6\varepsilon_r + 3.8\right)\left(w/h\right)\right] \qquad (4.5.4)$$

$$L = 0.44h\left[1 - 1.062\exp\left(-0.177\left(w/h\right)^{0.947}\right)\right] \qquad (4.5.5)$$

In the previous equations, "h" must be inserted in "mm," and "C" and "L" give a value in "pF" and "nH." The values given by these formulas have been proven to approximate very well the measured values, with a maximum error near some percent when $2 \leq \varepsilon_r \leq 13$ and $0.2 \leq$ w/h ≤ 6. The authors R. J. P. Douville and D. S. James[37] have obtained an optimum percent value for "m," given by:

$$m = 52 + 65\exp\left(-1.35w/h\right) \qquad (4.5.6)$$

which is applicable for $2.5 \leq \varepsilon_r \leq 25$ and w/h ≥ 25.

4.5.2 Stripline Case

The inductive "X_L" and capacitive "X_C" reactances for the elements of the equivalent circuit of Figure 4.5.3 for a noncompensated 90° bend are:

$$X_L/\zeta = 2D\left[0.878 + 2(D/\lambda)^2\right]/\lambda \qquad (4.5.7)$$

$$X_C/\zeta = -\lambda\left[1 - 0.114(2D/\lambda)^2\right]/2\pi D \qquad (4.5.8)$$

where:

$$D \overset{\perp}{=} bK(p)/K(p') \qquad (4.5.9)$$

$$p \overset{\perp}{=} \tanh(\pi w/2b) \qquad (4.5.10)$$

and "ζ" is the stripline characteristic impedance.

"$K(p)$" is the complete elliptic integral of first kind, defined in Appendix A8. The ratio $K(p)/K(p')$ can be calculated using the tabulated values of the elliptic integrals,[38] or using the approximated equations we give in Appendix A8.

4.6 OPEN END

Open ends are encountered any time a microstrip is open terminated. Typical devices where open ends are encountered are filters and matching stubs. An example is the low pass filter indicated in Figure 4.6.1. Due to its diffusion, this μstrip open end discontinuity has been studied by many authors.[39,40,41] The general representation of this discontinuity together with its equivalent circuit are shown in Figure 4.6.2b. To the open end capacitance it is sometimes associated with, an extra μstrip length "$\Delta\ell$" is attached to the original μstrip, permitting one to take into account the open end effect. This relationship is indicated in Figure 4.6.3. In the next points of study we will describe how to evaluate such quantities.

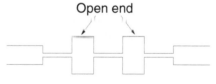

Open end

Figure 4.6.1

4.6.1 μStrip Case

According to a study by P. Silvester and P. Benedek,[40] the capacitance "C" of Figure 4.6.2, part b, can be obtained by:

$$\frac{C}{w} = \exp\left\{2.3026\sum_{i=1}^{i=5} a_i\left[\log\left(\frac{w}{h}\right)\right]^{i-1}\right\} \qquad (4.6.1)$$

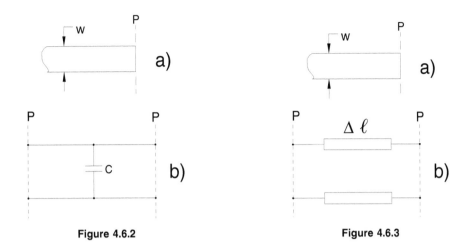

Figure 4.6.2   Figure 4.6.3

The previous equation gives a result in pF/m, and the coefficients "a_i" are functions of "ε_r" according to the following Table 4.6.1:

Table 4.6.1

	$\varepsilon_r = 1$	$\varepsilon_r = 2.5$	$\varepsilon_r = 4.2$	$\varepsilon_r = 9.6$	$\varepsilon_r = 16$	$\varepsilon_r = 51$
a_1	1.11	1.295	1.443	1.738	1.938	2.403
a_2	-0.2892	-0.2817	-0.2535	-0.2538	-0.2233	-0.222
a_3	0.1815	0.1367	0.1062	0.1308	0.1317	0.217
a_4	-0.0033	-0.0133	-0.026	-0.0087	-0.0267	-0.024
a_5	-0.054	-0.0267	-0.0073	-0.0113	-0.0147	-0.084

Once obtained from the previous equation, the capacitance $C_o \perp C/w$, in pF/m, this extra length is given by:

$$\Delta\ell \equiv v_0 \zeta C / \sqrt{\varepsilon_{re}} \qquad (4.6.2)$$

If we add this "$\Delta\ell$" to the original µstrip length, the capacitance "C_o" must not be added again to the end of this extra length.

The researchers M. Kirschning, R. H. Jansen, and N. H. L. Koster[42] have obtained a closed form expression for "$\Delta\ell$" through a curve fitting to full wave data analysis, according to:

$$\Delta\ell/h = abc/d \qquad (4.6.3)$$

$$a = 0.434907 \frac{\left(\varepsilon_{re}^{0.81} + 0.26\right)\left[(w/h)^{0.8544} + 0.236\right]}{\left(\varepsilon_{re}^{0.81} - 0.189\right)\left[(w/h)^{0.8544} + 0.87\right]} \qquad (4.6.4)$$

$$g = 1 + (w/h)^{0.371} / (2.358\varepsilon_r + 1) \qquad (4.6.5)$$

$$b = 1 + 0.5274 \, \text{atg}\left[0.084(w/h)^{1.9413/g}\right] / \varepsilon_{re}^{0.9236} \qquad (4.6.6)$$

$$c = 1 - 0.218e^{-7.5w/h} \qquad (4.6.7)$$

$$d = 1 + 0.0377 \, \text{atg} \left[0.067 \left(w/h \right)^{1.456} \right] \left[6 - 5e^{0.036(1-\varepsilon_r)} \right] \qquad (4.6.8)$$

The error produced by these equations, compared to the full wave analysis data, is lower than 2.5% for $\varepsilon_r \leq 50$ and $0.01 \leq w/h \leq 100$.

4.6.2 Stripline Case

An equivalent extra length "$\Delta\ell$" is associated with the stripline open end discontinuity, as indicated in Figure 4.6.3, given by:

$$\Delta\ell = \frac{\lambda_g}{2\pi} \, \text{acot} \left[\frac{4d + 2w}{d + 2w} \cot \left(\frac{2\pi}{\lambda_g} d \right) \right] \qquad (4.6.9)$$

$$d \overset{\perp}{=} b\ell n2 / \pi \qquad (4.6.10)$$

4.7 GAP

These types of discontinuities are encountered in filters and in DC blocks. An example is the band pass filter indicated in Figure 4.7.1. This discontinuity is quite diffused, and for the μstrip case, it has been studied or measured by several authors.[43,44,45] In Figure 4.7.2, part a, we represent the generic geometric structure of the gap, while part b shows the equivalent circuit.

Gaps

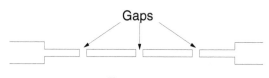

Figure 4.7.1

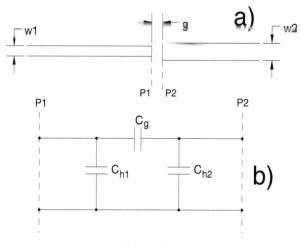

Figure 4.7.2

4.7.1 μStrip Case

According to work of the researchers M. Kirschning, R. H. Jansen, and N. H. L. Koster,[46] the element values in the network indicated in Figure 4.7.2 are given by:

$$\omega C_g = Q_1 \pi f h e^{-1.86 g/h} \left\{ 1 + 4.19 \left\{ 1 - \exp\left[-0.785 (h/w_1)^{0.5} w_2/w_1 \right] \right\} \right\} \qquad \text{in mS} \qquad (4.7.1)$$

$$\omega C_{h1} = 2\pi f C_{o1} (Q_2 + Q_3)/(1 + Q_2) \qquad \text{in mS} \qquad (4.7.2)$$

$$\omega C_{h2} = 2\pi f C_{o2} (Q_2 + Q_4)/(1 + Q_2) \qquad \text{in mS} \qquad (4.7.3)$$

where the quantities Q_1 through Q_4 are given by:

$$Q_1 \doteq 0.04598 (0.272 + 0.07\varepsilon_r) \left\{ 0.03 + (w_1/h) \wedge \left[1.23/(1 + 0.12 (w_2/w_1 - 1)^{0.9}) \right] \right\} \quad (4.7.4)$$

$$Q_2 \doteq 0.107 (w_1/h + 9) (g/h)^{3.23} + 2.09 (g/h)^{1.05} (1.5 + 0.3 w_1/h)/(1 + 0.6 w_1/h) \qquad (4.7.5)$$

$$Q_3 \doteq \exp\left[-0.5978 (w_2/w_1)^{1.35} \right] - 0.55 \qquad (4.7.6)$$

$$Q_4 \doteq \exp\left[-0.5978 (w_1/w_2)^{1.35} \right] - 0.55 \qquad (4.7.7)$$

In 4.7.2 and 4.7.3, "C_{o1}" and "C_{o2}" are the open end capacitances evaluated as in the previous section.

In the above equations the frequency and dimensions must be inserted respectively in GHz and mm, and are applicable for the following ranges:

$$0.1 \le w_1/h, w_2/h \le 3, \quad 1 \le w_2/w_1 \le 3, \quad d/h \ge 0.2, \quad 6 \le \varepsilon_r \le 13 \qquad (4.7.8)$$

The values produced by Equations 4.7.1 through 4.7.3, when compared to the full wave analysis method of the same authors,[47] have an error lower than 0.1 mS for frequencies up to 18 GHz and "f × h" product lower or equal to 12 GHz × mm.

Of course, as in the case of open end discontinuity, it is also possible to associate two extra lengths "$\Delta\ell_1$" and "$\Delta\ell_2$" with the two capacitances "C_{h1}" and "C_{h2}" respectively, inserting these capacitances into Equation 4.6.2.

4.7.2 Stripline Case

Only symmetrical gap analysis equations are available, i.e., relative to the structure indicated in Figure 4.7.3. The susceptances "B_g" and "B_h" of the elements in Figure 4.7.3 are:

$$\zeta B_h = \frac{1 + \zeta B_p \cot(\pi g/\lambda_g)}{\cot(\pi g/\lambda_g) - \zeta B_p} \qquad (4.7.9)$$

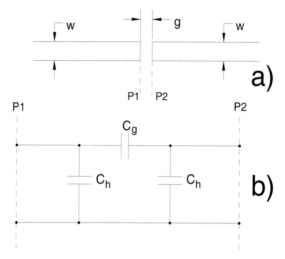

Figure 4.7.3

$$2\zeta B_g = \frac{1+\zeta\left(2B_s + B_p\right)\cot\left(\pi g/\lambda_g\right)}{\cot\left(\pi g/\lambda_g\right)-\zeta\left(2B_s + B_p\right)} - \zeta B_h \qquad (4.7.10)$$

where:

$$\zeta B_p \overset{\pm}{=} -\left(2\,b/\lambda_g\right)\ell n\left[\cosh\left(\pi g/2b\right)\right] \qquad (4.7.11)$$

$$\zeta B_s \overset{\pm}{=} \left(b/\lambda_g\right)\ell n\left[\coth\left(\pi g/2b\right)\right] \qquad (4.7.12)$$

4.8 CHANGE OF WIDTH

This kind of discontinuity is found in many devices like quarter wavelength transformers, multistep quarter wave directional couplers, and filters. An example is the four sections, $\lambda/4$ coupled lines, asymmetrical directional coupler* shown in Figure 4.8.1. This discontinuity, for the μstrip case, has also been studied by many authors,[48,49,50,51] due to its high diffusion. A lot of changes of width shapes are possible, but the symmetrical one indicated in Figure 4.8.2 is the simplest to build and can be considered a good approximation for any other type.

4.8.1 μStrip Case

Figure 4.8.3 indicates the suggested equivalent circuit of the symmetrical step shown in Figure 4.8.2. Simple formulas for the element's values in the equivalent network have been obtained by researchers,[52,53,54] according to the following equations:

$$L_{wi} \overset{\pm}{=} \zeta_i \sqrt{\varepsilon_{rei}}/v_0 \qquad (4.8.1)$$

* This type of directional coupler will be studied in the next Chapter 7.

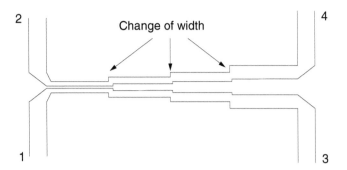

Figure 4.8.1

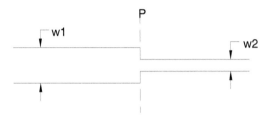

Figure 4.8.2

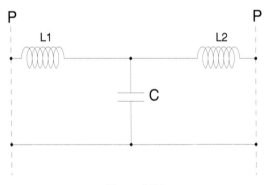

Figure 4.8.3

where i = 1,2 and "ζ_i" and "ε_{rei}" are respectively, the characteristic impedance and effective dielectric constant of a µstrip with width "w_i,"

$$L_s/h = 40.5\left(w_1/w_2 - 1\right) - 32.57\ln\left(w_1/w_2\right) + 0.2\left(w_1/w_2 - 1\right)^2 \quad \text{in nH/m} \quad (4.8.2)$$

$$L_i = L_{wi}L_s/\left(L_{wi}L_{w2}\right) \quad (4.8.3)$$

The equations above give an error of less than 5% for "w_1/w_2" ≥ 5 and $w_2/h = 1$. The general expression of capacitance, for $\varepsilon_r \le 10$ and $1.5 \le w_1/w_2 \le 3.5$, is given by:

$$C/\left(w_1w_2\right)^{0.5} = \left(4.3861n\varepsilon_r + 2.33\right)w_1/w_2 - 5.4721n\,\varepsilon_r - 3.17 \quad \text{in pF/m} \quad (4.8.4)$$

giving an error of less than 10%. For the case of alumina, i.e., $\varepsilon_r = 9.7$, and $3.5 \le w_1/w_2 \le 10$ we have the following equation:

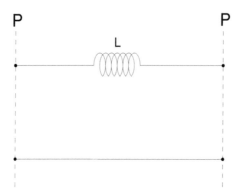

Figure 4.8.4

$$C/\left(w_1 w_2\right)^{0.5} = 56.46 \ln\left(w_1/w_2\right) - 44 \qquad \text{in pF/m} \qquad (4.8.5)$$

which gives a maximum error of 0.5% when compared to the measured values.

4.8.2 Stripline Case

Figure 4.8.4 shows the suggested equivalent circuit of the symmetrical step shown in Figure 4.8.2. The reactance "X_L" of the inductance is given by:

$$X_L = \left(2\zeta_1 D_1/\lambda_g\right)\ell n\left[\cosec\left(\pi D_2/2D_1\right)\right] \qquad (4.8.6)$$

where "D_1" and "D_2" are the application of equation 4.5.9 to the stripline of width "w_1" and "w_2."

4.9 "T" JUNCTIONS

The geometry of the discontinuity under study is indicated in Figure 4.9.1 and is called the "T-junction." This discontinuity is found mainly in matching networks and directional couplers,* and especially for the μstrip case it has been studied and measured by many authors.[55,56,57]

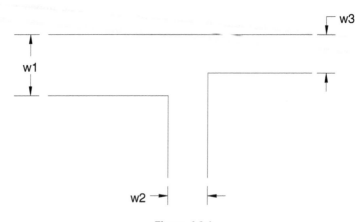

Figure 4.9.1

* Some directional couplers where "T-junctions" are used are the "branch line" and "rat-race," studied in Chapter 5.

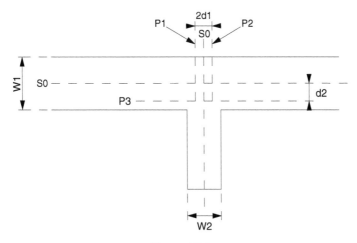

Figure 4.9.2

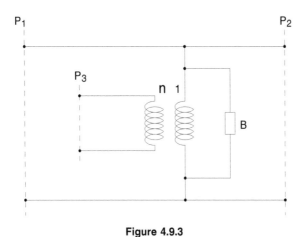

Figure 4.9.3

In contrast with the "bend" discontinuity previously introduced, in the present case more than one equivalent circuit is suggested. In addition, care must be taken to associate the equivalent circuit with the reference plane's position.

4.9.1 µStrip Case

For the particular case of symmetric "T" indicated in Figure 4.9.2 together with the reference planes "T_1," "T_2," and "T_3," the researcher E. O. Hammerstad[58,59] suggests the equivalent circuit indicated in Figure 4.9.3. Note that it is a dynamic equivalent circuit, i.e., the DC is not propagated by this equivalent circuit. The elements of this circuit are:

$$w_{ei} = 120\pi h/\zeta_{i0} \qquad \text{with } i = 1, 2 \tag{4.9.1}$$

where "ζ_{i0}" is the impedance of the µstrip "i" with the substrate replaced by the vacuum i.e., with $\varepsilon_r = 1$,

$$\lambda_g = \lambda/\sqrt{\varepsilon_{re}} \qquad (4.9.2)$$

$$n = sen\left(\pi w_{el}\zeta_1/\lambda_g\zeta_2\right)/\left(\pi w_{el}\zeta_1/\lambda_g\zeta_2\right) \qquad (4.9.3)$$

$$d_1 = 0.05 w_{e2} n^2 \zeta_{10}/\zeta_{20} \qquad (4.9.4)$$

$$d_2/w_{el} = 0.5 - 0.16\zeta_{10}\left[1 + \left(2w_{el}/\lambda g\right)^2 - 2\ln\left(\zeta_{10}/\zeta_{20}\right)\right]/\zeta_{20} \qquad (4.9.5)$$

$$B = -w_{el}\left(1 - 2w_{el}/\lambda_g\right)/\lambda_g\zeta_{20} \qquad \text{for } \zeta_{10} \le 0.5\zeta_{20} \qquad (4.9.6)$$

$$B = w_{el}\left(1 - 2w_{el}/\lambda_g\right)\left(3\zeta_{10}/\zeta_{20} - 2\right)\lambda_g\zeta_{10} \qquad \text{for } \zeta_{10} \ge 0.5\zeta_{20} \qquad (4.9.7)$$

It is important to observe that this model only describes the "T-junction" according to the reference plane indicated in Figure 4.9.2. So, the effect of the eventual length and termination of the lines connected at the reference planes have to be taken into account separately from this model. This is a typical procedure for any circuit simulation.

4.9.2 Stripline Case

With reference to Figure 4.9.4, the suggested equivalent network is indicated in Figure 4.9.5. The inductive "X_L" and capacitive "X_C" reactances for the elements of the equivalent circuit are:

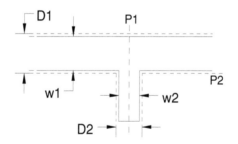

Figure 4.9.4

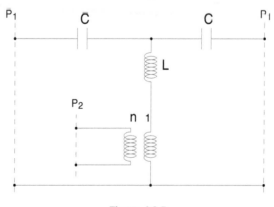

Figure 4.9.5

$$n' = \lambda_g \text{sen}\left(\pi D_2 / \lambda_g\right) / \pi D_2 \tag{4.9.8}$$

$$n = n'\left(D_2 / D_1\right)^{0.5} \tag{4.9.9}$$

$$X_c = -(0.785n)^2 D_2 / \lambda_g \tag{4.9.10}$$

$$X_L = -X_c/2 + \left\{\rho + \left(2D_1/\lambda_g\right)\left[\ell n2 + \pi D_2/6D_1 + 1.5\left(D_1/\lambda_g\right)^2\right]\right\}/n'^2 \quad \text{for } D_2 < 0.5D_1 \tag{4.9.11}$$

$$\rho = \left(2D_1/\lambda_g\right)\left\{\ell n\left[\cosec\left(\pi D_2/2D_1\right)\right] + 0.5\left(D_1/\lambda_g\right)^2 \cos^4\left(\pi D_3/2D_1\right)\right\} \tag{4.9.12}$$

$$X_L = -X_c/2 + \left(2D_1/n'^2\lambda_g\right)\left[\ell n\left(1.43D_1/D_2\right) + 2\left(D_1/\lambda_g\right)^2\right] \quad \text{for } D_2 > 0.5D_1 \tag{4.9.13}$$

where "D_1" and "D_2" are the application of Equation 4.5.9 to the stripline of width "w_1" and "w_2." Note that the reference planes are different between Figures 4.9.2 and 4.9.4, producing different equivalent circuits.

4.10 CROSS-JUNCTION

The geometry of this discontinuity is shown in Figure 4.10.1 and is called a "cross-junction." * Similar to the case of "T-junction," this discontinuity is mainly found in matching networks and filters, like the notch filter indicated in Figure 4.10.2. Nevertheless, the cross-junction has been studied and measured only for the µstrip case.[60,61]

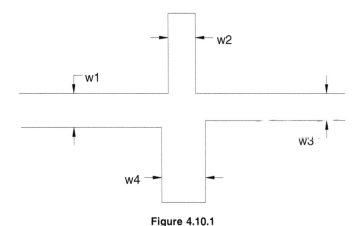

Figure 4.10.1

4.10.1 µStrip Case

In contrast with the case of "T-junction," there is no particular reason to have all four ports of different widths, i.e., different impedances, and for this reason the symmetric "X-junction," shown in Figure 4.10.3, has been studied more deeply than its fully asymmetrical counterpart.[62] With the reference planes indicated in Figure 4.10.3, we have[63] the suggested equivalent lumped element circuit as indicated in Figure 4.10.4 with elements value given by:

* We will indicate the "cross-junction" with "X-junction."

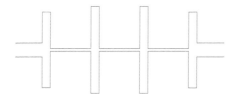

Figure 4.10.2

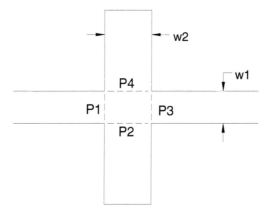

Figure 4.10.3

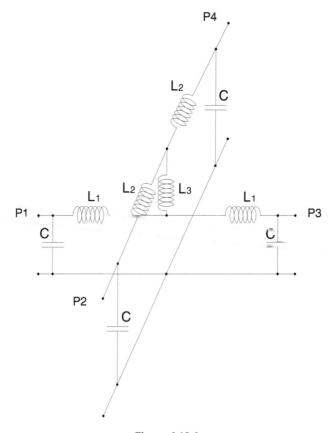

Figure 4.10.4

$$4c/w_1 = \left\{ \left[37.61 w_2 / h - 13.42 \left(w_2 / h \right)^{0.5} + 159.38 \right] \ln \left(w_1 / h \right) + \left(w_2 / h \right)^3 + 74\, w_2 / h + 130 \right\}$$

$$\left(w_1 / h \right)^{-1/3} + 0.5 h / w_2 - 60 - 0.375 w_1 \left(1 - w_2 / h \right) / h \qquad \text{in pF/m} \quad (4.10.1)$$

$$L_1 / h = \left\{ w_1 / h \left[165.6 w_2 / h + 31.2 \left(w_2 / h \right)^{0.5} - 11.8 \left(w_2 / h \right)^2 \right] - 32 w_2 / h + 3 \right\} \left(w_1 / h \right)^{-1.5}$$
$$\text{in nH/m} \quad (4.10.2)$$

$$L_2 / h = \left\{ w_2 / h \left[165.6 w_1 / h + 31.2 \left(w_1 / h \right)^{0.5} - 11.8 \left(w_1 / h \right)^2 \right] - 32 w_1 / h + 3 \right\} \left(w_2 / h \right)^{-1.5}$$
$$\text{in nH/m} \quad (4.10.3)$$

$$-L_3 / h = 337.5 + \left(1 + 7 h / w_1 \right) h / w_2 - \left(5 w_2 / h \right) \cos \left[\pi \left(1.5 h - w_1 \right) / 2 h \right] \qquad \text{in nH/m} \quad (4.10.4)$$

The previous equations were obtained as curves fitting the theoretical data, and are applicable for $\varepsilon_r = 9.9$; in particular, for capacitance "C" it is required that $0.3 \leq w_1/h \leq 3$ and $0.1 \leq w_2/h \leq 3$, while for the inductances it is required that $0.5 \leq (w_1/h, w_2/h) \leq 2$. In such a hypothesis, the error given by the previous equations is less than 5% with respect to the measured values.

4.10.1 Stripline Case

As we said before, no dedicated study of this discontinuity has been made. This is due to the high diffusion of μstrip networks and devices with respect to stripline counterparts. However, a first approximation of the loading of a transverse arm on a main line for a symmetric stripline X-junction (indicated in Figure 4.10.3) can still be obtained using the equivalent circuit for the stripline "T."

REFERENCES

1. T. K. Sarkar, Z. A. Maricevic, M. Salazar-Palma, Characterization of power loss from discontinuities in guided structures, *Microwave Symp.*, 2, 613, 1997.
2. E. J. Denlinger, Radiation from μstrip resonators, *IEEE Trans. on MTT*, 235, Apr, 1969.
3. T. Rozzi, G. Cerri, Radiation modes of open microstrip with applications, *IEEE Trans. on MTT*, 1364, June 1995.
4. G. Cerri, M. Mongiardo, T, Rozzi, Radiation from via-hole grounds in microstrip lines, *Microwave Symp.*, 1, 341, 1994.
5. L. J. van der Paw, The radiation of electromagnetic power by microstrip configurations, *IEEE Trans. on MTT*, 719, Sept. 1977.
6. M. D. Abouzahra, L. Lewin, Radiation from microstrip discontinuities, *IEEE Trans. on MTT*, 722, Aug. 1979.
7. M. D. Abouzahra, L. Lewin, Radiation from microstrip discontinuities, *IEEE Trans. on MTT*, 722, Aug. 1979.
8. L. Lewin, Spurious radiation from microstrip, *Proc. IEE*, 633, July 1978.
9. W. T. Lo, C. K. C. Tzuang, S. T. Peng, and C. H. Lin, Full-wave and experimental investigations of resonant and leaky phenomena of microstrip step discontinuity problems with and without top cover, *Microwave Symp.*, 1, 473, 1994.
10. N. Marcuvitz, *Waveguide handbook*, MIT Radiation Lab, McGraw-Hill, New York, 1951.
11. G. H. Robinson, Resonant frequency calculations for microstrip cavities, *IEEE Trans. on MTT*, 665, July 1971.

12. M. V. Schneider and B. S. Glance, Suppression of waveguide modes in strip transmission lines, *Proc. of the IEEE*, 1184, Aug. 1974.
13. G. D. Vendelin, Limitations on stripline "Q," *Microwave J.*, 63, May 1970.
14. E. J. Denlinger, A frequency dependent solution for microstrip transmission lines, *IEEE Trans. on MTT*, 30, Jan. 1971.
15. D. Vanhoenacker, I. Huynen, Prediction of surface wave radiation coupling on microwave planar circuits, *IEEE Trans. on MGWL*, 255, Aug. 1995.
16. G. D. Vendelin, Limitations on stripline "Q," *Microwave J.*, 63, May 1970.
17. R. Mittra, T. Itoh, A new technique for the analysis of the dispersion characteristics of microstrip lines, *IEEE Trans. on MTT*, 47, Jan. 1971.
18. R. Mittra, T. Itoh, Analysis of microstrip transmission lines, *Adv. in Microwaves*, 8, 67, 1974.
19. A. Farrar, A. T. Adams, Computation of propagation constants for the fundamental and higher order modes in microstrip, *IEEE Trans. on MTT*, 456, July 1976.
20. Z. Ma, and E. Yamashita, FDTD investigation of higher order mode leakage from three dimensional microstrip line structures, *Microwave Symp.*, 3, 1631, 1997.
21. D. Nghiem, J. T. Williams, D. R. Jackson, A. A. Oliner, Suppression of leakage on stripline and microstrip structures, *Microwave Symp.*, 1, 145, 1994.
22. S. Tsitsos, A. A. P. Gibson, A. H. I. McCormick, Higher order modes in coupled striplines, prediction and measurement, *IEEE Trans. on MTT*, 2071, Nov. 1994.
23. F. Mesa, R. Marques, Low frequency leaky regime in covered multilayered striplines, *IEEE Trans. on MTT*, 1521, Sept. 1996.
24. C. Di Nallo, F. Mesa, D. R. Jackson, Excitation of leaky modes on multilayer stripline structures, *IEEE Trans. on MTT*, 1062, Aug. 1998.
25. G. D. Vendelin, Limitations on stripline "Q," *Microwave J.*, 63, May 1970.
26. M. Kahrizi, T. K. Sarkar, Z. A. Maricevic, Dynamic analysis of a microstrip line over a perforated ground plane, *IEEE Trans. on MTT*, 820, May 1994.
27. M. D. Prouty, K. K. Mei, S. E. Schwarz, R. Pous, Y. W. Liu, Solving microstrip discontinuities by the measured equation of invariance, *IEEE Trans. on MTT*, 877, June 1997.
28. K. S. Oh, J. E. Schutt-Aine, R. Mittra, Computation of excess capacitances of various strip discontinuities using closed-form Green's functions, *IEEE Trans. on MTT*, 783, May 1996.
29. D. Bica, B. Beker, Analysis of microstrip discontinuities using the spatial network method with absorbing boundary conditions, *IEEE Trans. on MTT*, 1157, July 1996.
30. A. A. Oliner, Equivalent circuits for discontinuities in balanced strip transmission line, *IRE Trans. on MTT*, 134, March 1955.
31. H. M. Altschuler, A. A. Oliner, Discontinuities in the center conductor of symmetric strip transmission line, *IRE Trans. on MTT*, 328, May 1960. See also: Addendum to Discontinuities in the center conductor of symmetric strip transmission line, *IRE Trans. on MTT*, 143, March 1962.
32. D. S. James, R. J. P. Douville, Compensation of microstrip bends by using square cutouts, *Electron. Lett.*, 577, Oct. 1976.
33. R. Chada, K. C. Gupta, Compensation of discontinuities in planar transmission lines, *IEEE Trans. on MTT*, 2151, Dec. 1982.
34. A. F. Thomson, A. Gopinath, Calculation of microstrip discontinuity inductances, *IEEE Trans. on MTT*, 648, Aug. 1975.
35. A. J. Slobodnik Jr., R. T. Webster, Experimental validation of microstrip bend discontinuity models from 18 to 60 GHz, *IEEE Trans on MTT*, 1872, Oct. 1994.
36. M. Kirschning, R. H. Jansen, N. H. L. Koster, Measurements and computer aided modeling of microstrip discontinuities by an improved resonator method, *IEEE MTT Symp. Digest*, 495, 1983.
37. R. J. P. Douville, D. S. James, Experimental study of symmetric microstrip bends and their compensation, *IEEE Trans. on MTT*, 175, March 1978.
38. M. Abramowitz, I. A. Stegun, *Handbook of Mathematical Functions*, Dover, New York, 1970.
39. A. Farrar, A. T. Adams, Matrix methods for microstrip three dimensional problems, *IEEE Trans. on MTT*, 497, Aug. 1972.
40. P. Silvester, P. Benedek, Equivalent capacitances of microstrip open circuits, *IEEE Trans. on MTT*, 511, Aug. 1972.

41. J. C. Goswami, A. K. Chan, C. K. Chui, Spectral domain analysis of single and coupled microstrips open discontinuities with anisotropic substrates, *IEEE Trans. on MTT*, 1174, July 1996.

42. M. Kirschning, R. H. Jansen, N. H. L. Koster, Accurate model for open end effect of microstrip lines, *Electron. Lett.*, 123, Feb. 1981.

43. M. Maeda, An analysis of gap in microstrip transmission lines, *IEEE Trans. on MTT*, 390. June 1972.

44. P. Benedek, P. Silvester, Equivalent capacitances for microstrip gaps and steps, *IEEE Trans. on MTT*, 729, Nov. 1972.

45. N. G Alexopoulos, Shih-Chang Wu, Frequency-independent equivalent circuit model for microstrip open-end and gap discontinuities, *IEEE Trans. on MTT*, 1268, July 1994.

46. M. Kirschning, R. H. Jansen, N. H. L. Koster, Measurements and computer aided modeling of microstrip discontinuities by an improved resonator method, *IEEE MTT Symp. Digest*, 495, 1983.

47. N. H. L. Koster, R. H. Jansen, The equivalent circuit of the asymmetrical series gap in microstrip and suspended substrate lines, *IEEE Trans. on MTT*, 1273, Aug. 1982.

48. A. Farrar, A. T. Adams, Matrix methods for microstrip three dimensional problems, *IEEE Trans. on MTT*, 497, Aug. 1972.

49. R. Chada, K. C. Gupta, Compensation of discontinuities in planar transmission lines, *IEEE Trans. on MTT*, 2151, Dec. 1982.

50. A. Gopinath, A. F. Thomson, I. M. Stephenson, Equivalent circuit parameters of microstrip step change in width and cross junctions, *IEEE Trans. on MTT*, 142, March 1976.

51. C. Yinchao, B. Beker, Study of microstrip step discontinuities on bianisotropic substrates using the method of lines and transverse resonance technique, *IEEE Trans. on MTT*, 1945, Oct, 1994.

52. R. Garg, I. J. Bahl, Microstrip discontinuities, *Intern. J. Electron.*, 81, Jan. 1978.

53. K. C. Gupta, R. Garg, I. J. Bahl, *Microstrip Lines and Slotlines*, Artech House, Norwood, MA, 1979.

54. K. C. Gupta, R. Garg, R. Chada, *Computer Aided Design of Microwave Circuits*, Artech House, 1981. This book contains some revisited formulas already introduced in the book of K. C. Gupta, R. Garg, I. J. Bahl in Reference 53.

55. R. Chada, K. C. Gupta, Compensation of discontinuities in planar transmission lines, *IEEE Trans. on MTT*, 2151, Dec. 1982.

56. M. Dydyk, Master the T-junction and sharpen your MIC design, *Microwaves*, 184, May 1977.

57. R. W. Vogel, Effects of the T-junction discontinuity on the design of microstrip directional couplers, *IEEE Trans. on MTT*, 145, March 1973.

58. E. O. Hammerstad, Equations for microstrip circuit design, V Eur. µwave Conf., 1975.

59. K. C. Gupta, R. Garg, I. J. Bahl, *Microstrip Lines and Slotlines*, Artech House, Norwood, MA, 1979.

60. A. Gopinath, A. F. Thomson, I. M. Stephenson, Equivalent circuit parameters of microstrip step change in width and cross junctions, *IEEE Trans. on MTT*, 142, March 1976.

61. B. Easter, The equivalent circuit of some microstrip discontinuities, *IEEE Trans. on MTT*, 655, Aug. 1975.

62. R. J. Akello, B. Easter, I. M. Stephenson, Equivalent circuit of the asymmetric crossover junction, *Electron. Lett.*, 117, Feb. 1977.

63. K. C. Gupta, R. Garg, I. J. Bahl, *Microstrip Lines and Slotlines*, Artech House, Norwood, MA, 1979.

CHAPTER 5

Coupled Microstrips

5.1 GEOMETRICAL CHARACTERISTICS

An illustration of two coupled* µstrips is provided in Figure 5.1.1 in a cross-sectional view. In general, more than two µstrips** can be e.m. coupled, but we will only study the case indicated in Figure 5.1.1, i.e., where only two µstrip are e.m. coupled. In addition, we will assume that in the coupling region, µstrips widths, their spacing, and substrate height will remain constant, i.e., we will assume uniform coupling.***

In this chapter we will consider that coupling between µstrip is desired. Crosstalk between coupled µstrips has been investigated by the researchers Y. Qian and E. Yamashita.[1]

Coupled µstrips**** provide a very important network with which a lot of devices can be built such as directional couplers and filters. At lower frequencies "cµ" are employed to transmit digital information at a high rate of speed, typically hundreds of MHz where "ECL" devices are employed. "ECL" devices are in fact "differential logic," where the best noise immunity and high speed rate can be achieved just using coupled lines.

As we can see, this structure is composed by setting two µstrip hot conductors at a distance "s", one of width "w_1" and the other with width "w_2." Similar to the case of single µstrip, the substrate can now be a ferrimagnetic material. However, unless otherwise stated we will assume the "cµ" to be on a nonferrimagnetic substrate, covering this argument in Chapter 7.

In general, there is no particular reason to have different hot conductor widths, and the devices we will discuss in Chapter 7 that employ "cµ" are examples of this assertion. So, the following theory in this chapter will assume $w_1 \equiv w_2$,***** unless otherwise stated.

5.2 ELECTRIC AND MAGNETIC FIELD LINES

Some electric "e" and magnetic "h" field lines for the fundamental "qTEM" mode in microstrips are indicated in Figure 5.2.1, in a defined cross-section and a defined time. In the fundamental mode each hot conductor is equipotential. Note that as in the general theory of coupled lines, "cµ" supports two independent excitation modes, i.e., the "even" and the "odd" mode. See Chapter 1 for definitions of even and odd excitation modes. The e.m. field line disposition is obviously different for each mode. Due to this phenomenon, the effective relative dielectric constant is different for

---

* The concept of coupling, together with the general theory of coupled "TEM" lines, is explained in Chapter 1.
** µstrips are studied in Chapter 2.
*** The general theory of nonuniform coupled lines is covered in Chapter 1. In Chapter 7 we will study µstrip devices working on nonuniform coupling.
**** Coupled µstrip will be simply abbreviated with "cµ."
***** The case $w_1 \equiv w_2$ is named as symmetric "cµ."

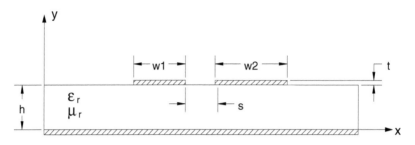

Figure 5.1.1

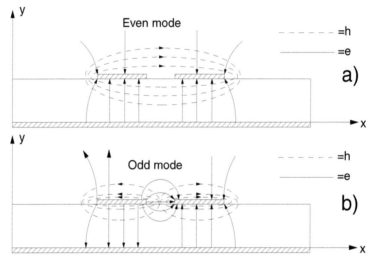

Figure 5.2.1

each mode. For this reason, we have an "even mode effective relative permitivity," indicated with "ε_{ree}," and an "odd mode effective relative permittivity," indicated with "ε_{reo}." The introduction of these quantities can be done similarly to the µstrip case studied in Chapter 2. With reference to Figure 5.2.2, in part a, the "cµ" with an even excitation has been enclosed in a box, with dimensions such that its effect on the field distribution can be neglected. In part b, the "cµ" is surrounded by a homogeneous, isotropic dielectric medium, with permitivity "ε_{ree}," so that the wave phase velocity is the same for cases in Figure 5.2.2 a and b. Of course, a similar discourse needs to be repeated for "cµ" with odd excitation, so that we can define an "ε_{reo}."

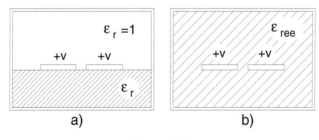

Figure 5.2.2

In general, it is the difference between the numbers of field lines in the air and substrate that creates the difference between the values of "ε_{ree}" and "ε_{reo}." If this number of lines was equal, as in the stripline case, there would not be any difference between "ε_{ree}" and "ε_{reo}." These quantities are considered as a constant in a quasi static group, while for the other groups they are frequency dependent functions, as they are in reality. In general, the determination of "$\varepsilon_{ree}(f)$" and "$\varepsilon_{reo}(f)$" is a very important point of any group of study. In fact, once we know "$\varepsilon_{re}(f)$," the even and odd characteristic impedances can be evaluated inserting this function in any of the quasi static impedance expressions that we will give in a later section. The result is that the "cμ" impedances will also be a function of frequency.

Similarly, we have an "even mode characteristic impedance," indicated with "ζ_e," and an "odd mode characteristic impedance," indicated with "ζ_o." These impedances are defined as the impedance of each μstrip, with the other μstrip present and excited with the proper polarity. In Chapter 1 we showed that $\zeta_e \geq \zeta_o$ and, in the present case, we have $\varepsilon_{ree} \leq \varepsilon_{reo}$.

Unfortunately, "cμ" are not t.l. supporting pure "TEM" waves, and the e.m. field disposition indicated in Figure 5.2.1 is only a good approximation for low frequency operation, let us say below some GHz. Due to the non- "TEM" propagation, "cμ" is a dispersive* structure and caution needs to be used when employing microstrip devices in wideband precision devices. However, due to their technological simplicity, "cμ" devices like directional couplers and filters are widely employed.

5.3 SOLUTION TECHNIQUES FOR THE ELECTROMAGNETIC PROBLEM

Three theory groups are mainly employed to study "cμ," similar to the case of μstrip studied in Chapter 3. They are:

1. *Quasi static group*, where the "cμ" is evaluated as being composed of the lines supporting a pure "TEM" mode. Examples[2,3] of methods used in this group are the solution of the Laplace equation[4] or conformal transformation method,[5,6] whose fundamental theory is studied in Appendix A1.
2. *Dispersion group*, where the "cμ" is evaluated as a particular coupling between a "TEM" and "TE" t.l.[7] or with other dispersion models.[8]
3. *Full wave group*, where no simplification is made and a full Maxwell's equations solution is found.[9]

The first two groups are quite simple to apply, while the third requires more analytical applications.

The next sections will discuss each one of the three groups.

5.4 QUASI STATIC ANALYSIS METHODS

These methods consider the evaluation of "cμ" properties as an electrostatic problem, so that it can be studied with rules and equations of electrostatics. In the following text we will indicate with the subscript "0," the quantities with the substrate replaced by the vacuum. It is assumed that each μstrip only supports a "TEM" wave, and even-odd excitation is applied to the structure. In this case, each μstrip has a characteristic impedance "ζ_m" and phase constants "β_m," with m = e,o to identify the mode, given by:

$$\zeta_m = \zeta_{0m}\left(C_{0m}/C_m\right)^{0.5} \pm \zeta_{0m}\Big/\left(\varepsilon_{rem}\right)^{0.5} \qquad \text{with } m = e,o \qquad (5.4.1)$$

---

* Dispersion is studied in Appendix A2.

$$\beta_m = \beta_{0m}\left(C_m/C_{0m}\right)^{0.5} \overset{\perp}{=} \beta_{0m}\left(\varepsilon_{rem}\right)^{0.5} \qquad \text{with m = e,o} \qquad (5.4.2)$$

Similar to the case of single µstrip studied in Chapter 2, we have:

$$\zeta_m = \zeta_v\varepsilon_0\left(C_mC_{0m}\right)^{-0.5} \equiv 1/\left[v_0\left(C_mC_{0m}\right)^{0.5}\right] \qquad \text{with m = e,o} \qquad (5.4.3)$$

where $\zeta_v \overset{\perp}{=} (\mu_0/\varepsilon_0)^{0.5} = 120\pi$. The effective relative dielectric constants "ε_{rem}" are defined as:

$$\varepsilon_{rem} \overset{\perp}{=} C_m/C_{0m} \qquad \text{with m = e,o} \qquad (5.4.4)$$

Among the quasi static methods [10] we will refer to work by the researcher J. A. Weiss,[11] based on the determination of Green's function* for the "cµ" structure. The geometry under study is shown in Figure 5.4.1. As unitary charge, it is assumed to be an infinitesimal width "dw" of the µstrip conductor, with theoretically infinite longitudinal length, carrying a unitary charge per u.l. This source is called the "line of charge." The hot conductors and distance "s" are divided in a number "M" of substrips. Consequently, each hot conductor is supposed to be composed of "M" lines of charge in parallel, so that the whole charge per unit area of the strip has a value "q_s." The solution of the problem is to find the Green's function for the line of charge, assuming a defined potential function inside the structure that satisfies the boundary conditions. The determination of this function requires some analytical practice, and for this reason we refer the reader to Weiss's article. Therefore, to show the analytical key points of this theory we would assume that the Green's function for a line source is known.

With reference to Figure 5.4.1, part a, let us indicate with "g_{ij}" the Green's function relative to the line of charge "$q_{\ell j}$" in position "x_j," working on the substrip in position "x_i." Applying the superposition principle, the potential "V" of each strip is:

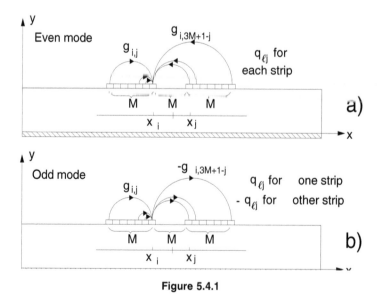

Figure 5.4.1

* The Green's function, (solution of Poisson's equation for a pulse charge), is studied in Appendix A1.

$$V = \sum_{j=1}^{j=M} q_{\ell j} \left(g_{i,j} \pm g_{i,3M+1-j} \right) \qquad (5.4.5)$$

where only values of "i" corresponding to hot "cµ" conductors need to be inserted in 5.4.5. In the previous equation "+" and "−" signs correspond to the even and odd excitations. The previous equation defines a system of "M" equations in the "M" unknowns "$q_{\ell j}$" which has a solution. Once found, the "$q_{\ell j}$" the charge "Q" for each strip is:

$$Q = \sum_{j=1}^{j=M} q_{\ell j} \qquad (5.4.6)$$

and consequently the capacity "C" per u.l. is:

$$C_m = Q_m/V \qquad (5.4.7)$$

where "Q_m" is obtained from 5.4.6 when 5.4.5 is evaluated once with the "+" sign, which corresponds to m ≡ e, and once with the "−" sign, which corresponds to m ≡ o. The procedure is then repeated again using vacuum as dielectric, obtaining:

$$C_{0m} = Q_{0m}/V \qquad (5.4.8)$$

Once "C_m" and "C_{0m}" are obtained, the even and odd impedances can be evaluated using the equations at the beginning of this section.

Similar to the case of single µstrip, an effective relative permeability "μ_{rem}" can be defined for a ferrimagnetic substrate, according to:

$$\mu_{rem} \overset{\pm}{=} L_m/L_{0m} \qquad (5.4.9)$$

where "L_{0m}" is the single µstrip equivalent inductance when the substrate is evaluated as $\mu_r = 1$, i.e., the substrate replaced by the vacuum. In particular, expression 5.4.1 is replaced with:

$$\zeta_m = \zeta_{0m} \left(\mu_{re}/\varepsilon_{re} \right)^{0.5} \qquad (5.4.10)$$

However, relative permeability is quite often a complicated function of material properties, such as the intensity of the eventual applied static magnetic field, the power of the RF signal, and others, as will be described in Appendix 7. So, expression 5.4.9 is actually only an approximation. It is evident, from the above calculation, that with this method the conductor thickness is assumed to be negligible. In practice, this hypothesis is approximated when t ≤ 0.1s and t ≤ 0.02h.

5.5 COUPLED MODES ANALYSIS METHOD

Among the types of dispersion group analysis, the coupled mode method is the simplest to study. It permits evaluation of the effective relative permittivity for the even and odd modes as a function of frequency. Consequently, the characteristic impedances "ζ_e" and "ζ_o" will be frequency dependent.

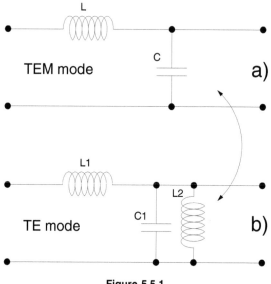

Figure 5.5.1

This method supposes that the propagation mode in a "cμ" can be obtained through a coupling between a "TEM" and "TE" lossless mode, as indicated in Figure 5.5.1 part a and b, respectively, with their equivalent t.l.* networks. This explanation of the microstrip propagation was first suggested by researchers H. J. Carlin and P. P. Civalleri.[12] We have already encountered such method of study in Chapter 2 for single μstrip. In the present case, both the even and odd modes are represented by a couple of t.l., as indicated in Figure 5.5.1. Of course, the element's values for each mode are different. In the following, we will refer to the case of lossless t.l. Using the subscript "m" to identify the modes "e" and "o," the elements of the equivalent t.l. have the following values:

$$L_m = \mu_0 d_m \tag{5.5.1}$$

$$C_m = \varepsilon_0 \varepsilon_{re0m}/d_m \tag{5.5.2}$$

$$L_{1m} = \mu_0 \tag{5.5.3}$$

$$L_{2m} = \mu_0/k_{tm}^2 \tag{5.5.4}$$

$$C_{1m} = \varepsilon_0 \varepsilon_{re0m} \tag{5.5.5}$$

where:

d_m is a parameter that depends on the transverse geometric dimensions of the "cμ"
ε_{re0m} is the static effective relative dielectric constant
k_{tm} is the "TE" mode cutoff wave number, a generalization of the same parameter defined in Chapter 2 for the single μstrip case, given by:

$$k_{tm}^2 = \frac{\left(\varepsilon_r - \varepsilon_{rem0}\right)\zeta_{am}^2}{4.143(60h)^2\left(0.5 + 0.001\zeta_{am}^{1.5}\right)} \tag{5.5.6}$$

---

* See Appendix A2 for equivalence between modes and t.l.

In 5.5.6 the parameter "ζ_{am}" is given by:

$$\zeta_{ae} = \zeta_e/2 \quad \text{and} \quad \zeta_{ao} = 2\zeta_o \tag{5.5.7}$$

To evaluate the effective relative dielectric constants, the series impedance "$[Z_s(\omega)]$" and parallel admittance "$[Y_p(\omega)]$" 4×4 matrices are written for each coupled t.l. for the "e" and "o" modes, and the following matrix equation is solved in "k":

$$\det\left(\left[Z_s(\omega)\right]\left[Y_p(\omega)\right] - k_m^2[I]\right) = 0 \tag{5.5.8}$$

where "$[I]$" is the 4×4 identity matrix. Note that the previous equation is a generalization of the scalar equation $Z_s Y_p = !k^2$ defined in Chapter 1. After the "k_m" value has been found, it is possible to extract the effective relative dielectric constant from the well-known relation* $k = j\omega(\mu\varepsilon_c)^{0.5}$. The resulting expression is:

$$\varepsilon_{rem}(\omega) = \varepsilon_{re0m} - \frac{\left(v_0 k_{tm}\right)^2}{2\omega^2} \pm \left[\left(C_m \varepsilon_{re0m}\right)^2 + \frac{\left(v_0 k_{tm}\right)^4}{4\omega^4}\right]^{0.5} \tag{5.5.9}$$

where the coefficient "C_m" is given by:

$$C_m = \left(\varepsilon_r - \varepsilon_{rem0}\right)/\varepsilon_{rem0} \qquad \text{with m = e,o} \tag{5.5.10}$$

and determines the values of the coupling capacitor "C_m" between the two t.l.s of Figure 5.5.1, i.e.:

$$C_m = c_m \varepsilon_0 \varepsilon_{rem0}/\sqrt{d_m} \qquad \text{with m = e,o} \tag{5.5.11}$$

The plus sign in 5.5.9 represents the fundamental "qTEM" even and odd modes, i.e., which propagate from DC, while the minus sign represents other modes, which are the solution of 5.5.8 but have finite cutoff frequencies.

In general, the effecive relative dielectric constants increase with frequency, which means the field is more and more contained inside the substrate.

5.6 FULL WAVE ANALYSIS METHOD

With this method, a solution of Maxwell's equations is found for the "cμ" structure enclosed in the box, applying the boundary conditions that the fields must satisfy. We have already used this general method in Chapter 2 for the single μstrip case. The same procedure can be applied here, starting, of course, with different boundary conditions. For this reason, in the present case of "cμ" we will only indicate the differences between this and the same procedure applied for single μstrip, sending the reader to Chapter 2 for the analytical expressions.

The enclosed "cμ" as indicated in Figure 5.6.1 is supposed to have cylindrical symmetry, and a Cartesian coordinate system is applied. Following the classical methods of study for structures with such a symmetry,** the electric "A" and magnetic "F" vector potentials are written as:

---

* See Chapter 1 for propagation constant and characteristic impedance relations.
** See Appendix A2 for a general discussion of guided propagation.

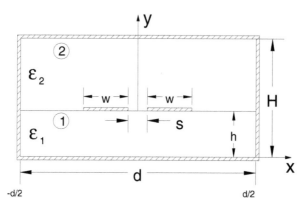

Figure 5.6.1

$$A(x,y,z) \doteq L(z)A_t(x,y) \tag{5.6.1}$$

$$F(x,y,z) \doteq L(z)F_t(x,y) \tag{5.6.2}$$

where "$L(z)$," (assumed to be in a lossless case and in a reflectionless propagation in the "z_0" direction), is given by:

$$L(z) = Ce^{-j\beta z} \tag{5.6.3}$$

and "A_t" and "F_t" verify the transverse wave equation.

Then, for every region $i = 1,2$ indicated in Figure 5.6.1 we assume the propagation mode to be composed of the sum of infinite "TM" and "TE" modes,* obtained using Equations A2.6.1 and A2.6.2. Therefore, the e.m. field inside the structure can be written with the same expressions we gave in Chapter 2.

The condition to consider in the expression of the field of all the possible "TE" and "TM" modes is performed expressing the transverse potential vectors with a series of infinite terms. Of course, the resulting expressions for "A_t" and "F_t" must verify the transverse wave equation and the boundary conditions for the structure in Figure 5.6.1. To this purpose, with reference to Figure 5.6.1, the e.m. field components must verify:

1. For $y = 0$ and $y = H$, with $|x| \le d/2$, i.e., top and bottom shields: $e_x = !0$ and $e_y - !0$;
2. For $x = \pm d/2$, with $0 \le y \le H$, i.e., the lateral shields: $e_z = !0$ and $e_y = !0$
3. For $y = h$, with $s/2 \le x \le w$, and $-w \le x \le -s/2$, i.e., the hot conductors: $e_x = !0$ and $e_z = !0$
4. For $y = h$, with $|x| \le s/2$, $-d/2 \le x \le -s/2 - w$, and $s/2 + w \le x \le d/2$, i.e., the dielectrics interface:
   a. $e_{x1} = !e_{x2}$ and $e_{z1} = !e_{z2}$, i.e., continuous tangential electric components
   b. $\varepsilon_1 e_{y1} = !\varepsilon_2 e_{y2} - q_s$, i.e., variation of "d"** normal component is equal to surface charge density "q_s"
   c. $h_{y1} = !h_{y2}$, i.e., continuous normal magnetic components
   d. $h_{x1} = !h_{x2} + i_z(x)$ and $h_{z1} = !h_{z2} - i_x(x)$, i.e., variation of "h" tangential component is equal to linear current "i."

So, the difference in boundary conditions with respect to the single μstrip are in conditions 3 and 4 above. Until this point, the relations are very simple and they are a direct application of the boundary conditions to a cylindrical guiding structure which supports an e.m. field. Analytic

---

* See Appendix A2 for general expressions of "TE" and "TM" modes.
** $d \perp \varepsilon e$ is the electric displacement vector.

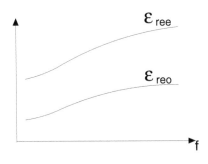

Figure 5.6.2

difficulties arise when we apply the boundary conditions 1 through 4 above to the field expressions. We refer the reader to Chapter 2 where a possible solution procedure[13] was outlined, which can still be adapted here.

The results of this procedure give an even and odd effective dielectric constant that increases with frequency, as indicated in Figure 5.6.2, and that is the same as the "ζ_e" and "ζ_o." Consequently, the even and odd phase velocities are different and the "cμ" is a dispersive structure, as the single μstrip we studied in Chapter 2.

5.7 DESIGN EQUATIONS

The most simple and widely used expressions for symmetric "cμ" synthesis are provided by researchers R. Garg and I. J. Bahl.[14,15] These equations are extracted using a quasi static analysis of the "cμ" structure, together with a curve fitting to measured values, which permits obtaining the coupled lines' capacitances. The capacitances involved are indicated in Figure 5.7.1. In detail, for each hot conductor:

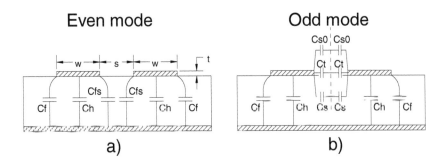

Figure 5.7.1

"C_f" is the fringing capacitance of the more distant sides of the "cμ"

"C_h" is the parallel capacitance of the hot conductors

"C_{fs}" is the fringing capacitance of the nearest sides of the "cμ"

"C_s" is the coupling capacitance, only considering the effect of "ε_r"

"C_{s0}" is the coupling capacitance, only considering the effect of "ε_0"

"C_t" is a coupling capacitance due the strip thickness "t," only considering the effect of "ε_0"

All these capacitances are relative to a unit length. With these notations, the even "C_e" and odd "C_o" capacitances for each line are given by:

$$C_e = C_h + C_f + C_{fs} \tag{5.7.1}$$

$$C_o = C_h + C_f + C_{s0} + C_s + C_t \tag{5.7.2}$$

The capacitances "C_h" and "C_t" come from the parallel plate capacitance, simply given by:

$$C_h = \varepsilon_0 \varepsilon_r \, w/h \tag{5.7.3}$$

$$C_t = 2\varepsilon_0 \, t/s \tag{5.7.4}$$

Fringing capacitance "C_f" is given by:

$$2C_f = \left(v_0 \zeta_s\right)^{-1} \sqrt{\varepsilon_{res}} - C_h \tag{5.7.5}$$

where "ζ_s" and "ε_{res}" are the impedance and effective relative dielectric constant for the single, uncoupled µstrip of Figure 5.7.1. These two quantities can be evaluated as indicated in Chapter 2. Fringing capacitance "C_{fs}" is given by:

$$2C_{fs} = C_f \Big/ \left[1 + A\left(h/s\right)\tanh\left(10s/h\right)\right] \left(\varepsilon_r/\varepsilon_{re}\right)^{-0.5} \tag{5.7.6}$$

where:

$$A = \exp\left[-0.1\exp\left(2.33 - 2.53\,w/h\right)\right] \tag{5.7.7}$$

Coupling capacitance "C_{s0}" is given by:

$$C_{s0} = \left(\varepsilon_0/\pi\right)\ln\left(2\frac{1+\sqrt{p'}}{1-\sqrt{p'}}\right) \qquad \text{for } 0 \le p^2 \le 0.5 \tag{5.7.8}$$

$$C_{s0} = \pi\varepsilon_0 \left[\ln\left(2\frac{1+\sqrt{p}}{1-\sqrt{p}}\right)\right]^{-1} \qquad \text{for } 0.5 \le p^2 \le 1 \tag{5.7.9}$$

where:

$$p \overset{\perp}{=} (s/h)/(s/h + 2w/h) \tag{5.7.10}$$

$$p' \overset{\perp}{=} \left(1 - p^2\right)^{0.5} \tag{5.7.11}$$

Coupling capacitance "C_s" is given by:

$$C_s = \left(\varepsilon_0 \varepsilon_r/\pi\right)\ln\left[\coth\left(\pi s/4h\right)\right] + 0.65 C_f \left[0.02 h\sqrt{\varepsilon_r}\big/s + 1 - 1/\varepsilon_r^2\right] \tag{5.7.12}$$

Once obtained "C_e" and "C_o" are obtained, i.e., "C_m" with m = e,o, we can simply calculate "ζ_m" and "ε_{rem}" by applying the relations of Section 5.4, i.e., Equations 5.4.3 and 5.4.4. The effect of strip thickness can be taken into account using the effective strip width* "w_e" which for the particular case of "cµ" is:

$$w_{ee} = w + \Delta w\left[1 - 0.5\exp\left(-0.69\Delta w/\Delta t\right)\right] \tag{5.7.13}$$

$$w_{eo} = w_{ee} + \Delta t \tag{5.7.14}$$

$$\Delta t = th/s\varepsilon_r \tag{5.7.15}$$

And Δw is given by 2.8.21 and 2.8.22.

In the ranges:

$$0.2 \leq w/h \leq 2; \quad 0.05 \leq s/h \leq 2; \quad \varepsilon_r \geq 1 \tag{5.7.16}$$

the previous formulas give a maximum error of 3% when the values are compared to the static rigorous analysis.[16]

Garg and Bahl have also given expressions to evaluate the effect of dispersion on "ε_{rem}" and "ζ_m," based on a study for single µstrip by the researcher W. J. Getsinger.[17] These are:

$$\varepsilon_{rem}(f) = \varepsilon_r - \frac{\varepsilon_r - \varepsilon_{rem}}{1 + \left(f/f_{pm}\right)^2 G_m} \tag{5.7.17}$$

The quantities "G_m" and "f_{pm}" are given by:

$$G_o = 0.6 + 0.018\zeta_o \quad \text{and} \quad f_{po} = \zeta_o/\mu_0 h \tag{5.7.18}$$

$$G_e = 0.6 + 0.0045\zeta_e \quad \text{and} \quad f_{pe} = 3\zeta_e/4\mu_0 h \tag{5.7.19}$$

An expression similar to 5.10.17 is given for "$\zeta_m(f)$" resulting in:

$$\zeta_m(f) = \zeta_{sm} - \frac{\zeta_{sm} - \zeta_{m0}}{1 + G_m\left(f/f_{pm}\right)^2} \tag{5.7.20}$$

where:

1. "G_m" and "f_{pm}" are given by 5.7.18 and 5.7.19
2. "ζ_{sm}" are the even and odd impedances of a side coupled stripline** with the same "w" and "s" as our "cµ," and planes spacing b = 2h, as indicated in Figure 5.7.2. These "ζ_{sm}" are simply given by:

$$\zeta_m = \frac{30\pi}{\sqrt{\varepsilon_r}} \frac{K\left(p'_m\right)}{K\left(p'_m\right)} \tag{5.7.21}$$

with:

* See Chapter 2 for the explanation of "effective strip width."
** Side coupled striplines will be studied in the next chapter.

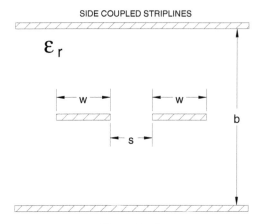

Figure 5.7.2

$$p_e \doteq \tanh\left(\pi w/2b\right)\tanh\left[\pi(w+s)/2b\right] \qquad (5.7.22)$$

$$p_o \doteq \tanh\left(\pi w/2b\right)\operatorname{ctgh}\left[\pi(w+s)/2b\right] \qquad (5.7.23)$$

and $p_m' \doteq (1 - p_m^2)$. The ratio "$K(p_m)/K(p_m')$" of the complete elliptic integrals of the first kind can be evaluated using the tabulated values of the elliptic integrals.[18] However, this ratio can be simply approximated as we have indicated in Appendix A8. The values obtained by the dispersion equations have been compared to the result of the full wave analysis,[19] resulting in an error of less than 5% for the same validity ranges given by 5.7.16.

The previous formulas are capable of taking into account the metal thickness and are very simple. In the cases where thickness can be neglected, the most accurate formulas for effective relative dielectric constants have been obtained by the researchers M. Kirschning and R. H. Jansen[20] through a process where curves are fit to full wave analysis data. Defining:

$$u \doteq w/h \quad \text{and} \quad g \doteq s/h \qquad (5.7.24)$$

the resulting equations for even and odd effective relative dielectric constants are:

$$\varepsilon_{ree} = 0.5\left(\varepsilon_r + 1\right) + 0.5 b_e\left(\varepsilon_r - 1\right)\left(1 + 10/v\right)^{\wedge}\left(-a_e\right) \qquad (5.7.25)$$

$$v = u\left(20 + g^2\right)/\left(10 + g^2\right) + g e^{-g} \qquad (5.7.26)$$

$$a_e = 1 + \ln\left\{\left[v^4 + (v/52)^2\right]/\left(v^4 + 0.432\right)\right\}/49 + \ln\left[1 + (v/18.1)^3\right]/18.7 \qquad (5.7.27)$$

$$b_e = 0.564\left[\left(\varepsilon_r - 0.9\right)/\left(\varepsilon_r + 3\right)\right]^{0.053} \qquad (5.7.28)$$

$$\varepsilon_{reo} = \left[0.5\left(\varepsilon_r + 1\right) + a_o + \varepsilon_{re}\right]\exp\left(-c_o g^{\wedge} d_o\right) + \varepsilon_{re} \qquad (5.7.29)$$

$$a_o = 0.7287\left[\varepsilon_{re} - 0.5\left(\varepsilon_r + 1\right)\right]\left[1 - \exp\left(-0.179u\right)\right] \qquad (5.7.30)$$

$$b_o = 0.747\varepsilon_r/(0.15 + \varepsilon_r) \tag{5.7.31}$$

$$c_o = b_o - [b_o - 0.207]e^{-0.414u} \tag{5.7.32}$$

$$d_o = 0.593 + 0.694e^{-0.562u} \tag{5.7.33}$$

where the quantity "ε_{re}" is the zero thickness effective relative dielectric constant for a single µstrip. This can be evaluated as we stated in Chapter 2, Equations 2.4.12 through 2.4.14.

Defining:

$$f_h \overset{\pm}{=} fh \tag{5.7.34}$$

with the frequency "f" in GHz and the substrate height in "mm," the frequency dependence of "ε_{rem}" becomes:

$$\varepsilon_{rem}(f_h) = \varepsilon_r - (\varepsilon_r - \varepsilon_{rem})/(1 + F_m(f_h)) \tag{5.7.35}$$

where:

$$F_e(f_h) = P_1 P_2 [(P_3 P_4 + 0.1844 P_7) f_h]^{1.5763} \tag{5.7.36}$$

$$P_1 = 0.27488 + [0.6315 + 0.525/(1 + 0.0157 f_h)^{20}] u - 0.065683 e^{-8.7513u} \tag{5.7.37}$$

$$P_2 = 0.33622[1 - \exp(-0.03442\varepsilon_r)] \tag{5.7.38}$$

$$P_3 = 0.363 e^{-4.6u}\{1 - \exp[-(f_h/38.7)^{4.97}]\} \tag{5.7.39}$$

$$P_4 = 1 + 2.751\{1 - \exp[-(\varepsilon_r/15.916)^8]\} \tag{5.7.40}$$

$$P_5 = 0.334 \exp[-3.3(\varepsilon_r/15)^3] + 0.746 \tag{5.7.41}$$

$$P_6 = P_5 \exp[-(f_h/18)^{0.368}] \tag{5.7.42}$$

$$P_7 = 1 + 4.069 P_6 g^{0.479} \exp(-1.347 g^{0.595} - 0.17 g^{2.5}) \tag{5.7.43}$$

$$F_o(f_h) = P_1 P_2 ((P_3 P_4 + 0.1844) f_h P_{15})^{1.5763} \tag{5.7.44}$$

$$P_8 = 0.7168\{1 + 1.076/[1 + 0.0576(\varepsilon_r - 1)]\} \tag{5.7.45}$$

$$P_9 = P_8 - 0.7913\{1 - \exp[-(-f_h/20)^{1.424}]\}\operatorname{atan}[2.481(\varepsilon_r/8)^{0.946}] \tag{5.7.46}$$

$$P_{10} = 0.242\left(\varepsilon_r - 1\right)^{0.55} \tag{5.7.47}$$

$$P_{11} = 0.6366\left[\exp\left(-0.3401 f_h\right) - 1\right]\text{atan}\left[1.263\left(u/3\right)^{1.629}\right] \tag{5.7.48}$$

$$P_{12} = P_9\left(1 - P_9\right)\big/\left(1 + 1.183 u^{1.376}\right) \tag{5.7.49}$$

$$P_{13} = 1.695 P_{10}\big/\left(0.414 + 1.605 P_{10}\right) \tag{5.7.50}$$

$$P_{14} = 0.8928 + 0.1072\left\{1 - \exp\left[-0.42\left(f_h/20\right)^{3.215}\right]\right\} \tag{5.7.51}$$

$$P_{15} = \text{abs}\left[1 - 0.8928\left(1 + P_{11}\right)P_{12}\,\exp\left(-P_{13}g^{1.092}\right)\big/P_{14}\right] \tag{5.7.52}$$

The function "abs( )" in the previous equation is a function that returns the absolute value of the argument "( )." The validity ranges of the above formulas are:

$$0.1 \le u \le 10 \qquad 0.1 \le g \le 10 \qquad 1 \le \varepsilon_r \le 18 \tag{5.7.53}$$

giving an error of less than 1.5% with respect to the full wave data analysis.

5.8 ATTENUATION

Similar to the case of single µstrip, losses in "cµ" are due to four causes:

1. Nonperfect conductivity of the conductors, or "conductor loss"
2. Nonzero conductivity of the dielectric, or "dielectric loss"
3. Substrate magnetic loss, if the substrate is a ferrimagnetic material
4. Radiation

Magnetic losses are mainly due to damping phenomena inside ferrimagnetic material and, if the signal frequency is of an appropriate value, to resonance absorption.*

Radiation losses in "cµ" have not been investigated as well as their µstrip counterpart. However, as a first analysis, we can use the results of Chapter 4 to evaluate the equivalent network elements' values for the discontinuity in a single µstrip of the "cµ" structure.

In this section we will study how to evaluate the first two causes of losses, which are directly related to the geometry of the "cµ" indicated in Figure 5.1.1. We consider the "cµ" as a t.l. only supporting a "TEM" mode, so that we can apply the theory developed by researcher H. A. Wheeler.[21,22] The procedure is similar to that used for the µstrip case in Chapter 2, and for this reason, we will only outline the differences with respect to µstrips at this time. With the subscript "m" we indicate the generic mode, i.e., m = e or m = o, and all the attenuation constants have dimensions [nepers/u.l.] unless otherwise stated.

First, let us consider the attenuation due the nonperfect conductivity of the conductors. In these conditions, we know** that at a depth "p," called the "penetration depth," inside the conductors and given by:

---

* See Appendix A7 for energy exchange phenomena between e.m. signal and ferrite.
** See Appendix A2 for e.m. energy penetration inside nonperfect conductors.

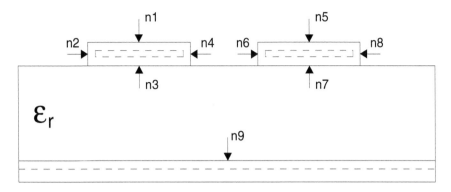

Figure 5.8.1

$$p \overset{\perp}{=} \left(\pi f \mu_c g\right)^{-0.5} \qquad [\text{meters}*] \qquad (5.8.1)$$

the field amplitudes are reduced by "1/e." In Equation 5.8.1 "f" is the signal frequency, and "g" and "μ_c" are the conductor conductivity and absolute permeability, respectively. The effect of each penetration can be regarded as the introduction of an additional series inductance and resistance** per u.l., indicated respectively with "L_{im}" and "R_{im}" and called respectively, "incremental inductance" and "incremental resistance" for the m = e or m = o mode. The situation is depicted in Figure 5.8.1, where the dashed line indicates the penetration depth. If "L_m" is the equivalent series inductance per u.l. of the lumped equivalent network of each µstrip of the "cµ," given by:***

$$L_m = \zeta_m / v_m = \zeta_m \left(\mu_{rem} \varepsilon_{rem}\right)^{0.5} / v_0 \qquad [\text{Henry/meter}] \qquad (5.8.2)$$

"L_{im}" and "R_{im}" are:

$$L_{im} = \frac{p \mu_{cr}}{2} \frac{\delta L_m}{\delta n} \qquad [\text{Henry/meter}] \qquad (5.8.3)$$

$$R_{im} \overset{\perp}{=} \omega L_{im} = \pi f p \mu_{cr} \frac{\partial L_m}{\partial n} \qquad [\Omega/\text{meter}] \qquad (5.8.4)$$

where "μ_{cr}" is the conductor relative permeability and "∂n" is an infinitesimal penetration inside the conductor, which is positive when the vector "$\underline{n}$" is directed inside it. If we define the quantity "R_s," called the "sheet resistance" for the conductor, as:

$$R_s \overset{\perp}{=} \pi f p \mu_c \equiv \left(\pi f \mu_c / g\right)^{0.5} \qquad [\Omega/\text{square}]**** \qquad (5.8.5)$$

Equation 5.8.4 becomes:

$$R_{im} = \frac{R_s}{\mu_0} \frac{\partial L_m}{\partial n} \qquad (5.8.6)$$

* Remember that unless otherwise stated, in this book we will use the MKSA units reference system.
** Since we assume the "cµ" only supports a "TEM" mode, we are referring to the simple low pass equivalent network for a line supporting a "TEM" mode. This argument is treated in Chapter 1.
*** See Chapter 1 for relationships between t.l. characteristics quantities.
**** See Appendix A2 for dimension unit of "conductor resistance."

We now have to take into account all the incremental inductances and resistances to obtain the whole additional inductance "L_{am}" and resistance "R_{am}." With reference to Figure 5.8.1 and using Equations 5.8.3 and 5.8.6, we have:

$$L_{am} = \sum_{j=1}^{j=9} \left(L_{im}\right)_j = \frac{1}{2\mu_0} \sum_{j=1}^{j=9} p_j \left(\mu_c\right)_j \frac{\partial L_m}{\partial n_j} \qquad (5.8.7)$$

$$R_{am} = \sum_{j=1}^{j=9} \left(R_{im}\right)_j = \frac{1}{\mu_0} \sum_{j=1}^{j=9} \left(R_s\right)_j \frac{\partial L_m}{\partial n_j} \qquad (5.8.8)$$

The conductor attenuation coefficient "α_{cm}"* is defined as:

$$\alpha_{cm} = W_{cm}/2W_{tm} \qquad (5.8.9)$$

where "W_{cm}" and "W_{tm}" are, respectively, the mean power dissipated in the conductor and the mean transmitted power, and are given by:

$$W_{cm} = R_{am}\left|i_m\right|^2, \quad W_{tm} = \zeta_m\left|i_m\right|^2 \qquad (5.8.10)$$

Using 5.8.8 and 5.8.10, Equation 5.8.9 becomes:

$$\alpha_{cm} = \frac{1}{2\mu_0\zeta_m} \sum_{j=1}^{j=9} \left(R_s\right)_j \frac{\partial L_m}{\partial n_j} \qquad (5.8.11)$$

From 5.8.2 we note that:

$$\frac{\partial L_m}{\partial n} = \frac{1}{v_0} \frac{\partial\left[\left(\mu_{rem}\varepsilon_{rem}\right)^{0.5}\zeta_m\right]}{\partial n} \perp \frac{1}{v_0} \frac{\partial \zeta_{zm}}{\partial n} \qquad (5.8.12)$$

where $\zeta_{zm} \perp \left(\mu_{rem}\varepsilon_{rem}\right)^{0.5}\zeta_m$. Equation 5.8.11 then becomes:

$$\alpha_{cm} = \frac{1}{\zeta_m 240\pi} \sum_{j=1}^{j=9} \left(R_s\right)_j \frac{\partial \zeta_{zm}}{\partial n_j} \qquad (5.8.13)$$

Since "ζ_m" and "ε_{rem}" are also functions of "w," "s," "t," and "h," as shown in the previous paragraph, the derivative "$\partial\zeta_{zm}/\partial n$" is:

$$\frac{\partial \zeta_{zm}}{\partial n} = \frac{1}{dn}\left(\frac{\partial \zeta_{zm}}{\partial s}ds + \frac{\partial \zeta_{zm}}{\partial w}dw + \frac{\partial \zeta_{zm}}{\partial t}dt + \frac{\partial \zeta_{zm}}{\partial h}dh\right) \qquad (5.8.14)$$

From Figure 5.8.1 we observe that:

---

* We are assuming a longitudinal variation of conductor attenuation with $e^{-\alpha z}$. See Chapter 1 for the fundamental theory of transmission lines.

$$ds = 2dn, \quad dt = -2dn, \quad dw = -2dn, \quad dh = 2dn \qquad (5.8.15)$$

and 5.8.13 becomes:

$$\alpha_{cm} = \frac{1}{\zeta_m 120\pi} R_s \left(\frac{\partial \zeta_{zm}}{\partial s} - \frac{\partial \zeta_{zm}}{\partial w} - \frac{\partial \zeta_{zm}}{\partial t} + \frac{\partial \zeta_{zm}}{\partial h} \right) \qquad (5.8.16)$$

Closed form formulas are not available, but the derivatives in the previous equation can be easily obtained numerically from the expressions of "ζ_m" presented in previous sections. The result is that "α_{co}" is always greater than "α_{ce}," and is four or five times the value in dB corresponding to "α_{co}."

Dielectric losses "α_{dm}" have been verified to be obtained for that of a single µstrip. In Chapter 2 we gave the closed form expression of such an attenuation constant, based on a study by M. V. Schneider, B. Glance, and W. F. Bodtmann[23] for a nongyromagnetic substrate. Applying this equation to the even and odd mode,[24] we have:

$$\alpha_{dm} = \frac{20\pi}{\ln 10} \frac{1/\varepsilon_{rem} - 1}{1/\varepsilon_r - 1} \frac{\tan \delta}{\lambda_0} \sqrt{\varepsilon_{rem}} \qquad [db/u.l.] \qquad (5.8.17)$$

where "$\tan\delta$" is the dielectric loss tangent.* Of course, the quantities "ε_r" and "ε_{rem}" are all relative to the real part of the substrate dielectric constant.** Similar to the case of single µstrip, in the most used substrates employed in MIC circuits, i.e., alumina and quartz, conductor losses are predominant. The situation is different in the case of semiconductor substrates employed in MMIC devices, like GaAs, where "α_d" is comparable to "α_c."

5.9 A PARTICULAR COUPLED MICROSTRIP STRUCTURE: THE MEANDER LINE

A particular type of "cµ" is the so-called "meander line" structure indicated in Figure 5.9.1. This structure is employed in phase shifters, as shown in Chapters 7 and 8 where phase shifters will be studied. The meander line can be thought of as being composed of a repeated single cell, still called a "meander line" or "C" section, indicated in Figure 5.9.2. In practice, the "C" section is built by connecting the two nearest ports in short circuit. This cell is characterized by:

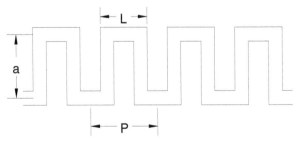

Figure 5.9.1

* See Chapter 1 for "dielectric loss tangent" definition.
** See Chapter 1 for complex permittivity definition.

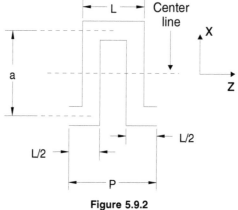

Figure 5.9.2

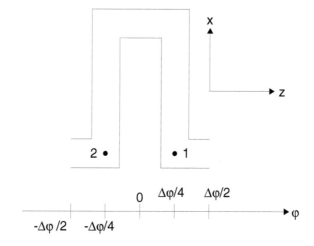

Figure 5.9.3

a. A width "a" transverse to the longitudinal axis "z," i.e., the mean value of the coupling length
b. A spatial repetition period "P" along "z"
c. A phase shift $\Delta\varphi$ along "z," per unit cell "P." This is a consequence of point a,

A characteristic of the "C" section is that for $\Delta\varphi = \pi$, the even and odd effective relative dielectric constants are equal. This can easily be shown using the even and odd excitation method in Figure 5.9.3. In this case, since the "cμ" are connected together at one side, whenever a generator is connected to one open extreme, the signal must arrive at the other extreme by shifting its phase of the proper value along the "cμ." So, using as connecting points those points labeled "1" and "2" in Figure 5.9.3, the even excitation will be identified by:

$$v_{1e} = ve^{j\Delta\varphi/4} \quad \text{and} \quad v_{2e} = ve^{-j\Delta\varphi/4} \qquad (5.9.1)$$

so that the phase difference "$\Delta\varphi_e$" between these signals is

$$\Delta\varphi_e = \Delta\varphi/2 \qquad (5.9.2)$$

in accordance with the phase shift for the points "1" and "2." The odd excitation will be identified by:

$$v_{1o} = ve^{j\Delta\phi/4} \quad \text{and} \quad v_{2o} = -ve^{-j\Delta\phi/4} \tag{5.9.3}$$

so that the phase difference "$\Delta\phi_e$" between these signals is:

$$\Delta\phi_o = \Delta\phi/2 + \pi \tag{5.9.4}$$

i.e., not evaluating the obvious phase shift of "π" due to the odd excitation, still a phase shift of $\Delta\phi/2$. Setting $\Delta\phi = \pi$ in 5.9.2 and 5.9.4, we note that $|\Delta\phi_o| \equiv |\Delta\phi_e|$. From the definition of effective relative dielectric constants, the equivalence between even and odd phase differences means that $\varepsilon_{ree} \equiv \varepsilon_{reo}$.

Among the methods of study of "C" section[25,26] we will refer to the quasi static method of Weiss,[27] which is similar to the method we used in Section 5.4 for the "cμ." The geometry under study is indicated in Figure 5.9.4. The procedure begins by finding a solution for the general "homogeneous wave equation":*

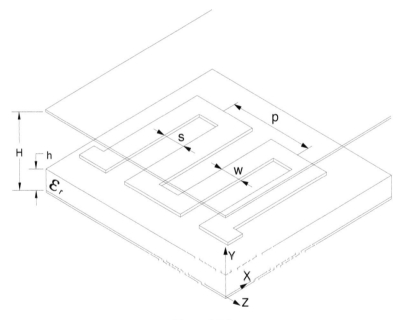

Figure 5.9.4

$$\nabla^2(F) = k^2 F \tag{5.9.5}$$

where "F" represents the electric or magnetic field and $k^2 \doteq -\omega^2\mu\varepsilon_c$. Due to the symmetry along the "x" axis, we will find a field,** "F," with a coordinate dependence given by:

$$F(x,y,z) \doteq V_t(y,z)e^{-jk_x x} \tag{5.9.6}$$

---

* See Appendix A2 for general definition of "wave equations."
** In the following, we will consider it obvious that the solution of the wave equation is, in this case, a field.

Due to the hypothesis of "TEM" mode, "V_t" will satisfy the transverse wave equation:

$$\nabla_t^2(V_t) = 0 \tag{5.9.7}$$

i.e., "V_t" is a solution of the Laplace* equation,** and for this reason it can be considered as a potential for the structure. Due the periodicity along "z," we assume:

$$v_t(y,z) \overset{\perp}{=} v_t(y,z)\exp(jz\Delta\varphi/p) \tag{5.9.8}$$

where "$v_t(y,z)$" will also be periodic along "z." In our rectangular coordinate system, Equations 5.9.7 and 5.9.8 are rewritten as:***

$$\frac{\partial^2 v_t}{\partial y^2} + \frac{\partial^2 v_t}{\partial z^2} + 2j\frac{\Delta\varphi}{p}\frac{\partial v_t}{\partial z} - \frac{\Delta\varphi^2}{p^2}v_t = 0 \tag{5.9.9}$$

Due the periodicity of "v_t," this function can be expandend in Fourier series, i.e.:

$$v_t(y,z) = \sum_{m=-\infty}^{m=+\infty} v_m(y)\exp\left(jm2\pi\frac{z}{p}\right) \tag{5.9.10}$$

Inserting this equation into:

a. Equation 5.9.8, we have:

$$v_t(y,z) = \sum_{m=-\infty}^{m=+\infty} v_m(y)\exp\left(j\frac{m2\pi + \Delta\varphi}{p}z\right) \tag{5.9.11}$$

b. Equation 5.9.9, we have the condition to be satisfied for the coefficients "$v_m(y)$" of Equation 5.9.10, i.e.:

$$\sum_{m=-\infty}^{m=+\infty} \frac{\partial^2 v_m(y)}{\partial y^2} - \sum_{m=-\infty}^{m=+\infty} \beta_m^2 v_m(y) = 0 \tag{5.9.12}$$

with "β_m" given by:

$$\beta_m \overset{\perp}{=} (m2\pi + \Delta\varphi)/p \tag{5.9.13}$$

Note that for every subscript "m," Equation 5.9.12 represents the "harmonic motion" equation,**** whose useful solution is:

---

* Pier Simon de Laplace, French mathematician, born in Beaumont en Auge in 1749 and died in Paris in 1827.
** See Appendix A1 for the Laplace equation.
*** Of course, for simplicity, we will write "v_t" for "$v_t(y,z)$."
**** The "Harmonic motion" equation was introduced in Chapter 1.

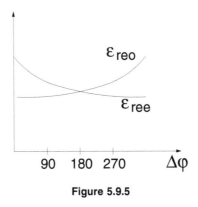

Figure 5.9.5

$$v_m(y) = c_m \operatorname{senh}\left(\left|\beta_m\right|y\right) + q_m \operatorname{senh}\left(\left|\beta_m\right|y\right) \tag{5.9.14}$$

With appropriate boundary conditions it is possible to use only one term of the previous equation, so that Equation 5.9.11 becomes:

$$v_{th}(y,z) = \sum_{m=-\infty}^{m=+\infty} c_m \exp\left(j\beta_m z\right) \operatorname{senh}\left(\left|\beta_m\right|y\right) \tag{5.9.15}$$

for $0 \le y \le h$, and

$$v_{tH}(y,z) = \sum_{m=-\infty}^{m=+\infty} d_m \exp\left(j\beta_m z\right) \operatorname{senh}\left[\left|\beta_m\right|(H-y)\right] \tag{5.9.16}$$

for $h \le y \le H$. In our case, the boundary conditions are the unity of the potential at the interface layer between air and dielectric, i.e.:

$$v_{th}(h,z) \equiv!\ v_{tH}(h,z) \tag{5.9.17}$$

and the difference between the normal components of the electric displacement* vector "D_y^-" and "D_y^+" at the strips is equal to the strip charge "q_s" and zero elsewhere, i.e.:

$$D_y^+ - D_y^- = q_s \qquad \text{with } \left[q_s\right] = \mathrm{Coul}/m^2 \tag{5.9.18}$$

Once the expression of the potential "v" for the structure is found, the problem is to find the coefficients "c_m" and "d_m" of the two previous equations. The problem has been solved by Weiss with a procedure similar to that used in Section 5.4 for the general quasi static study of "cμ." In practice, each hot conductor is divided into "M" substrips, each with charge "q_ℓ" for u.l. The determination of "c_m" and "d_m" permits the evaluation of Green's function "$g(y,z)$" for a substrip. Then, with a system of equations as indicated in Equation 5.4.5, the charge "q_ℓ" for each substrip is found. At this point, the procedure is exactly as explained in Section 5.4 using Equations 5.4.6 through 5.4.10, and finally obtaining the even and odd characteristics for the meander line. It is

* See Appendix A1 to find the definitions of other vector fields.

interesting is to note that a graph of "ε_{ree}" and "ε_{reo}," in Figure 5.9.5 shows that when $\Delta\varphi = \pi$, then $\varepsilon_{ree} \equiv \varepsilon_{reo}$.

The meander line is an example in planar transmission line technology of the general theory for the "slow wave structure,"[28,29] i.e., a structure where the e.m. field has a phase speed lower than the light speed in the vacuum. Another example of a commonly employed slow wave guiding structure is the helix inside a traveling wave tube.[30]* Other types of slow wave structures are possible as shown by some researchers.[31]

REFERENCES

1. Y. Qian, E. Yamashita, Characterization of picosecond pulse crosstalk between coupled microstrip lines with arbitrary conductor width, *IEEE Trans. on MTT*, 1011, Jun/Jul 1993.
2. D. Homentcovschi, R. Oprea, Analytically determined quasi static parameters of shielded or open multiconductor microstrip lines, *IEEE Trans. on MTT*, 18, Jan. 1998.
3. F. Sellberg, Simple determination of all capacitances for a set of parallel microstrip lines, *IEEE Trans. on MTT*, 195, Feb. 1998.
4. M. K. Krage, G. I. Haddad, Characteristics of coupled microstrip transmission lines-I, coupled mode formulation of inhomogeneous lines, *IEEE Trans. on MTT*, 217, April 1970 and Characteristics of coupled microstrip transmission lines-II, evaluation of coupled line parameters, *IEEE Trans on MTT*, 222, April 1970.
5. S. Akhtarzad, T. R. Rowbotham, P. B. Johns, The design of coupled microstrip lines, *IEEE Trans. on MTT*, 486, June 1975.
6. C. Wan, Analytically and accurately determined quasi static parameters of coupled microstrip lines, *IEEE Trans. on MTT*, 75, Jan. 1996.
7. H. J. Carlin, P. P. Civalleri, A coupled line model for dispersion in parallel coupled microstrips, *IEEE Trans. on MTT*, 444, May 1975.
8. W. J. Getsinger, Dispersion of parallel coupled microstrip, *IEEE Trans. on MTT*, 144, Mar. 1973.
9. R. H. Jansen, High speed computation of single and coupled microstrip parameters including dispersion, high order modes, loss and finite strip thickness, *IEEE Trans. on MTT*, 75, Feb. 1978.
10. T. G. Bryant, J. A. Weiss, Parameters of microstrip transmission lines and a coupled pairs of microstrip lines, *IEEE Trans. on MTT*, 1021, Dec. 1968.
11. J. A. Weiss, Microwave propagation on coupled pairs of microstrip transmission lines, *Adv. Microwaves*, 8, 295, 1974.
12. H. J. Carlin, P. P. Civalleri, A coupled line model for dispersion in parallel coupled microstrips, *IEEE Trans. on MTT*, 444, May 1975.
13. M. K. Krage, G. I. Haddad, Frequency dependent characteristics of microstrip transmission lines, *IEEE Trans. on MTT*, 678, Oct. 1972.
14. R. Garg, I. J. Bahl, Characteristics of coupled microstriplines, *IEEE Trans. on MTT*, 700, July 1979 See also, by the same authors, Correction to: Characteristics of coupled microstriplines, *IEEE Trans. on MTT*, 272, March 1980.
15. K. C. Gupta, R. Garg, R. Chada, *Computer Aided Design of Microwave Circuits*, Artech House, Norwood, MA, 1981. This book contains some revisited formulas introduced in the previous reference of R. Garg, I. J. Bahl.
16. T. G. Bryant, J. A. Weiss, Parameters of microstrip transmission lines and a coupled pairs of microstrip lines, *IEEE Trans. on MTT*, 1021, Dec. 1968.
17. W. J. Getsinger, Microstrip dispersion model, *IEEE Trans. on MTT*, 34, Jan. 1973.
18. M. Abramowitz, I. A. Stegun, *Handbook of Mathematical Functions*, Dover, New York, 1970.
19. R. H. Jansen, High speed computation of single and coupled microstrip parameters including dispersion, high order modes, loss and finite strip thickness, *IEEE Trans. on MTT*, 75, Feb. 1978.
20. M. Kirschning, R. H. Jansen, Accurate wide range design equations for the frequency dependent characteristic of parallel coupled microstrip lines, *IEEE Trans. on MTT*, 83, Jan. 1984.
21. H. A. Wheeler, Formulas for the skin effect, *Proc. of the IRE*, 30, 412, 1942.

* Traveling wave tubes are usually abbreviated "TWT."

22. R. Sturdivant, Transmission line conductor loss and the incremental inductance rule, *Microwave J.*, 156, Sept. 1995.

23. M. V. Schneider, B. Glance, W. F. Bodtmann, Microwave and millimeter wave hybrid integrated circuits for radio systems, *Bell System Tech. J.*, 1703, July-Aug. 1969.

24. B. R. Rao, Effect of loss and frequency dispersion on the performance of microstrip directional couplers and coupled line filters, *IEEE Trans. on MTT*, 747, July 1974.

25. E. G. Cristal, Analysis and exact synthesis of cascaded commensurate transmission line C-section all pass networks, *IEEE Trans. on MTT*, 285, June 1966.

26. J. A. Weiss, Dispersion and field analysis of a microstrip meander line slow wave structure, *IEEE Trans. on MTT*, 1194, Dec. 1974.

27. J. A. Weiss, Dispersion and field analysis of a microstrip meander line slow wave structure, *IEEE Trans. on MTT*, 1194, Dec. 1974.

28. S. Ramo, J. R. Whinnery, T. Van Duzer, *Fields and Waves in Communication Electronics*, John Wiley, NY, 1965.

29. A. F. Harvey, Periodic and guiding structures at microwave frequencies, *IRE Trans. on MTT*, 30, Jan. 1960.

30. A. S. Gilmour, Jr., *Principles of Traveling Wave Tubes*, Artech House, Norwood, MA, 1994.

31. Fei-Ran Yang, Yongxi Qian, R. Coccioli, T. Itoh, A novel low loss slow wave microstrip structure, *IEEE MGWL*, 372, Nov. 1998.

CHAPTER 6

Coupled Striplines

6.1 GEOMETRICAL CHARACTERISTICS

Coupled* striplines** are a very important network, employed in many devices like directional couplers, filters, and phase shifters. These devices in stripline configuration are bigger, but in general have better performances than the "$c\mu$" counterpart studied in Chapter 5. Also at lower frequencies they can be used to transmit digital information at a high rate of speed, typically hundreds of MHz, where "ECL" devices are employed. In general, a high density PCB will contain striplines and μstrips as well.

In contrast to "$c\mu$," the present case allows more than one way to couple two striplines, as indicated in Figures 6.1.1 through 6.1.3. Figure 6.1.1 represents a four layer structure, named "broadside coupled stripline" (BCS). In practice, two hot stripline conductors, which can be assumed to belong to two otherwise isolated offset striplines, are set near each other along the widest dimension "w." They share the same ground planes. A particular case of "BCS" is the "offset broadside coupled stripline" or "OBCS," indicated in Figure 6.1.2. In the case of the OBCS, the overlapping of hot conductors is not complete.

Figure 6.1.3 represents the "classical" coupled stripline structure, called the "side coupled striplines" or "SCS." This structure preserves the already known three layers' aspect for the single stripline. For all three configurations it is common practice to indicate with:

a. "w_i," the i-th hot conductor width
b. "s," the hot conductors separation
c. "t," the hot conductors thickness
d. "b," the ground plane spacing

In general, more than two striplines can be e.m. coupled, but we will only study the case indicated in Figure 6.1.1, where only two striplines are e.m. coupled.

Similar to the case of the single stripline, the dielectric material can be a ferrimagnetic one. However, unless otherwise stated we will assume the "cs" to be on a nonferrimagnetic substrate, and will cover this topic in Chapter 8.

In general, there is no particular reason to have different hot conductor widths, and the devices discussed in Chapter 8 that employ "cs" are an example of this assertion. So, the theory in this chapter will assume $w_1 \equiv w_2$, unless otherwise stated. In addition, we will assume that in the coupling region, stripline widths, their spacing, and backplane distance will remain constant, i.e., we will assume a uniform coupling.***

* The concept of e.m. coupling among t.l.s is explained in Chapter 1.
** Coupled μstrip will simply abbreviated with "cs."
*** The general theory of nonuniform coupled lines is explained in Chapter 1. In Chapter 8 we will study stripline devices working on nonuniform coupling.

BROADSIDE COUPLED STRIPLINES

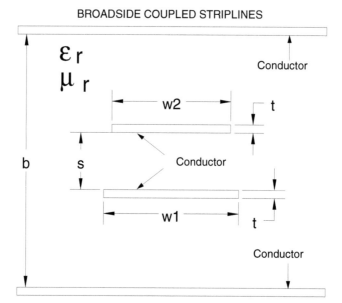

Figure 6.1.1

OFFSET BROADSIDE COUPLED STRIPLINES

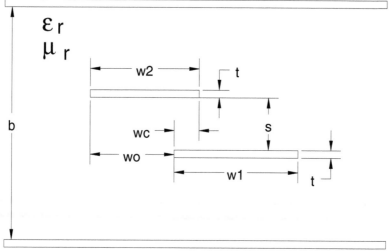

Figure 6.1.2

In this chapter we will assume that coupling between striplines is desired. Crosstalk between coupled striplines has been investigated by researchers Ponchak et al.[1]

6.2 ELECTRIC AND MAGNETIC FIELD LINES

Some electric "e" and magnetic "h" field lines in a defined cross-section and a defined time for the fundamental "TEM" mode in the three configurations of "cs" are indicated in Figures 6.2.1 through 6.2.3. In the fundamental mode each hot conductor is equipotential. Note that as in the general theory of coupled lines, cs supports two independent excitation modes, i.e., the "even" and

SIDE COUPLED STRIPLINES

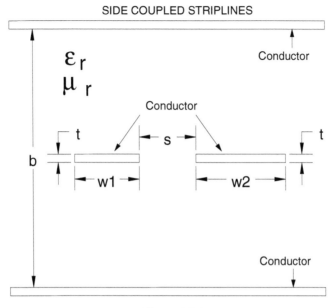

Figure 6.1.3

the "odd" mode.* This fields line comes directly from the local orthogonality between vector "e" and "h." The e.m. field lines disposition is obviously different for each mode.

Due to the homogeneity of the dielectric, even and odd phase constants are practically coincident in a wider bandwidth with respect to the "cμ" case. This characteristic made the "cs" very attractive with respect to the "cμ," especially in high precision and wide bandwidth devices, if "cs" has a bigger size and weight.

"BCS" and "OBCS" are mainly employed in devices where high coupling is required, for instance, inside the range –1 to –6 dB, while "SCS" are useful for a lower coupling.

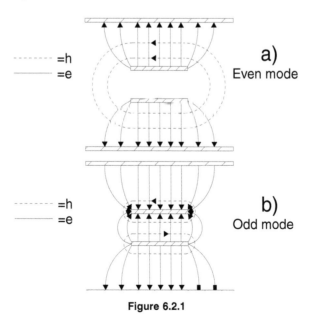

Figure 6.2.1

---

* See Chapter 1 for even and odd excitation mode definitions.

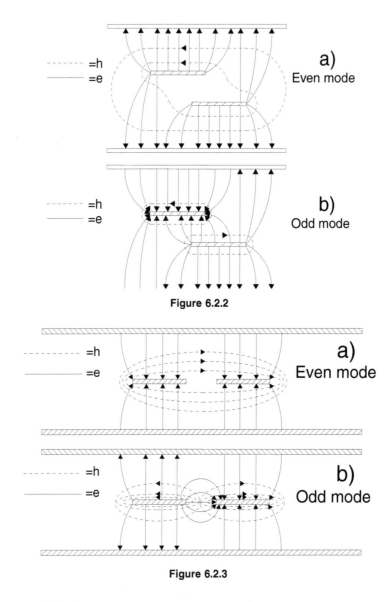

Figure 6.2.2

Figure 6.2.3

6.3 SOLUTION TECHNIQUES FOR THE ELECTROMAGNETIC PROBLEM

Due to the intrinsic ability to support a "TEM" mode, "cs" have been studied with quasi static methods. Of course, it is also possible to use full wave methods. The most employed quasi static methods are:

1. Solution of Laplace's equation[2]
2. Conformal transformation

The second method has been widely used. It is an application of the conformal transformation method, already applied to the single stripline, to the most general case of "cs," i.e., "BCS,"[3] "OBCS,"[4] or "SCS."[5,6,7] In contrast to the simple procedure for the single stripline, now for "even" and "odd" excitation we need to apply a transformation, resulting in quite a lengthy procedure. All the particular analytical steps can be found in the indicated references.

6.4 DESIGN EQUATIONS

All the following formulas are a result of the conformal transformation method applied to the "cs" structure. We will give such formulas for any "cs" we introduced in section 6.1.

6.4.1 SCS

The even "ζ_e" and odd "ζ_o" characteristic impedances for the "SCS" indicated in Figure 6.1.3, with the condition $w_1 \equiv w_2 \perp w$, are given by:[8]

$$\zeta_m = \frac{30\pi}{\sqrt{\varepsilon_r}} \frac{K(p'_m)}{K(p_m)} \qquad \text{with } m = e, o \tag{6.4.1}$$

with:

$$p_e \doteq \tanh(\pi w/2b)\tanh[\pi(w+s)/2b] \tag{6.4.2}$$

$$p_o \doteq \tanh(\pi w/2b)\operatorname{ctgh}[\pi(w+s)/2b] \tag{6.4.3}$$

$$p'_m \doteq (1 - p_m^2) \tag{6.4.4}$$

The operator "$K()$" is the complete elliptic integral of first type, already encountered in this text. This integral has been defined in Appendix A8, together with its closed form formulas with approximated ranges, which can also be used here.

Other used and simpler expressions have been obtained by Cohn[9] using the even "C_e" and odd "C_o" capacitance for each strip, with the condition $w/b \geq 0.35$. We have:

$$\zeta_e = E[w/b + \ln 2/\pi + \ln[1 + \tanh(\pi s/2b)]/\pi]^{-1} \tag{6.4.5}$$

$$\zeta_o = E[w/b + \ln 2/\pi + \ln[1 + \coth(\pi s/2b)]/\pi]^{-1} \tag{6.4.6}$$

where

$$E \doteq 94.15/\sqrt{\varepsilon_r} \tag{6.4.7}$$

These last two expressions can be easily inverted for synthesis purposes as a function of the strip voltage coupling "c_v" for a matched system with system reference impedance "Z_s." The resulting expressions are:

$$s/b = (2/\pi)\operatorname{atgh}(e^{\pi\delta E}) \tag{6.4.8}$$

$$w/b = (E/Z_s)\left(\frac{1-c_v}{1+c_v}\right)^{0.5} - (1/\pi)\ln[2(1+e^{\pi\delta E})] \tag{6.4.9}$$

where "δ" is given by:

$$\delta \overset{\perp}{=} -2c_v \Big/ \Big[Z_s \big(1 - c_v^2 \big)^{0.5} \Big]$$ (6.4.10)

Conductor thickness effect in the even and odd impedances has been evaluated by Cohn.[10] For $w/b \geq 0.35$ and $t/b \leq 0.1$ we have:

$$\zeta_e = \left\{ \frac{1}{\zeta[w/b;\, t/b]} - \frac{C_f[t/b]}{C_f[0]} \left[\frac{1}{\zeta[w/b;\, 0]} - \frac{1}{\zeta_e[w/b;\, 0;\, s/b]} \right] \right\}^{-1}$$ (6.4.11)

and the odd mode impedance for $s \geq 5t$ is:

$$\zeta_o = \left\{ \frac{1}{\zeta[w/b;\, t/b]} + \frac{C_f[t/b]}{C_f[0]} \left[\frac{1}{\zeta_o[w/b;\, 0;\, s/b]} - \frac{1}{\zeta[w/b;\, 0]} \right] \right\}^{-1}$$ (6.4.12)

while for $s \leq 5t$ is:

$$\zeta_o = \left\{ \frac{1}{\zeta_o[w/b;\, 0;\, s/b]} + \frac{1}{\zeta[w/b;\, t/b]} - \frac{1}{\zeta[w/b;\, 0]} - \frac{2\Delta C}{377} + \frac{2t}{377s} \right\}^{-1}$$ (6.4.13)

with "ΔC" given by:

$$\Delta C \overset{\perp}{=} \big(C_f[t/b] - C_f[0] \big) \big/ \varepsilon$$ (6.4.14)

For the previous formulas we have:

1. $\zeta[w/b;\, t/b]$ single stripline impedance, with the same "w," "t," and "b" of the actual structure. It can be evaluated as indicated in Chapter 3.
2. $\zeta[w/b;\, 0]$ single stripline impedance, with the same "w" and "b" of the actual structure, but with t = 0. It can be evaluated as indicated in Chapter 3.
3. $\zeta_m[w/b;\, 0;\, s/b]$ zero thickness "m" mode impedance, with the same "w," "s," and "b" of the actual structure, but with t = 0. They can be evaluated for example as indicated in Equations 6.4.5 and 6.4.6.
4. $C_f[t/b]$ is the single stripline fringing capacitance, with the same "w," "b," and "t" of the actual structure. This expression has been given in Chapter 3, Equation 3.5.12.
5. $C_f[0]$ is the fringing capacitance defined in point 4 above, evaluated with t = 0. We have:

$$C_f(0) = 0.03985\varepsilon_r \qquad \text{pf/cm}$$ (6.4.15)

6.4.2 BCS

These "cs" were first studied by S.B. Cohn,[11] and the geometric structure is indicated in Figure 6.1.1. With the conditions:

1. $w_1 \equiv w_2 \overset{\perp}{=} w$
2. $w/s \geq 0.35$
3. $w/(b - s) \geq 0.35$

the even "ζ_e" and odd "ζ_o" characteristic impedances are given by:

$$\zeta_e = G\big[(w/b)/(1-s/b) + C_{fe}/\varepsilon\big]^{-1} \tag{6.4.16}$$

$$\zeta_o = G\big[(w/b)/(1-s/b) + w/s + C_{fo}/\varepsilon\big]^{-1} \tag{6.4.17}$$

where:

$$G \doteq 188.3/\sqrt{\varepsilon_r} \tag{6.4.18}$$

$$C_{fe}/\varepsilon = 0.4413 + C \tag{6.4.19}$$

$$C_{fo}/\varepsilon = (b/s)C \tag{6.4.20}$$

$$C \doteq \big\{[-(s/b)/(1-s/b)]\ln(s/b) - \ln(1-s/b)\big\}/\pi \tag{6.4.21}$$

The expressions for "ζ_e" and "ζ_o" can easily be inverted for synthesis purposes, as we did before for the "SCS." The resulting expressions are:

$$s/b = \left(\frac{1-c_v}{1+c_v}\right)^{0.5}\left[\left(\frac{1-c_v}{1+c_v}\right)^{0.5} - \frac{0.4413Z_s}{G}\right] \tag{6.4.22}$$

$$w/b = \frac{G(1-s/b)s/b}{Z_s}\left(\frac{1+c_v}{1-c_v}\right)^{0.5} - (1-s/b)C \tag{6.4.23}$$

Conductor thickness effect in the even and odd impedances has been evaluated by Cohn.[12] In this case, he makes use of the total even "C_e" and odd "C_o" capacitances for each strip, which can be obtained from "ζ_e" and "ζ_o" by:*

$$C_m = (\mu\varepsilon)^{0.5}/\zeta_m \tag{6.4.24}$$

Of course, from the previous equation, if the capacitance is known, then we can evaluate the impedance.

For the odd mode, the total odd "C_o" capacitance to be used is:

$$C_o = C_o\big[w/b_0; s/b_0; 0\big] + 4\Delta C[t/s] \tag{6.4.25}$$

with "$\Delta C[t/s]$" and "b_0" given by:

$$\Delta C[t/s] = (\varepsilon/2\pi)\big[(1+t/s)\ln(1+t/s) - (t/s)\ln(t/s)\big] \tag{6.4.26}$$

$$b_0 \doteq b - 2t \tag{6.4.27}$$

---

* Of course, a pure "TEM" propagation mode is assumed. See Chapter 1 on relations among "ζ" and t.l. capacitances.

and "$C_o[w/b_0; s/b_0; 0]$" is the odd capacitance for the zero thickness "BCS" with "w" and "s" of Figure 6.1.1 and ground plane spacing given by 6.4.27. This capacitance is obtainable by the zero thickness "ζ_o" expression given above and using 6.4.24.

For the even mode,
if $t \leq 5s$, then:

$$C_e = C_e\left[w/b; s_e/b; 0\right] \qquad (6.4.28)$$

where:

$$s_e \stackrel{\perp}{=} s + 2t \qquad (6.4.29)$$

and "$C_e[w/b; s_e/b; 0]$" is the even capacitance for the zero thickness "BCS" with "w" and "b" of Figure 6.1.1 and hot conductors spacing given by 6.4.29. This capacitance is obtainable by the zero thickness "ζ_e" expression given above and using 6.4.24;

if $t \geq 5s$ then:

$$C_e = 0.5C\left[w/b; t_e/w\right]$$

where:

$$t_e \stackrel{\perp}{=} s + 2t$$

and "$C[w/b; t_e/b]$" is the capacitance for the single symmetric stripline with thickness "t_e" given by 6.4.29 and same "w" and "b" of Figure 6.1.1. This capacitance is obtainable by the "ζ" expression for a single stripline given in Chapter 3 and using 6.4.24.

6.4.3 OBCS

This structure was first studied by J. P. Shelton.[13] The resulting equations are synthesis oriented, which means that "ζ_o," "ζ_e," "ε_r," and the system reference impedance "Z_s" need to be known, and the dimensions of Figure 6.1.2 are determined. We will define $\rho \stackrel{\perp}{=} \zeta_e/\zeta_o$ and assume the "OBCS" is a matched "cs" for which $\zeta_e\zeta_o = Z_s^2$. Two sets of equations are given, where each dimension is normalized to $b = 1$. To remember this, we will add the subscript "n" to the dimensions whenever necessary. A general constraint is given as:

$$w_n/(1 - s_n) \geq 0.35 \qquad (6.4.30)$$

One set is for tight coupling, where the following additional constraint must be verified:

$$w_c/s \geq 0.7 * \qquad (6.4.31)$$

and the synthesis equations are:

---

* Since this equation is a ratio of two numbers, it is the same whether or not we use the normalized quantities.

$$A \stackrel{\perp}{=} \exp\left[\frac{60\pi^2\left(1-\rho s_n\right)}{Z_s\sqrt{\rho}\sqrt{\varepsilon_r}}\right] \tag{6.4.32}$$

$$B \stackrel{\perp}{=} 0.5\left[A - 2 + \left(A^2 - 4A\right)^{0.5}\right] \tag{6.4.33}$$

$$p \stackrel{\perp}{=} \frac{0.5(B-1)\left(1+s_n\right)+\left\{\left[\left(1+s_n\right)/2\right]^2(B-1)^2+4s_nB\right\}^{0.5}}{2} \tag{6.4.34}$$

$$r \stackrel{\perp}{=} s_n B/p \tag{6.4.35}$$

$$C_{fo} = \left\{\left(1/s_n\right)\ln\left[\frac{pr}{\left(p+s_n\right)\left(1+p\right)\left(r-s_n\right)\left(1-r\right)}\right] - \frac{2}{1-s_n}\ln\left(s_n\right)\right\}\Big/\pi \tag{6.4.36}$$

$$C_o = \frac{120\pi\sqrt{\rho}}{Z_s\sqrt{\varepsilon_r}} \tag{6.4.37}$$

$$w_n = 0.5s_n\left(1-s_n\right)\left(C_o - C_{fo}\right) \tag{6.4.38}$$

$$w_{o,n} = \left\{\left(1+s_n\right)\ln\left(p/r\right)+\left(1-s_n\right)\ln\left[\frac{\left(1+p\right)\left(r-s_n\right)}{\left(s_n+p\right)\left(1-r\right)}\right]\right\}\Big/2\pi \tag{6.4.39}$$

The other set of equations is for loose coupling, where the following constraint, together with that in Equation 6.4.30 must be verified:

$$2w_{o,n}\big/\left(1+s_n\right) \geq 0.85 \tag{6.4.40}$$

and the synthesis equations are:

$$\Delta C \stackrel{\perp}{=} \frac{120\pi(\rho-1)}{Z_s\sqrt{\varepsilon_r}\sqrt{\rho}} \tag{6.4.41}$$

$$k \stackrel{\perp}{=} \left[\exp\left(0.5\pi\Delta C\right)-1\right]^{-1} \tag{6.4.42}$$

$$a \stackrel{\perp}{=} \left[\left(\frac{s_n-k}{s_n+1}\right)^2+k\right]^{0.5}-\frac{s_n-k}{s_n+1} \tag{6.4.43}$$

$$q \stackrel{\perp}{=} k/a \tag{6.4.44}$$

$$C_{fo} = \left(2/\pi\right)\left\{\frac{1}{1+s_n}\ln\left[\frac{1+a}{a\left(1-q\right)}\right]-\frac{1}{1-s_n}\ln\left(q\right)\right\} \tag{6.4.45}$$

$$w_{c,n} = \left[s_n\ln\left(q/a\right)+\left(1-s_n\right)\ln\left(\frac{1-q}{1+a}\right)\right]\Big/\pi \tag{6.4.46}$$

$$C_{fai} = -(2/\pi)\left[\frac{1}{1+s_n}\ln\left(\frac{1-s_n}{2}\right)+\frac{1}{1-s_n}\ln\left(\frac{1+s_n}{2}\right)\right]$$ (6.4.47)

$$w_n = \left(1-s_n\right)^2\left(C_o - C_{fo} - C_{fai}\right)/4$$ (6.4.48)

From this study, a simple equation can be obtained in the case for max coupling, i.e., for $w_o = 0.$* We have:

$$\frac{1-\rho s_n}{\sqrt{\rho}} = \frac{Z_s\sqrt{\varepsilon_r}}{60\pi^2}\ln 4$$ (6.4.49)

However, for $w_o = 0$, the original Cohn's "BCS" equations given above have a greater range of validity, and those equations should be used. Thickness effect on these expressions has not been evaluated, but for $t \leq 0.1s$ it is believed that the error is negligible.

6.5 ATTENUATION

The sources of attenuation in a t.l. are well known and can be reviewed in Chapter 1 for the general case, or in Chapter 3 for the case of stripline, which applies to the "cs" as well.

Magnetic losses will not be evaluated here; this topic will be studied in Chapter 8 and Appendices A5 through A7.

In contrast to the "cμ" case, or other open t.l., in this case radiation losses can be neglected with respect to the other source of attenuation, because "cs" is a closed structure.

Assuming a pure "TEM" mode in the stripline, the evaluation of conductor losses can be performed applying Wheeler's[14,15] incremental inductance rule. This way of proceeding to evaluate the t.l. attenuation has been used throughout this text, and for this reason we will only give the formulas that are characteristics for the "cs" structure here. For the other concepts and common formulas it is possible, for example, to see the losses evaluation for the stripline case in Chapter 3. The following formulae apply equally to all three types of "cs" we have introduced. For example, we will refer to the "SCS" structure reported in Figure 6.5.1.

Indicating with a subscript "m" the generic mode m = e,o, the additional inductance "$L_{a,m}$" and resistance "$R_{a,m}$" are given by:

$$L_{a,m} = \sum_{j=1}^{j=10}\left(L_{i,m}\right)_j = \frac{1}{2\mu_0}\sum_{j=1}^{j=10}p_j\left(\mu_c\right)_j\frac{\partial L_m}{\partial n_j}$$ (6.5.1)

$$R_{a,m} = \sum_{j=1}^{j=10}\left(R_{i,m}\right)_j = \frac{1}{\mu_0}\sum_{j=1}^{j=10}\left(R_s\right)_j\frac{\partial L_m}{\partial n_j}$$ (6.5.2)

where:

a. "$L_{i,m}$" is the "incremental inductance" per u.l.
b. "$R_{i,m}$" is the "incremental resistance" per u.l.
c. "p" is the "penetration depth," [u.l.]
d. "μ_c" is the conductor absolute permeability

* Of course, for this limited case OBCS $\equiv$ BCS.

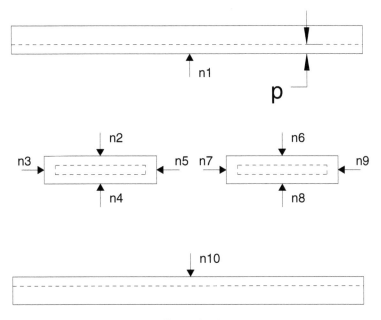

Figure 6.5.1

e. "R_s" is the conductor "sheet resistance," [Ω/square]*

The conductor attenuation coefficient "α_{cm}"** is defined as:

$$\alpha_{cm} = W_{cm}/2W_{tm} \tag{6.5.3}$$

where "W_{cm}" and "W_{tm}" are, respectively, the mean power dissipated in the conductor and the mean transmitted power, given by:

$$W_c = R_{am}\left|i_m\right|^2, \quad W_{tm} = \zeta_m\left|i_m\right|^2 \tag{6.5.4}$$

Consequently, the conductor attenuation constant does not depend on the additional inductance "L_a." Using 6.5.2 and 6.5.4, Equation 6.5.3 becomes:

$$\alpha_{cm} = \frac{1}{2\mu_0\zeta_m} \sum_{j=1}^{j=10} (R_s)_j \frac{\partial L_m}{\partial n_j} \tag{6.5.5}$$

Since we know*** that for a "TEM" t.l. we have:

$$L_m = \zeta_m/v = \zeta_m\left(\mu_r\varepsilon_r\right)^{0.5}/v_0 \tag{6.5.6}$$

then:

---

* See Appendix A2 for measurement unit of "conductor resistance."
** We are assuming a longitudinal variation of conductor attenuation with $e^{-\alpha z}$. See Chapter 1 for the fundamental theory of transmission lines.
*** See Chapter 1 for relations among t.l. characteristics.

$$\frac{\partial L_m}{\partial n} = \frac{(\mu_r \varepsilon_r)^{0.5}}{v_0} \frac{\partial \zeta_m}{\partial n} \tag{6.5.7}$$

Observing that $\mu_0 v_0 = \zeta_0 \equiv 120\pi$ and using Equation 6.5.7, Equation 6.5.5 becomes:

$$\alpha_{cm} = \frac{\sqrt{\mu_r \varepsilon_r}}{\zeta_m 240\pi} \sum_{j=1}^{j=10} (R_s)_j \frac{\partial \zeta_m}{\partial n_j} \tag{6.5.8}$$

The "ζ_m" is a function of "w," "t," "s," and "b," as was shown in the previous section, and so the derivative "$\partial \zeta_m / \partial n$" is:

$$\frac{\partial \zeta_m}{\partial n} = \frac{1}{dn} \left(\frac{\partial \zeta_m}{\partial w} dw + \frac{\partial \zeta_m}{\partial b} db + \frac{\partial \zeta_m}{\partial t} dt + \frac{\partial \zeta_m}{\partial s} ds \right) \tag{6.5.9}$$

From Figure 6.5.1 we observe that:

$$dw = -2dn, \quad db = 2dn, \quad dt = -2dn, \quad ds = 2dn \tag{6.5.10}$$

and 6.5.8 becomes:*

$$\alpha_{cm} = \frac{R_s \sqrt{\mu_r \varepsilon_r}}{\zeta_m 120\pi} \left(-\frac{\partial \zeta_m}{\partial w} + \frac{\partial \zeta_m}{\partial b} - \frac{\partial \zeta_m}{\partial t} + \frac{\partial \zeta_m}{\partial s} \right) \tag{6.5.11}$$

Of course, the value given by "α_{cm}" is in neper/meter.**

Simple closed form equations for "α_{cm}" are not available, and when needed, the derivative indicated in Equation 6.5.11 is numerically evaluated using a computer. In any case, it has been verified that "α_{cm}" is always greater then "α_{ce}," for a value near 10% the value of "α" in dB for "BCS" and "OBCS" and near 5% for "SCS."

Concerning the dielectric losses, we can use the general formula for dielectric losses in a "TEM" t.l.,*** resulting in:

$$\alpha_d = \frac{\pi \sqrt{\varepsilon_r} \tan \delta}{\lambda_0} [\text{Neper/u.1}] \quad \text{or} \quad \alpha_d = \frac{27.29 \sqrt{\varepsilon_r} \tan \delta}{\lambda_0} [\text{dB/u.1}] \tag{6.5.12}$$

where "tanδ" is the substrate "tangent delta" **** for g = 0 in the substrate. Of course, the quantity "ε_r" is relative to the real part of the substrate dielectric constant (see Chapter 1 for definition of permittivity). In general, for the typical substrates employed in "MIC," stripline circuit conductor losses are predominant.

* We assume that the top and bottom conductors have the same surface resistivity, as is usually the case.
** See Chapter 1 for attenuation constants dimensions.
*** See Appendix A2 for the procedure to have the attenuation constant "α_d."
**** See Chapter 1 for "tanδ" definition and conversion between Neper and dB.

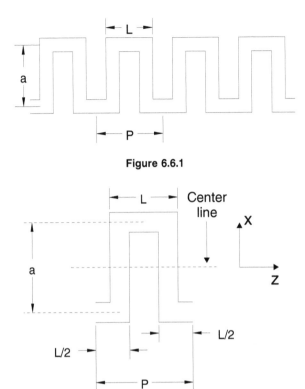

Figure 6.6.1

Figure 6.6.2

6.6 A PARTICULAR COUPLED STRIPLINE STRUCTURE: THE MEANDER LINE

A particular case of "cs" is the "meander line" structure indicated in Figure 6.6.1. This structure is employed in phase shifters, as we will see in Chapters 7 and 8 when we will study these and other devices. The meander line can be thought of as being composed of a repetition of a single cell, still named "meander line" or "C-section," indicated in Figure 6.6.2. In practice, the "C-section" is built connecting the two nearest ports of two "cs" with a short circuit. This cell is characterized by:

a. A width "a" transverse to the longitudinal axis "z," i.e., the mean value of the coupling length
b. A spatial repetition period "P" along "z"
c. A phase shift $\Delta\varphi$ along "z," per unit cell "P." This is a consequence of point a above.

We have already studied this network in Chapter 5, for the case of "cμ." Of course, with a little modification it is possible to use the theory we used in that chapter. However, here we will study this network using a theory directly applicable to the "cs" case, created by researchers J. T. Bolljahn and G. L. Matthaei.[16]

A typical construction of a meander line in stripline technology employs "SCS," as indicated in Figure 6.6.3. This structure can be thought of as a particular case of an array of coupled t.l., which, for the moment, we will assume to be of infinite extent. Hence, the analytical model is indicated in Figure 6.6.4. This is an infinite array of coupled lines, each one of characteristic impedance "ζ" and fed by a current generator* with modulus "I." The phase shift along the dimension "a," in the coupled case, will be indicated with "$\Delta\varphi$." For this reason, each "cg" is phase shifted by "$\Delta\varphi$" passing from one extreme "d" to "u."

---

* In the following, the current generators will be simply indicated with "cg."

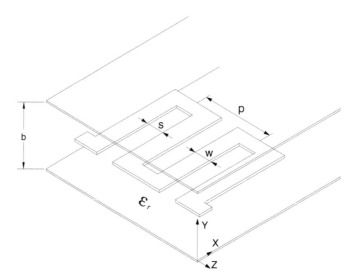

Figure 6.6.3

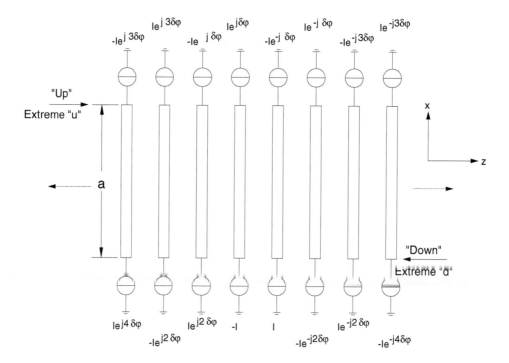

Figure 6.6.4

Since a "cg" is an open circuit for the line, these are open terminated. From the general theory of t.l.,* the voltage at the extreme "d" of a generic t.l. of position "p" in the array is that seen in 6.6.1, due to the presence there of only the generator,

$$V_{p,0;d} = I j \zeta \cot g (\beta a) \overset{\pm}{=} -I j \zeta \cot g (\theta) \qquad (6.6.1)$$

---

* See Chapter 1 for the foundations of the general theory of t.l.

"β" is the phase constant of a signal along a line of Figure 6.6.4, considered as uncoupled, and βa ≐ θ. So, in the previous equation, the subscripts "p," "0," and "d," respectively, denote a generic line "p," the fact that only a generator is present in the array, and that it is connected just to the line "p" at the extreme "d."

Now, let us define the coupling voltage coefficient "$c_{p,m}$" as the voltage at the line "p" when only a generator exists and it is connected to line "m," i.e.:

$$c_{p,m} = V_p/V_m \tag{6.6.2}$$

Note that for the complete geometric symmetry of the network, the coupling coefficients are so that:

$$C_{p,m} \equiv C_{m,p} \quad \text{and} \quad C_{p,p-k} \equiv C_{p,p+k}$$

Due to the linearity of the network, the total voltage at the extreme "d" of the line "p" is due to the following induced voltages:

1. "$V_{p,m;d+}$" of the generators at the extremes "d" of the lines at the right of the line "p":

$$V_{p.m:d+} = \sum_{m=p+1}^{m=\infty} c_{p.m} V_{m.0:d} \tag{6.6.3}$$

2. "$V_{p,m;d-}$" of the generators at the extremes "d" of the lines at the left of the line "p":

$$V_{p.m:d-} = \sum_{m=-\infty}^{m=p-1} c_{p.m} V_{m.0:d} \tag{6.6.4}$$

3. "$V_{p,0;du}$" of the generator at the extreme "u" of the line "p." From the forward transmission matrix of a t.l.:*

$$V_{p.0:du} = -Tj'_s \cos\theta c 0 \tag{6.6.5}$$

4. "$V_{p,m;du+}$" of the generators at the extremes "u" of the lines at the right of the line "p"

$$V_{p.m:du+} = \sum_{m=p+1}^{m=\infty} c_{p.m} V_{m.0:du+} \tag{6.6.6}$$

5. induced voltage "$V_{p,m;du-}$" by the generators at the extremes "u" of the lines at the left of the line "p"

$$V_{p.m:du-} = \sum_{m=-\infty}^{m=p-1} c_{p.m} V_{m.0:du-} \tag{6.6.7}$$

---

* See Chapter 1 for the transmission matrix of a t.l.

These relationships have to be evaluated with the proper phases, since, due to the phase shift along the length "A," each "cg" is offset by "$\Delta\varphi$" in phase. So, the total voltage "$V_{p,d}$" at the extreme "d" of a generic line "p" is given by:

$$V_{p,d} = V_{p,0;d}\left[1 - c_{p,p+1}e^{-j2\Delta\varphi} + c_{p,p+2}e^{-j2\Delta\varphi} - c_{p,p+3}e^{-j4\Delta\varphi} + \dots(\text{down}-\text{right})\right.$$

$$- c_{p,p-1} + c_{p,p-2}e^{j2\Delta\varphi} - c_{p,p-3}e^{j2\Delta\varphi} + c_{p,p-4}e^{j4\Delta\varphi} - \dots(\text{down}-\text{left})\Big]$$

$$+ V_{p,0;du}\left[-e^{j\Delta\varphi} + c_{p,p+1}e^{-j\Delta\varphi} - c_{p,p+2}e^{-j3\Delta\varphi} + \dots(\text{up}-\text{right})\right.$$

$$+ c_{p,p-1}e^{j\Delta\varphi} - c_{p,p-2}e^{j\Delta\varphi} + c_{p,p-3}e^{j3\Delta\varphi} - \dots(\text{up}-\text{left})\Big]$$

(6.6.8)

Now the meander line can be obtained from Figure 6.6.4 practically as indicated in Figure 6.6.5, i.e., alternatively connecting two extremes in short circuit. For example, it means that the potentials "$V_{p,d}$" and "$V_{p-1,d}$" must be equal. Using this condition and the previous equation it is possible to have an important relationship between "$\Delta\varphi$" and "θ," i.e.:

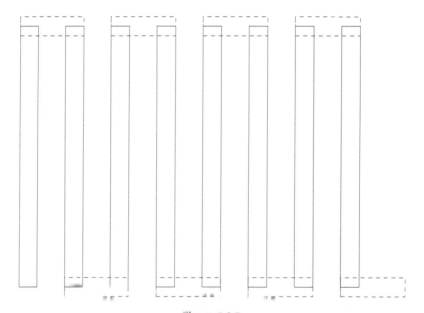

Figure 6.6.5

$$\cos(\theta) = N(\Delta\varphi)/D(\Delta\varphi) \tag{6.6.9}$$

where:

$$N(\Delta\varphi) \overset{+}{=} \left(1 - 2c_{p,p+1} + c_{p,p+2}\right)\cos\Delta\varphi + \left(c_{p,p+2} - 2c_{p,p+3} + c_{p,p+4}\right)\cos 3\Delta\varphi + \dots \tag{6.6.10}$$

$$D(\Delta\varphi) \overset{+}{=} 1 - c_{p,p+1} - \left(c_{p,p+1} - 2c_{p,p+2} + c_{p,p+3}\right)\cos 2\Delta\varphi - \left(c_{p,p+3} - 2c_{p,p+4} + c_{p,p+5}\right)\cos 4\Delta\varphi$$

(6.6.11)

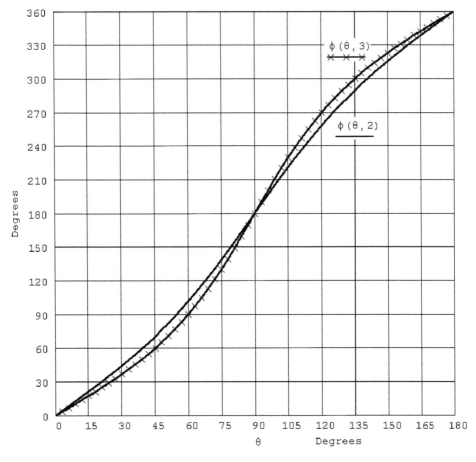

Figure 6.6.6

From equation 6.6.9 we can obtain the phase shift "φ" for a "C-section," as a function of only a coupling value "c" between two coupled lines whose graph is depicted in Figure 6.6.6. Here we have represented only two values for the coupling. This is a well-known graph for the phase shifter designer.*

The input impedance evaluation can be done with the network indicated in Figure 6.6.7. Here, the infinite array of coupled lines is limited at one side, which we will consider as the input and indicate with subscript "1; d." Of course, a new network does not satisfy the original requirements of an infinite number of coupled lines. However, it has been verified in practice that the coupling coefficient after the third or fourth line is so small as to be neglected in most practical cases of phase shifters.** So, if the meander line is longer than nearly ten coupled lines, it can be considered as an infinity of coupled lines at a distance from the input or output wider than the three nearest lines. In Figure 6.6.7, in this case to simplify the notation, we have named the bottom and upper "cg" as "$I_{1;d}$," "$I_{2;d}$," ... and "$I_{1;u}$," "$I_{2;u}$," ... respectively. Note that due to the connections among lines, just to have a phase shifter, we can write:

---

* Stripline phase shifter design is studied in Chapter 8.
** The meander line is mainly used in phase shifters, as we will show in Chapter 8.

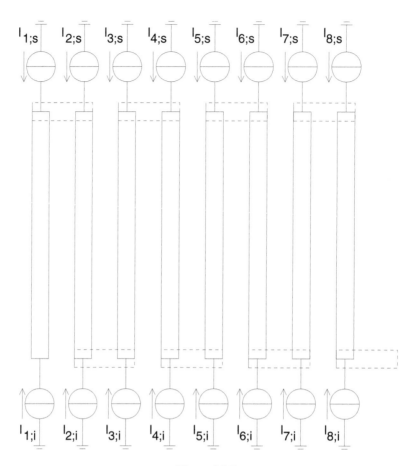

Figure 6.6.7

$$V_{p;u} = !V_{p+1;u} \quad \text{and} \quad V_{p;u} = ! - I_{p+1;u} \quad \text{for } p = 1,3,5... \tag{6.6.12}$$

$$V_{p;d} = !V_{p+1;d} \quad \text{and} \quad V_{p;d} = ! - I_{p+1;d} \quad \text{for } p = 2,4,6... \tag{6.6.13}$$

Then, the input impedance "Z_i" is found by the definition equation $Z_i = V_{1;d}/I_{1;d}$, with the application of the previous two conditions. Of course, voltages and current can be found in a manner similar to that used to obtain Equation 6.6.8. Considering only five coupled lines significant and omitting the subscript "d" for simplicity, the input impedance is:

$$Z_i = \zeta \left(1 - c_{1,2} - c_{1,3} + c_{1,4}\right) \left(\frac{1 - c_{1,2}}{1 - 2c_{1,2} + c_{1,3}} + \frac{-c_{1,3} + c_{1,4} + c_{1,5} - c_{1,6}}{1 - 2c_{1,2} + 2c_{1,4} - c_{1,5}} \right) \tag{6.6.14}$$

This equation is general, and is applicable to any "TEM" t.l. in a meander line structure. The final expression is, however, dependent on the technology employed, since the coupling coefficients are also.

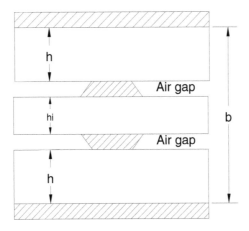

Figure 6.7.1

6.7 PRACTICAL CONSIDERATIONS

As with any other planar t.l., "cs" are also affected by the undercut,* which needs to be compensated for when high coupling values are required. As a result, the hot conductors are, in general, of trapezoidal shape.

"SCS" is a three layer t.l. structure, obtained by overlaying two separate "PCBs;" one "PCB" has only the ground layer, while the other "PCB" has one ground layer and one layer of desired tracks. From this technological point of view it is equal to the single stripline.

"BCS" or "OBCS" is a four layer structure. It is usually built with a sandwich of three "PCBs." The two external "PCBs" have only the ground planes, while the internal "PCB" has the required tracks. This situation is indicated in Figure 6.7.1. The dishomogeneity introduced by the air gap can often be neglected, since typical values of air gap when copper** conductors are employed are inside the range 20 to 40μm, and 3 to 8μm when gold conductors are used. In addition, for μwave devices, the air gap tends to decrease departing from the hot conductor, since the typically employed dielectrics are of a soft type.

"Cs" are more insensitive than "cμ" to lateral ground planes of a metallic enclosure since the e.m. field is strongly contained near the center conductor and the top–bottom ground planes.

REFERENCES

1. G. E. Ponchak, D. Chen, J. G. Yook, L. P. B. Katehi, Characterization of plated via hole fences for isolation between stripline circuits in LTCC packages, *IEEE MTT Int. Symp.*, 3, 1831, 1998.
2. J. L. Allen, M. V. Estes, Broadside coupled strips in a layered dielectric medium, *IEEE Trans. on MTT*, 662, Oct. 1972.
3. S. B. Cohn, Characteristic impedances of broadside coupled strip transmission line, *IRE Trans. on MTT*, 633, Nov. 1960.
4. J. P. Shelton, Jr., Impedances of offset parallel coupled strip transmission line, *IEEE Trans. on MTT*, 7, Jan. 1966.
5. S. B. Cohn, Shielded coupled strip transmission line, *IRE Trans. on MTT*, 29, Oct. 1955.
6. W. J. Getsinger, Coupled rectangular bars between parallel plates, *IRE Trans. on MTT*, 65, Jan. 1962.

---

* See Chapter 1 for undercut phenomenon.
** In practice, at μwave frequencies, copper conductors are gold plated.

7. J. D. Horgan, Coupled strip transmission lines with rectangular inner conductors, *IRE Trans. on MTT*, 92, April 1957.

8. S. B. Cohn, Shielded coupled strip transmission line, *IRE Trans. on MTT*, 29, Oct. 1955.

9. S. B. Cohn, Shielded coupled strip transmission line, *IRE Trans. on MTT*, 29, Oct. 1955.

10. S. B. Cohn, Shielded coupled strip transmission line, *IRE Trans. on MTT*, 29, Oct. 1955.

11. S. B. Cohn, Characteristic impedances of broadside coupled strip transmission line, *IRE Trans. on MTT*, 633, Nov. 1960.

12. S. B. Cohn, Thickness corrections for capacitive obstacles and strip conductors, *IRE Trans. on MTT*, 638, Nov. 1960.

13. J. P. Shelton, Jr., Impedances of offset parallel coupled strip transmission line, *IEEE Trans. on MTT*, 7, Jan. 1966.

14. H. A. Wheeler, Formulas for the skin effect, *Proc. of the IRE*, 30, 412, 1942.

15. R. Sturdivant, Transmission line conductor loss and the incremental inductance rule, *Microwave J.*, 156, Sept. 1995.

16. J. T. Bolljahn and G. L. Matthaei, A study of the phase and filter properties of arrays of parallel conductors between ground planes, *Proc. of the IRE*, 299, March 1962.

CHAPTER 7

Microstrip Devices

7.1 SIMPLE TWO PORT NETWORKS

Many lumped networks used in electronic fields have their microstrip technology counterparts, usually called distributed elements. For instance, inductors, capacitors, filters, and tranformers may be built with microstrips. It is very important to note that the greatest difference between a lumped network and its distributed counterpart is that a great number of the latter have the so-called "returns" while the former do not. The "returns" are frequency bands where the behavior of the network is quite similar to that obtained in the desired bandwidth. Consequently, caution must be used when employing a distributed network. In the course of the book we will show networks that posses returns.

The choice to build or not build these distributed devices depends on frequency value and the available space. Using the frequency as the technology selection criteria, we may say that below two or three hundred MHz, the lumped devices are preferred. From near 300 MHz to one or one and a half GHz, both technologies may be used, but most commonly the inductive elements are substituted with microstrip counterparts. From near one and a half GHz and above, all the lumped elements indicated are realized with distributed elements, with the capacitors as the only lumped elements that may still be used up to 4 or 5 GHz.

If we use the board space as the technology selection criteria, we may say that below one GHz, the lumped devices are preferred. We may easily recognize this assumption with an example. Let us suppose we want to realize an impedance transformer from 50 Ω to 100 Ω. As we know from the transmission line theory studied in Chapter 1, with the coordinate origin in the load "Z_ℓ" position, if we move toward the generator along a distance $(2n + 1)\lambda/4$, the following relation holds:

$$Z_\ell Z\left[(2n + 1)\lambda/4\right] = \zeta_t^2 \tag{7.1.1}$$

where "λ" is the signal wavelength in the microstrip, "n" is an integer number, and "ζ_t" is the impedance of the microstrip connected between the load and the generator. For n = 1, as it is usually done, we have $\zeta_t = 70.71 \ \Omega$. Using a signal at f = 100 MHz and a "FR4" substrate with $\varepsilon_r = 4.7$, as usually employed at these frequencies, the "$\lambda/4$" transformer is 42.19 cm long, a dimension that cannot be considered for today's small electronic circuits.

A general characteristic of microstrip elements is the higher thermal stability when compared to the lumped devices. For instance, a typical lumped inductor has an inductance variation with temperature near 100 ppm/°C at least, while for a microstrip inductor this value is restricted to near 5 to 10 ppm/°C, usually depending on the dielectric employed.

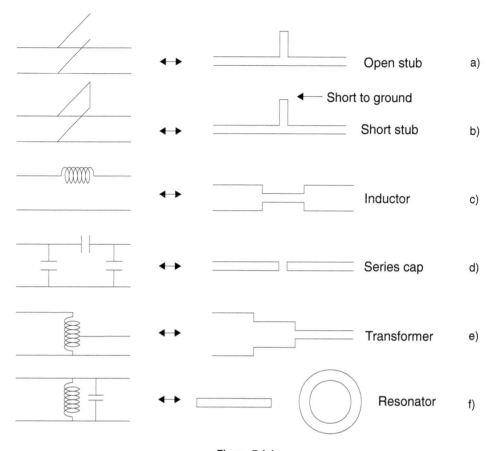

Figure 7.1.1

Let us start by showing which kind of networks are usually built with microstrip technology. Figures 7.1.1 and 7.1.2 represent some lumped networks, or the ideal network, and their microstrip counterparts. To simplify the drawings, we have represented the microstrips only with the hot conductor* and from a top view, without indicating the substrate. It is very important to remind the reader that we studied more exact correspondences between microstrip discontinuities and the equivalent lumped networks in Chapter 4. So, what we are going to explain in this section is only a simple representation of typical networks that can be built with microstrips, while the complete equivalent networks may be found in Chapter 4.

We can now begin to speak about the single networks that appear in the figures.

a. Stubs — Figure 7.1.1a comes directly from the transmission line theory. Once the matching stub impedance is defined, usually using the Smith chart,** it may be realized working on the microstrip hot conductor "w" and/or substrate height "h,"*** since the dielectric constant "ε_r" of the substrate cannot be easily changed, as desired, inside the same substrate. In the microwave region, "h" is also very difficult to change as desired, since the typical values are 128, 254, 508, and 635 micrometers, and consequently the change in "h" is always discontinuous, a situation that should be avoided in the microwave region. The height "h" can easily be changed in RF boards as indicated in Figure 7.1.3 where the number of layers may be a multiple of two. In Figure 7.1.3

---

* See Chapter 2 for definitions on microstrips.
** See Chapter I for definition and use of the Smith chart.
*** We assume the reader knows the definitions of microstrips, as indicated in Chapter 2.

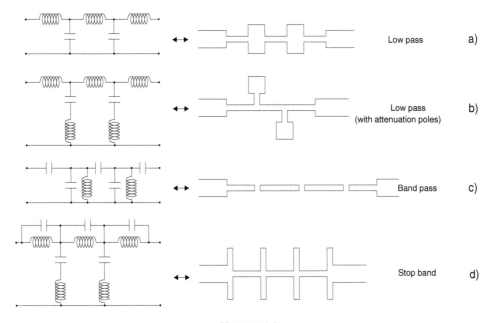

Figure 7.1.2

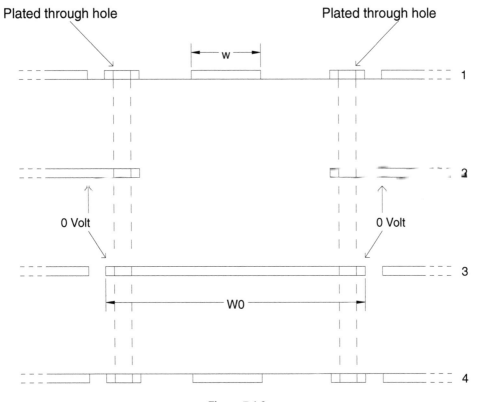

Figure 7.1.3

we have assumed that layer "2" is the ground for the whole board. If the impedance value required by the stub is too high to be realized with ground to layer "2," then the ground can be moved to layer "3" using metallized holes that connect layer "2" to a ground strip in layer "3," which works as a cold conductor for the microstrip stub. The width "w0" of the cold conductor should not be lower than 3w* to avoid error in the realization of the impedance values.

What was said about the open stub is equally applicable to the short stub indicated in Figure 7.1.1b. The connection to ground is realized with a metallized hole, both in RF and in microwave printed circuit board and, in this last case, the connection to ground is simply called "via."

An interesting result can be extracted from the previous discussion if we remember from Chapter 1 that the impedance "Z_0" of an open stub with length "ℓ" is:

$$Z_0 = -j\zeta_t \cot g(\beta \ell) \tag{7.1.2}$$

while the impedance "Z_s" of a short circuited stub of the same length is:

$$Z_s = -j\zeta_t \operatorname{tg}(\beta \ell) \tag{7.1.3}$$

where:

1. "β" is the phase constant along the microstrip, i.e.:

$$\beta = \frac{2\pi f}{c}\left(\varepsilon_{re}\right)^{0,5} \tag{7.1.4}$$

where "f" is the signal frequency, "c" the speed of light, and "ε_{re}" the effective** dielectric constant of the microstrip.

2. "ζ_t" is the impedance of the microstrip realizing the stub.

Note that since "Z_0" and "Z_s" are only imaginary, they are reactances. We know that the reactance "X_ℓ" of an inductor of inductance "L" is $X_\ell = \omega L$, and the reactance "X_c" of a capacitor of capacitance "C" is $X_c = -1/\omega C$. Comparing these formulas with 7.1.2 and 7.1.3 we may conclude that if $\beta \ell < \pi/2$ then the open stub is equivalent to a capacitor of capacitance:

$$C_o = \left[\zeta_t \omega \cot g(\beta \ell)\right]^{(-1)}$$

while the short stub is equivalent to an inductor of inductance:

$$L_s = \left[\zeta_t \operatorname{tg}(\beta \ell)\right]/\omega$$

Note that "C_o" and "L_s" are functions of frequency, while true ideal capacitors and inductors are not. If $\beta \ell \ll \pi/2$, the variability with frequency can quite often be tolerated and, in many matching networks, even desired. Note that to have high values of "L_s" we need high values of "θ," while the contrary holds when we need high values of "C_o." High values of "θ" mean small microstrip width "w," and/or high values of "h," and/or small values of "ε_r." If we assume the small available value of "ε_r," as we said in the previous item a, "h" is usually limited in its range of variability. The same happens for "w," which is limited by the technology used. In RF boards,

* See Chapter 2 for proper dimensioning of microstrip lines.
** See Chapter 2 for microstrip parameter definitions.

the minimum reliable width "w" is near 100 μm, while in microwave circuits the minimum value of "w" is near 10 μm.

b. *Series INDUCTORS* — This situation is indicated in Figure 7.1.1c. The existence of a series inductor may be understood if we approximate the "qTEM" fundamental mode of the microstrip* with a pure "TEM." In this case the impedance "ζ" of the microstrip is:

$$\zeta = (L/C)^{(1/2)} \tag{7.1.5}$$

If we roughly evaluate the capacitance "C" of the microstrip using the known "parallel plate formula," i.e.:

$$C = \varepsilon_r \varepsilon_{re} \frac{\ell w}{h} \tag{7.1.6}$$

once we have defined the impedance realized for the microstrip, we can easily extract from the previous formula the equivalent inductance, that is:

$$L = \zeta^2 C \tag{7.1.7}$$

Note that high values of "L" correspond to realized high values of "ζ," as needed in the previous item a to have a high value of shunt inductance. So, the same limitations on "h" and "w" hold in this case.

c. *Series CAPACITORS* — This case is represented in Figure 7.1.1d. Since this situation is exactly what we have studied in Chapter 4, Section 4.2, we send the reader to that chapter for the study of this layout example.

d. *TRANSFORMERS* — This situation is represented in Figure 7.1.1e. Note that a more correct name for this network should be "autotransformer," since no DC isolation is performed between input and output of the microstrip network. Nevertheless, this network is called "transformer" or "λ/4 transformer." The general theory for this device is treated in Chapter 1, so we send the reader there to see how these networks may be synthesized.

e. *RESONATORS* — As represented in Figure 7.1.1f, any piece of microstrip of length "ℓ" can be evaluated as a resonator, that is, like a system that is able to have oscillation modes inside it. Microstrip resonators are usually built with straight lines or circular rings.
For the straight resonators, the resonance is possible when the following condition holds:

$$\beta \ell = s\pi \tag{7.1.8}$$

with s = 1,2,3,... and "β" defined as in 7.1.4. Note that the previous condition is true both for open end resonators and short circuited ends. Once we have defined the frequency of the signal to which the microstrip straight resonator has to resonate, the length of the microstrip will be:

$$\ell = s \frac{c}{2f \left(\varepsilon_{re}\right)^{0.5}} \tag{7.1.9}$$

* See Appendix A2 for propagation mode definition and Chapter 2 for application to microstrips.

An interesting observation can be made if we assume a given length "ℓ" of the microstrip and try to obtain the frequency where it resonates. This can easily be done if we extract "f" from the previous equation, obtaining:

$$\ell = s\frac{c}{2\ell\left(\varepsilon_{re}\right)^{0.5}}$$

(7.1.10)

From this relation we may observe that a given length "ℓ" of microstrip has infinite frequencies of resonance, since "s" is an integer number greater than or equal to one. Of course, as studied in Chapter 3, if the operating frequency increases, then the radiation effects in the microstrip also increase, so that the quality factor* of the resonator decreases. For this reason, above near 40 GHz, microstrip resonators are seldom used.

For the ring resonator, the condition for resonance is:

$$\beta 2\pi r_{m} = s2\pi$$

(7.1.11)

where "r_{m}" is the medium radius of the ring. Once the frequency of the signal to which the microstrip ring resonator has to resonate is defined, the medium radius of the ring will be:

$$r_{m} = s\frac{c}{2\pi f\left(\varepsilon_{re}\right)^{0.5}}$$

(7.1.12)

Of course, the ring resonator also possesses infinite frequencies of resonance.

Microstrip straight resonators are used in filters, as we will see in item f below, or to measure the dielectric constant of the substrate. In this last case, the ring resonator that is not affected by the end discontinuities is preferred.** To evaluate the effective dielectric constant using a ring resonator, it is enough to measure an s-nth resonance frequency "$f_{r,s}$" and extract "$\varepsilon_{re,s}$" by the previous formula, obtaining:

$$\varepsilon_{re,s} = \left[\frac{sc}{2\pi r_{m}f_{r,s}}\right]^{2}$$

(7.1.13)

The measure of the resonance frequency is usually made using a network analyzer coupled to the ring resonator with two 50Ω microstrip lines, which couple to the resonator through gaps.***

Of particular importance is the "Q" of the resonators. This value is dependent on its geometry, on the losses of the dielectric used as substrate, and on the losses in the conductor. If we indicate with "Q_c" the quality factor of the conductor, with "Q_d" the quality factor of the dielectric, and with "Q_r" the quality factor of the resonator, the total quality factor "Q_0" fulfills the following relation:

$$Q_0^{(-1)} = Q_c^{(-1)} + Q_d^{(-1)} + Q_r^{(-1)}$$

(7.1.14)

The expression of "Q_r" is dependent on the geometry of the resonator, and this argument has been studied by many researchers.[1,2,3,4] For the straight resonator with open ends and long $\lambda/2$, E. J. Denlinger[5] gives the following formula for "Q_r:"

* See Appendix A4 for definition of the "quality factor" of a resonator.
** See Chapter 4 for microstrip discontinuities.
*** See Chapter 4 for microstrip gap.

$$Q_r \overset{\perp}{=} Q_{r0} = \frac{\zeta}{240\pi\rho_{\epsilon 0}\left(h/\lambda_0\right)^2} \qquad (7.1.15)$$

where:

$$\rho_{\epsilon 0} = \frac{\epsilon_{re}+1}{\epsilon_{re}} - \frac{\left(\epsilon_{re}-1\right)^2}{2\epsilon_{re}^{(1.5)}} \ln\left[\frac{\epsilon_{re}^{(0.5)}+1}{\epsilon_{re}^{(0.5)}-1}\right] \qquad (7.1.16)$$

and "λ_0" is the free space wavelength.

For the case of a straight resonator with short circuited ends and long $\lambda/2$, we have:

$$Q_r \overset{\perp}{=} Q_{rs} = \frac{\zeta}{240\pi\rho_{es}\left(h/\lambda_0\right)^2} \qquad (7.1.17)$$

where:

$$\rho_{es} = 3 - \frac{1}{\epsilon_{re}} - \left(3+\frac{1}{\epsilon_{re}}\right)\frac{\epsilon_{re}-1}{2\epsilon_{re}^{(0.5)}} \ln\left[\frac{\epsilon_{re}^{(0.5)}+1}{\epsilon_{re}^{(0.5)}-1}\right] \qquad (7.1.18)$$

For the ring resonator, L. J. Van der Pauw[6] gives the following expression for "Q_r":

$$Q_r = \frac{\zeta\epsilon_{re}}{120\pi^3\left(h/\lambda_0\right)^2\left[1-4/\left(3\epsilon_r\right)+8/\left(15\epsilon_r^2\right)\right]} \qquad (7.1.19)$$

The values of "Q_c" and "Q_d" do not, of course, depend on the geometry of the resonators, and they are given by:

$$Q_c = \frac{\pi\epsilon_{re}^{(0.5)}}{\alpha_c\lambda_0} \qquad (7.1.20)$$

$$Q_d = \frac{\pi\epsilon_{re}^{(0.5)}}{\alpha_d\lambda_0} \qquad (7.1.21)$$

where "α_c" and "α_d" are the attenuation constants* per u.l. of the conductor and dielectric, respectively.

The theoretical value of the resonators is usually higher than what can actually be obtained, due to substrate dishomogeneity, conductor roughness, and mechanical tolerances. It has been proven that the practical values of the "Q_0" are 50 to 80% of the theoretical ones. To give some values, we may say that with a low loss substrate, like alumina or silica quartz, values of "Q_0" near 200 can be reached for frequencies of 5 or 6 GHz.

---

* See Chapter 1 for definitions on the attenuation constants.

f. FILTERS — It is absolutely necessary to inform the reader that filter theory is one of the broadest theories of network synthesis. To give all the fundamental information for this very useful theory, it would be necessary to write at least another book the size of this one. In addition, a good knowledge of mathematical analysis is required. Filter theory and technology are not the subjects of this book. Since in this book we will study networks that also have filtering capabilities, Appendix A4 introduces the reader to the fundamental concepts of filter theory, which are enough for our networks, but not for pure filters. So, in this item we will show typical filters realized in microstrip technology just to give the reader the capability to recognize which type of filter the microstrip topology will realize. Good general network filter synthesis may be found in several references,[7,8,9] together with specialized filter technology.

Microstrip filters of the type shown in Figure 7.1.2 are realized joining together, in an opportune way, the single networks represented in Figure 7.1.1. It is a characteristic of all transmission line filters to have "returns," that is, bands of frequencies, higher than the desired one, where the network has a behavior similar to the one for which it is designed in the desired frequency band. These effects are not present in a lumped filter network. For instance, the parallel coupled microstrip bandpass filter indicated in Figure 7.1.2c and designed for a center frequency "f_c," will give similar bandpass shape at center frequencies "f_{cs}" given by:

$$f_{cs} = (2s + 1) f_c \qquad (7.1.22)$$

For instance, if the desired center frequency is at 6 GHz, the network will give returns to 18 GHz, 30 GHz, ... according to Equation 7.1.22. Of course, when the frequency increases, the intrinsic losses, dispersion, and radiation also increase, and the return shape is worse than the designed one at the original frequency.

g. BALUN — The word "balun" is created from the two words "BALanced" and "UNbalanced" and, as the words suggest, it is used to move the e.m. energy from an unbalanced line to a balanced one.* Balanced microstrips are usually used in mixers[10,11] and to feed microstrip antennas, and for this reason baluns are important devices.[12,13] A simple microstrip balun is indicated in Figure 7.1.4a. To simplify, in this figure we have only represented the two conductors of the lines, without indicating the substrate between them. As we see, the connection to ground of the cold conductor of the microstrip is abruptly removed at a desired axial coordinate, where it begins the balancing procedure. This is usually done tapering for a length "ℓ," equal to an odd number of quarter wavelengths, the bottom conductor of the microstrip, or both the conductors. Usually, the tapering shape is a microwave art, and it has been originally realized with a "cut and try" procedure. The shape of the tapering is very important for operating bandwidth and return loss values.

The synthesis procedure can start, at first approximation, with looking at the balun as a matching line between the load impedance value "Z_ℓ" and the source generator impedance "Z_g," as indicated in Figure 7.1.4b. This procedure is a typical problem of the "$\lambda/4$ transformer." In this case, as seen in Chapter 1, the impedance "ζ_t" of the $\lambda/4$ transformer is:

$$\zeta_t = (Z_\ell Z_g)^{0.5} \qquad (7.1.23)$$

The practical situation looking at the balanced line side of Figure 7.1.4a is indicated in Figure 7.1.5. This kind of guiding structure is called a "balanced broadside coupled line," or "broadside coupled line," or "suspended substrate broadside coupled line." We will refer to the first type of structure and will abbreviate with "BBCL." This structure is clearly not a planar transmission line, essentially due to the mechanical construction required to hold the "BBCL." In addition, to

* See Chapter 1 for definitions of balanced and unbalanced lines.

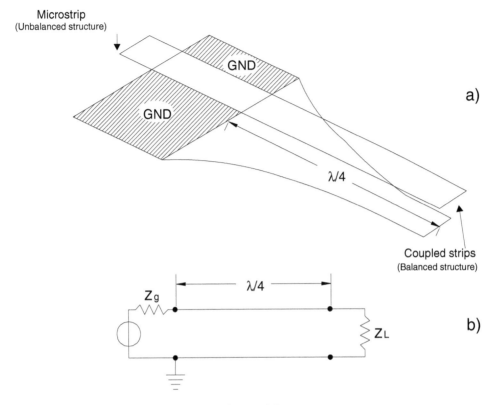

Figure 7.1.4

have good balancing characteristics from the balun of Figure 7.1.4a, it is extremely important that the even impedance "ζ_e" of the "BBCL" is realized at as high a value as possible, at least 10 times the odd impedance "ζ_o."[14] This means that the walls of the metal box shielding the balun, which also create the ground connection for the microstrip line feeding the balun, must be far away from the "BBCL," at least 10 times "w." If this condition is satisfied, we may use the easier formulas given by Wheeler,[15] as indicated in Chapter 2 where we studied microstrip theory. If the dimensions "a" and "b" may not be evaluated as infinite when compared to "w" and "h," some researchers[16] have studied the transmission line characteristics of the boxed structure indicated in Figure 7.1.5, giving formulas for the even and odd impedances, which are also functions of the thickness "t" of

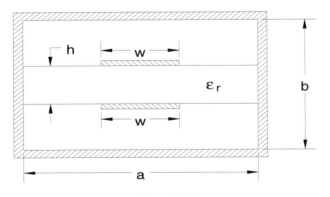

Figure 7.1.5

the balanced conductors. It has been proven that when t/b ≤ 0.02, the effect of "t" may be neglected. The typical operating bandwidth of microstrip baluns are near one octave and a half.

7.2 DIRECTIONAL COUPLERS

In the previous section we studied typical microstrip networks for which it is simple to associate the equivalent lumped network. In other words, looking at the dimensions of the lines that create the network, we can easily understand where a capacitor or an inductor may be associated for the lumped equivalent network. However, for the transformer this association is not so simple, and we need to have knowledge of transmission line theory as discussed in Chapter 1.

In this section we will study other very important microstrip networks that are building blocks of every RF and microwave system, for which it is not so simple to associate the equivalent lumped network. As was the case in the transformer study of the previous section, to have a deep understanding of directional couplers we need to have a good knowledge of transmission line theory and to know how networks may be analytically represented. As said, Chapter 1 is available for transmission line fundamentals, but we also suggest the reader see Appendix A3 where the main mathematical representations of networks are indicated.

As we have outlined in Appendix A3, we want to remind the reader that any directional coupler, regardless its realization, has two important parameters that characterize the coupler, i.e., the isolation "I" and the directivity "D," which are combined with the coupling "c_v" by the well-known formula:

$$\left|D\right| = \left|I\right| - \left|C_v\right|$$ (7.2.1)

where all the quantities are in dB. In the following points of study we will give typical values of isolation for the various couplers.

In all the figures we will show later, we have only drawn the hot conductor of the microstrip network, for simplicity purposes. Of course, the substrate and the ground conductor below it must be assumed to be effectively present in the practical device.

In addition, the microstrip network will be considered in this paragraph as a line that can support a pure "TEM" mode, unless otherwise noted. We know, after the reading of Chapters 2 and 3, that this assumption is not true, but it will help us to obtain some design formulas quite simply.

Directional couplers are used where power division or addition is required, or where a sample of the signal must be analyzed. In the next points we will study all the characteristics of these important networks.

An important class of directional couplers is one in which these devices are realized with coupled lines. These couplers are said to be "backward couplers" or "contraflow couplers" because the coupled port is the one that is geometrically nearest to the input one, and the coupled signal exits from this port. Opposite to this class is one in which the coupled port is not the one geometrically nearest to the input one, and the devices belonging to this class are named to "forward couplers" or "coflow couplers." As a general rule, all coupled line directional couplers are naturally "backward couplers." However, "forward couplers" are also possible with coupled μstrips, but they are seldom used due to the longer extension when compared to the contraflow counterpart. In the following we will study all such devices.

7.2.1 Branch Line

In Figure 7.2.1 parts a and b are two representations of the branch line coupler. The network in part b is sometimes called a "ring hybrid," but of course it is a branch line and the working

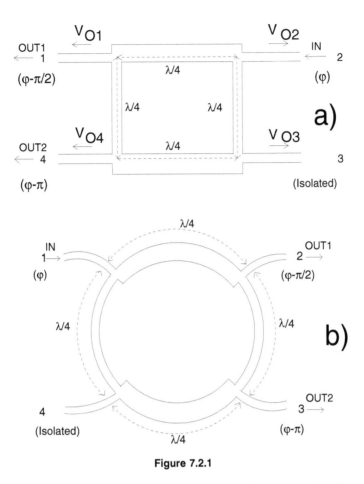

Figure 7.2.1

principle is the same. Referring for simplicity to Figure 7.2.1a, two parallel lines are connected together in two sections, which are a quarter wavelength apart, through two pieces of line a quarter wavelength long. The reference wavelength is that guided by the microstrip, at the center frequency of the desired bandwidth.

The working principle of the branch line coupler may be easily explained using the superposition effects theory, due to the linearity of the network. So, with reference to Figure 7.2.2a, we first feed the network with two generators of equal amplitude and in phase. We will call this situation "even excitation." With this feeding, we note that at the longitudinal symmetry line, we have a voltage maximum, that corresponds to having an infinite impedance at these points. This means that with even excitation, the branch line is equivalent to a network composed of half of the original branch line, with the transversal lines $\lambda/8$ long and open terminated. Then, with reference to Figure 7.2.2b, we feed the network with two generators of equal amplitude but 180° out of phase. We will call this situation "odd excitation." With this feeding, we note that at the longitudinal symmetry line, we have a voltage zero that corresponds to having a zero impedance at these points. This means that with an odd excitation, the branch line is equivalent to a network composed with half of the original branch line, with the transversal lines $\lambda/8$ long and short circuit terminated. Due to the symmetry and the series/shunt configuration of the branch line coupler, it is typical to use "chain matrices,"* also called "ABCD matrices," to study how it works. We know that for a shortened stub of electrical length "θ" and of line impedance "ζ," the ABCD matrix is:

---

* See Appendix A3 for chain matrix, or ABCD matrix, definition.

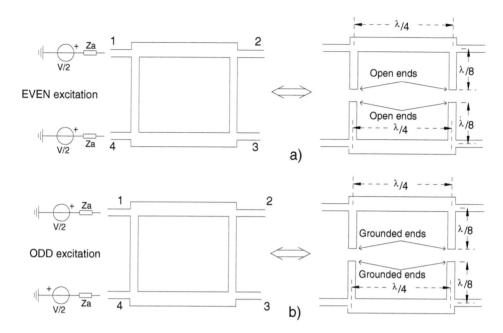

Figure 7.2.2

$$ABCD_{ss} = \begin{bmatrix} 1 & 0 \\ j(1/\zeta)\,tg(\theta) & 1 \end{bmatrix} \qquad (7.2.2)$$

while for an opened stub of the same characteristics:

$$ABCD_{so} = \begin{bmatrix} 1 & 0 \\ -j(1/\zeta)\,ctg(\theta) & 1 \end{bmatrix} \qquad (7.2.3)$$

and for a piece of transmission line of the same characteristics:

$$ABCD_{tl} = \begin{bmatrix} \cos(\theta) & j\zeta\,sen(\theta) \\ (j/\zeta)\,sen(\theta) & \cos(\theta) \end{bmatrix} \qquad (7.2.4)$$

When the elements of the ABCD matrix are known, we may easily obtain the quantities necessary to have the voltage phasors at each port. In particular, the voltage phasors "V_2" and "V_3" at ports "2" and "3" can be obtained from the transmission coefficient "t_{21}" of the matrix "T," * while "V_4" may be obtained from the reflection coefficient "s_{11}" at port "4." As seen in Appendix A3, these quantities are related to the ABCD matrix by the following relations:

$$s_{21} = \frac{2}{A + B + C + D} \qquad (7.2.5)$$

$$s_{11} = \frac{A + B - C - D}{A + B + C + D} \qquad (7.2.6)$$

---

* See Appendix A3 for transmission matrix definition.

The output voltage phasors, normalized to the input ones at the ports of the complete network, may be obtained as the superposition of the previous quantities relative to the even and odd excitation. In particular, we have:

$$V_{o1} = 0.5s_{11,e} + 0.5s_{11,o} \qquad (7.2.7)$$

$$V_{o2} = 0.5s_{21,e} + 0.5s_{21,o} \qquad (7.2.8)$$

$$V_{o3} = 0.5s_{21,e} - 0.5s_{21,o} \qquad (7.2.9)$$

$$V_{o4} = 0.5s_{11,e} - 0.5s_{11,o} \qquad (7.2.10)$$

So, to evaluate the phasors "V_{oi}" with i = 1, 2, 3, 4 we have to obtain the $ABCD_e$ and $ABCD_o$ matrices for the even and odd excitations. To do this, we start the evaluation of this matrix from the center on the "T" connection between the series and shunt arm of Figure 7.2.1a, proceed along the series line, and terminate at the next center on the "T" connection of the same side. From the start to the end we encounter a shunt line, a series line, and still another shunt arm. We now assume we are at center frequency because in this manner, the frequency dependence may not be considered, and, in addition, interesting relationships between arm impedances can easily be found. Proceeding with the even and odd excitation, the shunt arm will be $\lambda/8$ long, as previously mentioned, while the series line will be $\lambda/4$ long. Now, keeping this fact in mind and remembering from Appendix A3 that the ABCD matrix of a network is given by the products of the ABCD matrix of the components; indicating with $\zeta_{n\ell} \doteq 1/Y_{n\ell}$ and $\zeta_{nt} \doteq 1/Y_{nt}$, respectively, the normalized* impedance of the series or longitudinal, and shunt or transversal arm of the branch line coupler, and using 7.2.2 through 7.2.4 evaluated at center frequency we have:

$$ABCD_e = \begin{bmatrix} -\zeta_{nl}Y_{nt} & j\zeta_{nl} \\ -j(\zeta_{nl}Y_{nt}^2 + Y_{nl}) & -\zeta_{nl}Y_{nt} \end{bmatrix} \qquad (7.2.11)$$

$$ABCD_o = \begin{bmatrix} \zeta_{nl}Y_{nt} & j\zeta_{nl} \\ -j(\zeta_{nl}Y_{nt}^2 + Y_{nl}) & \zeta_{nl}Y_{nt} \end{bmatrix} \qquad (7.2.12)$$

By definition, an ideal directional coupler must be matched and perfectly directive. This means that feeding at any port, no signal must be reflected at this port, and no signal must appear as output at the adjacent port. Supposing to feed at port "1" of Figure 7.2.1b, we must have $V_{O1} = !0$ and $V_{O4} = !0$. This means, from 7.2.7 and 7.2.10, that:

$$s_{11,e} + s_{11,o} = !0 \qquad (7.2.13)$$

Evaluating these expressions using 7.2.6 we have the condition that for perfect matching and directivity we must have:

$$Y_{n\ell}^2 - Y_{nt}^2 = !1 \qquad (7.2.14)$$

Evaluating "V_{o2}" and "V_{o3}" using 7.2.8 and 7.2.9 and the previous condition, we have:

---

* See Chapter 1 for impedance and admittance normalizations.

$$V_{o2} = j\zeta_{n\ell} \tag{7.2.15}$$

$$V_{o3} = \frac{\left(Y_{n\ell}^2 - 1\right)^{0.5}}{Y_{n\ell}} \tag{7.2.16}$$

From the previous two equations we may note that the signal phasors "V_{o2}" and "V_{o3}" at the output of the ports "2" and "3" are in quadrature, due the presence of "j" in "V_{o2}." The case when $Y_{n\ell} = \sqrt{2}$ and $Y_{nt} = 1$ is very interesting. In this case, the expression 7.2.14 is satisfied and from the two previous equations we have $|V_{o2}| = 1/\sqrt{2} \equiv |V_{o3}|$. This means that with this condition the branch line coupler is a 3 dB power divider and, of course, is perfectly matched and directive. To give some values of impedance, using the classical 50Ω line as input and output connection lines, it results that the 3 dB coupler will have $\zeta_{n\ell} = (50/\sqrt{2})\Omega$ and $\zeta_{nt} = 50\Omega$. The capability of this network to reach a 3 dB splitting with such an easily realized value of microstrip impedance makes it very useful. We will show later that other networks can be built to reach a 3 dB of coupling, but the impedance values of the microstrips are very high for this technology, i.e., near 130Ω or more.

We now want to see the behavior of the branch line coupler vs. frequency. To do this, we repeat the previous procedure used to create 7.2.11 and 7.2.12, but we will use 7.2.2 through 7.2.4 without any imposition on the value of "θ." The result is:

$$ABCD_e = \begin{bmatrix} \varepsilon + j\beta Y_{nt} t & \beta \\ \left(\varepsilon + j\beta Y_{nt} t\right) j Y_{nt} t + \tau + j\varepsilon Y_{nt} t\right) & \varepsilon + j\beta Y_{nt} t \end{bmatrix} \tag{7.2.17}$$

and

$$ABCD_o = \begin{bmatrix} \varepsilon + j\beta Y ntt & \beta \\ \left(\varepsilon - j\beta Y_{nt}/t\right) j Y_{nt}/t + \tau - j\varepsilon Y_{nt}/t\right) & \varepsilon - j\beta Y_{nt}/t \end{bmatrix} \tag{7.2.18}$$

In the two previous equations we have used the following abbreviations:

$$tg(\theta) \doteq t$$

$$\cos(2\theta) \equiv \frac{1}{1+t^2} \cdot \frac{t^2}{1+t^2} \doteq f_i$$

$$sen(2\theta) \equiv \frac{2t}{1+t^2}$$

$$j\zeta_{n\ell} sen(2\theta) \doteq \beta$$

$$jY_{n\ell} sen(2\theta) \doteq \tau$$

The graphs of the normalized output signals "V_{o2}," "V_{o3}," and "V_{o4}," at ports "2," "3," and "4" relative to Figure 7.2.1b are given in Figure 7.2.3, where the input port is port "1." In this figure, the values above and below "0" are linear and dB, respectively, and in the case of a 3 dB coupler. We have used this notation because the signal at the output of the isolated port is theoretically exactly zero at center frequency, and consequently, the value in dB will be $-\infty$. In the abscissae we have indicated the normalized frequency "f_n," given by the ratio of the general variable frequency

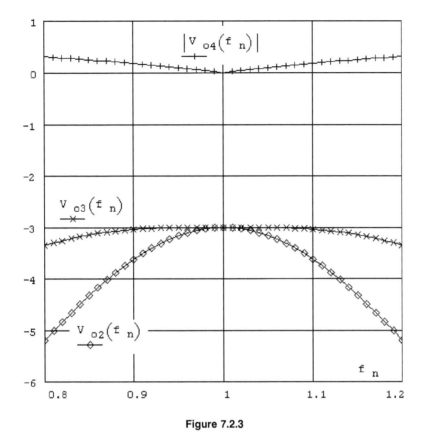

Figure 7.2.3

and that frequency where each arm of the "square" of the branch line is $\lambda/4$ long. In Figure 7.2.4 we have indicated the phases of the signals "V_{o2}," "V_{o3}," and "V_{o4}." From this figure we may recognize how the phase difference between the output signals "V_{o2}" and "V_{o3}" differ exactly of 90° at center band.

The counterpart of the high simplicity to reach the 3 dB coupling is the relatively narrow operating bandwidth. Limiting the useful bandwidth to that where the directivity is higher than 15 dB, the bandwidth of practical microstrip branch line couplers is near 15%. To increase the bandwidth sometimes two or three branch lines are connected in tandem,[17] a situation that permits reaching an operating bandwidth of 20%. To study such tandem devices we may apply the method used for the single branch line coupler. As said before, other devices may be used to increase the operating bandwidth. These are the Wilkinson and Lange networks, which will be discussed later.

7.2.2. "Rat Race" or "Magic T"

The geometry of the rat race coupler is indicated in Figure 7.2.5. This network is also called the "magic T," analogous to the "magic T" built-in waveguide technology and shown in Appendix A3. It is created using a microstrip ring with $3\lambda/2$ of circumference. On half of this ring, four ports are connected, each one $\lambda/4$ in distance from the previous one. The "rat race" device has two important characteristics:

1. Exactly in phase power distribution between one input port and two output ports. The fourth port is isolated.
2. Exactly 180° out of phase power distribution between one input port and two output ports. The fourth port is isolated.

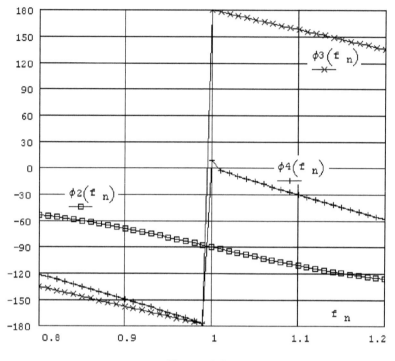

Figure 7.2.4

The two previous characteristics depend on the port chosen as input.

The situation indicated in characteristic 1 is reached when the internal ports, i.e., "2" or "3" are used as inputs. In this case, the two ports nearest to the input one are the outputs, and the remaining port is isolated. For instance, if we enter in port "2," ports "1" and "3" are the in phase outputs and "4" is isolated.

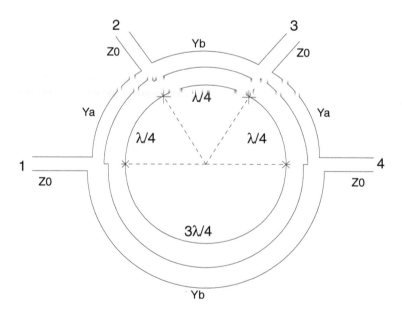

Figure 7.2.5

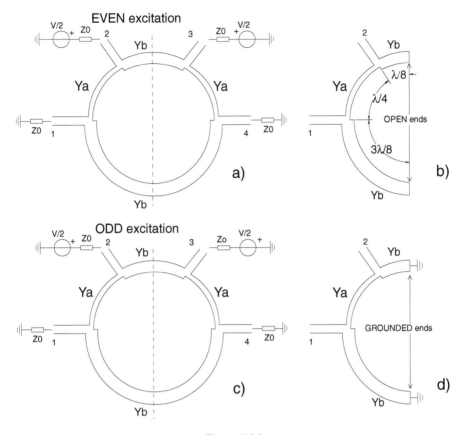

Figure 7.2.6

The situation indicated in characteristic 2 is reached when the two ports, which are diametrically opposed, i.e., "1" or "4," are used as input. In this case, choosing one input port, the two adjacent ports are the 180° phase difference output ones, while the remaining port is isolated. For instance, if we enter in port "4," ports "1" and "3" are the 180° phase difference outputs ones, while port "2" is isolated.

To study analytically how the rat race works, we can apply the superposition effects principle, as we did in the section 7.2.1 for the branch line. In this case, the situation relative to the even and odd excitation is indicated in Figure 7.2.6, assuming the original excitation is applied to port "2." In parts a and c, we have indicated with a dashed line, the symmetry axes where the network may be separated in two equal parts, each one indicated in parts b and d of the figure, depending on the type of excitation.

To obtain the matrices (ABCD$_e$) and (ABCD$_o$) for these networks, we have to customize Equations 7.2.2 through 7.2.4 to the length indicated in Figure 7.2.6 parts b and d. Note that the only difference between the chain matrices of a stub $\lambda/8$ long and of another stub $3\lambda/8$ long is the sign in the "C" element, where the stub $3\lambda/8$ long has the negative sign. With this note, we have:

$$\text{ABCD}_e = \begin{bmatrix} \zeta_{an} Y_{bn} & j\zeta_{an} \\ j\left(\zeta_{an} Y_{bn}^2 + Y_{an}\right) & -\zeta_{an} Y_{bn} \end{bmatrix} \qquad (7.2.19)$$

and

$$\text{ABCD}_o = \begin{bmatrix} -Z_{an}Y_{bn} & jZ_{an} \\ j\left(Z_{an}Y_{bn}^2 + Y_{an}\right) & Z_{an}Y_{bn} \end{bmatrix} \tag{7.2.20}$$

where the subscript "n" added to "Y_a," "Y_b," and "ζ_a," which appears in Equations 7.2.19 and 7.2.20 and indicates the normalization of these quantities to the reference admittance or impedance. To proceed with the evaluation of the electrical properties of the "magic T," we now suppose, to begin with, that we need a perfectly matched device. This means that feeding at any port, no signal must be reflected at this port. Assuming a feed at port "2," as indicated in Figure 7.2.6, we must have:

$$s_{11} \equiv 0.5s_{11,e} + 0.5s_{11,o} =! \; 0$$

Evaluating this expression using 7.2.6 we have the condition that for perfect matching and directivity we must have:

$$Y_{bn}^2 + Y_{an}^2 =! \; 1 \tag{7.2.21}$$

Of course, similar relations to Equations 7.2.7 through 7.2.10 may be used for our case, obtaining:

$$V_{o1} = 0.5s_{21,e} + 0.5s_{21,o} \tag{7.2.22}$$

$$V_{o2} = 0.5s_{11,e} + 0.5s_{11,o} \tag{7.2.23}$$

$$V_{o3} = 0.5s_{11,e} - 0.5s_{11,o} \tag{7.2.24}$$

$$V_{o4} = 0.5s_{21,e} - 0.5s_{21,o} \tag{7.2.25}$$

Using Equations 7.2.24, 7.2.6, and 7.2.19 through 7.2.21, the normalized phasor for the output at port "3" is:

$$V_{o3} = -jY_{bn} \tag{7.2.26}$$

and similarly, using Equations 7.2.22, 7.2.5, and 7.2.19 through 7.2.21, the normalized phasor at port "1" is:

$$V_{o1} = -jY_{an} \tag{7.2.27}$$

From the two previous equations we recognize that when feeding at port "2," the signals at ports "1" and "3" are in phase, since they are purely imaginary numbers, but in general they have different amplitude. The isolation of port "4" may be easily seen using 7.2.25, 7.2.5, and 7.2.19 through 7.2.21, obtaining $V_{o4} = 0$. Now, remembering that the power at port "i" is proportional to $|V_{oi}|^2$, from 7.2.26 and 7.2.27 it follows that:

$$\left|V_{o3}\right|^2 + \left|V_{o1}\right|^2 =! \; 1 \tag{7.2.28}$$

that is, the sum of the power exiting from port "1" and port "3" must be equal to the power that enters into port "2." This is a consequence of the lossless hypothesis on the network under study. Always applying the concept of power, if we denormalize 7.2.26 and 7.2.27 we have:

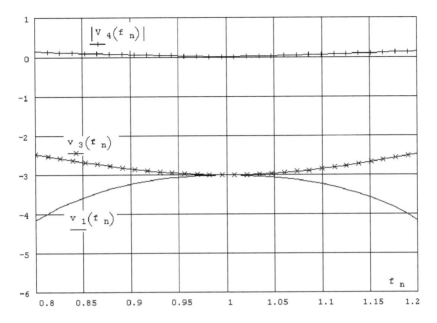

Figure 7.2.7

$$W_3/W_2 = \left(Y_b/Y_0\right)^2 \qquad\qquad (7.2.29)$$

$$W_1/W_2 = \left(Y_a/Y_0\right)^2 \qquad\qquad (7.2.30)$$

So, if we want to create a power divider we must have $W_3/W_2 = !0.5$ and $W_1/W_2 = !0.5$, and consequently from the two previous equations we need to create the line admittances so that:

$$Y_b \equiv Y_a =! Y_0/\sqrt{2}$$

To study the behavior of the rat race hybrid vs. frequency, we may follow the same procedure used for the branch line coupler. We assume that we insert the signal in port "2," as we have just done for our study. The results for the case of a 3 dB splitter are depicted in Figures 7.2.7 and 7.2.8. Comparing Figure 7.2.7 with Figure 7.2.3 we may see that the bandwidth where the splitting is closer to 3 dB is now slightly higher than the corresponding case of the branch line coupler. In fact, if we limit the useful bandwidth to that where the directivity is higher than 15 dB, the bandwidth of a practical microstrip rat race 3 dB divider is near 18%. In Figure 7.2.7 the values above and below "0" are linear and dB, respectively, and in the abscissae we have indicated the normalized frequency "f_n," given by the ratio of the general variable frequency and that frequency where each line among the ports "1" through "4" of the rat race is $\lambda/4$ long.

The phase of the signals at the ports are indicated in Figure 7.2.8, where we may see how the signals at the ports "1" and "3" are exactly equiphase at center frequency, and delayed 90° with reference to the input port "2."

It is important to say that all the previous formulas may be simply modified in the subscripts to study the case when the signal is applied to port "3"; it is enough to replace subscripts "2" with "3," "1" with "2," and "3" with "4."

The practical limitations on the minimum width of the microstrips, typically 15μm due to radiation phenomena* and effects of fabrication tolerances,** limit the use of this network to a

---

* See Chapter 4 for theory of higher order effects of microstrip lines.
** See Chapter 2 for information on fabrication tolerances of microstrip lines.

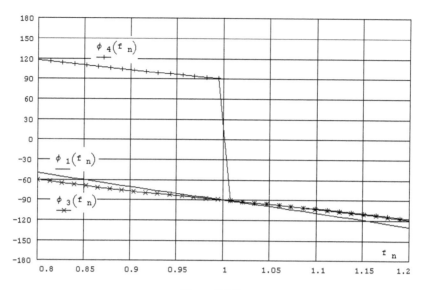

Figure 7.2.8

maximum power ratio of 6/1, i.e., near 8 dB. Using expressions 7.2.29 and 7.2.30 the reader can easily recognize how higher values of the splitting ratio lead to impractically small values of the width of the microstrip hot conductor. Different configurations for the rat race coupler, which permit a better matching and isolation, have been investigated by some researchers.[18,19]

7.2.3 "In-Line" or "Wilkinson"

This type of directional coupler (and with the "Lange" type, which we will discuss later) is one of the most famous and most used of all the planar transmission line directional couplers.[20,21,22] The original study and experiments of this type of network are due to researcher E. J. Wilkinson[23] who studied and created a divider by 8 network in coaxial technology. The planar transmission line counterpart is indicated in Figure 7.2.9a and, in this simple form, it is always a power divider by two. As we can see, the input line of impedance "ζ_0" is divided into two lines, each of them of impedance "ζ," with a shape that reminds one of a "ring." These two lines are a quarter of a wavelength long for the designed operating frequency and, after this length, they join together again through a resistor of value "R." At this point, two lines of impedance "ζ_0" depart, which creates the output of the network. Using the value of 50Ω as the reference impedance "R_a," the impedance value of the lines for the planar Wilkinson power divider are indicated in Figure 7.2.9b. We will show next how these values of impedances and resistances are calculated.

The geometrical aspect of the Wilkinson power divider suggests some qualitative topics of its electromagnetic properties to be treated.

First of all we want to invite the reader to pay attention to the fact that the Wilkinson network is a four port network, and not a three port device as it could appear at first sight. In fact, the fourth port of this device is balanced and is connected to the resistor "R." In addition, this device is theoretically perfectly matched at center frequency at all the three used ports "1," "2," and "3" indicated in Figure 7.2.9, as we will show later. Then, since we know from Appendix A3 that it is not possible to realize a three port device, which, among other thing, is perfectly matched at all the three ports, the Wilkinson power divider must be a four port one.

Second, due to the nature of codirectional flow of the energy, the Wilkinson network is also called an "in-line" power divider, with the signals at the outputs ports, which are equiphase.

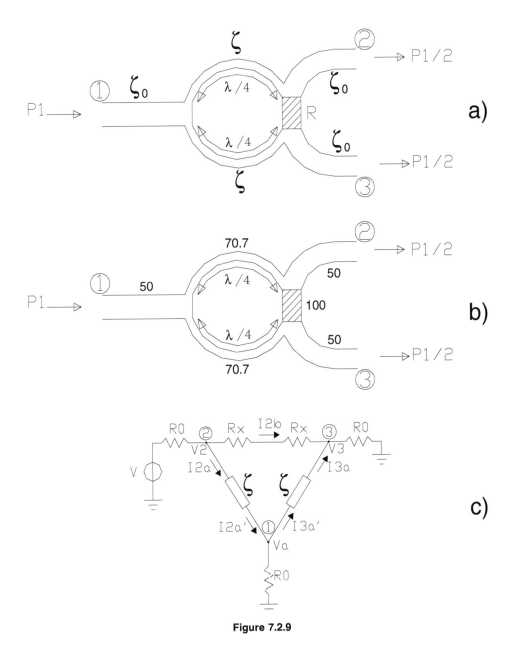

Figure 7.2.9

Third, since the input signal at port "1" is split in two equal paths, then at the terminals of the resistor "R," indicated in Figure 7.2.9a, there is no voltage drop. As a consequence, this resistor never dissipates energy.

Fourth, there is no dramatic reason that the splitting mechanism should not substantially work at other frequencies different from which it is designed. We will show later that the signals at the outputs are always very near −3 dB, with the input signal as reference, also at frequencies far enough from the design frequency. The counterpart of this wideband splitting is that isolation between the output ports the matching deteriorate quickly as the frequency changes from the optimum one. Nevertheless, the Wilkinson power divider has the widest operating band of all the networks studied before.

After these simple considerations, it is important to say that the particular shape of the "λ/4" lines, which resemble a circle, is due to the fact that optimum performances are obtained when these two lines are electromagnetically decoupled, that is when the even "ζ_e" and the odd "ζ_o" characteristic impedances of this structure are coincident. The reader is referred to Chapter 5, where the coupled microstrips network topic is studied and "ζ_e" and "ζ_o" are defined. For the moment, it is enough to remember that the two "$\lambda/4$" lines must be as decoupled as possible. From this consideration the ring shape follows.

Although this device can be studied with the superposition principle, that is with "even" and "odd" excitations as we did in sections 7.2.1 and 7.2.2, due to the nongeometrical four port topology of this device, we will study this network with the original theory of Wilkinson. We will only specialize this theory to the simple microstrip network shown in Figure 7.2.9a. We will use the even and odd excitations soon when we will analyze the bandwidth performance of the Wilkinson divider.

Let us start to represent Figure 7.2.9a with the schematic shown in Figure 7.2.9c. The resistor of value "R" in part a is divided in two resistors in series, each one of value "R_x," in part c. We now want to determine the condition which "ζ," "ζ_o," and "R_x" must satisfy so that the network in Figure 7.2.9c is matched at all the ports, and ports "2" and "3" are isolated between them. If we remember the forward transmission matrix "T_f" seen in Chapter 1, for the transmission line "1–2" in Figure 7.2.9c, with "V_a" and "I'_{2a}" as input quantities, we may write:

$$V_2 = V_a \cos\theta + j\zeta I'_{2a} \, sen\theta \tag{7.2.31}$$

$$I_{2a} = jYV_a \, sen\theta + I'_{2a} \cos\theta \tag{7.2.32}$$

that, for $\theta = \pi/2$ become:

$$V_2 = j\zeta I'_{2a} \tag{7.2.33}$$

$$I_{2a} = jYV_a \tag{7.2.34}$$

Similar relations may be obtained for line "3–1." Considering "V_3" and "I_{3a}" as input quantities, we may write for $\theta = \pi/2$:

$$V_a = j\zeta I_{3a} \tag{7.2.35}$$

$$I'_{3a} = jYV_3 \tag{7.2.36}$$

Applying the Kirchhoff* current law we have:

$$I'_{2a} =! V_a/R_0 + I'_{3a} \qquad \text{for node "1"} \tag{7.2.37}$$

$$I_{3a} =! V_3/R_0 - I_{2b} \qquad \text{for node "3"} \tag{7.2.38}$$

$$I_{2a} + I_{2b} =! (V - V_2)/R_0 \qquad \text{for node "2"} \tag{7.2.39}$$

The voltage between nodes "2" and "3" is according to Ohm law, i.e.:

$$V_2 - V_3 = 2R_x I_{2b} \tag{7.2.40}$$

* Gustav Robert Kirchhoff, German physicist, born in Koenigsberg in 1824 and died in Berlin in 1887.

Performing a system of equations with 7.2.35, 7.2.38, and 7.2.40 we may write:

$$jV_a R_0/\zeta + \left(1 + R_0/2R_x\right)V_3 - \left(R_0/2R_x\right)V_2 = 0 \qquad (7.2.41)$$

Performing a system of equations with 7.2.34, 7.2.39, and 7.2.40 we may write:

$$jV_a R_0/\zeta + \left(1 + R_0/2R_x\right)V_2 - \left(R_0/2R_x\right)V_3 = V \qquad (7.2.42)$$

Performing a system of equations with 7.2.33, 7.2.36, and 7.2.37 we may write:

$$V_a + j\left(R_0/\zeta\right)V_3 + j\left(R_0/\zeta\right)V_2 = 0 \qquad (7.2.43)$$

We now impose the condition that port "3" be isolated from port "2," in the situation of Figure 7.2.9 c. It means we must set $V_3 = !0$. With this hypothesis, we have:

$$jV_a R_0/\zeta = !\left(R_0/2R_x\right)V_2 = 0 \quad \text{from 7.2.41} \qquad (7.2.44)$$

$$jV_a R_0/\zeta + \left(1 + R_0/2R_x\right)V_2 = ! V \quad \text{from 7.2.42} \qquad (7.2.45)$$

$$V_a = ! - j\left(R_0/\zeta\right)V_2 \quad \text{from 7.2.43} \qquad (7.2.46)$$

Inserting the previous equation:

in 7.2.44, we have:

$$\left(R_0/\zeta\right)^2 = ! R_0/2R_x \qquad (7.2.47)$$

in 7.2.45, we have:

$$1 + R_0/R_x = ! V/V_2 \qquad (7.2.48)$$

We now impose the condition that port "2" be matched. It means that:

$$V_2/\left(I_{2a} + I_{2b}\right) = ! R_0 \qquad (7.2.49)$$

Doing a system of equations with 7.2.39, 7.2.48, and 7.2.49 it follows that:

$$R_x = ! R_0 \qquad (7.2.50)$$

and, inserting the 7.2.50 in 7.2.47, we have:

$$\zeta = ! R_0\sqrt{2} \qquad (7.2.51)$$

The two previous equations show the relations that must be satisfied between the impedance and resistance of the network shown in Figure 7.2.9a, to assure that it be perfectly matched to all ports and with ports "2" and "3" isolated. Note that for the symmetry of the device, if port "2" is matched and isolated from port "3," then port "3" is also isolated from "2" and matched. Note that since ports "2" and "3" are matched, port "1" is also matched. Remembering the effect of the $\lambda/4$ transformer studied in Chapter 1, each matched load "R_0" connected to ports "2" and "3" is transformed toward port "1" in a load of value "$2R_0$," with the result that the parallel of these resistors is still a matched load of value "R_0."

As a consequence of the perfect matching, isolation, and the lack of power dissipation in "R," if we assume an ideal lossless transmission line in the Wilkinson network shown in Figure 7.2.9, then the following relation must be satisfied at center frequency:

$$W_2 =! \ W_3 =! \ W_1/2 \qquad\qquad (7.2.52)$$

The importance of the resistor "R" is to assure the termination of the fourth balanced port of the network, so assuring the isolation of the output ports. For this reason, the resistor "R" is also called or known as the "isolation resistor."

Wilkinson's original study, which we have just shown, is oriented to synthesis of the device, and no theoretical information about acceptable bandwidth is given. Useful information about this important requirement can be obtained with a study of the device oriented to analysis. This study can be performed with the even and odd mode excitation procedure, just like we did to study the previous "branch line" and "rat race" devices. Let us examine Figure 7.2.10. In parts a and c, we have indicated the even and odd excitations at ports "2" and "3." The superposition of these two

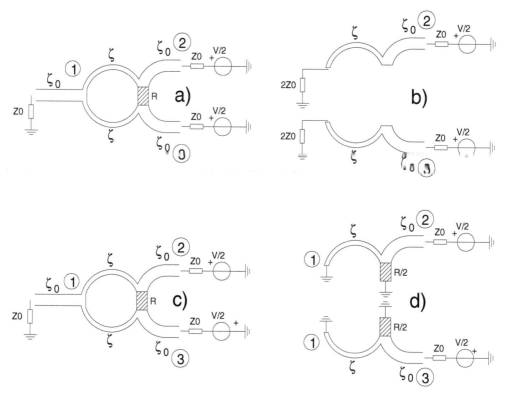

Figure 7.2.10

excitations is equivalent to one excitation to port "2." In parts b and d we have indicated the equivalent networks for even and odd excitations, respectively. Note that for the even excitation, no current flows in the resistor "R," and consequently the network in part a can be divided as in part b with the "$2\zeta_0$" load required since when the two networks in part b are connected together at the input, the resultant load is still "Z_0," as in part a. For odd excitation, note that at one half of the resistor "R," there is a short circuit, since for this excitation the generators have equal amplitude but phase reversal. For the same reason, at the input of the network, an equivalent short circuit is present. This situation is represented in part d. Note that from Figure 7.2.10 part b and d, the output phasor "V_{32}" at port "3" when the input is at port "2" is:

$$V_{32} = 0.5s_{11,e} - 0.5s_{11,o} \tag{7.2.53}$$

For the reciprocity of this network, the signal exiting at port "2" with the excitation at port "1" is equivalent to the signal exiting at port "1" with the excitation at port "2." This last one can be obtained just using the excitation in Figure 7.2.10. Note that only the even excitation creates an output signal at port "1" and from the conservation energy theorem* we may write:

$$V_{12} \equiv V_{21} = \left[0.5\left(1 - \left|s_{11,e}\right|^2\right)\right]^{0.5} \tag{7.2.54}$$

From the previous two equations it is clear that all the electrical characteristics of the network can be evaluated knowing the reflection coefficients for the even and odd excitations. To determine these quantities is very simple. In fact, for the even excitation, the input impedance "Z_{ie2}" at port "2" is that of a length "θ" of transmission line of impedance "ζ" terminated with a load of impedance "Z_ℓ" of value "$2\zeta_0$." From Chapter 1 we know that this impedance is:

$$Z_{ie2} = \frac{Z_\ell \zeta + j\zeta^2 \tan(\theta)}{\zeta + jZ_\ell \tan(\theta)} \equiv Z_{ie3} \overset{\perp}{=} Z_{ie} \tag{7.2.55}$$

and the corresponding reflection coefficient is:

$$s_{11,e} = \frac{Z_{ie} - \zeta_0}{Z_{ie} + \zeta_0} \tag{7.2.56}$$

For the odd case we note that the input impedance "Z_{io2}" at port "2" is the parallel of a resistor with value "R/2" and the impedance of a shorted stub of impedance "ζ" and length "θ." From the theory seen in Chapter 1 we may write:

$$Z_{io2} = \frac{j(R/2)\zeta \tan(\theta)}{R/2 + j\zeta \tan(\theta)} \equiv Z_{io3} \overset{\perp}{=} Z_{io} \tag{7.2.57}$$

and the correspondent reflection coefficient is:

$$s_{11,o} = \frac{Z_{io} - \zeta_0}{Z_{io} + \zeta_0} \tag{7.2.58}$$

---

* Remember that we are considering a lossless, reciprocal "TEM" mode network.

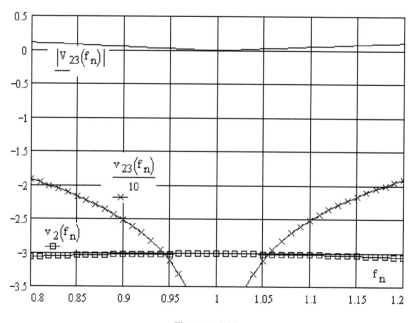

Figure 7.2.11

Inserting the values determined with the synthesis of Wilkinson in Equations 7.2.55 through 7.2.58 and applying 7.2.53 and 7.2.54 we have the result indicated in Figures 7.2.11 through 7.2.13. As in sections 7.2.1 and 7.2.2, for these figures we have used the normalized frequency "fn" as the independent variable, and the values above and below "0" are linear and dB, respectively. In addition, the lowercase variable "$v_{32}(f_n)$" is evaluated in dB while the uppercase one is a linear value.

In Figure 7.2.12 we made a larger sweep for "f_n," where we may see the interesting result that the Wilkinson divider is always capable of dividing the input power near to three dB also very far

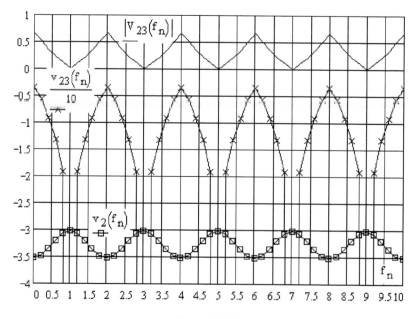

Figure 7.2.12

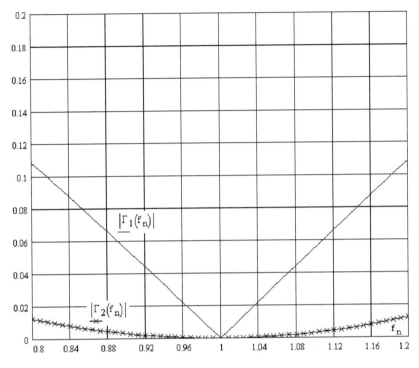

Figure 7.2.13

from the designed frequency, but the isolation is no more acceptable. The useful bandwidth for an isolation of 20 dB is near 44%. In Figure 7.2.13 we have indicated the reflection coefficients at ports "1" and "2." We may see how port "2" (and consequently port "3") is much less reflective than port "1" outside the designed frequency.

Note that the study we have made until now assumes in the case of lossless devices and perfect "TEM" * transmission lines. While the first hypothesis is never satisfied for any practical device, the "TEM" hypothesis is only approximated by microstrips, as was said in Chapter 3. In addition, the effects of discontinuities in the device have not been evaluated, as the input bifurcation or the step in width where the resistor "R" is connected. For this purpose, the reader can consult Chapter 4, where microstrip discontinuities are treated. As a result of the physical realization in microstrip technology of the Wilkinson power divider, we have losses, isolation, power division by two, and matching degrade with the increasing of the design frequency. Limiting the bandwidth to the value where isolation is at least 20 dB, the typical bandwidth for a microstrip Wilkinson power divider indicated in Figure 7.2.9b is near 38%, i.e., n = 1.5, until center frequencies of approximately 10 GHz.

To increase the bandwidth, it is possible to create the power divider using more rings, as indicated in Figure 7.2.14a. The synthesis of these devices has been performed by researcher S.B. Cohn,[24] under the hypothesis of "TEM" mode propagation. For the case of Figure 7.2.14a, the synthesis of the network can be performed directly using Cohn's expressions, i.e.:

$$R_1 = \frac{2\zeta_1\zeta_2}{\left[\left(\zeta_1 + \zeta_2\right)\left(\zeta_1 - \zeta_2 \cot^2\left(\varphi\right)\right)\right]^0} \tag{7.2.59}$$

where:

* See Appendix A2 for propagation mode definition and Chapter 2 for applications to microstrips.

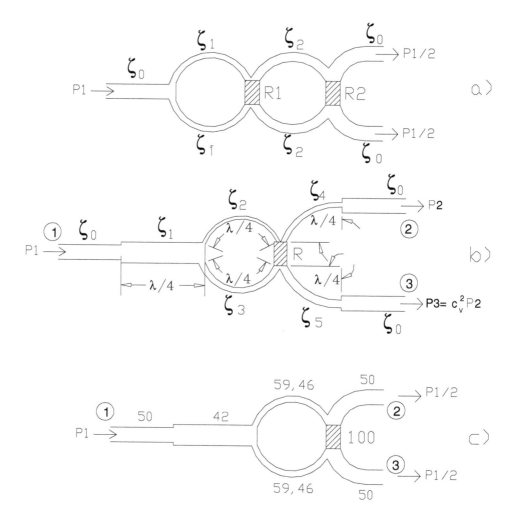

Figure 7.2.14

$$\varphi = (\pi/2)\left[1 - \frac{f_h/f_\ell - 1}{(f_h/f_\ell + 1)\sqrt{2}}\right] \qquad (7.2.60)$$

$$R_2 = \frac{2R_1(\zeta_1 + \zeta_2)}{R_1(\zeta_1 + \zeta_2) - 2\zeta_2} \qquad (7.2.61)$$

where F_h and F_ℓ are the higher and lower frequencies, respectively, of the operating bandwidth.

In the three previous equations, all the "R_i" and "ζ_i" are normalized to the reference impedance, i.e., 50Ω practically. The analysis of this network can be performed in the same manner with regard to Figure 7.2.10. The most important characteristics of the network in Figure 7.2.13a is that it can be designed for a bandwidth operation of Chebyshev shape,* contrary to the original Wilkinson divider in which the bandwidth shape is fixed. As we know, once a ripple is fixed in the bandwidth, the Chebyshev shape has the widest bandwidth or, conversely, once a bandwidth is fixed the Chebyshev shape has the lowest ripple. This characteristic can be deduced from Figure 7.2.10a,

---

* See Appendix A4 for Chebyshev polynomials and their characteristics.

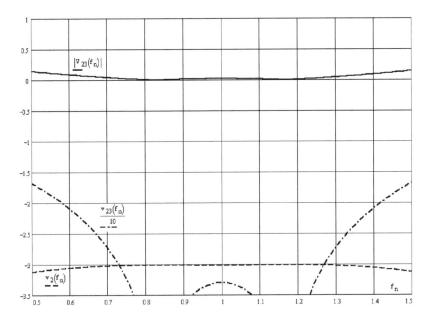

Figure 7.2.15

where the analysis using the even excitation is simply a matching problem between a load of value "$2\zeta_0$" and a source of impedance "ζ_0." We have seen in Chapter 1 how this problem can be resolved with a Chebyshev design. Consequently, the multiring Wilkinson divider can be designed with "ζ_1" and "ζ_2" according to a Chebyshev shape. For instance, let us attempt to create a two step Wilkinson 3 dB divider in an operating bandwidth of B = 1, with a return loss ripple for the prototype equal to 0.025. From the theory studied in Chapter 1 and referring to Figure 7.2.14a, we have $\zeta_1 = 1.203$ and $\zeta_2 = 1.654$. Then, from 7.2.59 through 7.2.61 we have $R_1 = 5.2$ and $R_2 = 1.882$. Iteratively applying the same theory given for 7.2.55 through 7.2.58 to the case of Figure 7.2.14a, for this example we have the results shown in Figure 7.2.15. We see how each electrical parameter has a Chebyshev shape, and how there is an increase in the operating bandwidth with respect to the single section divider shown in Figure 7.2.11. In fact, a two section microstrip Wilkinson divider has a useful bandwidth near 70% for center frequency up to 10 or 15 GHz, where "useful bandwidth" means the frequency region where the return loss "s_{11}" is higher or equal to 15 dB.

Of course, a still higher operating bandwidth can be reached if a higher ripple is accepted for the quarter wavelength transformer prototype, since it is a characteristic of the Chebyshev design.

A higher number of sections does not have a simple synthesis procedure like the single or two section case. The Wilkinson power divider can be found working in operating bandwidths of 10:1 or more, which are realized with six or more sections. These devices are synthesized using computer optimization procedures, until a quasi-Chebyshev behavior is reached.

A very important variation of the Wilkinson divider has been studied by Parad and Moynihan,[25] and the resultant network is indicated in Figure 7.2.14b. The important characteristic is that this network is able to divide the input power, not only equally for the two outputs, but also with different values of signal power at the two outputs. In any case, the higher signal power is present at the port connected to the branch where the lower impedance is present. So, in Figure 7.2.14 the signal at port "3" has higher power than signal at port "2." In addition, as in the Wilkinson divider, the resistor "R" never dissipates energy.

The theoretical study of this network can be performed in the same manner we used for the simple Wilkinson power divider, i.e., with the even and odd mode excitations. According to this theory, Parad and Moynihan give the following design formula for the network in Figure 7.2.14b:

$$|c_v|^2 \doteq W_3/W_2 \tag{7.2.62}$$

$$\zeta_1 = \left[|c_v| / \left(1 + |c_v|^2\right) \right]^{1/4} \tag{7.2.63}$$

$$\zeta_2 = |c_v|^{3/4} \left(1 + |c_v|^2\right)^{1/4} \tag{7.2.64}$$

$$\zeta_3 = \left(1 + |c_v|^2\right)^{1/4} / |c_v|^{5/4} \tag{7.2.65}$$

$$\zeta_4 = \sqrt{|c_v|} \tag{7.2.66}$$

$$\zeta_5 = 1 / \sqrt{|c_v|} \tag{7.2.67}$$

$$R = \left(1 + |c_v|^2\right) / |c_v| \tag{7.2.68}$$

Of course, all the "R" and "ζ" in Equations 7.2.62 through 7.2.68 are normalized to the system reference impedance, i.e., 50 Ohm. It is important to say that this kind of network is suitable for a ratio between the output power until 8 dB maximum, since for higher values the microstrip impedances become physically impractical. Note that in the previous formulas, there is no theoretical limitation on the value of "c_v," and for this reason the device in Figure 7.2.14b can also be synthesized for equal power output. In this case we simply have $cv^2 = 1$, and 7.2.62 through 7.2.68 become:

$$|c_v|^2 = 1, \quad \zeta_1 = 2^{-0.25}, \quad \zeta_2 = 2^{0.25} \equiv \zeta_3, \quad \zeta_4 = 1 \equiv \zeta_5, \quad R = 2$$

The denormalized values to 50 Ohm for the previous formula give the network in Figure 7.2.14c. Due to the additional transformers introduced in the network shown in Figure 7.2.14b with respect to the single section Wilkinson network shown in Figure 7.2.9a, the unequal power divider has an operating bandwidth near 30%, which is considerably lower than the single section Wilkinson power divider.

It is interesting to observe that both such networks are dividers, i.e., only two used outputs are present. It is possible to realize planar in-line multioutput dividers, as shown in Figure 7.2.16 for an in-line six output divider. All the lines with impedance "ζ" are a quarter wavelength long at the center of the operating bandwidth. Unlike the Wilkinson network, this kind of hybrid does not perform theoretical infinite isolation and matching at the ports, if these values can be reasonably optimized to an acceptable value.[26,27] Also, although the network in Figure 7.2.16 is very attractive from a theoretical point of view, it is seldom used in practice, and a cascade of Wilkinson power dividers are preferred when more than two outputs are required.

7.2.4 Step Coupled Lines

The simplest DC decoupled directional coupler is indicated in Figure 7.2.17. It is simply realized setting at a distance "s," two microstrip lines of width "w" and length "ℓ," which is a quarter wavelength long at center frequency. After the length "ℓ," the two lines are set more distant than possible, to remove any type of coupling after "ℓ." The characteristic of this coupler is that the coupled port is geometrically nearest to the input one; the direct port is that which lies on the same line of the input port, while the isolated port is the remaining one. This type of coupler belongs to

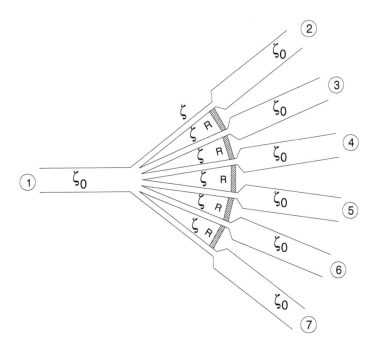

Figure 7.2.16

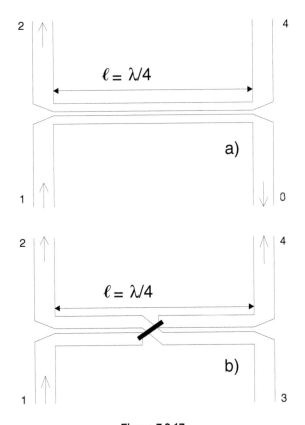

Figure 7.2.17

the class of "backward couplers." This situation is completely different from the previous couplers we have studied, where the energy flow of the coupled signal has the same direction as the input one. So, with reference to Figure 7.2.17a, if the input signal is connected to port "1," then the coupled port is port "2" while port "3" is the direct port. In Figure 7.2.17b, the direct port has been reversed using an "air bridge," i.e., a connection that passes over another line. Of course, it is always a "backward coupler." The network in part b is seldom used in practice, but it is a building block of a very important microstrip coupler we will discuss later, i.e., the Lange coupler.

The electromagnetic situation indicated in Figure 7.2.17 is very different from the previous situations we have studied. In fact, we are now dealing with coupled lines. Coupled μstrip theory is studied in detail in Chapter 5, and now we will review only the concepts necessary to understand the working principle of this coupler. Every time two lines come in proximity, each one changes the characteristic impedance it had when it was alone, and the new situation must be studied as a lone electromagnetic ambient. The impedance of each line is dependent on the amplitude and phase of the signal that is present on the other line. Of course, the two lines must be terminated with an equal impedance "Z_ℓ." Two impedances are defined:

a. The "even" impedance "ζ_e," which is defined as the characteristic impedance that a line presents to ground when two equiphase, equiamplitude, and equal impedance generators are connected to each line.

b. The "odd" impedance "ζ_o," which is defined as the characteristic impedance that a line presents to ground when two out of phase, equiamplitude, and equal impedance generators are connected to each line.

Theoretically, two lines are never electromagnetically isolated, since the electromagnetic field produced by each line is zero only at the infinite, but in microstrip practice, two lines can be evaluated as electromagnetically decoupled when the distance "s" between their nearest boundary is at least three times their width "w."

Assuming the situation indicated in Figure 7.2.17a, a center band voltage coupling coefficient "c_{0v}" is defined as:

$$c_{0v} \triangleq E_2/E_1 \quad \text{when } \ell = \lambda/4 \tag{7.2.69}$$

where "E_1" and "E_2" are the electric fields at the input port "1" and output port "2." The coupling coefficient is, in general, a complex number, with a modulus value between 0 and one. From the coupled line theory* it is possible to show that the following two relations hold:

$$\zeta_e = Z_\ell \left(\frac{1 + |c_{0v}|}{1 - |c_{0v}|} \right)^{0.5} \tag{7.2.70}$$

$$\zeta_o = Z_\ell \left(\frac{1 - |c_{0v}|}{1 + |c_{0v}|} \right)^{0.5} \tag{7.2.71}$$

From 7.2.70 and 7.2.71 it follows that:

$$Z_\ell \equiv \left(\zeta_e \zeta_o \right)^{0.5} \tag{7.2.72}$$

* See Chapter 1 for coupled line theory fundamentals.

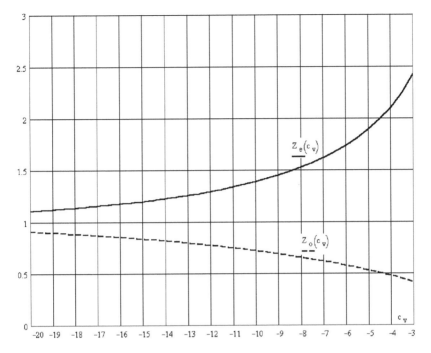

Figure 7.2.18

In Figure 7.2.18 we have reported the dependence of "ζ_e" and "ζ_o" vs. "c_v," with "c_v" evaluated in dB and $Z_\ell = 1$ Ohm. In microstrip technology it is very difficult to realize the coupler in Figure 7.2.18 with "c_v" higher than -6 dB because the values of "ζ_e" and "ζ_o" are so different, and in particular "ζ_e" so high that spacing "s" smaller than 10 micrometers should be realized. The only solution to realize a -3 dB coupled line coupler is to employ the Lange coupler, which we will study later.

The reader who wants to create the coupler in Figure 7.2.17 needs first to define the required coupling value. Then, from Figure 7.2.18, find the corresponding values of "ζ_e" and "ζ_o," and finally find the values of "s" and "w" for the defined substrate.

The study of the networks in Figure 7.2.17 can be performed using the well-known superposition principle, as was done in the previous items studied. The situation is indicated in Figure 7.2.19. In part a of the figure is represented the situation obtained with the superposition of the even excitation, indicated in part b, and with the odd excitation indicated in part c. We may write:

$$V_1 = V_{1e} + V_{1o} \tag{7.2.73}$$

$$V_2 = V_{2e} + V_{2o} \equiv V_{1e} - V_{1o} \tag{7.2.74}$$

$$V_3 = V_{3e} + V_{3o} \tag{7.2.75}$$

$$V_4 = V_{4e} + V_{4o} \equiv V_{3e} - V_{3o} \tag{7.2.76}$$

From the previous four equations we see that there are four system variables, i.e., "V_{1e}," "V_{1o}," "V_{3e}," and "V_{3o}." These variables can be found very easily. Using the subscript "m" to indicate the mode of excitation we have:

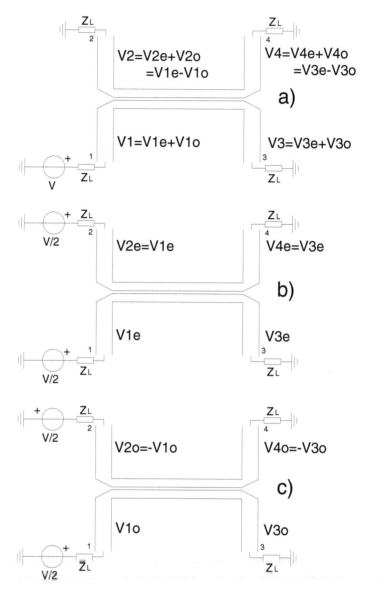

Figure 7.2.19

$$V_{1m} = \frac{V}{2} \frac{Z_{1m}}{Z_{1m} + Z_\ell} \quad \text{with } m = e, o \tag{7.2.77}$$

and the current at port "1" will be:

$$I_{1m} = \frac{V}{2} \frac{1}{Z_{1m} + Z_\ell} \tag{7.2.78}$$

The quantity "Z_{1m}" is the impedance we see at the input of port "1" with excitation "m." From the general line theory seen in Chapter 1, we have:

$$Z_{1m} = \frac{Z_m Z_\ell + jZ_m^2 \, tg(\theta)}{Z_m + jZ_\ell \, tg(\theta)} \qquad \text{with } m = e,o \qquad (7.2.79)$$

where "θ" is the electrical length of the coupled region of the coupler in Figure 7.2.17. The voltage and current at port "3" can be found using the "forward transmission matrix" defined in Chapter 1. So we can write:

$$\begin{bmatrix} V_{3m} \\ I_{3m} \end{bmatrix} = \begin{bmatrix} \cos(\theta) & -jZ_m \, sen(\theta) \\ -jY_m \, sen(\theta) & \cos(\theta) \end{bmatrix} \begin{bmatrix} V_{1m} \\ I_{1m} \end{bmatrix} \qquad \text{with } m = e,o \qquad (7.2.80)$$

Since we have determined all variables, we may evaluate the direct transmission "d_v," the coupling "c_v" and the isolation "i_v." We have:

$$d_v \triangleq \frac{V_3}{V/2} \equiv \frac{V_{3e} + V_{3o}}{V/2} = \left(\frac{Z_{1e}}{Z_{1e} + Z_\ell} + \frac{Z_{1o}}{Z_{1o} + Z_\ell} \right) \cos\theta - j\left(\frac{Z_e}{Z_{1e} + Z_\ell} + \frac{Z_o}{Z_{1o} + Z_\ell} \right) sen\theta \qquad (7.2.81)$$

$$c_v \triangleq \frac{V_2}{V/2} \equiv \frac{V_{1e} - V_{1o}}{V/2} = \frac{Z_{1e}}{Z_{1e} + Z_\ell} + \frac{Z_{1o}}{Z_{1o} + Z_\ell} \qquad (7.2.82)$$

$$i_v \triangleq \frac{V_4}{V/2} \equiv \frac{V_{3e} + V_{3o}}{V/2} = \left(\frac{Z_{1e}}{Z_{1e} + Z_\ell} - \frac{Z_{1o}}{Z_{1o} + Z_\ell} \right) \cos\theta + j\left(\frac{Z_o}{Z_{1o} + Z_\ell} - \frac{Z_e}{Z_{1e} + Z_\ell} \right) sen\theta \qquad (7.2.83)$$

In Figure 7.2.20 we have represented the dependence with the electrical length "θ" of the modulus of the above equations, for the case of a coupling factor of −10 dB and "θ" sweeping from zero to 2π. The values above "1" on the ordinate axis are in linear value, while the values below "1" are in dB. We have used the linear values for this figure to show that the value of the isolation is theoretically always zero. This means that the quarter wavelength coupled lines directional coupler is able to isolate a port outside the frequency where it has been dimensioned and exactly for every real frequency from zero to infinite. Note that this result is only theoretically valid since discontinuities in the structure and the nonperfect "TEM" propagation mode in the microstrip generate signals in the theoretically isolated port. Typical isolation values are near 15 or 20 dB for frequencies up to 20 GHz, decreasing to 10 or 15 dB for higher frequencies.

In Figure 7.2.21 we have reported the dependence with the electrical length "θ" of the phase of the above equations. We see that for $\theta = (2n + 1) \, 90°$, where "n" is an integer, the signal at port "2" is equiphase with that at port "1" while the signal at port "3" is 90° delayed. It is also interesting to note that the phase difference between signals at port "2" and "3" is always near 90°. The modulus of these signals, i.e., the coupling, does not have the wideband characteristic we can recognize in Figure 7.2.20.

The expressions for "$c_v(\theta)$" and "$d_v(\theta)$" can also be reported as functions of "c_{0v}," using 7.2.79, 7.2.70, and 7.2.71. We have:

$$d_v(\theta) = \frac{\left(1 - |c_{0v}|^2\right)0.5}{\left(1 - |c_{0v}|^2\right)^{0.5} \cos\theta + jsen\theta} \qquad (7.2.84)$$

$$c_v(\theta) = \frac{j|c_{0v}| \, sen\theta}{\left(1 - |c_{0v}|^2\right)^{0.5} \cos\theta + jsen\theta} \qquad (7.2.85)$$

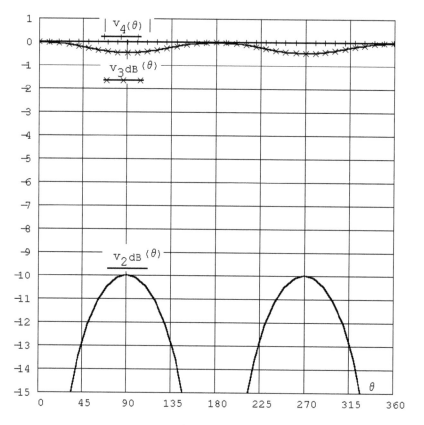

Figure 7.2.20

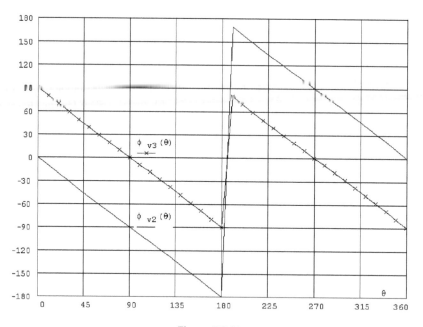

Figure 7.2.21

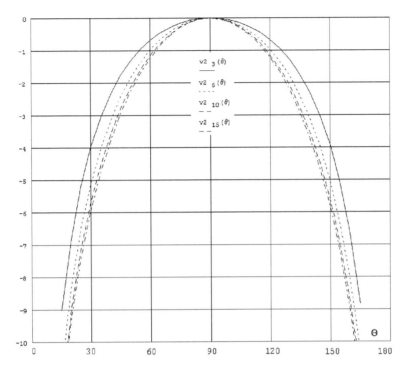

Figure 7.2.22

The reader can easily understand that defining an operating bandwidth for such a coupler is quite subjective. In addition, the bandwidth where the coupling is near its center frequency value is a function of the coupling value as indicated in Figure 7.2.22. In this figure every coupling vs. "θ" has been normalized to its maximum value, reached for $\theta = \pi/2$. We see how the increase in the coupling increases the bandwidth where the coupling is near its maximum value. The most used parameter to define a bandwidth is the isolation, and if a value not lower than 15 dB is taken as reference, the bandwidth is near 40% for coupling values below -10 dB. Assuming we could build a 3 dB microstrip coupler of the type under study, the operating bandwidth could reach 70%.

As we did for the Wilkinson power divider, if we increase the number of coupled sections, it is possible to have an equiripple shape of the coupling, and so to use a bandwidth definition applied to equal ripple coupling. Two types of multisection coupled line coupler are possible, the symmetrical and asymmetrical, as indicated in Figure 7.2.23. Only the asymmetrical coupler can be synthesized using Chebyshev polynomials. We will also show later that if the symmetrical couplers have an equiripple shape of the coupling, they cannot be synthesized with Chebyshev polynomials. This means that asymmetrical couplers are more bandwidth efficient compared to symmetrical versions, when a layout space and coupling ripple are fixed.

Note also that although for every section we have reported both the even and odd impedances, it is enough to indicate only one of these, since the Equation 7.2.72 must be satisfied. The study of this network can be done in the same manner used for the single section coupler. But another simple method exists, which is very suitable to be used with a computer. With this method, an "ABCD," or chain, matrix* is associated with every section of coupled line. Then the whole "ABCD" matrix is simply evaluated by multiplying each single section matrix. Finally, using the conversion formulas, it is possible to evaluate all the desired scattering parameters.

* See Appendix A3 for definition of "ABCD," or "chain," matrix.

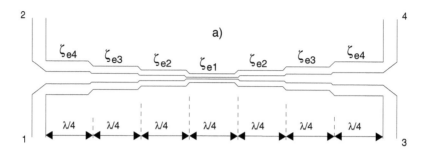

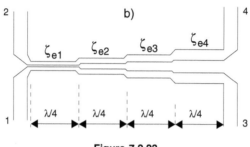

Figure 7.2.23

Let us do an example for a three section symmetrical coupler. For the terminal "t" coupled lines, the chain matrix $(C_m)_t$ will be:

$$(C_m)_t = \begin{bmatrix} \cos\theta & j\hat{Z}_{mt}\,\text{sen}\,\theta \\ j\hat{Y}_{mt}\,\text{sen}\,\theta & \cos\theta \end{bmatrix} \tag{7.2.86}$$

and for the center "c" section:

$$(C_m)_c = \begin{bmatrix} \cos\theta & j\hat{Z}_{mc}\,\text{sen}\,\theta \\ j\hat{Y}_{mc}\,\text{sen}\,\theta & \cos\theta \end{bmatrix} \tag{7.2.87}$$

where m = e,o and "$\hat{Z}_{mt}$," "$\hat{Y}_{mt}$," "$\hat{Z}_{mc}$," and "$\hat{Y}_{mc}$," are the normalized impedances and admittances. The chain matrix of the whole coupler will be:

$$(C_m)_c = \begin{bmatrix} A_m & B_m \\ C_m & D_m \end{bmatrix} = (C_m)_t (C_m)_c (C_m)_t \tag{7.2.88}$$

Now, let us examine Figure 7.2.24, which is Figure 7.2.19 from another point of view. We may write:

$$\Gamma_1 = 0.5s_{11,e} + 0.5s_{11,o} \tag{7.2.89}$$

$$\Gamma_2 = 0.5s_{11,e} - 0.5s_{11,o} \tag{7.2.90}$$

$$\Gamma_3 = 0.5s_{21,e} + 0.5s_{21,o} \tag{7.2.91}$$

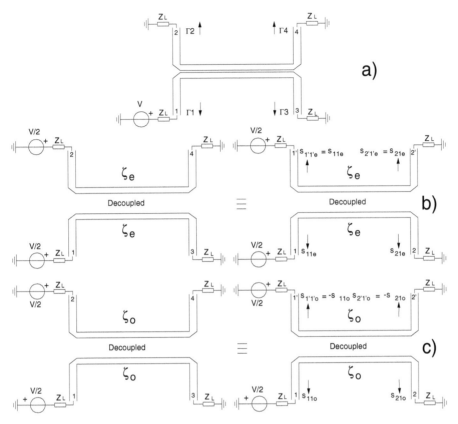

Figure 7.2.24

$$\Gamma_4 = 0.5s_{21,e} - 0.5s_{21,o} \tag{7.2.92}$$

where the generic "Γ_i" is the voltage reflection coefficient at port "i." From the previous equations, we may recognize that to study the network in Figure 7.2.24 we only need the scattering parameters "$s_{11,e}$," "$s_{11,o}$," "$s_{21,e}$," and "$s_{21,o}$." The same conclusions hold for our three element coupler, and from the conversion formulas given in Appendix A3 we have:

$$s_{11,m} = \frac{A_m + B_m - C_m - D_m}{A_m + B_m + C_m + D_m} \tag{7.2.93}$$

$$s_{21,m} = \frac{2}{A_m + B_m + C_m + D_m} \tag{7.2.94}$$

Resolving the expressions 7.2.88 through 7.2.94 for the case $\zeta_{et} = 56.05$ and $\zeta_{ec} = 107.35$ we have the graphs of Figure 7.2.25. It is important to inform the reader that in the case of multisection step coupled line couplers, the determination of the impedance values of the coupled lines is not so simple as the single step coupler. We will return later to this topic. In Figure 7.2.25 we may see how these values of "ζ_{et}" and "ζ_{ec}" give a directional coupler with -6 dB of coupling, with a coupling ripple of ± 0.2 dB. It is interesting to show that the phase difference between signals at ports "2" and "3" is always of $\pi/2$, as indicated in Figure 7.2.26. This phase difference is the same in all symmetrical directional couplers, regardless the number of sections. Note also that if the two

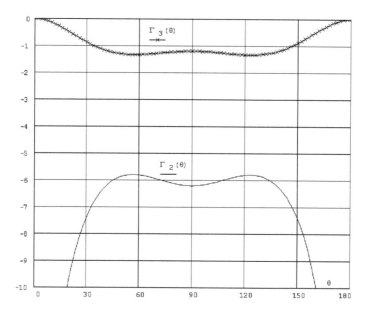

Figure 7.2.25

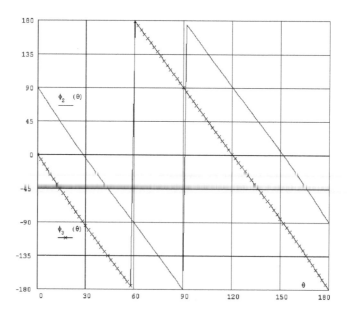

Figure 7.2.26

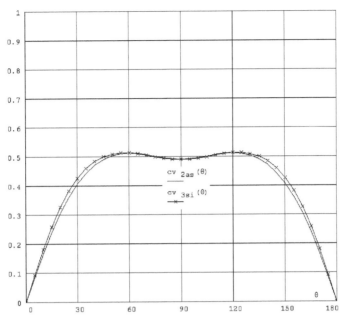

Figure 7.2.27

ring Wilkinson power divider and the coupler under study have two completely different layout aspects, both these networks are able to follow a Chebyshev characteristic, which means that an equiripple behavior exists. From this point of view, such networks are completely equal to any other network that can be synthesized using Chebyshev polynomials.

It is interesting to observe that an asymmetrical directional coupler with "n" sections, with "n" an even number, has approximately the same bandwidth as a symmetrical coupler with "n + 1" sections. If "n" is an odd number, then the bandwidth of the asymmetrical directional coupler is approximately equal to the bandwidth of a symmetrical coupler with "n + 2" sections. This can be shown in Figure 7.2.27, where we have drawn the linear coupling coefficient "cv_{2as}" for a two section asymmetrical coupler and "cv_{3si}" for a three section symmetrical 6 dB coupler. We see how approximately the same bandwidth can be reached with proper design. This means the asymmetrical coupler is shorter when a bandwidth and a ripple are fixed.

A large difference exists between the symmetrical and asymmetrical coupler. While the former always has a phase difference of $\pi/2$ between the signals at the direct and coupled ports, the asymmetrical coupler has no such characteristic. This result is shown in Figure 7.2.28, where with "ϕ_2" and "ϕ_3" we have respectively represented the phase of the signals at ports "2" and "3" for the two section asymmetrical coupler whose coupling coefficient "cv_{2as}" is represented in Figure 7.2.27.

With the chain matrix procedure we can study all multisection couplers. For instance, in figures 7.2.29 and 7.2.30 we have respectively represented the coupling "cv" and direct "dv" coefficients, the phase "ϕ_c" for coupled and "ϕ_d" direct coefficient for a five section asymmetrical coupler. In this example, the coupling has been synthesized for −6 dB and ripple ±0.2 dB.

The determination of the impedance values of the coupled line sections is not a simple task,[28] compared to the synthesis of the previous networks we have studied. For this purpose it is necessary to have great confidence in the general theory of electrical networks synthesis. It is not possible to treat a theory of this size in our book, since typical books on this topic[29,30,31] are composed of nearly one thousand pages. For these reasons, we will only study the most important steps in synthesizing such multisection couplers. However, for the reader who has familiarity with electrical

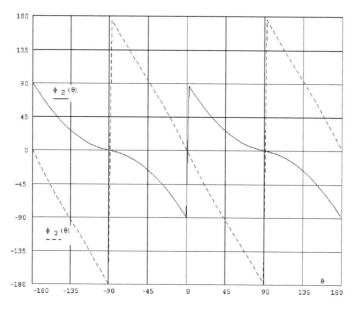

Figure 7.2.28

network synthesis, in Appendix A4 we have reported the fundamental concepts of this important branch of electrical engineering.

The first step in network synthesis is to find a mathematical function that is able to represent the desired transfer function shape, the coupling in our case. Finding this is not, in general, a simple thing, but in our case we may simplify this task using an analytical equivalence between the step transmission line transformers, which we have studied in Chapter 1, and our network. This situation is indicated in Figure 7.2.31, where the step transformer is a symmetric network composed of a series of step-up and step-down transformers, or vice versa, in order to realize a symmetric network. In fact, it is possible to show[32] that the square $|\Gamma|^2$ of the voltage reflection coefficient of the step

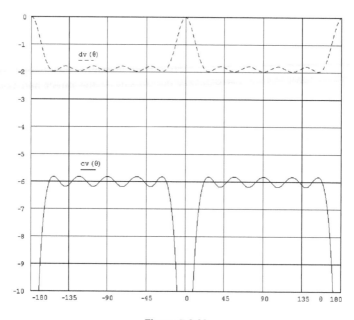

Figure 7.2.29

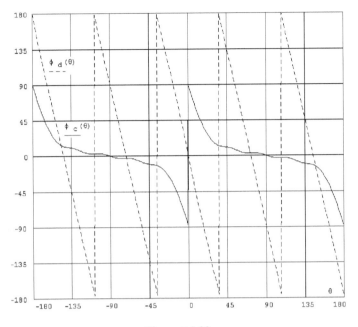

Figure 7.2.30

transformer has the same analytical form as the coupling "c_v" for the symmetrical coupler. So, it is possible to write:

$$c_w = ! \, w_2 / w_1 \equiv | \Gamma |^2 \tag{7.2.95}$$

and consequently, from the energy conservation theorem:

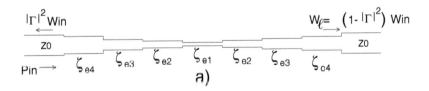

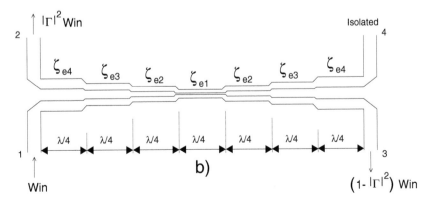

Figure 7.2.31

$$d_w =! w_3/w_1 \equiv 1 - |\Gamma|^2$$

Further help comes from a theorem applicable to symmetric networks, as in the coupler family under study. It is known from the filter network theory that a symmetric "n" section transmission line network is characterized as having an insertion loss function "L(θ)" given by:

$$L(\theta) \overset{\perp}{=} W_{in}/W_\ell =! 1 + \left[P_n(sen\theta)\right]^2 \qquad (7.2.96)$$

where "θ" is the line's electrical length and "$P_n(sen\theta)$" is an odd polynomial of degree "n." So the order "n" of the polynomial is an odd number, as of course the number "n" of sections of the symmetrical coupler must be. From the insertion loss function 7.2.96, we can obtain the square of the modulus of the voltage reflection coefficient "$\Gamma(\theta)$" from:

$$\left|\Gamma(\theta)\right|^2 =! \frac{L(\theta) - 1}{L(\theta)} \qquad (7.2.97)$$

which in our case becomes:

$$\left|\Gamma(\theta)\right|^2 \equiv \frac{\left[P_n(sen\theta)\right]^2}{1 + \left[P_n(sen\theta)\right]^2}$$

As it is known from network theory, and also from the previous points of study, a preferable shape for "c_v" is the Chebyshev type. Crystal and Young[33] have shown that symmetrical couplers can be synthesized with an equiripple shape for "c_v" but not responding to the Chebyshev type. The asymmetrical couplers, which we will study later, can instead be synthesized with Chebyshev polynomials, and this is the reason why asymmetrical directional couplers with "n" sections have approximately the same bandwidth as a symmetrical coupler with "n + 1" or "n + 2" sections, depending on whether "n" is an even or odd number. In fact, it is known from network theory that any electrical network synthesized with a Chebyshev polynomial has the widest useful bandwidth of any other synthesis when a ripple in the bandwidth is fixed. Crystal and Young have used their own equiripple polynomials $P_n(x)$, which approach the value 1 for values of "x" in the interval 0 through 1. In Figure 7.2.32 we have indicated four odd polynomials used for the synthesis. As in the case of Chebyshev polynomials, in this case, if we accept a higher value of coupling ripple, the useful bandwidth increases for a fixed number of sections.

These two researchers have found the normalized values of the coupled line impedances for the most used values of coupling "c_v," coupling peak ripple "δ_p," and fractional bandwidth "B." In Tables 7.2.1 and 7.2.2 we report their results for three and five sections.

The reason that only the even impedances are reported in the previous tables is due to the fact that for the multistep couplers indicated in Figure 7.2.31b, the known formula 7.2.72 must be satisfied. In addition, since the "n" section coupler is symmetrical, it is necessary to specify only (n + 1)/2 values of impedance. It is also important to remember that all the "ζ_{ei}," with i = 1,2,3, in the previous tables are normalized to the system reference impedance. Of course, as it happens in all electrical networks where an equiripple behavior can be found, the useful bandwidth for such couplers is defined as the frequency interval where the equiripple value of "c_v" is assured. The reason why a value of c_v = −8.34 dB is inserted in the tables will be discussed later when we discuss the 3 dB coupled line coupler. The reader should remember that in microstrip thin film technology, it is not possible to reach a coupling higher than −6 dB using only tandem of coupled lines, and for this reason the presence of c_v = −3 dB in the previous table could appear strange.

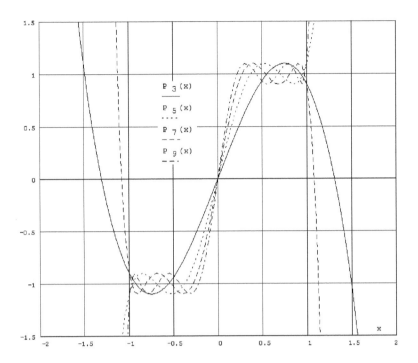

Figure 7.2.32

Table 7.2.1 Three Sections, Symmetrical

c_v(dB)	δ_p (dB)	ζ_{e1}	ζ_{e2}	B%
	0.1	1.171	3.26	100
−3	0.3	1.24	3.543	127
	0.5	1.3	3.785	141
	0.1	1.103	2.094	92
−6	0.3	1.136	2.19	117
	0.5	1.164	2.265	130
	0.1	1.074	1.719	89
−8.34	0.3	1.097	1.773	114
	0.5	1.115	1.815	127
−10	0.5	1.091	1.621	126
−20	0.5	1.027	1.16	124
	1	1.035	1.172	141

Table 7.2.2 Five Sections, Symmetrical

c_v(dB)	δ_p (dB)	ζ_{e1}	ζ_{e2}	ζ_{e3}	B%
	0.1	1.078	1.373	3.976	132
−3	0.3	1.136	1.495	4.401	153
	0.5	1.189	1.592	4.742	162
	0.1	1.045	1.22	2.38	125
−6	0.3	1.074	1.28	2.52	145
	0.5	1.098	1.323	2.62	155
	0.1	1.032	1.157	1.89	123
−8.34	0.3	1.043	1.180	1.934	135
	0.5	1.068	1.225	2.022	152
−10	0.5	1.054	1.179	1.772	150
−20	0.5	1.016	1.051	1.191	149
	1	1.024	1.063	1.206	162

But we have reported this value in light of stripline transmission line technology, which we will study in a following chapter. Tables 7.2.1 and 7.2.2 and the following Tables 7.2.3 through 7.2.5 assume a pure "TEM" propagation mode and for this reason these tables are valid for striplines too. In addition, we will see that using striplines, we can reach a 3 dB coupling with broadside coupled striplines. So, all these tables can also be used for striplines.

An asymmetrical step coupled lines directional coupler shown in Figure 7.2.23b has the advantage that it can be synthesized using Chebyshev polynomials for the loss function "$L(\theta)$." These kinds of couplers have been studied by many researchers, but one of the most complete studies was done by R. Levy,[34] who reported the synthesis procedure for these networks. Levy has also

Table 7.2.3 Two Sections, Asymmetrical

c_v(dB)	δ_p (dB)	ζ_{e1}	ζ_{e2}	B%
	0.13	3.129	1.321	100
−3	0.29	3.289	1.423	120
	0.48	3.445	1.532	133
	0.03	1.987	1.148	66.6
−6	0.19	2.06	1.205	100
	0.44	2.129	1.268	120
	0.04	1.641	1.108	66.6
−8.34	0.22	1.685	1.149	100
	0.5	1.726	1.194	120
−10	0.53	1.573	1.157	120
−20	0.58	1.145	1.047	120
	0.96	1.156	1.057	133

Table 7.2.4 Three Sections, Asymmetrical

c_v(dB)	δ_p (dB)	ζ_{e1}	ζ_{e2}	ζ_{e3}	B%
	0.08	3.525	1.639	1.138	120
−3	0.27	3.767	1.863	1.252	143
	0.51	3.983	2.074	1.38	155
	0.11	2.224	1.387	1.092	120
−6	0.24	2.275	1.45	1.128	133
	0.58	2.365	1.568	1.206	150
	0.13	1.779	1.275	1.063	120
−8.34	0.27	1.808	1.318	1.094	133
	0.45	1.835	1.358	1.123	143
−10	0.48	1.656	1.287	1.1	143
−20	0.53	1.168	1.082	1.03	143
	1	1.175	1.096	1.044	155

Table 7.2.5 Four Sections, Asymmetrical

c_v(dB)	δ_p (dB)	ζ_{e1}	ζ_{e2}	ζ_{e3}	ζ_{e4}	B%
	0.1	3.921	2.047	1.379	1.111	143
−3	0.25	4.097	2.252	1.518	1.188	155
	0.43	4.256	2.438	1.656	1.275	164
	0.16	2.379	1.603	1.241	1.075	143
−6	0.38	2.446	1.706	1.325	1.126	155
	0.64	2.505	1.796	1.405	1.183	164
	0.18	1.865	1.417	1.175	1.056	143
−8.34	0.43	1.902	1.483	1.234	1.094	155
	0.73	1.934	1.539	1.289	1.135	164
−10	0.45	1.708	1.384	1.189	1.077	155
−20	0.5	1.179	1.106	1.055	1.023	155
	1.2	1.189	1.126	1.078	1.043	169

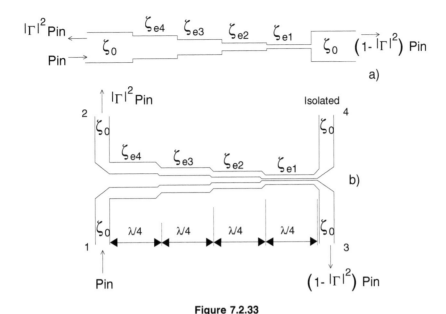

Figure 7.2.33

shown that an analytical equivalence between an asymmetrical network of quarter wavelength transmission lines and an asymmetrical step coupler exists, in the same manner as Young has shown for symmetrical step couplers. This situation is indicated in Figure 7.2.33. With this analytical equivalence we may use 7.2.95 and 7.2.97 for the synthesis procedure, to evaluate the coupling function of the coupler. For these kinds of asymmetrical couplers, Levy has shown that the insertion loss function "L(θ)" may be written as:

$$L(\theta) = 1 + \beta^2 - h^2 T_n^2\left(\cos\theta/\cos\theta_0\right) \tag{7.2.98}$$

where "β," "h," and "θ_0" are three constants that are related to the mean value and ripple of "L(θ)," while $T_n(\cos\theta/\cos\theta_0)$ in the Chebyshev polynomial of the first kind and order "n." Also in this case, the order "n" corresponds to the number of coupled sections used in the coupler.

We report Tables 7.2.3 through 7.2.5 for two, three, and four sections of coupled lines asymmetrical coupler as reported by Levy in his tables.[15] We have used the notation to indicate with the subscript "1," the most coupled section, and consequently the higher subscript number will indicate the section with the smallest coupling.

Of course, in Tables 7.2.3 through 7.2.5 we have only given the normalized even impedances of the lines since for this coupler the general relation $Z_\ell \equiv (\zeta_e\zeta_o)^{0.5}$ holds. The reader should remember that in practice, a number higher than three or four for the coupled sections is seldom used. In fact, the theoretical increase of bandwidth is limited by the increasing source of discontinuity in each step of the coupler, as we have seen in Chapter 4. The increase in the number of discontinuities has the undesirable effect of decreasing the isolation that is a very important parameter in any directional coupler. Of course, this effect is common for both the symmetrical and asymmetrical coupler. It is for this reason that another type of coupler, based on coupled lines has been developed, as we are going to study in the next section.

Maximum isolation values for these microstrip coupled lines couplers are near −25 dB at 10 GHz, regardless the value of coupling. So, to have a good directivity, which is the most desirable characteristic of a directional coupler, from 7.2.97 it follows that the coupling cannot be much lower than −15 dB or −20 dB, otherwise the coupler will have a very small directivity.

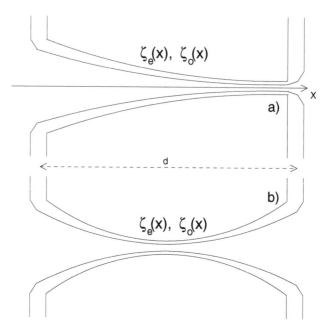

$\zeta_e(x),\ \zeta_o(x)$

a)

d

b)

$\zeta_e(x),\ \zeta_o(x)$

Figure 7.2.34

7.2.5 Tapered Coupling

To decrease the discontinuities effects due to step transitions of the step coupled lines directional couplers, the step coupling has been replaced with a continuous increase and/or decrease of the coupling. This means that the distance between the microstrips decreases and/or increases in a continuous way. The networks that realize this type of coupling are indicated in Figure 7.2.34. Part a shows the asymmetrical version, while part b shows the symmetrical network. These networks are backward couplers like the previous ones, but in addition, they do not have higher order operating bands for the coupling. These couplers are also indicated as "nonuniform coupled line directional couplers," but in our study we will refer to these networks as "tapered coupling directional couplers" or simply "tapered couplers." The analytical formula that gives the distance "s (x)" between the conductors, with "x" the longitudinal variable, can theoretically be of any type, but always continuous. An often used shape is the exponential one.

In general, these networks have higher isolation than the step coupled line directional couplers we have just studied.[36,37] When a "c_v" is defined, the tapered coupling couplers are a little longer and the highest coupling between the lines is a little higher than the length and the highest coupling value of the previous couplers we studied.

The most evident difference with the step coupled couplers is that now there must exist a mathematical equation that describes the variation of "c_v" with the longitudinal variable "x." This function must be continued inside the values $|x| < d$, where "d" is the absolute length of the coupler. For the tapered coupling directional couplers, the general relations 7.2.70 and 7.2.71 still hold, with the difference that now the center band coupling is a function of "x." So we may write:

$$\zeta_e(x) = Z_\ell \left[\frac{1 + |c_{0v}(x)|}{1 - |c_{0v}(x)|} \right]^{0.5} \tag{7.2.99}$$

$$\zeta_o(x) = Z_\ell \left[\frac{1 - |c_{0v}(x)|}{1 + |c_{0v}(x)|} \right]^{0.5} \tag{7.2.100}$$

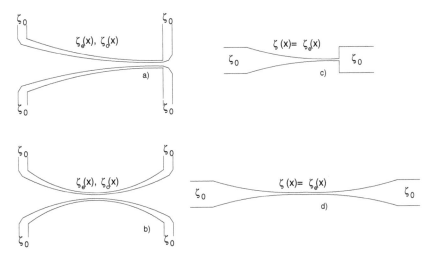

Figure 7.2.35

Actually, for these kinds of couplers, the concept of center band coupling is not general since these couplers can also have a theoretically high pass shape for the coupling. In such a case, a center band does not exist, and "$c_{0v}(x)$" assumes the value "$c_v(x)$" of the coupling in the high pass band. We will show later how these couplers may have the high pass shape for the coupling.

For the tapered coupling directional couplers, researcher C. B. Sharpe[38] has reported an analytical equivalence with a nonuniform transmission line in the same manner as L. Young and R. Levy have reported for symmetrical and asymmetrical step coupled lines directional couplers. This equivalence is indicated in Figure 7.2.35, for symmetrical and asymmetrical cases.

So, expressions 7.2.95 and 7.2.97 still hold for our case, and we may concentrate our study on the simple transmission line networks indicated in Figure 7.2.35c and d. In this case, from the theory of nonuniform transmission lines studied in Chapter 1, the voltage reflection coefficient "$\Gamma(\beta)$" for the general nonuniform transmission line can be written as:

$$\Gamma(\beta) = \int_0^d e^{-2j\beta x}\Gamma(x)\,dx \equiv \Gamma\left(\frac{2\pi}{v}f\right) \tag{7.2.101}$$

where "v" is the speed of the electromagnetic wave along the line. Note that we assume our study to be a case of pure "TEM" wave, while we know that microstrips only approximate such a mode. Later we will see the consequences of the "qTEM" microstrip propagation mode for such a coupler.

In addition to the previous equation, to study a tapered coupling directional coupler we may use the theory developed by B.M. Oliver,[39] which is one of the first theories regarding the general electromagnetic coupling between lines. Starting from the telegraphists equations, that we studied in Chapter 1, B. M. Oliver has shown that the resulting coupling value "$c_v(\beta)$" of the coupler is related to the punctual coupling "$C_v(x)$" through the equation:

$$c_v(\beta) = j\beta \int_0^d e^{-2j\beta x}C_v(x)\,dx \tag{7.2.102}$$

Both the 7.2.101 and 7.2.102 can be used for the overall coupling, and the choice depends on what distribution $f(x)$ we specify for the coupler.

Concerning the voltage reflection coefficient "$\Gamma(x)$" along the equivalent line, it can be simply evaluated as seen in Chapter 1, and is given by:

$$\Gamma(x) \doteq \frac{d\left[\ln\left((\zeta_n(x))\right)\right]}{2dx} \tag{7.2.103}$$

where "$\zeta_n(x)$" is the normalized impedance of the equivalent line, which is equivalent to the normalized even impedance expression "$\zeta_e(x)$" along the coupler. Remembering the equivalence represented by 7.2.95, Equation 7.2.101 also gives the voltage coupling coefficient "$c_v(\beta)$" for the tapered coupler. In addition, the normalized line impedance "$\zeta_n(x)$," which appears in 7.2.103 is equivalent to the even normalized line impedance "$\zeta_{ne}(x)$" of the coupler. So, given a "$\Gamma(x)$" we can obtain the "$\zeta_{ne}(x)$" through a simple integration and subsequent exponentation of the previous relation.

It is important to remember from Chapter 1 that 7.2.101 and 7.2.102 have been obtained under an assumption of small reflections along the tapered line. Typically these equations are accurate for coupling below -10 dB. Researcher C. P. Tresselt, who has performed one of the most important studies on the tapered couplers,[40] has shown that the previous expression can be modified to include the case of higher coupling, obtaining:

$$c_v(\beta) = \frac{\left|\int_0^d \Gamma(x)e^{-j2\beta x}dx\right|}{\sqrt{\left|1+\int_0^d \Gamma(x)e^{-j2\beta x}dx\right|^2}} \tag{7.2.104}$$

After the expression 7.2.101, which is strongly related to the coupled transmission lines theory given in Chapter 1, it is very simple to show how the tapered couplers have a high pass behavior for the coupling. First of all, let us assume a symmetric tapered coupler. Then, shift the origin of the "x" axis to the middle of the coupler so that its length is from $x = -l/2$ up to $x = l/2$. Then, normalize each half of the coupler so that its length is from $x = -1$ up to $x = 1$. Then, use a function "$\Gamma(x)$" given by:

$$\Gamma(x) \equiv \begin{array}{ll} \Gamma_M(x) \doteq -0.201(1+x)/2x & \text{for } -1 \le x < 0 \\ \Gamma_p(x) \doteq -0.201(1-x)/2x & \text{for } 0 < x \le 1 \end{array} \tag{7.2.105}$$

This function is depicted in Figure 7.2.36. Inserting 7.2.104 in 7.2.101 we have the graph in Figure 7.2.37, where $\theta \doteq \beta d/\pi$. Note the coupling has a high pass shape, which is completely different from any other directional coupler we have studied. Equation 7.2.105 cannot be physically synthesized since for $x = 0$ this expression assumes an infinite value. For this purpose, two considerations need to be made: first of all, nothing exists in physics that is infinite; in addition, from 7.2.103 it follows that the infinite value of "$\Gamma(x)$" also means that "ζ_e" is infinite, with the consequence that a local coupling value of "1" exists. In fact, from 7.2.99 it follows that:

$$\left|c_{0v}(x)\right| = \frac{\zeta_{en}^2(x)-1}{\zeta_{en}^2(x)+1} \tag{7.2.106}$$

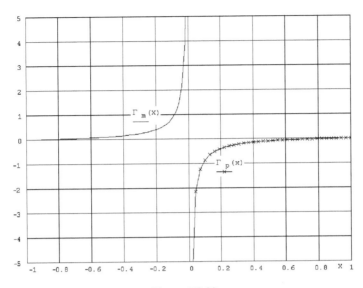

Figure 7.2.36

from which we can see that when the normalized even impedance "$\zeta_{en}^2(x)$" is infinite, then $|c_{0v}(x)| = 1$. So, it is clear that we need to truncate "$\Gamma(x)$" to some realizable value. In Figure 7.2.38 we have the coupling shape "cl_θ" when the function 7.2.105 is modified as:

$$\Gamma(x) \equiv \begin{array}{l} \Gamma_m(x) \doteq -0.201(1+x)/2(x-g) \quad \text{for } -1 \leq x < 0 \\ \Gamma_p(x) \doteq -0.201(1-x)/2(x+g) \quad \text{for } \ 0 < x \leq 1 \end{array} \tag{7.2.107}$$

with g = 0.005, so that $\Gamma(0) = 20$. For comparison we have also reported the coupling "c_θ" when we use the function 7.2.105. From this figure we see how using the limited function, the coupling has a strong negative slope, which causes a decrease in the −10 dB coupling value. It is instead

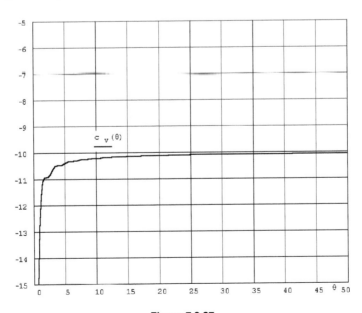

Figure 7.2.37

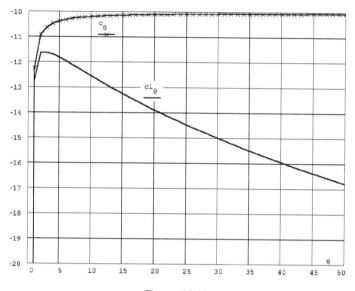

Figure 7.2.38

possible to have a good band pass shape with proper weighting of the limited function "$\Gamma(x)$." This situation is indicated in Figure 7.2.39, where with "cc_θ" we have reported the coupling with an opportunely weighted "$\Gamma(x)$" given by 7.2.107 with g = 0.001. To this value of "g," integrating and exponentiating 7.2.103, corresponds a value of normalized even impedance $\zeta_{en}(x) = 3.28$, which is the practical upper limit for microstrip technology. Note as the coupling "cl_θ" for the limited function 7.2.107 with g = 0.001 is better than the case reported in Figure 7.2.38 when g = 0.005, but a deep slope always exists.

From these figures we should recognize that in a tapered coupling symmetrical directional coupler a perfect high pass behavior for the coupling is only theoretical, and only band pass shapes

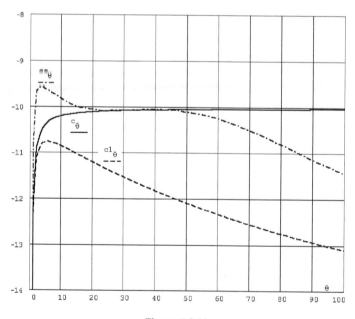

Figure 7.2.39

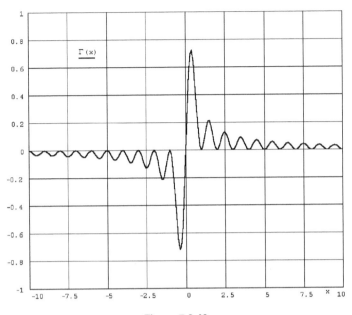

Figure 7.2.40

are physically realizable. We will show later that with asymmetrical couplers it is instead possible to have a good high pass shape for the coupling.

In order to realize a band pass equal ripple shape for the coupling of symmetrical tapered coupling couplers, Tresselt has found a family of functions "$\Gamma(x)$" which are very suitable to the purpose. One of them is the function:

$$\Gamma(x) = A \frac{\text{sen}^2(kx)}{kx} \tag{7.2.108}$$

where "k" is a constant to transform in angle the length "x," and "A" is a scale factor. The previous function is represented in Figure 7.2.40, with $k = \pi$ and $A = 1$. Note that the previous functions also have theoretically an infinite length, with the consequence that the symmetrical directional coupler will have a maximum local coupling value equal to one. So, also in this case we have to limit 7.2.108. In Figure 7.2.41 we have reported the linear value of the coupling "$|c_4(\beta)|$" and "$|c_8(\beta)|$," when we have limited to 4 and 8, respectively, the total lobes number of 7.2.108, i.e., the sum of left half plane and right half plane lobes. For this case $A = 0.98$. Note that the shape of the coupling in the pass band is not an equiripple, but if the number of the lobes used is increased, the pass band shape assumes an equiripple aspect. To improve the equiripple shape, Tresselt has performed a weighting of 7.2.108 for any lobe used, which produces a very good result.

Practical applications of the band pass coupling shape tapered couplers have shown that such devices have their lower reciprocal distance "s" and width "w," which is somewhat smaller than the value of "s" and "w" for the case of step couplers with the same coupling value and operating bandwidth. Also the overall length of the tapered case is somewhat longer than its stepped coupler counterpart.

Another method of studying the tapered coupling directional couplers is based on the use of ABCD matrices. In this case, the length "d" of the coupler is divided in many segments for which the variation of impedance along "x" may be evaluated as constant. In other words, with this method the coupler is evaluated as composed of "N" stepped directional couplers, as we studied in section 7.2.4, with the difference that their lengths need not be a quarter wavelength long. C. P. Tresselt

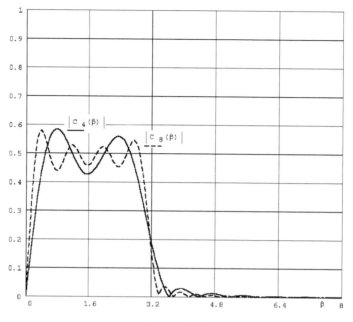

Figure 7.2.41

has used[41] this method to study some asymmetrical tapered coupling directional couplers. Using the even impedance distribution function:

$$\zeta_e(x) = e^{[0.024192 + 0.7822\,(x/d)2]}$$

which corresponds to a coupling value "c_v" of −8.34 dB, we obtain the result given in Figure 7.2.42. We see how the asymmetrical tapered coupler has a high pass shape, and in our case, this behavior is obtained for a maximum coupling value of 0.668, which is a reachable value in microstrip

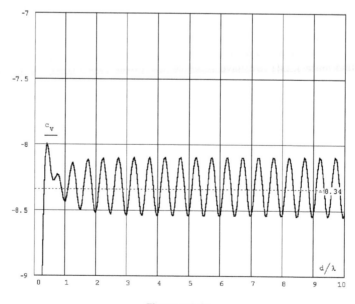

Figure 7.2.42

technology. Of course, radiation phenomena and the non- "TEM" propagation mode in the microstrip will practically limit the high pass shape of the coupling. With this method it is very important to use a large number of sections to form the whole coupler, usually some hundreds, otherwise the error in the result cannot be acceptable with respect to the exact result obtained with the integral analysis given by 7.2.101. For example, the graph in Figure 7.2.44 has been obtained with N = 300 subdivision of the asymmetrical tapered coupler.

Another important aid in designing tapered line directional couplers comes from F. Arndt,[42,43] who has studied the case of a high pass shape of the coupling in an asymmetrical coupler. He has applied the integral theory given in 7.2.102 to coupling functions "$C_v(x)$" given by an n-th order polynomial, i.e.:

$$C_v(x) = \prod_{m=0}^{n} a_m x^m \qquad (7.2.109)$$

where the coefficients "a_m" are functions of the desired operating bandwidth coupling. Arndt has given some tables for the case of n = 6, where we can find the coefficients "a_i" of the previous equation, for the most used coupling values. We report some of his results in the following Table 7.2.6:

Table 7.2.6 Six Order, Asymmetrical, Tapered Polynomial-Coupling, Directional Coupler

c_v(dB)	δ_p (dB)	a_0	a_1	a_2	a_3	a_4	a_5	a_6
−3	0.08	0.943	−0.494	−0.969	−3.911	11.094	−9.326	2.693
	0.34	0.944	−0.38	−0.645	−1.625	3.534	−2.12	0.407
−6	0.23	0.8	−0.752	−0.791	0.438	1.507	−1.686	0.524
	0.4	0.8	−0.682	−0.617	0.193	1.227	−1.199	0.344
−8.34	0.17	0.669	−0.889	−0.487	1.106	−0.043	−0.583	0.246
	0.42	0.669	−0.759	−0.324	0.555	0.184	−0.426	0.147
−10	0.26	0.575	−0.814	−0.159	0.7	−0.184	−0.187	0.091
	0.6	0.575	−0.687	−0.105	0.376	−0.025	−0.134	0.05
−20	0.25	0.198	−0.354	0.141	0.058	−0.023	−0.027	0.013
	0.42	0.198	−0.323	0.118	0.04	−0.008	−0.024	0.01
	0.82	0.198	−0.275	0.088	0.014	−0.005	−0.013	0.003

In this table we have indicated the peak value of the coupling ripple with "δ_p", with respect to the mean value "c_v(dB)." Using the values in the previous table and expression 7.2.109, in Figure 7.2.43 we have reported the coupling functions "$C_6(x_n)$" and "$C_{20}(x_n)$," respectively, for the case of 6 dB and 20 dB of coupling, for "δ_p" near 0.25 dB in both cases. With "x_n" we have indicated the normalized length of the asymmetrical coupler indicated in Figure 7.2.34a. Characteristic of Arndt's polynomials is that the resulting "$c_v(\beta)$" has an almost equal ripple shape in the high pass band. As an example, in Figure 7.2.44 we have indicated the coupling "$c_{20}(\beta)$" for the case −20 dB of coupling and 0.25 dB of ripple peak, whose polynomial coefficients are given in Table 7.2.6. The phase "$\phi_{20}(\beta)$" behavior of the signal at the coupled port is interesting, indicated in Figure 7.2.45.

Due to small discontinuities in the structure, the consequent high directivity, and the absence of higher unwanted operating bandwidth, the tapered couplers are the preferred choice when high performance microstrip directional couplers are required.

Of course, a practical limitation in the operating bandwidth exists, due to dispersion and radiation phenomena in the microstrip structure. The most detrimental cause for bandwidth reduction is the difference between the even "β_e" and odd "β_o" phase constants in coupled µstrips, as we studied in Chapter 5, and consequently the different phase velocities. A method to equalize such velocities has been suggested by S. Uysal and H. Aghvami,[44] resulting in the structure indicated in Figure 7.2.46, where the coupled side of the µstrips are wiggled. With a tandem configuration of two of these couplers, each one with cv = −8.34 dB, they have built a 3 dB coupler in the

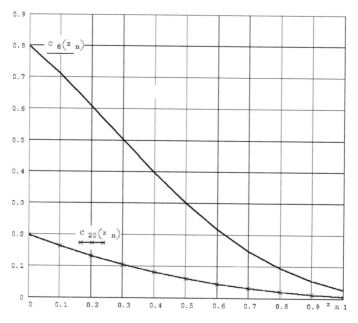

Figure 7.2.43

operating bandwidth of 2 to 18 GHz with good characteristics. Of course, wiggled lines can also be used for step couplers.[45]

Tapered coupling couplers have also been investigated for forward coupling by P. Ikalainen and G. Matthaei.[46] Employing asymmetric width coupled µstrips in a structure similar to that indicated in Figure 7.2.35b, they have built a 3 dB forward coupler with an operating bandwidth of 57%. However, due to the longer extension compared to other 3dB couplers, as will be seen in the next section, this structure could be attractive in the millimeter wave region.

7.2.6 Interdigital or "Lange"

The "Lange coupler" is the only compact, wideband, microstrip directional coupler that is able to easily reach −3 dB of coupling or even more. This network was proposed for the first time by J. Lange,[47] and it is perhaps one of the most used microstrip networks. Its physical realizations are indicated in Figure 7.2.47, where the network in part a is the original one developed by Lange.

We see that a number "n" of coupled lines with length $\lambda/4$ are connected together at each extreme. In the case of part a of the figure, the two $\lambda/8$ lines form one $\lambda/4$ line for those frequencies where the length of the connecting wires may be neglected. The connections are indicated in Figure 7.2.47 with solid thick lines and are called "air bridges." The number "n" of coupled lines is usually an even number, typically four or six, although in theory it is possible to use three lines.[48] The Lange coupler is also called an "interdigital coupler" or simply "hybrid." With the hypothesis of neglecting the length of the air bridges, the Lange coupler may be thought of as many quarter wavelength directional couplers connected in parallel. So, to explain the path of a signal entering the coupler we may use the theory studied in section 7.2.4. As a consequence of this coincidence, the Lange coupler is a backward type. For example, and with reference to Figure 7.2.47a, a signal entering in port "1" has the coupled port "2" and the direct port "4," while port "3" is isolated. In addition, phase shift between signals at direct and coupled ports is 90°. For this reason, the Lange coupler is also called a "3 dB quadrature hybrid."

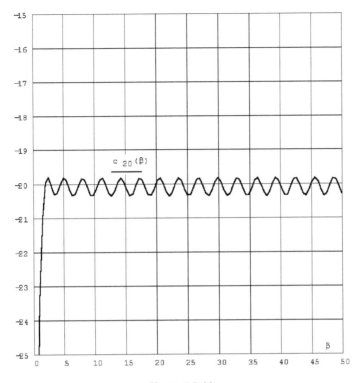

Figure 7.2.44

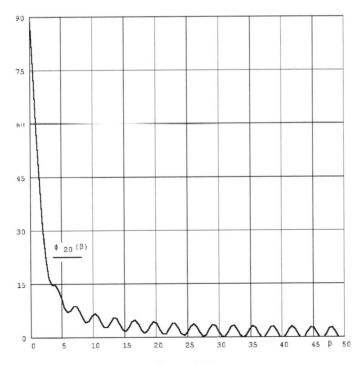

Figure 7.2.45

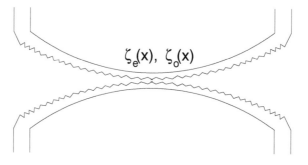

$\zeta_e(x), \ \zeta_o(x)$

Figure 7.2.46

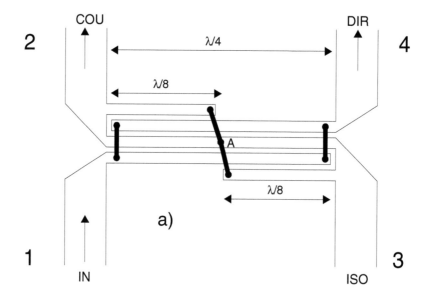

a)

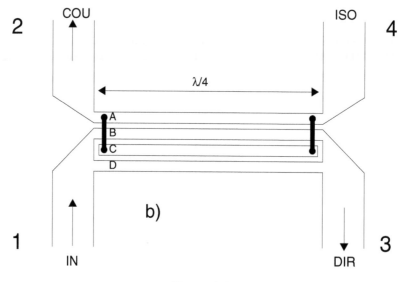

b)

Figure 7.2.47

The network in Figure 7.2.47b still belongs to the Lange coupler family shown in part a of the figure. It was first suggested by R. Waugh and D. Lacombe.[49] It is usually called an "unfolded Lange coupler," and the difference between these two networks is that they exchange the isolated and direct port, as we see in Figure 7.2.47 parts a and b. For both the versions, each of the coupled lines is called a "finger" in the normal practice.

Due to the multicoupling effect between the fingers, with the Lange coupler it is possible to reach 3 dB of coupling or even more with the consequence that this network is the only "3 dB quadrature microstrip coupler" employed in practice in wide bandwidth circuits. In fact, for the four fingers network and 2.5 dB of center band coupling, it is possible to reach operating bandwidth of 100%, i.e., 3:1 between the limits of the frequency band. Theoretically, coupling increases with the number of fingers, but in practice only four fingers are used, especially for center frequencies above 10 GHz. The theoretical increase of coupling is limited by the increase in the number of discontinuities encountered using a higher number of fingers. As we will see later, with the Lange coupler it is also possible to have coupling lower than –3 dB, but in this case other networks can be simply used for this purpose.

One of the simplest studies on the Lange coupler was made by Wen Pin Ou.[50] In this method the equivalence of the coupler is used with an array of coupled transmission lines, as indicated in Figure 7.2.48a. Since we think that it is important to have a theoretical view of the Lange coupler, we will discuss such a method of study. Let us suppose that coupling between the coupled lines of Figure 7.2.48a only exists among adjacent lines, so that, for example, line "2" is only affected by lines "1" and "3." Each line has its own admittance "Y," which depends on its position in the array. If a line would be alone and short circuited at one extreme it would have an input admittance "$Y(\theta)$" given by:

$$Y(\theta) = -jY_\ell \cot g(\theta) \overset{+}{_{-}} Y_\ell P(\theta)$$

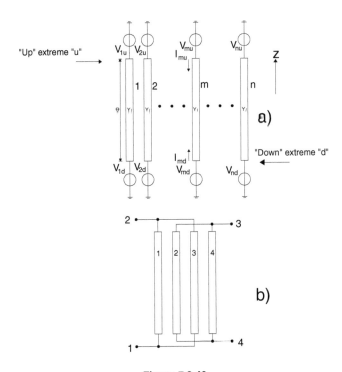

Figure 7.2.48

where "Y_ℓ" is the admittance of the line. The fact that the lines in Figure 7.2.48a are not an infinite distance apart changes this situation, and the input admittance "$Y_{mm}(\theta)$" for the generic line "m" is:

$$Y_{mm}(\theta) \overset{\perp}{=} Y_{mm} P(\theta)$$

where "Y_{mm}" depends on the position in the array of the line "m." In fact, for the first "1" and last "n" line, the input admittance is:

$$Y_{11} \equiv Y_{nn} =! \; Y_{\ell\ell} \qquad\qquad (7.2.110)$$

For any other line "m," that is with m ≠ 1 or m ≠ n, we have:

$$Y_{mm} = Y_{\ell\ell} + Y_c^2 / Y_{\ell\ell} \qquad\qquad (7.2.111)$$

where "Y_c" is the mutual admittance between two adjacent lines, that is:

$$Y_{m,m+1} \equiv Y_{m,m-1} =! \; Y_c \qquad\qquad (7.2.112)$$

Now, let us evaluate the current entering in one extreme, for example the "d" extreme. This current will be affected by the "d" generators and "u" generators. Since the system under study is a linear one, we can use the superposition effects principle.

Let us start to evaluate the system when only the lower generators are present. The current "I_{md}" entering in the lower extreme of the line "m" will be due to three terms:

I1. The current "$I_{mmd}(\theta)$" due to the generator "V_{md}":

$$I_{mmd}(\theta) = V_{md} \, Y_{mm}(\theta) \qquad\qquad (7.2.113)$$

I2. The current "$I_{+md}(\theta)$" due to the generator "$V_{(m+1)d}$":

$$I_{+md}(\theta) = V_{(m+1)d} \, Y_c(\theta) \qquad\qquad (7.2.114)$$

where "$Y_c(\theta)$" is given by:

$$Y_c(\theta) \overset{\perp}{=} Y_c \, P(\theta)$$

I3. The current "$I_{-md}(\theta)$" due to the generator "$V_{(m-1)d}$" current, which is equal to the previous "$I_{+md}(\theta)$," that is:

$$I_{-md}(\theta) = V_{(m-1)d} \, Y_c(\theta) \equiv I_{+md}(\theta) \qquad\qquad (7.2.115)$$

Then, let us continue to evaluate the effect on the current "I_{md}" caused by the upper generators. This effect can be evaluated using the transmission matrix, studied in Chapter 1. There will be three currents due to the upper generators:

I4. The current "$I_{mmdu}(\theta)$" due to the generator "V_{mu}":

$$I_{mmdu}(\theta) = jV_{mu}Y_{mm}(\theta)\,\mathrm{sen}\,\theta + I_{mmu}\cos\theta$$

Since $I_{mmu} \equiv -I_{mmd}$, using 7.2.113 the previous equation becomes:

$$I_{mmdu}(\theta) = jV_{mu}Y_{mm}(\theta)\cos ec\,\theta \overset{\perp}{=} V_{mu}Y_{mm}(\theta)T(\theta) \qquad (7.2.116)$$

I5. The current "$I_{+mdu}(\theta)$" due to the generator "$V_{(m+1)u}$":

$$I_{+mdu}(\theta) = V_{(m+1)u}Y_cT(\theta) \qquad (7.2.117)$$

I6. The current "$I_{-mdu}(\theta)$" due to the generator "$V_{(m-1)u}$," current which is equal to the previous "$I_{+mdu}(\theta)$," that is:

$$I_{-mdu}(\theta) = V_{(m-1)u}Y_cT(\theta) \equiv I_{+mdu}(\theta) \qquad (7.2.118)$$

So, the current "$I_{md}(\theta)$" at the lower extreme "d" of the generic line "m" is:

$$I_{md}(\theta) = I_{mmd}(\theta) + I_{+md}(\theta) + I_{-md}(\theta) + I_{mmdu}(\theta) + I_{+mdu}(\theta) + I_{-mdu}(\theta) \qquad (7.2.119)$$

From 7.2.119 it is possible to evaluate the current "$I_{mu}(\theta)$" at the upper extreme "u" of the generic line "m" only changing "d" with "s," obtaining:

$$I_{mu}(\theta) = I_{mmu}(\theta) + I_{+mu}(\theta) + I_{-mu}(\theta) + I_{mmud}(\theta) + I_{+mud}(\theta) + I_{-mud}(\theta) \qquad (7.2.120)$$

With 7.2.118 and 7.2.119 we have completely determined the coupled line system indicated in Figure 7.2.48a. This theory is the general one, which we can apply to the network indicated in Figure 7.2.48b, which represents the case of the four fingers unfolded Lange coupler. Note that as a consequence of the physical connections between the lines, the admittance "Y_p" to ground of each line is:

$$Y_p = 2Y_{\ell\ell} + Y_c^2/Y_{\ell\ell}$$

and the overall coupling admittance "Y_a" of each line will be:

$$Y_a = 3Y_c$$

The two previous formulas can be simply adapted to the case of "n" coupled lines, with "n" an even number, always connected as indicated in Figure 7.2.48b so that we always have a four port network. In such a case "Y_p" and "Y_a" become:

$$Y_p^{(n)} = (n/2)Y_{\ell\ell} + (n/2-1)Y_c^2/Y_{\ell\ell} \qquad (7.2.121)$$

$$Y_a^{(n)} = (n-1)Y_c \qquad (7.2.122)$$

For the case of such a network, the relation for perfect isolation and matching is:

$$Y^2 =! \, Y_p^{(n)2} - Y_a^{(n)2} \tag{7.2.123}$$

where "Y" is the system reference load admittance connected to each port. So, remembering what we said to obtain the previous points I1 through I6, for the case of Figure 7.2.48b we can write the following relation:

$$
\begin{bmatrix} I_1 \\ I_2 \\ I_3 \\ I_4 \end{bmatrix} =
\begin{bmatrix}
Y_p P(\theta) & Y_p T(\theta) & Y_a T(\theta) & Y_a P(\theta) \\
Y_p T(\theta) & Y_p P(\theta) & Y_a P(\theta) & Y_a T(\theta) \\
Y_a T(\theta) & Y_a P(\theta) & Y_p P(\theta) & Y_p T(\theta) \\
Y_a P(\theta) & Y_a T(\theta) & Y_p T(\theta) & Y_p P(\theta)
\end{bmatrix}
\begin{bmatrix} V_1 \\ V_2 \\ V_3 \\ V_4 \end{bmatrix} \tag{7.2.124}
$$

Setting $\theta = \pi/2$ in the previous matrix and designating port "1" as the input port we can obtain:

$$|c_v|^2 = P_4/P_1 = Y_a^2/Y_p^2 \tag{7.2.125}$$

$$|d_v|^2 = P_2/P_1 = 1 - P_4/P_1 \tag{7.2.126}$$

$$P_3/P_1 = 0 \tag{7.2.127}$$

from which we can recognize that the network connected as in Figure 7.2.48 b. The quantities "$Y_{\ell\ell}$" and "Y_c" can be related to the even "Y_e" and odd "Y_o" admittance using the following relations:

$$Y_{\ell\ell} = 0.5\left(Y_o + Y_e\right) \tag{7.2.128}$$

$$Y_c = -0.5\left(Y_o - Y_e\right) \tag{7.2.129}$$

Inserting the two previous equations in 7.2.121 and 7.2.122, 7.2.123 and 7.2.125 become:

$$Y^2 = \frac{\left[(n-1)Y_o^2 + Y_o Y_e\right]\left[(n-1)Y_e^2 + Y_o Y_e\right]}{\left(Y_o + Y_e\right)^2} \tag{7.2.130}$$

$$|c_v| = \frac{(n-1)\left(Y_o^2 - Y_e^2\right)}{(n-1)\left(Y_o^2 + Y_e^2\right) + 2Y_o Y_e} \tag{7.2.131}$$

After this study, it should be clear that the air bridges that connect the coupled lines on Figure 7.2.48 in parallel must be evaluated with zero length in order to study the Lange coupler with this method. In such a case all the lines connected in parallel are equipotential at the connection points. In particular, with Figure 7.2.48a as reference, the two $\lambda/8$ lines are connected in series with a zero length connection. Note that point "A" belongs to the $\lambda/4$ line, which is in parallel to the series of the two $\lambda/8$ lines, and since these lines are equipotential for any transversal section, the connection at the point "A" has no electromagnetic effect. So, the connection of the air bridge at point "A" only has the effect of creating a mechanically stronger connection with the wires. This discussion is obviously valuable only until that frequency where the wire connections have a length which is no more than a 1/10 of the quarter wavelength coupled lines.

From Equations 7.2.130 and 7.2.131, R. M. Osmani[51] has obtained other formulas that are among the first and simplest dedicated to the design of a Lange coupler. According to Osmani, from the previous formula we can obtain the relationships between "ζ_e," "ζ_o," and "Z" for any adjacent pair of lines in the array:

$$\zeta_o = Z\sqrt{\frac{1-|c_v|}{1+|c_v|}}\,\frac{(n-1)(1+q)}{\left(|c_v|+q\right)+(n-1)\left(1-|c_v|\right)} \tag{7.2.132}$$

$$\zeta_e = \zeta_o\,\frac{|c_v|+q}{(n-1)\left(1-|c_v|\right)} \tag{7.2.133}$$

where:

$$q \overset{\perp}{=} \sqrt{|c_v|^2+\left(1-|c_v|^2\right)(n-1)^2} \tag{7.2.134}$$

Equations 7.2.132 and 7.2.133 have been depicted in Figure 7.2.49 for coupling values between 3 dB and 6 dB in abscissae and for four and six coupled lines. It is interesting to note that for the Lange coupler, which is formed with $\lambda/4$ long coupled lines, the general formula $\zeta_e\zeta_o = Z^2$ does not hold as we can see from 7.2.132 through 7.2.134.

From Figure 7.2.49, once the coupling "c_v," the reference system admittance "Z," and the number "n" of coupled lines are defined, it is possible to evaluate the even and odd admittance and synthesize the coupler.

Although the Lange coupler is the most used 3 dB wideband microstrip directional coupler, its physical realization always requires quite a small line width "w" and coupling spacing "s." For example, a Lange coupler centered at 12 GHz with a BW% = 100% and a center band coupling

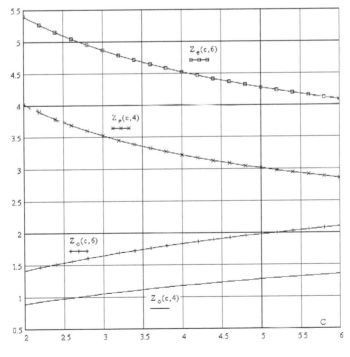

Figure 7.2.49

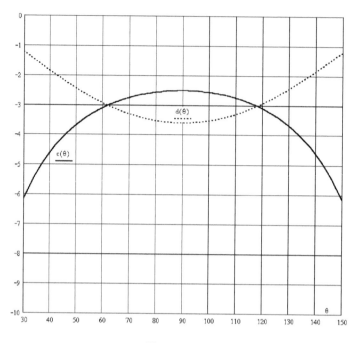

Figure 7.2.50

of 2.5 dB, requires coupled microstrip dimensions near s = 18 μm and w = 35 μm, with a conductor thickness t = 4μm and alumina substrate of height h = 508 μm.

The shape of the coupling vs. frequency of the Lange coupler and the phase relation between the output signals are the same as the single quarter wavelength coupled line coupler that we studied in section 7.2.4. So, also in this case the equations for coupled "c_v" and direct "d_v" ports are the same as given in 7.2.84 and 7.2.85, which we report here:

$$d_v(\theta) = \frac{\left(1 - |c_{0v}|^2\right)0.5}{\left(1 - |c_{0v}|^2\right)^{0.5}\cos\theta + j sen\theta} \tag{7.2.84}$$

$$c_v(\theta) = \frac{j|c_{0v}|sen\theta}{\left(1 - |c_{0v}|^2\right)^{0.5}\cos\theta + j sen\theta} \tag{7.2.85}$$

The graphs for coupled and direct ports are depicted in Figure 7.2.50 for the case of a center band coupling c_{0v} = −2.5 dB. Of course, in all these discussions the radiative phenomena, the non-"TEM" propagation mode and losses have not been taken into consideration. In practice, all these undesirable effects will cause some bandwidth reduction and increasing attenuation with frequency.

7.3 SIGNAL COMBINERS

If we recall the conditions used in the previous section to study the directional couplers, i.e., lossless devices and "TEM" propagation mode, all the previous networks also work as signal combiners. This hypothesis will be considered as true in the rest of this section, unless otherwise stated. This means that when two signals enter in two ports, some combination of these signals must appear at the other ports, with the condition that all the power entering must be equal to all the power leaving the network. In addition, we will assume that the ports are perfectly matched.

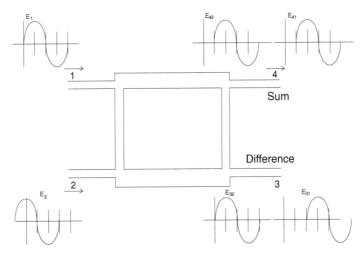

Figure 7.3.1

Some particular notes are needed to better understand how they work. Therefore, we will analyze the networks in the previous section from the point of view of their ability to realize a signal combining. A common rule for any power combiner we are going to study is that the applied signals must be connected to the isolated ports of each directional coupler used as a combiner. In such a case, the two generators are not influenced by each other.

7.3.1 Branch Line

Let us examine Figure 7.3.1. Let us suppose that two signals that are 90° out of phase and at the same frequency enter at the two adjacent ports of the branch line coupler. From the study of the previous section we know that the adjacent ports of this coupler are decoupled at center frequency. So, these two signals do not alter each other.

The rigorous analysis of the branch line coupler is indicated in the previous section and it can be used for the case of the branch line working as a combiner. A qualitative interpretation of its behavior as a signal combiner can be given representing the phases of each signal along the path, as indicated in Figure 7.3.1. So, the signal "E_1" entering port "1" comes out at port "4" as "E_{41}" and at port "3" as "E_{31}." The same applies to signal "E_2," which exits at port "4" as "E_{41}" and at port "3" as "E_{31}." Assuming the network is lossless and propagating pure "TEM" modes, the output signals add at port "4" and subtract at port "3." If, in addition, the signals "E_1" and "E_2" have the same amplitude and the network is supposed to be lossless, then at port "3" there will be no signal and at port "4" there will be available the sum of the available power of the signals "E_1" and "E_2." This situation is represented in Figure 7.3.2, where we have also indicated the power as a function of "θ" exiting at all the ports, where "θ" is the electrical length of the branch line arms. For this figure we have assumed a 3 dB branch line with the signals applied to the network being of the same amplitude. The $\pi/2$ factor that appears in the expressions in the ordinate is the phase offset between the two input signals according to Figure 7.3.1. In any case, the output port of the branch line combiner is always in front of the port where the most delayed signal is connected.

7.3.2 "Rat Race" or "Magic T"

The study of the "rat race" working as a signal combiner can be performed using the complete analysis of the "magic T" that was given in the previous section. A qualitative interpretation of its

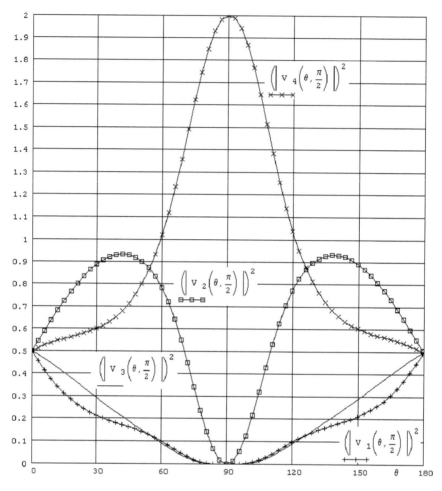

Figure 7.3.2

behavior as a signal combiner can be given representing the phases of each signal along the network, as indicated in Figure 7.3.3. The generators are of course applied to the isolated ports, in this case ports 1 through 3, but also ports 2 to 4 could be used. We see that depending on the phase difference "$\Delta\varphi$" of the signals "E_1" and "E_3" the output and isolated ports can be "2" and "4" when $\Delta\varphi = 0$ or "4" and "2" when $\Delta\varphi = \pi$. In Figure 7.3.4 we have drawn the powers exiting at ports "2" and "4" for the case of $\Delta\varphi = 0$ or $\Delta\varphi = \pi$, while in Figure 7.3.5 we have drawn the powers exiting at all ports. We see that when the electrical distance "θ" between ports 1 through 4 is equal to 90°, then there is no reflection against the generators, and they are completely isolated between them.

Also note that when the signals have a phase offset of "π" the operating bandwidth is a little wider than the case of $\Delta\varphi = 0$. Of course, for both Figures 7.3.4 and 7.3.5 we have assumed a 3 dB rat race device with the signals applied to the network being of the same amplitude.

7.3.3 "In Line" or "Wilkinson"

The case of the Wilkinson directional coupler working as a power combiner is represented in Figure 7.3.6. With the hypothesis of a lossless and pure supporting "TEM" mode network and remembering from the previous section that all the ports are perfectly matched and ports "2" and

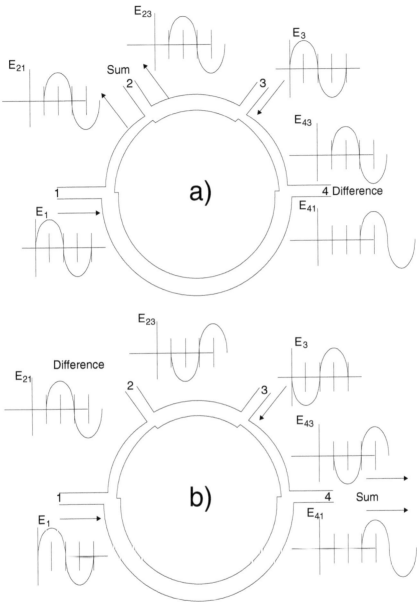

Figure 7.3.3

"3" isolated at center frequency, we find that the sum of the power entering appears at the sum port "1." Of course, from this result the two input signals must be in phase, otherwise a fraction of the power will be dissipated in the resistor "R." An evaluation of the power lost "P_w" in the isolation resistor is given in Figure 7.3.7, as a function of the phase offset "$\Delta\varphi$" between the input signals. Here we have assumed that the signals at ports "2" and "3" have equal power "P1." We see that when $\Delta\varphi = \pi$ the sum of the available power of the generators is lost in "R" and consequently no power will be available at port "1." In Figure 7.3.7 with $P_w(\Delta\varphi,1,1)$ we have indicated the power lost in "R" when both the input signals are connected, while with $P_w(\Delta\varphi,1,0)$ we have indicated the power lost in "R" when only one input signal is connected. In this last case, one half of the input power is lost in "R" and consequently the other half exits from port "1."

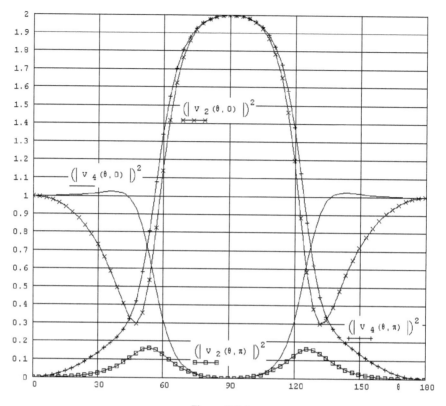

Figure 7.3.4

Particular attention must be paid to this fact when dimensioning the size of the isolation resistor in a Wilkinson adder. In other words, "R" must be able to dissipate all the power occurring in a possible phase offset between the inputs or when one input is no longer present.

7.3.4 Step Coupled Lines

The situation of a single step coupled line directional coupler working as a signal adder is indicated in Figure 7.3.0. Assuming that the network has been designed for 3 dB of coupling and remembering what we said in section 7.2.4 we find that the power available at ports "2" and "4" have the shape indicated in Figure 7.3.9. Here we have assumed that the input signals have the same amplitude and phase offset of 90°. We see that when the electrical length "θ" of the coupled lines is equal to 90° there is only one output from which all the power is available. If we increase the frequency, power appears at the output ports. Note that since the coupled line coupler is perfectly matched at all frequencies, no reflection coefficient exists at the input ports. This means that the sum of the power at the output ports is equal to the sum of the powers entering the network. This situation is very different from the previous cases studied in sections 7.3.1 and 7.3.2. From Figure 7.3.9 we also see that when the frequency of the signals is three times the frequency where the coupling length is 90° long, the sum and difference ports reverse.

We now want to come back to Tables 7.2.1 and 7.2.2 where we gave the impedances to build multistep directional couplers with a coupling value of –8.34 dB. To this purpose, let us examine Figure 7.3.10. With "c" and "d" we have indicated the modulus of the "coupling" and "direct coefficient." For the moment, "c" and "d" are unknowns. We know that "c" and "d" must satisfy the condition:

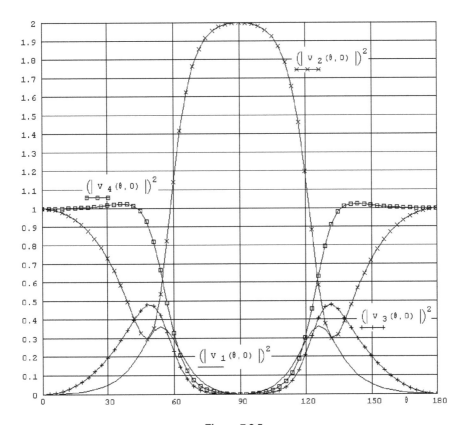

Figure 7.3.5

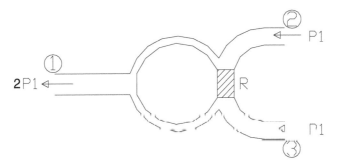

Figure 7.3.6

$$|c|^2 + |d|^2 = 1 \qquad (7.3.1)$$

Now, we want the signals at ports "1" and "3" to have the same amplitude, i.e.:

$$2|c|\,|d| = ! |d|^2 - |c|^2 \qquad (7.3.2)$$

Combining 7.3.1 and 7.3.2 we have:

$$1 - 2|c|^2 = 2|c|\left(1 - |c|^2\right)^{0.5} \qquad (7.3.3)$$

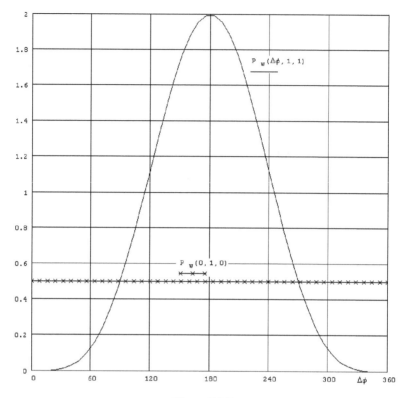

Figure 7.3.7

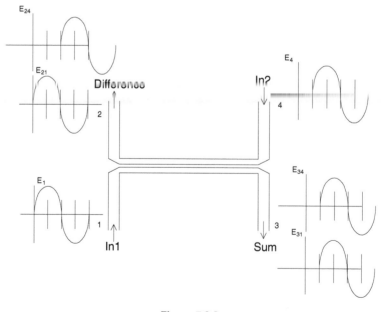

Figure 7.3.8

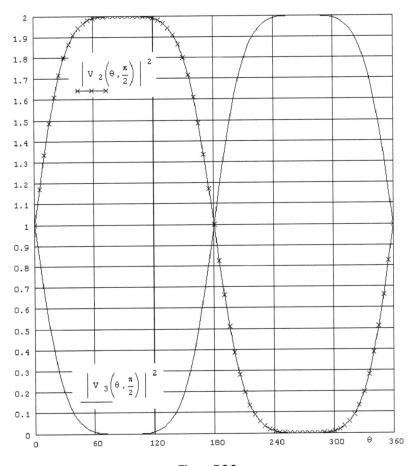

Figure 7.3.9

This equation has four solutions, i.e.:

$$\left| c_1 \right| = 0.924 = - \left| c_2 \right| \text{ and } \left| c_3 \right| = 0.383 = - \left| c_4 \right| \tag{7.3.4}$$

While "c_1" and "c_2" give a coupling value of –0.686 dB, which cannot be reached in microstrip "MIC" technology, "c_3" and "c_4" give a coupling value of –8.34 dB, which is quite simple to realize. From the fact that all the power entering in port "4" must be equal to the sum of the output powers at ports "1" and "3," and from the condition in 7.3.2 it follows that the network in Figure 7.3.10 is equivalent to a 3 dB directional coupler. This is the reason why we have inserted the design values for –8.34 dB of coupling in Tables 7.2.1 and 7.2.2.

The reader should recognize that this explanation is also true for the multistep symmetrical directional couplers and Lange couplers, while multistep asymmetrical directional couplers are seldom used, due to the fact that they are not 90° out of phase between the outputs. At this point it is important to remember that since the Lange coupler permits the construction of 3 dB directional couplers, the network shown in Figure 7.3.11 is seldom used.

A typical application of 3 dB 90° directional couplers is to give a sufficient match to a source when we also need to use reflective devices. Let us examine Figure 7.3.11. In part a of the figure the two shunt diodes give a high reflection coefficient "s_{11}" when they are biased forward. The two

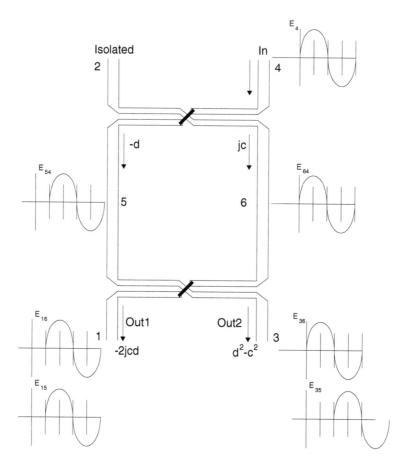

Figure 7.3.10

reflected signals "$s_{11}V_u$" and "$s_{11}V_d$" in part b of the figure, sum at the terminated port of the input hybrid while subtracting at the input port, with the consequence that the reflections do not appear at the input. Of course, this is theoretically true only if all the ports of the hybrid are matched, while in practice this is not the case because the diodes are reflecting. Nevertheless, the hybrid still gives an isolation, and a value of 10 dB for the overall input return loss is possible in the hybrid operating bandwidth.

After we have seen the working principles of the multistep and Lange directional couplers as power combiner-divider networks, we want to show a network that employs all the concepts we have studied until now. In practice this network is not used, but it is an example of how the desire for wider bandwidth 3 dB directional couplers has stimulated researchers.

The original network is indicated in Figure 7.3.12. It was designed by G. Kemp and others.[52] A three step symmetrical directional coupler has been modified so that the inner section is composed of two Lange couplers connected in tandem. One Lange hybrid works as a signal divider, the other as signal adder. With this network the researchers have overcome the difficulty of creating with microstrips, the high coupling value of the inner section for wide bandwidths, as we can see in Tables 7.2.1 and 7.2.3. In fact, performing the same calculations used for the series connections of the two coupled line directional coupler shown in Figure 7.3.10, and indicating with "c" the coupling value for the single Lange coupler in Figure 7.3.12, we have the tandem connections of the two hybrids giving a resultant coupling value of "c_s" if:

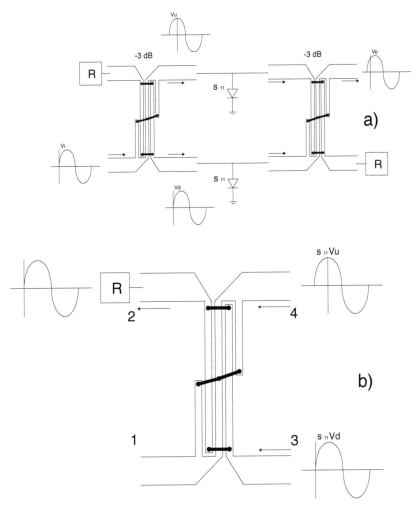

Figure 7.3.11

$$|c|^2 = \frac{1 - \left(1 - |c_s|^2\right)^{0.5}}{2} \qquad (7.3.5)$$

Other solutions are analytically possible, but they give values of "c" higher than "c_s."

As an example, let us attempt to build a 3 dB, 0.1 dB ripple, B = 100%, three step directional coupler. From Table 7.2.1 we have for the highest coupling region, ζ_{e2} = 3.26. To this value of "ζ_e" corresponds a value for the odd impedance of ζ_{o2} = 0.307, and consequently a coupling value of c_s = 0.828. Using this value of "c_s" in 7.3.4 we obtain c = 0.469, or c = −6.583 dB which is the coupling value for any hybrid that appears at the center of the coupler shown in Figure 7.3.12. As was said before, this type of network is seldom used because the use of a Kemp type coupler results in a bandwidth reduction with respect to the theoretical band of the original step coupled coupler. This bandwidth reduction has been evaluated as being near 10% by J. L. B. Walker.[53] This network is an example of how very desirable a wideband 3 dB − 90° directional coupler is.

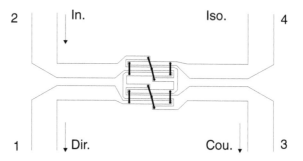

Figure 7.3.12

7.4 DIRECTIONAL FILTERS

Directional filters are widely used networks in systems where a lot of narrowband signals need to be treated. They have the ability to simultaneously realize a band pass filter in one direction and a stop band filter in another direction. Variations of this type of directional filter, also in waveguide technology, can be found in references 54,55,56.

Examples of applications are in signal mux-demux and phase lock systems.

7.4.1 Resonant Ring

The most used microstrip directional filter is indicated in Figure 7.4.1, and is called a "resonant ring" or "resonant loop" directional filter. As we can see, two single step coupled directional couplers are connected "in parallel" to form something similar to an inner ring with two coupled lines. This network must satisfy the condition that the length "ℓ = !λ/4" of each directional coupler plus the length "D1" and "D2" of each matched connecting line be an integer multiple "n" of wavelength in the microstrips. Usually D1 = D2 $\doteq$ D, and for analytical simplicity, notation in our study we will assume that D $\equiv$ λ/4.

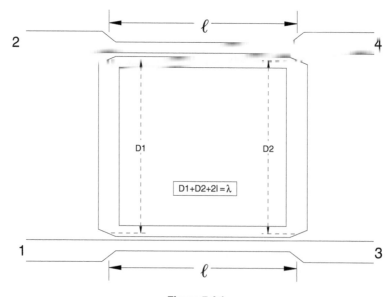

Figure 7.4.1

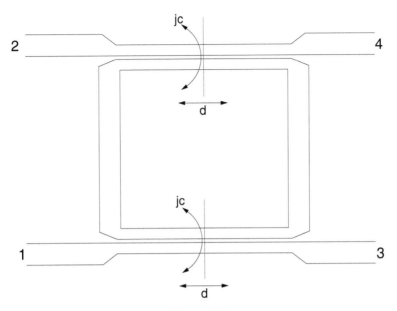

Figure 7.4.2

So, let us start indicating with "θ" the electrical length of the two couplers, and consequently βD ≡ θ. Let us place the phase reference planes at the middle of the directional coupler as indicated in Figure 7.4.2 by the dashed lines and assume as reference phase that corresponding to the direct signal "d_{0v}." With this notation, every signal crossing the coupling region transversely and longitudinally will be given by "$j|c_{0v}|$" and "$|d_{0v}|$," respectively.

The signal "V_2" at port "2" will be given by the sum of infinite terms traveling along the ring. For instance, the first three terms "$V'_{2,0}$," "$V'_{2,1}$," and "$V'_{2,2}$" of the sum, for the zero, first, and second rounds in the loop are:

$$V'_{2,0} = -\left|c_{0v}\right|^2 \exp(j2\theta); \quad V'_{2,1} = -\left|c_{0v}\right|^2\left|d_{0v}\right|^2 \exp(j6\theta); \quad V'_{2,2} = -\left|c_{0v}\right|^2\left|d_{0v}\right|^4 \exp(j10\theta) \quad (7.4.1)$$

It is simple to recognize that the generic term "$V_{2,k}$" of the sum can be written as:

$$V_{2,k} = -\left|c_{0v}\right|^2 \exp(j2\theta)\sum_{m=0}^{m=k}\left[\left|d_{0v}\right|^2 \exp(j4\theta)\right]^m \quad (7.4.2)$$

and consequently:

$$V_2 = \lim_{k\to\infty} V_{2k} \quad (7.4.3)$$

The series in 7.4.2 when $k \to \infty$, as also required from our study and expressed in 7.4.3, is a case of a canonical one, the geometric series, whose result is well known and can be found in many mathematics books.[57] As a consequence, Equation 7.4.3 has the result:

$$V_2 = \frac{-\left|c_{0v}\right|^2 \exp(j2\theta)}{1 - \left|d_{0v}\right|^2 \exp(j4\theta)} \quad (7.4.4)$$

Also the signal "V_3" at port "3" will be given by the sum of infinite terms traveling along the ring. For instance, the first three terms "$V'_{3,0}$," "$V'_{3,1}$," and "$V'_{3,2}$," of the sum, for the zero, first, and second rounds in the loop are:

$$V'_{3,0} = |d_{0v}|; \quad V'_{3,1} = -|c_{0v}|^2 |d_{0v}| \exp(j4\theta); \quad V'_{3,2} = -|c_{0v}|^2 |d_{0v}|^3 \exp(j8\theta) \quad (7.4.5)$$

The reason why only odd exponents exist for "d" in 7.4.5 depends on the fact that when the signal travels in the loop and exits from port "3," it only exceeds the 2 – 4 reference plane but not the 1 – 3 one. In any case, for every travel in the loop, a phase change of "4θ" exists.

It is simple to recognize that the generic term "$V_{3,k}$" of the sum can be written as:

$$\text{for } k = 0: \ V_{3,0} = |d_{0v}|; \quad \text{for } k \geq 1: \ V_{3,k} = -c_{0v}^2 |d_{0v}| \exp(j4\theta) \sum_{m=0}^{m=k-1} \left[|d_{0v}|^2 \exp(j4\theta) \right]^m \quad (7.4.6)$$

and consequently:

$$V_3 = \lim_{k \to \infty} V_{3,k} \quad (7.4.7)$$

From the previous equation and the well-known sum of the geometric series we have:

$$V_3 = |d_{0v}| - \frac{|c_{0v}|^2 |d_{0v}| \exp(j4\theta)}{1 - |d_{0v}|^2 \exp(j4\theta)} \quad (7.4.8)$$

Note that 7.4.4 and 7.4.8 have been obtained as if they were at the center band of the coupler, and the phase term "$e^{(jx\theta)}$" only gives the phase change along the loop. To have complete solutions for 7.4.4 and 7.4.8 with "θ" as a variable, it is only necessary to replace "$|c_{0v}|$" and "$|d_{0v}|$" with "$|c(\theta)|$" and "$|d(\theta)|$" given by 7.2.84 and 7.2.85. The resulting shape of "$|V_2|$" and "$|V_3|$" is indicated in Figure 7.4.3, with $c_{0v,dB} = -3$.

Interesting is the case when losses inside the loop are evaluated. To show what happens, let us indicate with "$|\Gamma_{01}|$" and "$|\Gamma_{0L}|$" the two center band voltage coupling values of the two quarter wavelength couplers and indicate with "u" the u.l. attenuation factor. With these notations, Equation 7.4.8 becomes:

$$V_3 = |d_{01}| - \frac{|c_{01}|^2 |d_{02}| \exp(j4\theta) e^{(-4\alpha\ell)}}{1 - |d_{01}| |d_{02}| \exp(j4\theta) e^{(-4\alpha\ell)}} \quad (7.4.9)$$

where "ℓ" is the length of each arm of the ring, which for simplicity we have already assumed to be of equal length. At resonance, i.e., at $\theta = \pi/2$, we want $V_3 = 0$. Inserting these two conditions into the previous equation, we have the following relation that must be satisfied:

$$e^{(-8\alpha\ell)} =! \frac{1 - |c_{02}|^2}{1 - |c_{01}|^2} \quad (7.4.10)$$

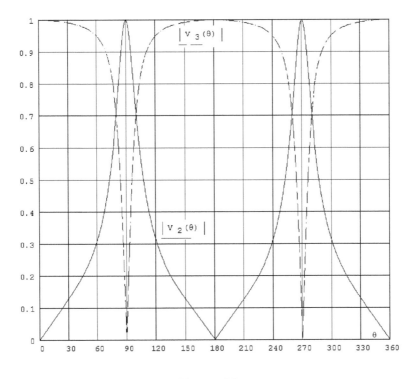

Figure 7.4.3

Due to the presence of losses, the two couplings cannot be equal. Unfortunately, the previous condition does not agree with the condition $|c_{01}| = |c_{02}|$ for maximum "V_2" at resonance. So, some compromise is necessary.

It is interesting to observe that if we studied this device using transmission line theory, it could also be regarded as a filter. In fact, the ring inside the two lines "1 – 3" and "2 – 4" can be evaluated as a resonant circuit, and with the two coupling regions as two coupling capacitances toward the resonant circuit, one at the input and one at the output. From the general filter theory, whose fundamentals are reported in Appendix A4, we know that the pass bandwidth of a band pass filter is narrower when the coupling capacitances, to the resonant elements are smaller. If we regard the coupling region as capacitances as stated, we should note that when decreasing the coupling, the band pass of the directional filter will decrease. This is exactly what happens for this ring directional coupler. The situation is represented in Figure 7.4.4 where the shape in dB of "V_2" and "V_3" is drawn for three values of coupling, i.e., –3, –6, and –9 dB. Always from the analogy with the general filter theory, the bandwidth of this device is that of a single resonant circuit and its value is dependent on the "Q" of the resonant element. In this case, bandwidth is near 10% for frequencies up to 15 GHz while it increases for higher center frequency.

Some modifications of this device can be performed remembering from filter theory that bandwidth and rejection of a filter can be increased if we increase the number of resonant elements. This is exactly what we can do if we realize the device indicated in Figure 7.4.5. Two full wavelength resonant rings are coupled to form a double ring directional filter. This device can be studied as was the case of single ring, or by the use of general filter theory where various pass band shapes can be performed. A theory for a directional ring filter with generic number "n" of resonant rings can be found in reference 58. Multiring directional filters are seldom used in practice since if some bandwidths that need to be filtered are wider than some percent, a multiplex filter is preferred.

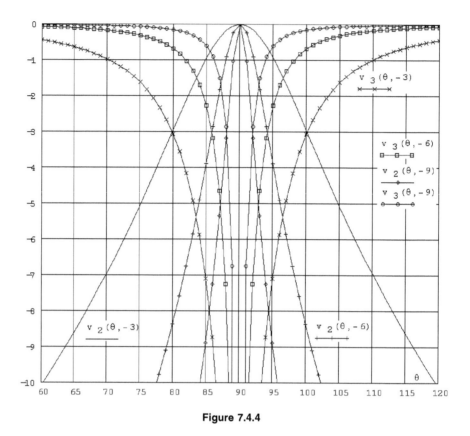

Figure 7.4.4

It is interesting to observe from Figure 7.4.3 how this device has the same behavior for odd harmonics of the designed center frequency, a characteristic in common with the step coupled directional couplers studied in section 7.2.4.

7.4.2 Transverse Resonant Lines

Another type of microstrip directional filter is indicated in Figure 7.4.6, the "transversal resonant lines" directional filter. Two parallel lines, "1 – 3" and "2 – 4," are coupled to two half wavelength

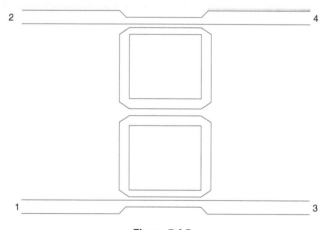

Figure 7.4.5

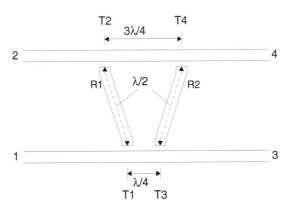

Figure 7.4.6

lines usually called "transversal resonators" and indicated with "R1" and "R2." For the lines, the coupling distance to "R1" and "R2" is different so that for one line, the distance is $\lambda/4$ and for the other, the distance is $3\lambda/4$. This type of network is not as widely used as the "resonant ring" type, but sometimes it is employed in low microwave regions, let us say up to some GHz. Its study is a good example of the application of the general transmission line theory discussed in Chapter 1 and the "even-odd" technique widely used to study all symmetrical, reciprocal, passive networks.

Let us start our study using the even-odd excitation method, as indicated in Figure 7.4.7. Assuming that port "1" is the input, let us apply an even excitation at reference planes "T1" and "T2" as indicated in part b of Figure 7.5.7. We can recognize that at resonator "R2" these two excitations arrive with opposite phase, and "R2" can resonate while "R1" cannot. For the odd excitation indicated in Figure 7.4.7c, the situation is reversed, i.e., "R1" can resonate while "R2" cannot. Note that at resonance the middle of any resonator is at zero potential.

Now let us suppose that at resonance a complete reflection exists at the coupling regions, due to a reported zero impedance by the resonator. Later we will see the conditions for such a supposition. With this hypothesis, Figure 7.4.8 describes the symbolic waveforms of the reflected signals. In this figure, the referement time is the same as that used in Figure 7.4.7. For the even case, indicated in Figure 7.4.8a, "$E_{T4'}$" and "$E_{T3'}$" represent the signals reflected by the resonator "R2," while "E_{T1R}" and "E_{T2R}" represent the signals reflected by the resonator "R2" and coming back at port "1" and "2." For convenience, in part a we have drawn with dashed line the waveform of "E_{T2R}" up to the referement time, just to show the steady state situation. In Figure 7.4.8b "$E_{T1'}$" and "$E_{T2'}$" represent the signals reflected by the resonator "R1" and coming back at port "1" and "2" respectively. From Figure 7.4.8 we can recognize how at complete reflections, that is at resonance, the reflections add at port "2" and cancel at port "1." This means that a signal entering at port "1" at resonance will be completely transferred to port "2," and nothing will exit from ports "3" and "4." Conversely, a signal entering at port "1" out of resonance will be completely transferred to port "3" and nothing will exit from the other ports.

It is now interesting to study if and when a complete reflection can exist at the reference planes "T1" → "T4." To this purpose, Figure 7.4.9 represents the equivalent circuit of our network at the resonance, i.e., when the middle of the resonators is at zero potential, between ports "1" and "3." For the preceding discussion, this equivalent network is also applicable to ports "2" and "4." Ports "1" and "3" are of course properly terminated. Note that being "ℓ" the length of the resonators, now the length of the stub in Figure 7.4.9 is "$\ell/2$," and its electrical length will be indicated with "θ." The impedance of any line is set equal to "Z." The coupling region of the resonators to main lines has been simplified to a single capacitance "C." We studied in Chapter 4 that as our coupling regions can be approximated, a gap between microstrips has a good approximation in a "π" of

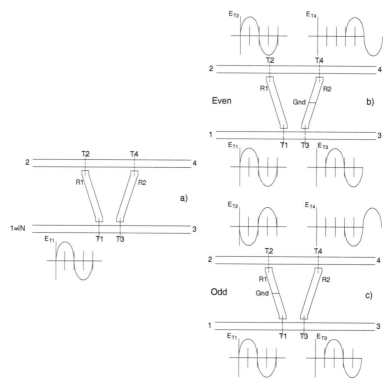

Figure 7.4.7

capacitors. The series capacitor is a reduced equivalent circuit, which is enough for our study. With reference to Figure 7.4.9, we can write:

$$Z_i = j\zeta_0 tg\theta \tag{7.4.11}$$

$$Z_t = -j/\omega C + j\zeta_0 tg\theta \doteq jX \tag{7.4.12}$$

$$Y = Y_0 + Y_t = Y_0 - j/X \tag{7.4.13}$$

and the voltage reflection coefficient "ρ" looking from port "1" to "T1" will be:

$$\rho \doteq \frac{Y_0 - Y}{Y_0 + Y} \equiv \frac{j}{2Y_0 X - j} \tag{7.4.14}$$

A complete reflection exists when $|\rho| = 1$, and from 7.4.14 it happens when $X = 0$ or:

$$tg\theta = 1/\zeta_0 \omega c \tag{7.4.15}$$

Since $\theta = (\omega/v)\ell$, with "v" being the speed of e.m. energy in the microstrip, it is possible from the previous equation to have the length "ℓ" of half resonator when a resonance frequency "f_r" is chosen. In fact we have:

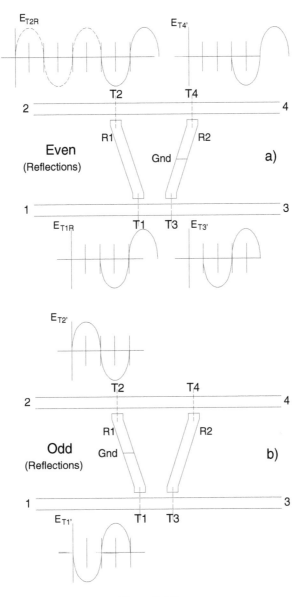

Figure 7.4.8

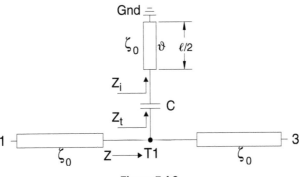

Figure 7.4.9

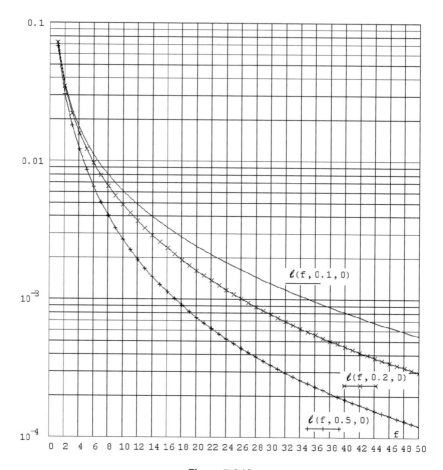

Figure 7.4.10

$$\ell_n = \frac{v}{2\pi f_r}\, atg\!\left(\frac{1}{2\pi f_r \zeta_0 C}\right) + n\,\frac{v}{2f_r} \quad \text{with } n = 0,1,2\ldots . \tag{7.4.16}$$

Figure 7.4.10 represents the previous equation vs. frequency in GHz with the capacitance "C" in pF and "n" as a parameter. The capacitance has been set equal to 0.1, 0.2, and 0.5 pF while "n" is always equal to zero. The ordinate is the length "ℓ" in meters of half resonator.

Using frequency as a dependent variable, Equation 7.4.15 is implicit since frequency appears in both terms. In fact we have:

$$\frac{2\pi f}{v}\,\ell = atg\!\left(\frac{1}{2\pi f C \zeta_0}\right) + n\pi \quad \text{with } n = 0,1,2\ldots . \tag{7.4.17}$$

For our study Equation 7.4.17 can be solved in "f" graphically in a very easy manner. The situation is indicated in Figure 7.4.11, where $g_1(f,\ell) \doteq 2\pi f\ell/v$ and $g_2(f,C,n) \doteq atg(1/2\pi f C\zeta_0) + n\pi$. The intersection points between these functions give the resonant frequency of our device. For this figure, values of "ℓ" are in mm, "C" in pF, and "f" in GHz.

The frequency response of the transversal resonant lines directional filter is the same as that shown in Figure 7.4.4, but there is not the simple tuning of bandwidth reachable with the resonant

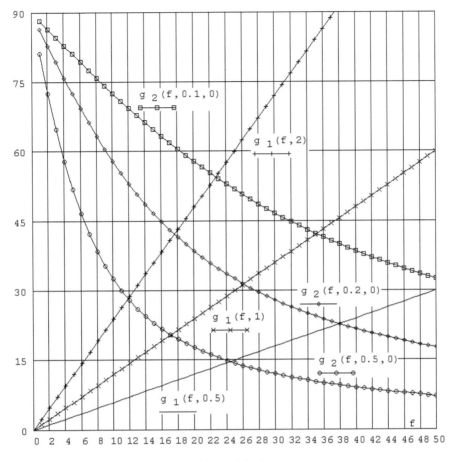

Figure 7.4.11

ring directional filter. In addition, difficulties of tuning and difficulties of synthesizing a proper coupling gap make this network usable in the low microwave region where, of course, the resonant ring directional filter can also be used.

7.5 PHASE SHIFTERS

Quite often in transmission systems it is required that two signals have a precise phase difference between them. Examples are signals to be sent in a mixer or signals used in interforemetric measurements. We already have seen in previous sections that some power dividers give 90° of phase offset between output signals, and this characteristic can be used to build phase shifters.[59] In this section we will show that such dividers are not the only networks that can be used to create a phase shift between signals.

7.5.1 Coupled Lines or "Schiffman"

The most well-known and used family of phase shifter networks are the Schiffman types shown first by B. M. Schiffman.[60] They all are 90° phase shifters and all have in common the same theoretical arguments, i.e., how the phase of a signal changes in a coupled line. Due to its widespread use, this device has been widely discussed in literature.[61,62]

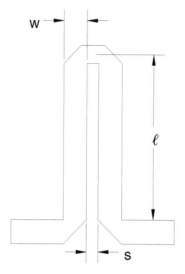

Figure 7.5.1

The fundamental cell of Schiffman's shifters is indicated in Figure 7.5.1. This shape is called a "C section" or "meander line" in the literature. It consists of a length "ℓ" of coupled µstrips, connected in short circuit at an extreme. Coupled microstrips have been studied in detail in Chapter 5, but we don't need the complete theory to understand the Schiffman shifter. It is enough to remember the introduction to coupled line theory given with regard to quarter wavelength step coupled line directional couplers in Section 7.2. Of course, the length of the connection between the two coupled microstrips must be considered in determining the proper length "ℓ" of coupling region. This is usually done in an empirical way, tuning the length "ℓ," which we will now describe, until the proper phase shift is reached. Also in this case, unless otherwise stated, to simplify our study, we will assume that the microstrip supports a pure "TEM" mode in an omogeneous dielectric medium. With this hypothesis, phase change "φ" along the length "ℓ" of coupled microstrips is given by:

$$\varphi = a\cos\left(\frac{\zeta_e/\zeta_0 - tg^2(\theta)}{\zeta_e/\zeta_0 + tg^2(\theta)}\right) \tag{7.5.1}$$

where "θ" is the electrical length of a single isolated microstrip of length "ℓ." Figure 7.5.2 reports the previous equation vs. the electrical length "θ," with $\rho \triangleq \zeta_e/\zeta_0$ as a parameter. Note that since the voltage coupling coefficient "c_v" is given by:

$$|c_v| \equiv \frac{\zeta_e - \zeta_o}{\zeta_e + \zeta_o}$$

it follows that:

$$|\rho| \equiv \frac{1 + |c_v|}{1 - |c_v|} \tag{7.5.2}$$

Coupled microstrips are, of course, realized so that the matching condition $\zeta_e\zeta_o = Z^2$ is respected, "Z" being the system reference impedance.

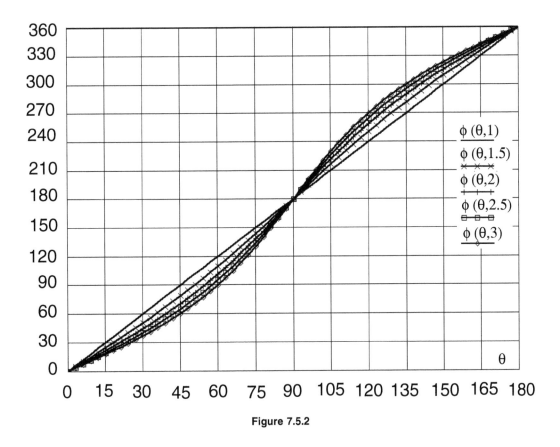

Figure 7.5.2

As known from the theory of coupled lines, the value $\rho = 1$ means completely uncoupled lines. From Figure 7.5.2 we see that as the coupling between the microstrips increases, the phase change along the network assumes a pseudosinusoidal shape along the line for $\rho = 1$.

The working principle of Schiffman shifters is all inside this figure. For example, measure the phase difference between a signal that exits from a coupled line of length "ℓ" and an isolated line of length "3ℓ." A possible network is indicated, for example, in Figure 7.5.3 where all the components are already known. This figure is only an example since the number of rings for a Wilkinson divider can also be greater than that shown in this figure. The electrical performance of such a network, with respect to the phase difference "$\Delta\varphi$" between the two signals at ports "2" and "3," is indicated in Figure 7.4.4 for $\rho = 3.7$ and $\ell = \lambda/4$. The value of $\ell = \lambda/4$ is always used in all Schiffman's phase shifters and this value will be assumed for these networks, unless otherwise noted.

We see how the phase difference stays for a bandwidth $\Delta\theta = \theta_2 - \theta_1$ inside a window "Δa" centered at $\Delta\varphi = 90°$. In this case the operating bandwidth is 3:1, with $\Delta a = 20°$. Operating bandwidth can be increased if a greater "Δa" is accepted, which physically traduces in a higher value of "ρ." However, $\rho = 3.7$ is near the maximum value for coupled microstrips in typical microstrip technology, and Figure 7.4.4 represents the maximum "$\Delta\theta$" for the network indicated in Figure 7.5.3. The Schiffman shifter in Figure 7.5.3 is based on a single coupled line plus a single line and is called "type A."

More complicated shifters can be obtained, with the aim of decreasing the ripple "Δa" without noticeably reducing the bandwidth. Schiffman suggests using a correction network composed of a coupled line and a single line of lengths "$n\ell$" and "$2n\ell$," respectively. From what we have said

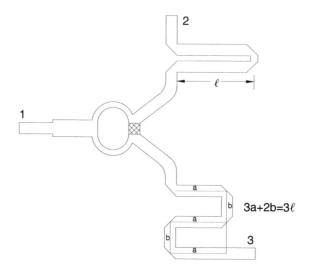

Figure 7.5.3

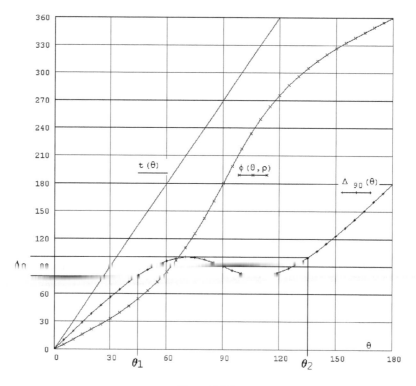

Figure 7.5.4

about the theory of Schiffman's shifter, such a structure has a phase difference around $\Delta\varphi = 90°$, with a ripple depending on "ρ." Since there are two ways to connect such a correction network to a "type A" shifter, two new phase shifters are possible, called "type B" and "type "C.""

A type "B" phase shifter is indicated in Figure 7.5.5 where the value "n" of the correction network has been chosen equal to "3." In this figure, and in this paragraph, the required power divider in front of the shifter is omitted for simplicity. An example of "Δa" reduction by this network is represented in Figure 7.5.6 by its phase characteristics. In this figure, for the "type A"

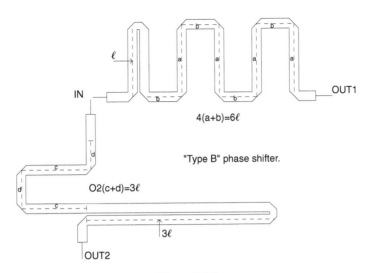

$$4(a+b)=6\ell$$

"Type B" phase shifter.

$$O2(c+d)=3\ell$$

$$3\ell$$

Figure 7.5.5

network $\rho = 3$ has been set, while for the error correction we have $\rho = 1.18$. Note that the "type A" shifter has a ripple of $\Delta a = \pm 4.8°$ in BW = 2.33:1, while the "type B" network has $\Delta a = \pm 0.8°$ max. in approximately the same BW.

The "type C" phase shifter is indicated in Figure 7.5.7 where the value "n" of the correction network has been chosen equal to "2." The original Schiffman's example of "Δa" reduction by this

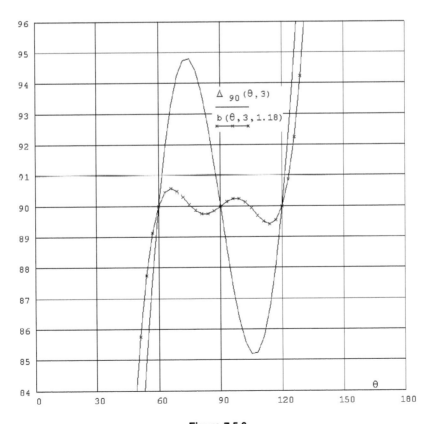

$$\frac{\Delta_{90}(\theta, 3)}{b(\theta, 3, 1.18)}$$

Figure 7.5.6

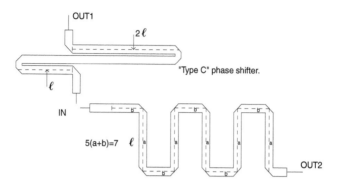

Figure 7.5.7

network is represented in Figure 7.5.8 by its phase characteristics. In this figure, for the "type A" network $\rho = 5.83$ has been set, while for the error correction $\rho = 2.35$. The value $\rho = 5.83$ is clearly not realizable in microstrip technology, but the "type C" network can of course be used with other values of "ρ" to compensate for the "Δa" for the "type A" shifter. Note that the "type A" shifter has a ripple of $\Delta a = \pm 22.7°$ in BW = 4.6:1, while the "type C" network has $\Delta a = \pm 5°$ max. in approximately the same BW.

Another compact Schiffman's shifter is the "type F" indicated in Figure 7.5.9. In this network, two coupled microstrip sections are employed in a series configuration, and phase is compared to

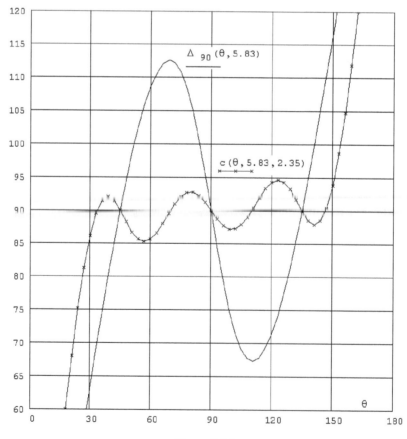

Figure 7.5.8

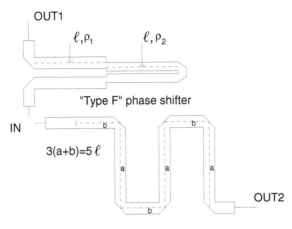

Figure 7.5.9

that of a single microstrip. Phase change "φ_2" along two sections of coupled microstrips connected in series is given by:

$$|\varphi_2| = a\cos\left(\frac{\rho_1 - tg^2\left\{atg\left[(\rho_1/\rho_2)^{0.5}tg(\theta_2)\right] + \theta_1\right\}}{\rho_1 + tg^2\left\{atg\left[(\rho_1/\rho_2)^{0.5}tg(\theta_2)\right] + \theta_1\right\}}\right) \qquad (7.5.3)$$

where "θ_1" and "θ_2" are the electrical lengths of two isolated microstrips with the same physical length "ℓ_1" and "ℓ_2" of each coupled section while "ρ_1" and "ρ_2" are the ratio of ζ_e/ζ_o for each coupled section. The "type F" phase shifter is realized with $\ell_1 = \ell_2$. The original Schiffman's example of "Δa" reduction by this network is represented in Figure 7.5.10 by its phase characteristics. In this figure, for the "type A," network $\rho = 6.2$ has been set while for "type F," $\rho_1 = 1.6$ and $\rho_1 = 6.2$. The value $\rho = 6.2$ is clearly not realizable in microstrip technology, but the "type F" network can of course be used with other values of "ρ" to compensate for the "Δa" for the "type A" shifter. Note that the "type A" shifter has a ripple of $\Delta a = \pm 24.4°$ in BW = 4.8:1 while the "type F" network has $\Delta a = \pm 2.8°$ max. in a BW = 3.2:1.

At this point it is interesting to see an example of the synthesis of the most simple Schiffman network, i.e., the "type A" shifter. We assume that the ripple "Δa" and the operating bandwidth "BW" are known. The first thing to do is to evaluate whether "Δa" and "BW" can be made with a "type A" network. In our example we will assume this is possible. Then, applying the procedure explained in Figure 7.5.4 with some values of "ρ," the proper value of this parameter is found. Now, since the matching condition $\zeta_e\zeta_o = Z^2$ must be satisfied and the value of "ρ" is known, it follows that "ζ_e" and "ζ_o" are determined. Now, using the microstrip coupled theory studied in Chapter 5, the value of width "w" and spacing "s" of the coupled microstrip region can be obtained, when the height "h" and relative dielectric constant "ε_r" are known.

The original Schiffman work only describes 90° phase shifters, but with these networks, other mean values of phase can also be reached. An interesting work on this subject has been developed by B. Schiek and J. Köhler.[63] Figure 7.5.11 indicates, for instance, the phase characteristic of a 45° phase shifter, realized with a coupled line section with $\rho = 3.2$ and a reference line of length $\ell = 5\lambda/8$. The resulting bandwidth is BW = 3.2 and the phase amplitude window is $\Delta a = \pm 4.5°$.

Figure 7.5.12 indicates the phase characteristic of a 180° phase shifter, realized with two coupled line sections, both with $\rho = 2.82$. The resulting bandwidth is BW = 2.1 and the phase amplitude window is $\Delta a = \pm 10.6°$.

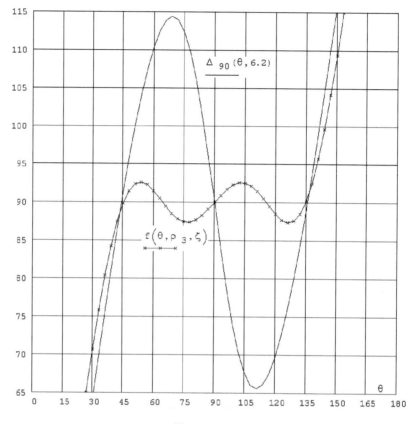

Figure 7.5.10

In their study, Schiek and Köhler have also given a procedure to increase the matching to 50Ω of the coupled line sections. In fact, from the coupled µstrips theory studied in Chapter 5 it is possible to understand that the dishomogeneity of the dielectric medium for the microstrip networks causes the difference between phase constants "β_e" and "β_o." This fact is responsible for the high frequency mismatch of the structure. The method used by Schiek and Köhler is to divide the single coupled line section of length of $\ell = \lambda/4$ in two coupled sections in series, each of length $\ell = \lambda/8$. This method allowed them to build a $180°$ shifter with a matching of better than 20 dB in the band of 4 to 8 GHz.

7.5.2 Transverse Stubs or "Wilds"

Another type of simple phase shifter has been reported by R.B. Wilds[64] and is indicated in Figure 7.5.13. The phase shifting is measured between the output of an isolated reference microstrip of length $\lambda/2 + x$, where "x" will be defined next, and the output of a microstrip, called the "crossed line" of length $\lambda/2$ with two shunt stubs of equal length $\lambda/8$, one open and one short circuited at the middle of its length. The reference line has an impedance "ζ" equal to the system reference impedance, the two stubs have equal impedance "ζ_t," while the other lines have equal impedance "ζ_ℓ." The length "x" is chosen to be proportional to the desired phase shifting "$\Delta\varphi$," so that $\beta x = \Delta\varphi$ with "β" the phase constant of the signal along the reference line. Figure 7.5.13 assumes the feeding element of the device, for instance a Wilkinson divider. Wilds has shown that "ζ_ℓ" and "ζ_t" are in relation to the mean value "$\Delta\varphi_m$" of "$\Delta\varphi$" through the following equation:

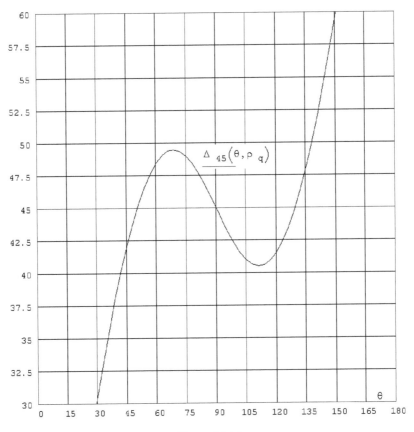

Figure 7.5.11

$$90/\Delta\varphi_m = \zeta_t/\zeta_\ell \qquad\qquad (7.5.4)$$

The network in Figure 7.5.13 can easily be studied using "ABCD" matrices, as were used, for instance, in the analysis of the branch line coupler. When the resulting matrix of the crossed line is obtained, it is simple to compare the elements of the ABCD matrix of the reference line with those of the ABCD matrix of the crossed line.

The phase characteristic of this phase shifter is indicated in Figure 7.5.14, as a function of the electrical angle "θ" where x = λ/4, ζ = 50 and $\zeta_\ell \equiv \zeta_t = 31\Omega$. The resulting operating bandwidth "BW" is near 2:1. In any case, the "BW" of this kind of shifter is limited near 3:1, and cannot reach the value of 4:1 (or more) of some Schiffman's networks. The limitation is due to the value of the return loss more than the high value of the ripple in the phase difference. Note that while the Schiffman shifter is theoretically always matched, since the matching condition $\zeta_e\zeta_o = Z^2$ is satisfied, Wilds' shifter uses resonant lines and transformers for matching. So, we can expect some resonances in the "s_{11}" value of the Wilds' shifter.

A typical shape for "s_{11}" is shown in Figure 7.5.15, where we can see how a perfect matching only exists at some frequencies, or if an equiripple bandwidth can be found. It is interesting to show how at center frequency, the network in Figure 7.5.13 is always matched. In fact, at center frequency the shortened stub reports to the connection point an admittance value of $Y_s = -jY_t$ while the open stub reports an admittance value $Y_o = jY_t$. So, the sum of these two admittances is zero, and the resultant situation is that of a λ/2 line of impedance "ζ_ℓ" that connects the load to the generator. From the general transmission line studied in Chapter 1, we know that a line λ/2 long

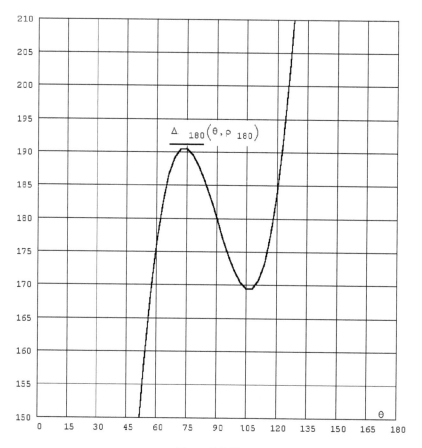

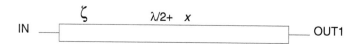

Figure 7.5.12

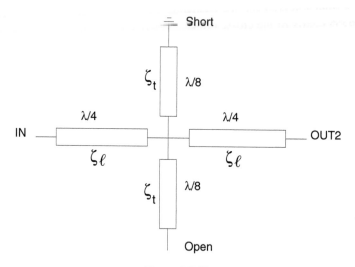

Figure 7.5.13

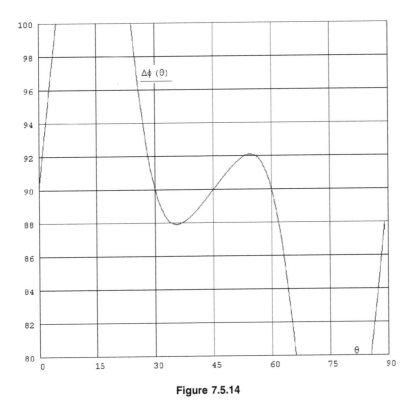

$\Delta\phi$ (θ)

θ

Figure 7.5.14

connects the load to the generator, regardless of its impedance. Consequently, at center frequency the network in Figure 7.5.13 is always matched. Out of resonance, only some values of "ζ_ℓ" and "ζ_t" give a reasonable bandwidth. To this purpose, Table 7.5.1 indicates the normalized values of "ζ_ℓ" and "ζ_t" for 90° phase shifters, with bandwidths ranging from 2:1 up to 3:1.

Figures 7.5.4 and 7.5.15 show that center frequency of the network corresponds to the frequency where the stubs are λ/8 long. In any case, Wilds' shifters give a better matching than the Schiffman counterpart, with the same "BW" of course, due to the most simple construction.

7.5.3 Reflection Type

The Lange coupler discussed in section 7.2 is well suited to be used as a building block for a phase shifter. This may be easily understood remembering how a practical 3 dB/90° splitter, usually called "hybrid," works as a divider or an adder. This was studied in sections 7.5.2 and 7.5.3.

Let us start to examine Figure 7.5.16 where a practical 3 dB/90° Lange coupler divider is indicated. Assuming:

1. Infinite isolation of hybrid
2. Ports perfectly matched

   The waveforms indicated in the figure represent the signal situation when the coupler is fed at only one port, port "1" in the figure, i.e., it is working as a divider. We see that signal output at ports "2" and "3" are in quadrature between them, while port "4" is isolated and no signal exits from this port.

   The signal situation when the Lange coupler works as a power adder is indicated in Figure 7.5.17. Assuming that the same previous hypotheses 1 and 2 hold, we see that feeding the coupler at two adjacent ports with two signal generators of equal amplitude but with 90° phase offset, the signal appears at only one port of the remaining two, while the last one is isolated. In the case of Figure 7.5.17, port "2" is the isolated one, since the signals at this port are 180° out of phase between them.

A mixed situation between Figures 7.5.16 and 7.5.17 can be created if the output signals "E_{21}" and "E_{31}" of Figure 7.5.16 are allowed to enter again at these ports. To realize this situation, a complete reflection must be created at ports "2" and "3." In this case, a signal entering at port "1" divides equally at ports "2" and "3," reflecst back, enters again in these ports, and finally a signal exits from port "4" while at port "1" no signal appears. If, in addition to the previous two hypotheses, we assume:

3. A lossless device,
4. Complete reflection at ports "2" and "3,"

Then we have the result that the output signal at port "4" has the same amplitude as the input one at port "1," but with a phase offset that is dependent on the phase of the reflection coefficient created at ports "2" and "3" to realize the complete reflection. Note that it is very important that a complete reflection be realized at these ports, since any resistive load different from zero or infinite will cause energy absorption, with the consequence that the output signal will be attenuated.

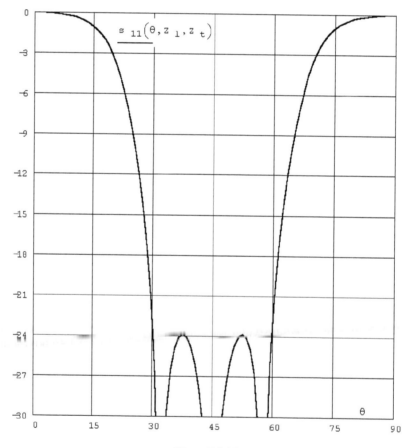

Figure 7.5.15

With reference to Figure 7.5.18 indicating with "E_1" the signal at the input port "1" and with "$\Gamma(Z)$" the reflection coefficient created by the load "Z," the signal output "E_4" to port "4" is given by:

$$E_4 = jE_1\Gamma(Z) \tag{7.5.5}$$

As previously said, it is necessary to have a complete reflection at the coupled and direct ports of the hybrid. Any pure reactance, working in a defined bandwidth, can perform a good approximation to a complete reflection. In addition, if we want to create a tunable phase shifter we need

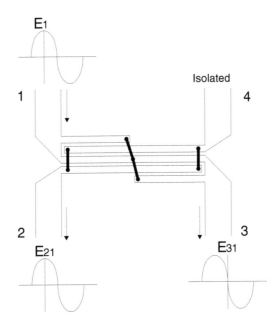

Figure 7.5.16

Table 7.5.1

B_j	B_g	ζ_t	ζ_ℓ	s_{11M},dB	$\Delta a°$
2.2:1	2.1:1	0.6	0.6	−18.3	±2
2.5:1	2.6:1	0.6	0.55	−15.3	±3.2
3:1	3.2:1	0.6	0.48	−10.1	±5.5

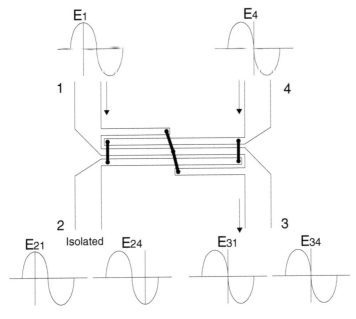

Figure 7.5.17

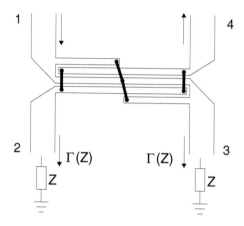

Figure 7.5.18

to change the value of this reactance in an easy manner. In the microwave frequency region, good tunable reactance may be obtained with varactor diodes, whose capacitance value "C" is dependent on the applied voltage "V."

A block diagram of a phase shifter using a hybrid and ideal capacitors is indicated in Figure 7.5.19. Ports "1" and "4" are the input and isolated ports, or vice versa, and ports "2" and "3" are the coupled and direct ports, or vice versa, of the splitter. In this case, choosing a reference value "V_r" of the applied control voltage to the varactors from 7.5.5 we find that if we want the output signal changes only for a phase "φ" at the control voltage "V," with respect to the output signal for "V_r," the reflection coefficients must be related by:

$$\Gamma(V) = e^{j\varphi}\,\Gamma(V_r) \tag{7.5.6}$$

If the previous equation holds, signal "E_4" will not be affected by amplitude variation when varying the control voltage, but only its phase will change.

Also in this case of ideal hybrid and ideal capacitors, the bandwidth of the phase flatness of the tuneable phase shifter is limited by the frequency dependence of the capacitive reactance. This can be simply shown by evaluating the reflection coefficient "$\Gamma(C)$" created by the capacitor "C" of Figure 7.5.19, that is:

$$\Gamma(C) = \frac{-jX_c - Z_h}{-jX_c + Z_h} \tag{7.5.7}$$

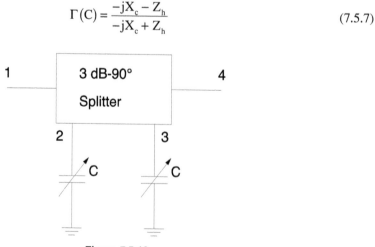

Figure 7.5.19

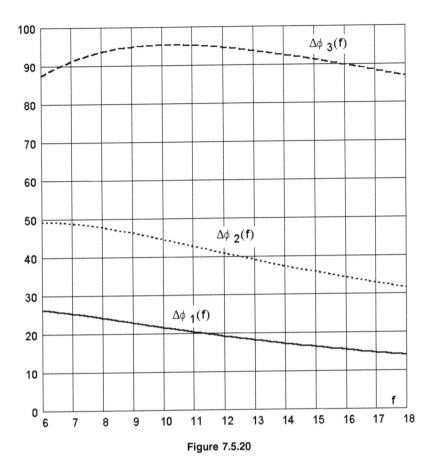

Figure 7.5.20

where $X_c = 1/\omega C$ and "Z_h" is the impedance presented by the hybrid to the capacitor. Then, applying 7.5.6 with "V" and "V_r" replaced by two values of capacitance, it is possible to evaluate the differential phase shifting inserted on the signal. The graphical results are represented in Figure 7.5.20, where in abscissae we have frequency in GHz and in ordinates the phase shifting in degree. Here, we have chosen three values of phase shifting at 12 GHz, nearly 20°, 40,° and 90°. In this figure the smaller values of capacitance are 0.5pF, 0.33pF, and 0.12pF, respectively, while the higher value of capacitance, that is the reference value, is 0.8pF. We see how a desired phase shifting can be obtained just by changing the values of the capacitors, but the phase flatness is a compromise with the operating bandwidth. Using some reactance compensation it is possible to flatten the phase shifting somewhat. Note that all the components in the calculations have been considered ideal.

Another source of error is due to the limited value of isolation of the hybrid. Typical values are near 20 dB for the Lange coupler working in the 6 to 18 GHz band, or 25 dB using high technological accuracy. Adapting the results of some researchers[65] to the case of Figure 7.5.19, the maximum peak phase error "ε_{pk}" due to the finite isolation "I" of the hybrid is given by:

$$\varepsilon_{pk} = \left(1 + 3\left|\Gamma(Z)\right| - 0.25\sin\pi\left|\Gamma(Z)\right|\right) a\sin\left(I/\left|\Gamma(Z)\right|\right) \tag{7.5.8}$$

In our case of phase shifter we desire $|\Gamma(Z)| = 1$, as previously assumed in point 4. The graphical result of Equation 7.5.8 is given in Figure 7.5.21 for the case of $|\Gamma(Z)| = 1$, where in abscissae we have isolation in dB and in ordinate the peak phase shifting error in degree. Note as with 20 dB of isolation the maximum peak phase error is near 23°. With care in choosing a matching network

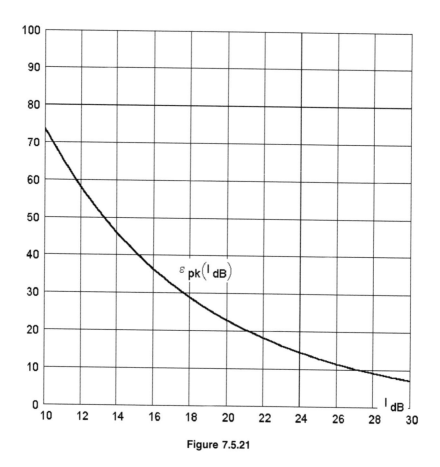

Figure 7.5.21

between the reflective elements of Figure 7.5.18 and the hybrid, it is possible to reduce this cause of error. Unfortunately, these matching networks have a narrower bandwidth with respect to the Lange hybrid and, in addition, can perform good phase precision only for some values of $\Gamma(Z)$. As a result, for a wide bandwidth phase shifter some compromise is needed between phase flatness, tunability, and available layout area. Due to the simple construction and small required area, a reflection phase shifter is well suited to be realized in MMIC technology.[66] Of course, the same discourse can be applied to any 3 dB/90° divider, as for example the branch line divider. However, this last device has a narrower band in comparison to the Lange coupler.

7.6 THE THREE PORT CIRCULATOR

Very interesting transmission properties can be obtained when microstrips come near, or are realized on, ferrimagnetic materials.* The most commonly used material for this purpose is called "ferrite," but effectively under this name there are a lot of ferrimagnetic materials** each of which is best suited for a particular application and/or operating bandwidth. In this section we will use the term "ferrite" to mean a generic ferrimagnetic material, unless we indicate an otherwise certain, specified material.

A microstrip circulator is a three port device using ferrite in some part, with the characteristic that for any input "I" port there is only one output "O" port; in addition, (and it is the main characteristic that makes the circulator unique) exchanging "I" with "O" there is no transmission.

---

* See Appendices A5, A6, and A7 for energy interactions between the e.m. field and ferrimagnetic materials.
** See Appendix A7 for definitions of ferrimagnetic materials.

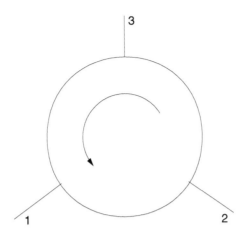

Figure 7.6.1

In other words, connection between two ports is circular with a well-defined direction.* Since the interest in these devices is very high, circulators are not only stand-alone devices, but they have also been inserted in MMIC.[67,68]

With this definition, the circulator is represented with a symbol indicated in Figure 7.6.1. If port "1" is the input, then port "3" is the output; if port "2" is the input, then port "3" is the output; finally, if port "3" is the input, then port "1" is the output. So, the "rotation" of the signal inside the circulator is possible only in the direction "1 → 2 → 3 → 1," and not in the opposite direction, i.e., "1 → 3 → 2 → 1." A physical realization of a microstrip circulator is indicated in Figure 7.6.2. In part a it is indicated in a cross-sectional view, while part b indicates a top view of the device. Three microstrips are connected together with a circular disk, so that the lines are 120° from the nearest line. This structure is constructed on ferrite as a substrate, at least beneath the circular disk and part of the microstrips. Finally, a magnet must properly and uniformly magnetize the ferrite through a static magnetic field "$\underline{H}_{dc}$." Usually, a microstrip circulator is a "drop-in" device, i.e., a stand-alone component that is inserted in another microstrip circuit, whichever it is. The device indicated in Figure 7.6.2 is also called a "junction circulator."

A lot of theory, especially regarding the physics of ferrimagnetic materials, is necessary to explain why this device works as a circulator. The reader who is not clear about the energy interaction phenomena between the e.m. field and ferrite is strongly encouraged to read at least Appendix A7 where we have reported the fundamental concepts on this topic. These concepts will also be necessary for the coming sections where we will discuss other microstrip devices working through energy interactions with ferrimagnetic materials. In addition, Appendices A5 and A6 contain the fundamental concepts required to understand Appendix A7. However, when necessary, we will recall the most important topics regarding the interaction between the e.m. field and ferrite that are necessary to explain the circulator operation mode.

From the theory developed in Chapter 2, we know that the fundamental propagation mode of the microstrip is the "qTEM." Here we will assume that this mode can be replaced with a pure "TEM." So, before the connection between the microstrip and the circular disk, "$\underline{h}$" is linearly polarized and parallel to "w" of the microstrip. Consequently, the internal static magnetic field "$\underline{H}_s$" is orthogonal to "$\underline{h}$." From this fact, and from the fact that the wave is propagating in a direction orthogonal to "$\underline{H}_s$," we know** that the "TEM" mode divides itself into "TE" and "TM" modes having circular magnetic polarization, that in our case is as perfect as the wave is near the center of the disk. In this circumstance, the ferrite presents a permeability "$\mu_{eq\perp}$."

---

* We will see later how the circular direction between ports can be changed.
** See Appendix A7 for the theory.

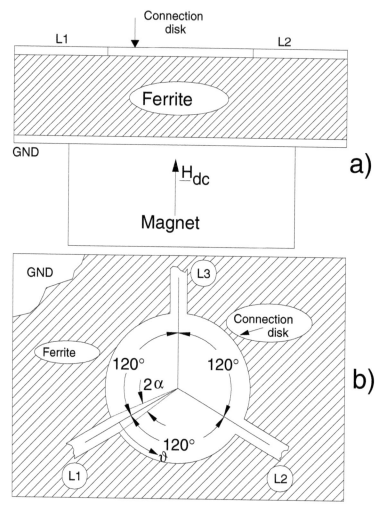

Figure 7.6.2

After these simple fundamental notes regarding the energy interaction between ferrite and e.m. waves, we can begin to study the electromagnetic structure extracted from that of the circulator. This situation is indicated in Figure 7.6.3. Two conductor disks are separated by a ferrite cylinder of height "h."* For the present case of the microstrips, we assume that all the "RF" field energy is contained inside this structure. With this hypothesis, we don't need to evaluate effective parameters like permeability or permeability effective values.**

The analytical study to find the solution of Maxwell's equations for this geometry is quite involved with mathematics. In fact we have to resolve these equations in a medium with tensorial permeability, applying the boundary condition to a system with a cylindrical coordinate system. Some theories to explain the circulator behavior can be found in the literature.[69,70,71,72,73,74,75,76,77] Here, we will show the fundamental steps that are necessary to understand the operation of the junction circulator. We will suppose that the e.m. field has no dependence on the coordinate "z," i.e., we assume that the value of "h" is much smaller than the wavelength of the signal inside the ferrite.

---

* In this text "h" is used throughout to indicate substrate height. Here it will also be used to indicate the magnetic field of the wave. We think that the objects are so different that no confusion will arise.
** See Chapter 2 for definitions of microstrip effective parameters.

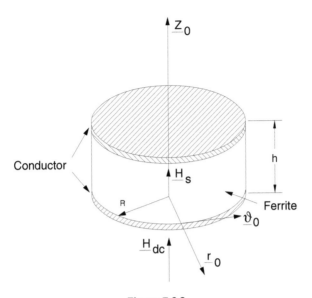

Figure 7.6.3

To find the natural oscillation frequencies of this structure, we have to resolve Maxwell's equations. So, the wave equation* for the electrical field "e_z" for the structure indicated in Figure 7.6.3 can be written as:

$$\frac{\partial^2 e_z}{\partial r^2} + \frac{\partial e_z}{r\partial r} + \frac{\partial^2 e_z}{r^2 \partial \theta} - k_{te\perp}^2 e_z =! \, 0 \qquad (7.6.1)$$

where "$k_{te\perp}^2$" and "$\mu_{eq\perp}$" are given by **:

$$k_{te\perp}^2 \doteq -\omega^2 \mu_0 \varepsilon \mu_{eq\perp} \qquad (7.6.2)$$

$$\mu_{eq\perp} \doteq \frac{\mu_p^2 - \mu_\ell^2}{\mu_p} \qquad (7.6.3)$$

Analogous with the case of rectangular coordinates, to find a solution of 7.6.1 we apply the "separation variable" method. We assume the field "e_z" to be the product of a function of "r" only and a function of "θ" only, i.e., $e_z(r,\theta) \doteq f_r(r) f_\theta(\theta)$. With this hypothesis, the solution of 7.6.1 is:

$$e_z(r,\theta) = J_n\left(k_{te\perp} r\right)\left[a_n e^{(-jn\theta)} + b_n e^{(jn\theta)}\right] \qquad (7.6.4)$$

where "$J_n(k_{te\perp} r)$" is the Bessel function of the first kind and of order "n," and "a_n" and "b_n" are two constants to be determined when the boundary conditions are defined. Then, inserting this expression into Maxwell's equation:

$$\underline{\nabla} \otimes \underline{h} = j\omega \varepsilon \underline{e} \qquad (7.6.5)$$

---

* See Appendix A2 for the "Wave equation."
** See Appendix A7 for definitions of these quantities.

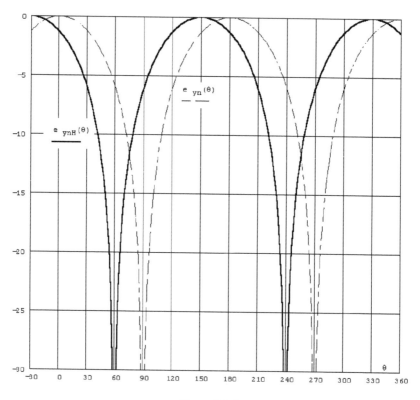

Figure 7.6.4

The value of "$\gamma_{T\mu}$" is known and in the "MKSA" unit system is $-2.21(10)^5$ (rad./sec)/(A.t/meter). Quite often it is more useful to speak about frequencies rather than angular frequencies, and in technical literature "$\gamma_{T\mu}$" is given the value $-3.52(10)^4$ Hz/(A.t/m) or -2.8 MHz/Oe in "CGS" system unit. When we use frequencies instead of pulsations we will use the "frequency gyromagnetic ratio," indicated with "$\gamma_{T\mu}$" and defined as $\gamma_{T\mu} \doteq 2\pi\gamma_{T\mu'}$.

It is common practice in magnetism to use the "CGS" unit system, where magnetic fields are measured in Oersted (Oe) and magnetic flux density in Gauss. The relations between these quantities and the corresponding quantities in "MKSA" are:

$$1 \text{ Oe} = 10^3/4\pi \text{ A.t}/\text{m} \tag{7.6.11}$$

$$1 \text{ Gauss} = 10^{-4} \text{ Web}/\text{m}^2 \tag{7.6.12}$$

where "A.t/m" means "Ampere.turn/meter" and "Web/m²" means "Weber/square meters."

Returning to Figure 7.6.5, in practice values of $\sigma > 1$ are associated with operations above resonance,* while for $\sigma < 1$ we are below resonance. Since the required value of "H_{dc}" when $\sigma < 1$ is lower than the value for $\sigma > 1$, the below resonance circulator operation is preferred. The values of "$r(\sigma,p)$" near $\sigma = 0$ can be reached theoretically in two ways, as can be recognized by the definition of "σ" in 7.6.10: one for $H_s = 0$ and the other for $\omega = \infty$. Of these two possibilities, only $H_s = 0$ can be physically obtained. So, at left of $\sigma = 0$ the values of "σ" should be associated with an operating case of zero static magnetic field; if required, "ω" can also be varied. Just to recall what we have studied in Appendix A7, the ferrite operating at such a low DC bias** field induces

---

* See Appendices A6 and A7 for resonance definitions and associated theory.
** The "DC" magnetic field will also be simply called the "bias field."

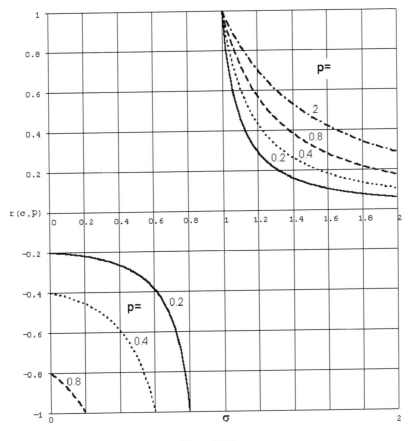

Figure 7.6.5

losses in the passing e.m. wave. Appendix A7 has formulas for the case of an unsaturated DC magnetic field, which characterizes the ferrite in such situations well. In the present study, instead, all the formulas are relative to a DC magnetic field saturating the ferrite, which is the operating situation of the circulator. It is important to observe that the same absolute value of "$r(\sigma,p)$" can be obtained above or below resonance. However, since the sign changes for this value, progressive and regressive modes exchange places.

The consequence of this dependence of the "RF" fields on the ferrite "DC" magnetic status (as indicated in Figure 7.6.5) is that with a particular value of "H_{dc}," the stationary wave pattern of Figure 7.6.4 rotates. When the rotation is near 30° a null of the field can be placed at a port, causing the isolation of that port. This situation is indicated in Figure 7.6.4 with the solid line. Of course, as a consequence of the change of sign (but same value) of "$r(\sigma,p)$" for above or below resonance, the circulation direction changes if we operate above or below resonance. We have the same change in circulation direction if we reverse the direction of application of "H_{dc}."

After this brief explanation of the circulator working principle, we want to find the field components inside this device when we use the boundary conditions where no generator applied. This is the case relative to the "natural oscillation frequencies." These boundary conditions are:

$$h_\theta = !0 \quad \text{for} \quad r = R \qquad\qquad (7.6.13)$$

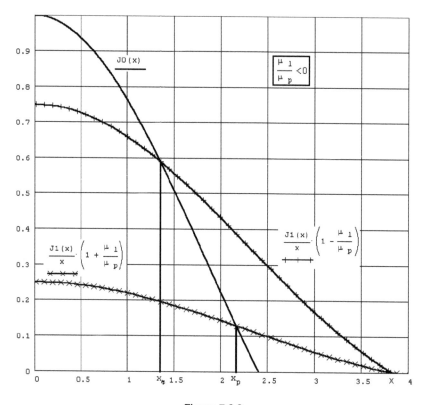

Figure 7.6.6

Inserting this condition in 7.6.7 and 7.6.8, for the progressive and regressive modes we have, respectively:

$$J_{n-1}\left(k_{te\perp}R\right) = ! \left[nJ_n\left(k_{te\perp}R\right)\big/k_{te\perp}R\right]\left(1+\mu_\ell/\mu_p\right) \tag{7.6.14}$$

$$J_{n-1}\left(k_{te\perp}R\right) = ! \left[nJ_n\left(k_{te\perp}R\right)\big/k_{te\perp}R\right]\left(1 - \mu_\ell/\mu_p\right) \tag{7.6.15}$$

The fact that these equations give the resonant frequencies depends on the definition in 7.6.4 of "$k_{te\perp}$," where the frequency appears and from the fact that "R" is known. The solutions of the two previous equations can be easily obtained graphically, after a value of "n" has been defined. In Figure 7.6.6 we have represented the case of $n = 1$, $\mu_\ell/\mu_p = -0.5$, i.e., below resonance operation, and with $x \perp k_{te\perp}R$. Note that for below resonance operation, i.e., $\sigma < 1$, the progressive mode has a higher resonant frequency than the regressive one, while for $\sigma > 1$ the opposite is true. The difference between the resonant frequencies of the two modes depends on the value of the ratio "μ_ℓ/μ_p," whose shape is a function of the applied internal DC magnetic field "H_s" as indicated in Figure 7.6.5. Operatively, it has been discovered that circulation is present when the frequencies of these two modes are nearly the same. This situation is called "mode degeneracy." So, indicating with "f_p" and "f_r," respectively, the resonance frequency of the progressive and regressive modes, the circulation frequency "f_c" is something near the value of the following equation:

$$f_c \approx \left(f_p + f_r\right)\big/2 \tag{7.6.16}$$

The nonexact correspondence between the values of the two members of equation 7.6.16 depends on the approximations we did in considering only the case of n = 1 and a perfect "TEM" mode supported by the microstrip.

It is interesting to study the case when a generator is applied to the structure in 7.6.3 in the manner indicated in Figure 7.6.2 for our microstrip network. With the generator connected at port 1, the boundary condition becomes:

$$h_\theta \equiv h_1 \text{ at input port, i.e., for } -\alpha < \theta < \alpha \qquad (7.6.17)$$

$$h_\theta \equiv h_1 \text{ at output port, i.e., for } 120° - \alpha < \theta < \alpha + 120° \qquad (7.6.18)$$

$$h_\theta = 0 \text{ otherwise} \qquad (7.6.19)$$

and similar relations for the field "e_z." Inserting the three previous conditions in 7.6.6 we find that the resonant frequencies are given by:

$$J_{n-1}\left(k_{te\perp} R\right) =! J_n\left(k_{te\perp} R\right)/\left(k_{te\perp} R\right) \qquad (7.6.20)$$

This equation can easily be solved as we did for the evaluation of the natural oscillation frequencies, i.e., graphically. Since every Bessel function of the first kind and of every order is known, the first root for Equation 7.6.20 for n = 1 is:

$$k_{te\perp} R = 1.84 \qquad (7.6.21)$$

Practical circulator synthesis is always a "cut and try" procedure, where experience and theory work closely together. For example, it is often necessary to adjust the input port impedances to that of the connecting lines. Experience suggests that conductor disk diameter "d" is smaller than magnet diameter "D," according to:[79]

$$d = 0.8D \qquad (7.6.22)$$

where "D" is given by:

$$D = \frac{3.3c}{\omega\left(\varepsilon_r \mu_{eq\perp}\right)^{0.5}} \qquad (7.6.23)$$

and "c" is the speed of light in the vacuum. Note that 7.6.22 is also a consequence of the fact that a uniform DC magnetic bias is required in all the ferrite inside the conductor disk. If the magnet were of the same diameter as the disk, some DC magnetic disuniformity would exist in the periphery.

Practical synthesis design starts with the definition of the operating fractional bandwidth "B" and VSWR, which are closely related, and then we define the value of "μ_ℓ/μ_p."[80] In fact, from VSWR we extract the angle parameter "τ" from:

$$\tau \stackrel{\pm}{=} a\cos\left[(VSWR)^{-0.5}\right] \qquad (7.6.24)$$

From this value, we have the loaded quality factor "Q_ℓ"* using:

* See Appendix A4 for quality factor definitions.

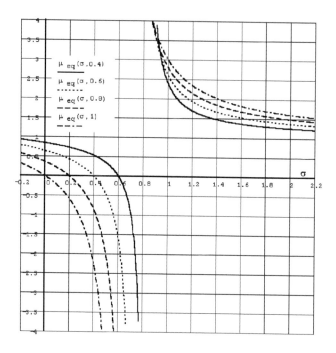

Figure 7.6.7

$$Q_\ell = \tan(\tau)/B \qquad (7.6.25)$$

and the value of "μ_ℓ/μ_p" is:

$$\left|\mu_\ell/\mu_p\right| = 0.71/Q_\ell \qquad (7.6.26)$$

Usually, when the value of $|r(\sigma,p)| = |\mu_\ell/\mu_p|$ from the previous equation, a value of "p" is chosen so that the corresponding value of "σ" is as small as possible so as to have the smaller value of "H_{dc}." Note that the disk diameter can also be obtained with this procedure. In fact, once the value of "σ" and "p" are defined, we can extract the value of "$\mu_{eq\perp}$" from Figure 7.6.7. After "$\mu_{eq\perp}$" is known, from 7.6.2 we extract the value of "$k_{te\perp}$," and from 7.6.21 we have the value "R" of the disc conductor.

It is important at this point to inform the reader that other methods are possible for studying the circulator operation. For instance, some researchers[81,82] have used the "s" parameter method, the empirical circular rotating waves method,[83] or the phase shifter method.[84] We have decided to study the circulator with the classical Maxwell's equation method for two reasons. First, it is the only method applicable to all electromagnetic situations even if other methods are simpler to use, and second, this is a practical application of the electromagnetic theory in Appendix A2.

The circulator performances are strongly related to their operating bandwidths. These devices can operate on narrow bandwidth (for "B," something below 10%), a moderate bandwidth ("B" near 20%), or a wide bandwidth ("B" near 70%). In any case, insertion loss is below 2 dB and isolation is higher than 15 dB. Narrow band devices can have an improvement of these values also of 50%. Center frequencies "f_c" are limited by size for frequencies $f_c < 1$ GHz, and for $f_c > 25$ GHz, by microstrip dispersion phenomena,* ferrite losses, and tolerances accuracy. For example, a typical circulator working at 1 GHz has a disk diameter $d \approx 2.5$ mm, while $d \approx 2.5$ cm for $f_c = 15$ GHz. The need for this type of device in any transmission-reception unit has moved the research

* See Chapter 4 for higher order effects in microstrip.

to produce circulators with $f_c \approx 90\text{GHz}$.[85] The power handling capability of these devices is dependent on the ferrite mass; peak power of some hundreds of watts and CW power of some tenths of watts are typical.

In general, for the operation of this device it is not required that the biasing magnet always be present mainly for two reasons: first, because it is quite heavy, and second, because the employed ferrite has low saturation magnetization, so that it can simply be permanently magnetized. This last operation mode is usually called "latched operation." Experiments on latched circulators have been made by some researchers[86,87] with good results. This operation mode has the characteristic of being more sensitive to external magnetic fields.

In this section we have tried to simply explain the operating principle of one of the most attractive and commonly used microwave nonreciprocal passive devices. However, this is only a fraction of the quantity of scientific disciplines required to understand all the phenomena involved in the circulation. For instance, the influence of the chemical composition of ferrite and its shape regarding the electrical performances have been ignored here. Nevertheless, chemistry is as important as physics, mathematics, and electromagnetism to really understand the circulator operating principle.

7.7 FERRIMAGNETIC PHASE SHIFTERS

Ferrimagnetic phase shifters can be realized in two configurations: reciprocal and nonreciprocal. The former type have nonferrimagnetic counterparts, as we studied in section 7.5, while the latter type is unique. However, every ferrimagnetic phase shifter has a unique characteristic, i.e., the possibility of changing the phase shifting continuously, changing the intensity of the applied "DC" magnetic field. Also these devices are attractive in microwave communication systems, and for this reason studies to increase performance are continuing.[88,89]

7.7.1 Reciprocal

Microstrip reciprocal ferrimagnetic phase shifters are characterized by the fact that phase shifting is insensitive to the wave direction of propagation inside the structure. In contrast to the fixed phase shifter studied in Section 7.5, where for one input there exist two outputs, here phase difference is always evaluated using the same signal path. Of course, the reason why a phase difference exists lies in the energy interaction between the e.m. wave and static magnetic field "H_{dc}." In practice, the phase difference is evaluated changing the direction of "H_{dc}" in two positions: the first one corresponding to the maximum interaction between the "RF" magnetic field "h" and "H_{dc}," and the second one corresponding to the minimum interaction. A typical top view of a reciprocal ferrimagnetic phase shifter is indicated in Figure 7.7.1.

The microstrip hot conductor is folded many times with a serpentine shape. This folded structure is usually called the "meander line." In this case, the meander line is realized so that the coupling between any two lines can be neglected.* Let us suppose that a conductor coil is realized passing the winding inside the slots "F2 – F2'." If a current is now sent in the wire, a static magnetic field "H_{dc}" is generated,** whose field lines can be approximated by the lines indicated with "H_{dc}" in Figure 7.7.1. Of course, the direction of the field along these lines depends on the current direction in the coil. If the conductor coil is inserted in the holes "F1 – F1'" and a current is sent inside it, then the generated static magnetic field is rotated 90° with respect to the lines represented in the figure. For any microstrip parallel to "H_{dc}" the magnetic situation for the coil in "F2 – F2'" is indicated in Figure 7.7.2 in a crossed view. In this figure we have also indicated the "RF" magnetic field lines of the microstrip, and we have assumed that the static magnetic field is entering in the

---

* See Chapter 2 for uncoupled microstrip conditions.
** See Appendix A5 for relations between currents and magnetic fields.

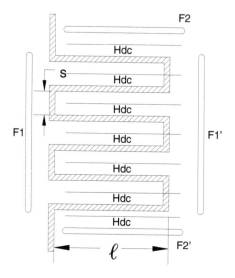

Figure 7.7.1

paper. If we approximate the wave supported by the microstrip to a "TEM" in a homogeneous ferrimagnetic media, this is the same electromagnetic situation covered in Appendix A7. With this hypothesis for the wave magnetic field "h" and the static magnetic field "H_{dc}" we can write:

$$\underline{h} = h_x \underline{x}_0 \tag{7.7.1}$$

$$\underline{H}_{dc} = H_z \underline{z}_0 \tag{7.7.2}$$

where "H_z" is the internal saturating static magnetic field. This is a case where a "TEM" linear polarized wave propagates inside an isodirectional magnetized ferrite, with orthogonality between "RF" and "DC" magnetic fields.

Studies made by some researchers[90] have shown that in the practical case of residual magnetization* the relative permeability "μ_r" for the situation indicated in Figure 7.7.2 is:

$$\mu_r = \mu_{eq\perp}\left[1 - (h/w)^{0.8}\left(\mu_\ell/\mu_p\right)^\perp \ln\left(1 + 1/\mu_{eq\perp}\right)/7\right]^{(-1)} \tag{7.7.3}$$

where all the magnetic quantities are already defined in the previous sections or in Appendix A7.

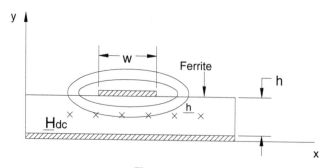

Figure 7.7.2

---

* See Appendix A7 for secondary effects for a magnetized ferrite.

So, when "H_{dc}" is directed as indicated in Figure 7.7.1, the previous equation gives the appropriate phase constant for the "TEM" wave. Instead, when "H_{dc}" is directed at 90° with respect to the magnetic lines indicated in Figure 7.7.1, i.e., the coil is inserted in the slots "F1 – F1′," the ferrite presents a relative permeability equal to 1, and the phase constant is:

$$\beta_{z0} = \omega\left(\mu_0\varepsilon_0\varepsilon_{re}\right)^{0.5} \tag{7.7.15}$$

Then, with a fixed length "ℓ" of the phase shifter, i.e., the length of the longer parallel microstrips in Figure 7.7.1, the phase shifting "$\Delta\theta$" is:

$$\Delta\theta \propto n\left(\beta_z - \beta_{z0}\right)\ell \tag{7.7.16}$$

where "n" is the number of parallel microstrips. The quantity "$\Delta\theta$" is usually called "differential phase shift." This parameter in the most general case can change sign in two circumstances:

a. there is a change in phase constant when changing the direction of propagation. For ferrimagnetic phase shifters, this condition also assumes that the direction of "$\underline{H}_{dc}$" is fixed for the two directions of propagation.
b. there is a change in phase constant when changing the direction of "$\underline{H}_{dc}$." For ferrimagnetic phase shifters, this condition also assumes that the direction of propagation is fixed for the two directions of the static magnetic field.

In the present case, condition b holds.

A parameter that is frequently used to classify phase shifters is the "figure of merit." It is the available phase change for each dB of attenuation introduced by the phase shifter when it changes the phase.

With these concepts in mind, it is clear that the device in Figure 7.7.1 refects the following Table 7.7.1:

Table 7.7.1

Coil in:	$\Delta\theta$
F2 – F2′	Maximum
F1 – F1′	Minimum

It is not required that the current into the winding always be present for this device to operate, mainly for two reasons: first, because some amount of energy is required to permit the current flow and it cannot always be available; second, because the employed ferrite has low saturation magnetization, so that it can simply be permanently magnetized. This last operation mode is usually called "latched operation." Experiments on microstrips ferrimagnetic reciprocal phase shifters have been made by G. T. Roome and H. A. Hair[93] using latching operation. As a result, this device has a narrow bandwidth, near some percent, with a figure of merit of 40°/dB at 2 GHz. Since other ferrimagnetic phase shifters exist that can give a higher figure of merit, as we will see in the next section, the ferrimagnetic reciprocal phase shifter is not widely used.

7.7.2 Nonreciprocal

Nonreciprocal ferrimagnetic phase shifters can be built using the following two distinct physical phenomena:

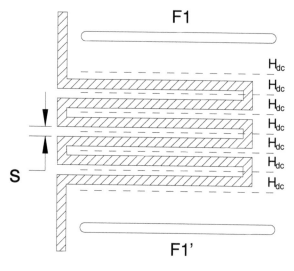

Figure 7.7.3

1. The interaction between biased ferrite and an "RF" magnetic field with circular polarization
2. Impenetrability of ferrite, a physical phenomena called "field displacement"

a. Using a Circular Polarized "RF" Magnetic Field

In contrast to the previous case in 7.7.1, the nonreciprocal ferrimagnetic phase shifter has the characteristic of presenting to the wave traveling inside it, different phase constants for each direction of propagation. Alternatively, for a fixed direction of propagation there is a change of phase constant for each change in the application direction of static "$\underline{H}_{dc}$" magnetic field.

The microstrip construction of this device is very similar to its reciprocal counterpart, and a possible microstrip realization is indicated in Figure 7.7.3.

We see how the meander line is now more compact so that e.m. interactions between adjacent lines cannot be avoided. However, in our case this interaction is desirable, as we will show. So, a significant difference exists between the symmetrical and asymmetrical ferrimagnetic shifters. the former uses an uncoupled meander line only to decrease the required dimensions, the latter must use a coupled microstrip structure just to work properly. In fact, this structure presents inside it points with circular polarization of the "RF" magnetic field, which can be used for our phase shifting problem. The slots named "F1 – F1'" in Figure 7.7.3 need to insert the winding of a coil to create a "DC" magnetic field "$\underline{H}_{dc}$" that biases the device. The direction of "$\underline{H}_{dc}$" is indicated in Figure 7.7.3 where we can see how it is aligned with the coupling region of the meander line. The reason why in the present case, "$\underline{H}_{dc}$" also needs to be so directed, can be understood by observing Figure 7.7.4. Part a of the figure represents the hot conductor of Figure 7.7.3, where we have imposed the condition:

$$2\ell + d = ! \; \lambda/4 \qquad\qquad (7.7.17)$$

Part b of Figure 7.7.4 indicates a transverse view of part a taken from the section A_A. We see how inside the ferrite, somewhere in the middle of the microstrips separation "s," points "C_j" exist where the "RF" magnetic field "h" is circular polarized. The loci of the "RF" magnetic field for any µstrip are represented by the elliptic dashed lines, as was explained in Chapter 2. The reason for the existence of these "C_j" points lies in Equation 7.7.17, which also means that at center frequency between each middle of the meander line, a 90° phase offset exists. Figure 7.7.4 indicates

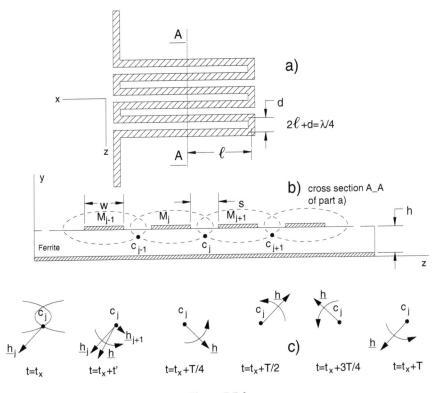

Figure 7.7.4

how such polarization comes out. Starting at a generic time "t_x" where only a vector "$\underline{h}_j$" exists, after a time "t'" the vector "$\underline{h}_{j+1}$" starts to appear: the resulting vector "$\underline{h}$" rotates. After a time "$t_x + T/4$," with "T" the period of the signal passing in the meander line, the vector "$\underline{h}_{j+1}$" is at maximum while "$\underline{h}_j$"is zero: the resulting vector "$\underline{h}$" rotates 90° with respect to the position at time "t_x." This behavior continues, so that "$\underline{h}$" is circular polarized. The direction of such polarization depends on the direction of propagation; so, if we have the left circular polarization of Figure 7.7.4c when the signal travels from left to right, a right circular polarization arises when the signal travels from right to left.

Of course, to consider the e.m. coupling composed only by coupling a line with position "j" with lines of positions "j – 1" and "j + 1" is a simplification. However, experiments show that not much error arises using such a method, and at points "C_j," a circular polarization exists.

We know from the theory of e.m. energy inside ferrimagnetic materials* that when a "TEM" circular polarized wave travels inside an isodirectional magnetized ferrite, it meets two permeabilities called "concordant" "μ_c" and "discordant" "μ_d," given by:

$$\mu_c \overset{\perp}{=} \mu_p + \mu_\ell \tag{7.7.18}$$

$$\mu_d \overset{\perp}{=} \mu_p - \mu_\ell \tag{7.7.19}$$

The shapes of the previous equations for zero losses vs. angular frequency "ω" are given in Figure A7.5.2 (from Appendix A7), which we report here for simplicity. Associated with the two previous permeabilities there are two phase constants "β_c" and "β_d" given by:

* See Appendix A7 for energy exchange between e.m. waves and ferrimagnetic materials.

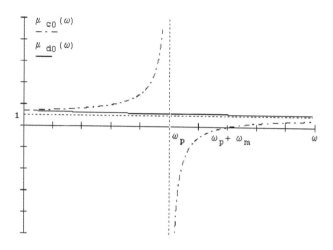

Figure A7.5.2

$$\beta_c \doteq \omega\left(\mu_0 \mu_c \varepsilon\right)^{0.5} \tag{7.7.20}$$

$$\beta_d \doteq \omega\left(\mu_0 \mu_d \varepsilon\right)^{0.5} \tag{7.7.21}$$

where $\varepsilon \perp \varepsilon_0\,\varepsilon_{re}$ So:

1. If we approximate the wave passing in the meander line as a "TEM" one,
2. If we statically magnetize the device as indicated in Figure 7.7.3,

then a fixed direction of "$\underline{H}_{dc}$" the wave will have a phase constant "β_c" in one direction of propagation and a phase constant "β_d" in the opposite direction of propagation. In such a case the differential phase shift "$\Delta\theta$" will be:

$$\Delta\theta \propto n\left(\beta_c - \beta_d\right)2\ell \tag{7.7.22}$$

since "2ℓ" is the length of a meander line. Of course, the same differential phase shift can be obtained if we fix a direction of propagation and change the direction of "$\underline{H}_{dc}$." It is important to remember that in our case of microstrip technology the practical values of differential phase shift are quite different from those obtained from Equation 7.8.6. This is mainly because wave propagation in a microstrip is not in homogeneous media. So, also in this case it is necessary to evaluate an effective permeability "μ_{re}." To do this, it is possible to use the equations we gave in section 7.7.1, with attention to evaluating two effective permeabilities, one "μ_{cre}" relative to the case when the wave encounters a "μ_c," and one "μ_{dre}" when the wave encounters a "μ_d." Note that in this case it is also possible to use this device with a residual, or "latching" magnetization, similar to the reciprocal phase shifter.

As we said before, in the present case the coupled meander line is a particular example of coupled microstrip theory that we have already encountered in this chapter.* The general theory of the meander line is more deeply studied in Chapter 5, where we treat just about all the coupled µstrip types. The reader is directed to that chapter for a closer study of this coupled lines network, especially concerning the impedance matching of this device. This is always a critical task because

* See sections 7.24 through 7.2.6, 7.3.4, 7.4.1, and especially section 7.5.1 where Schiffman's shifter is studied.

high values of "Δθ" come with high values of VSWR, which need to be eliminated. So, the study of optimum performance of the meander line phase shifter is quite often done with measurements and manual optimization, starting from an initial device theoretically synthesized. For instance, after having found an optimum shifting, it is often necessary to adjust the input port impedances to the system reference impedance. This is usually done using the final "λ/4" lines, working as impedance transformers.*

Experiments[94] made with these devices have given figures of merit near 300°/dB and differential phase shift of 140°/cm[95] for an "H_{dc}" biased operation, or 200°/dB for a latching operation. These values are almost three times the reachable values for the reciprocal phase shifter we studied previously. This means that interactions between ferrite and wave circular polarization are a highly effective phenomena. Operating bandwidth is always quite narrow and comparable to the reciprocal phase shifter, i.e., near some percent. Caution must be used with the maximum operating frequency, independent of the ferrite properties. According to meander line theory,[96] stop band behavior of this structure can happen at those frequencies where, with reference Figure 7.8.2, the length "2ℓ" satisfies:

$$\beta 2\ell = n2\pi \qquad \text{with } n = 1,2,3\ldots \qquad (7.7.23)$$

Actually, the coupled meander line microstrip phase shifter is a widely used device, which is preferred to its reciprocal counterpart.

b. Using "Field Displacement"

The "field displacement" is an effect produced by ferrite when particular conditions exist among:

1. "DC" magnetic field strength
2. Direction of "DC" magnetic field
3. Signal frequency
4. Direction of "RF" magnetic field

When these four conditions hold, the field displacement effect consists of a ferrite "RF" signal impenetrability. We will study the field displacement theory, effects, and associated waveguide devices in Appendix A7. Here we will see how this theory can be applied to microstrips and how to build phase shifter devices. One of the first studies on microstrip field displacement effect was performed by M. E. Hines,[97] who has applied Maxwell's equations to the microstrip structure. We will follow this method, which is an application of the general theory developed in Appendices A2 and A7.

Examine Figure 7.7.5. Here a microstrip built on a ferrimagnetic substrate is biased transversely to the direction of propagation with a static field "H_{dc}." Let us suppose that dimension "w" is much higher than dimension "h," i.e.,

$$w \gg h \qquad (7.7.24)$$

This condition makes it possible to study propagation along "x," in addition to the obvious propagation along "z." With this hypothesis, and assuming

5. to be in an homogeneous media
6. to evaluate only the propagation along "$\underline{z}_0$,"
   the "RF" fields "h" and "e" have the following expressions, in accordance with the theory given in Appendix A7:

---

* See Chapter 1 for "λ/4" lines used as impedance transformers.

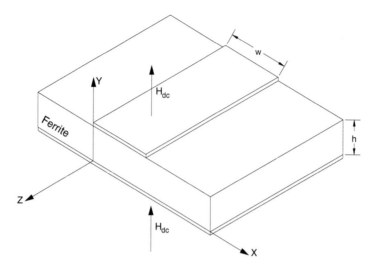

Figure 7.7.5

$$h_x(x,z) = \left[P \frac{-\mu_\ell k_x - j\mu_p k_z}{\omega\mu_0\left(\mu_\ell^2 - \mu_p^2\right)} e^{(-k_x X)} + R \frac{\mu_\ell k_x - j\mu_p k_z}{\omega\mu_0\left(\mu_\ell^2 - \mu_p^2\right)} e^{(k_x X)} \right] e^{(-k_z Z)} \qquad \text{(A7.14.3)}$$

$$h_z(x,z) = \left[P' \frac{j\mu_p k_x - \mu_\ell k_z}{\omega\mu_0\left(\mu_\ell^2 - \mu_p^2\right)} e^{(-k_x X)} + R' \frac{-j\mu_p k_x - \mu_\ell k_z}{\omega\mu_0\left(\mu_\ell^2 - \mu_p^2\right)} e^{(k_x X)} \right] e^{(-k_z Z)} \qquad \text{(A7.14.4)}$$

$$e_y(x,z) = \left[e^+ e^{(-k_x X)} + e^- e^{(k_x X)} \right] e^{(-k_z Z)} \qquad \text{(A7.14.9)}$$

with the condition:

$$k_x^2 + k_z^2 = \left(1 - m^2 \right) \mu_0 \mu_{eq} \varepsilon \qquad \text{(7.7.25)}$$

where "P," "P'," "R," "R'," "e+," and "e-" are generic constants with the dimensions of Volt/m. For a better understanding of changes of sign of these quantities as a function of direction of propagation or direction of "$\underline{H}_{dc}$," A7.14.3 and A7.14.4 can be rewritten as:

$$h_x(x,z) = \frac{1}{\omega\mu_0\mu_{eq\perp}} \left[\left(\frac{\mu_\ell}{\mu_p} k_x + jk_z \right) P e^{(-k_x X)} + \left(\frac{-\mu_\ell}{\mu_p} k_x + jk_z \right) R e^{(k_x X)} \right] e^{(-k_z Z)} \qquad \text{(7.7.26)}$$

$$h_z(x,z) = \frac{1}{\omega\mu_0\mu_{eq\perp}} \left[\left(\frac{\mu_\ell}{\mu_p} k_z - jk_z \right) P' e^{(-k_x X)} + \left(\frac{-\mu_\ell}{\mu_p} k_z + jk_x \right) R' e^{(k_x X)} \right] e^{(-k_z Z)} \qquad \text{(7.7.27)}$$

Now we impose the conditions of having attenuation along "$\underline{x}_0$" and no attenuation along "z." This results in* "k_x" being real and "k_z" being imaginary. So we set:

---

* See Appendix A2 for relations among propagation constants.

$$k_x \overset{\perp}{=} \alpha_x \tag{7.7.28}$$

$$R = R' = e^- = 0 \tag{7.7.29}$$

$$k_z \overset{\perp}{=} j\beta_z \tag{7.7.30}$$

With these three conditions, the "RF" field components become:

$$h_x(x,z) = \left(1/\omega\mu_0\mu_{eq\perp}\right)\left[\left(\mu_\ell/\mu_p\right)\alpha_x - \beta_z\right]Pe^{(-\alpha_x X)}\ e^{(-k_z Z)} \tag{7.7.31}$$

$$h_z(x,z) = \left(j/\omega\mu_0\mu_{eq\perp}\right)\left[\left(\mu_\ell/\mu_p\right)\beta_z - \alpha_x\right]P'e^{(-\alpha_x X)}\ e^{(-k_z Z)} \tag{7.7.32}$$

$$e_y(x,z) = e^+e^{(-\alpha_x X)}\ e^{(-j\beta_z Z)} \tag{7.7.33}$$

Now, with reference Figure 7.7.5, we can recognize how propagation outside the dimension "w" depends on the fields "e_y" and "h_z." So if we impose:

$$h_z = 0 \quad \text{for } x = 0 \text{ and } x = w \tag{7.7.34}$$

it means we don't want propagation outside "w."

Inserting expressions 7.7.28 through 7.7.30 and 7.7.34 in 7.7.31 through 7.7.33 we have "α_x" and "β_z" must verify the following equations:

$$\alpha_x = \left(\mu_\ell/\mu_p\right)\omega\left(\mu_0\mu_p\varepsilon\right)^{0.5} \tag{7.7.35}$$

$$\beta_z = \omega\left(\mu_0\mu_p\varepsilon\right)^{0.5} \tag{7.7.36}$$

Combining the two previous equations it follows that:

$$\alpha_x \equiv \left(\mu_\ell/\mu_p\right)\beta_z \tag{7.7.37}$$

It is interesting to evaluate the wave impedance* "θ" along "z." Then, inserting 7.7.35 and 7.7.36 in 7.7.31 and 7.7.33 we have:

$$\theta = \left(\mu_0\mu_p/\varepsilon\right)^{0.5} \tag{7.7.38}$$

Since "μ_p" can also assume negative values for some bias field "H_{dc}," when it happens we have an imaginary impedance that corresponds to an impenetrability of ferrite to "RF" fields.

The graph of "μ_p" is indicated for the case of zero ferrite losses in Figure A7.4.1b reported here for simplicity. The electromagnetic situation when "θ" assumes a negative value is indicated in Figure 7.7.6. The nonreciprocity of the field displacement phenomena can be easily understood knowing** that "μ_ℓ" changes sign if "$\underline{H}_{dc}$" changes direction of application, while "β_z" changes

---

* See Appendix A2 for wave impedance definitions.
** See Appendix A7 for descriptions of permeability quantities and their signs.

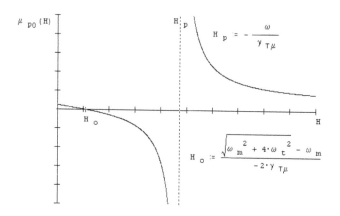

Figure A7.4.1b

sign if we change direction of propagation.* Since the sign of "α_x" depends on the product of "$\mu_\ell \beta_z$," as given by 7.7.37, we can conclude that if we change the direction of propagation or direction of "$\underline{H}_{dc}$," then we have the change of the side of "w" where the wave propagates, while two changes of sign of course do not change the propagation side. Note that in Figure 7.7.6 the "initial state" of the field displacement effects is represented. In other words, it exactly represents the side where the wave propagates when $\underline{H}_{dc} \equiv H_y \underline{y}_0$ and the direction of propagation is along "$\underline{z}_0$." The required value of "H_{dc}" for proper field displacement operation has been evaluated as the value that slightly saturates the ferrite and also generates a negative value of "μ_p."

Note that all the theory we have discussed here is founded on an assumption of homogeneous propagation media. Microstrip technology does not satisfy such an assumption. So, for a rigorous theoretical analysis a lot of theory needs to be developed to use some effective permeability in the field displacement microstrip devices. As a first approximation it is possible to use the equations we gave previously in Section 7.7.1. However, in this case the construction of a final field displacement phase shifter is a goal reached with a lot a tuning on various prototypes.

After explanation of the field displacement microstrip effect, it is simple to understand that a nonreciprocal phase shifter can be easily realized with an opportune dielectric loading of one edge of the field displacement area. This situation can be observed in Figure 7.7.7. In part a we have represented a hot microstrip conductor built on a ferrite substrate when, at a desired point, the width is increased many times with respect to the substrate height. We will call the area where

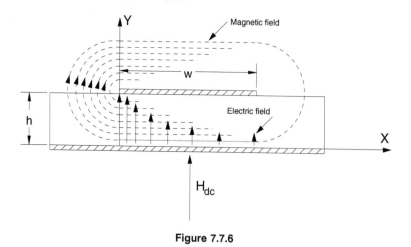

Figure 7.7.6

* See Appendix A2 for propagation constants and their signs.

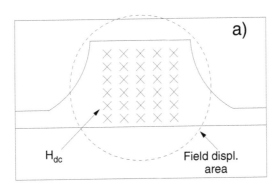

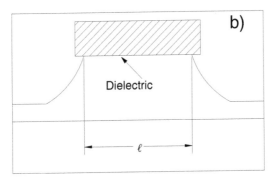

Figure 7.7.7

w >> h the "field displacement area." Note that since this area is a zone of low impedance, typically 20 to 30 Ohm, some matching to the reference system impedance of 50 Ohm is necessary. This is usually done by tapering the "field displacement area" width to the dimension of the 50 Ohm lines, or using $\lambda/4$ transformers.*

In part b we presented the case where a side of the field displacement area is loaded with a bar of insulating material with dielectric constant "$\varepsilon_{r\ell}$." We can see how in the direction of propagation where energy is concentrated on the loaded side, the propagation constant "β_ℓ" will be similar to 7.7.36. After explicating the effective quantities and considering the dielectric loading, 7.7.36 becomes:

$$\beta_\ell \approx \omega\left(\mu_0\mu_{pre}\varepsilon_{re\ell}\varepsilon_{r\ell}\right)^{0.5} \tag{7.7.39}$$

In this equation, "μ_{pre}" is the effective permeability obtained using "μ_p" in the formulas of section 7.7.1, while "$\varepsilon_{re\ell}$" is the effective permeability. In the opposite direction of propagation, or when "H_{dc}" changes direction, the propagation constant "β_0" will be:

$$\beta_0 \approx \omega\left(\mu_0\mu_{pre}\varepsilon_{re}\right)^{0.5} \tag{7.7.40}$$

where we have introduced the effective quantities. So the differential phase shift "$\Delta\theta$" will be:

$$\Delta\theta \propto \left(\beta_\ell - \beta_0\right)\ell \tag{7.7.41}$$

where "ℓ" is the dielectric loading bar length.

---

* See Chapter 1 for $\lambda/4$ transformers.

Typical differential phase shift values are near 50°/cm, with a figure of merit of 50°/dB. These values are smaller than those obtainable with coupled meander line phase shifters, and for this reason the latter devices are the preferred, especially when a high figure of merit is required.

7.8 FERRIMAGNETIC ISOLATORS

Microstrip ferrimagnetic isolators are mostly found as terminated circulators. From the theory we gave in Section 7.6, it is evident that if we terminate a port, the three port circulator becomes an isolator. However, isolators can also be built with other physical phenomena. These are the field displacement isolator and the resonance isolator, which will now be discussed.

7.8.1 Field Displacement Isolator

In the previous section we studied field displacement phenomena. With that theory in mind, it is clear that a device built as indicated in Figure 7.7.7 and with the dielectric bar replaced with absorptive material works as an isolator.[98] In the direction of propagation where energy is concentrated on the loaded side, the signal will encounter additional losses introduced by the loading bar, while in the opposite direction of propagation, or when "$\underline{H}_{dc}$" changes direction, the signal will pass practically unattenuated.

M.E. Hines[99] has built a microstrip isolator working from 6 GHz to 12 GHz with a reverse attenuation greater than 20 dB and a direct insertion loss lower than 1.5 dB.

7.8.2 Resonance Isolator

Any ferrimagnetic device has the ability, under particular conditions, to insert attenuation in a signal passing inside it. If the frequency "ω_p" given by:

$$\omega_p \overset{\perp}{=} -\gamma_{j\mu}H \tag{7.8.1}$$

where "H" is the internal static magnetic field is the same as the signal frequency, then a potential energy phenomena absorption exists.* This absorption is maximum if the signal has a circular polarization direction coincident with the spin electron magnetic moment precession motion. The absorption is very selective in frequency, and for this reason the resonance isolators are very narrow band devices, typically with a "BW" near 10% of center frequency.

Based on this concept, G. R. Harrison, G. H. Robinson, B. R. Savage, and D. R. Taft[100] have built a microstrip isolator using the circular polarization of the magnetic field in the air–dielectric boundary region near the edge of the hot conductor. The circular polarization lies in a longitudinal plane, i.e., aligned to the direction of propagation. Their device is indicated in Figure 7.8.1. Near the edge of the microstrip is posed a bar of ferrite biased transversely by "$\underline{H}_{dc}$." In the direction where the "RF" magnetic circular polarization is coincident with the induced electron spin magnetic moments precession motion inside the ferrite, the wave gives its energy to the ferrite, which starts to warm up. In the opposite direction of propagation, or when "$\underline{H}_{dc}$" changes direction, the signal passes along the line quite unattenuated. G. R. Harrison, G. H. Robinson, B. R. Savage, and D. R. Taft report on a microstrip resonance isolator centered at 6.1 GHz, with a resonance attenuation near 30 dB and an insertion loss in the unattenuated direction lower than 1 dB.

With the concepts of resonance absorption, it is simple to understand that the coupled meander line phase shifter can also become an isolator. To do that, it is necessary to bias the device with

* See Appendix A7 for the theory of energy absorption at ferrimagnetic resonance.

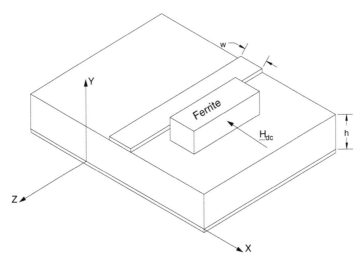

Figure 7.8.1

an "$\underline{H}_{dc}$" so that 7.8.1 holds, and for a signal with "ω_p" passing in one direction will be maximally attenuated, while in the opposite direction, or when "$\underline{H}_{dc}$" changes direction, the signal will have the minimal attenuation.

7.9 COMPARISON AMONG FERRIMAGNETIC PHASE SHIFTERS

After studying the most used ferrimagnetic phase shifters in the previous sections, we think it is useful to do a comparison among these devices, so that readers can simply choose the phase shifter most suited to their purpose. As discussion topics we will use the phase shifter characteristics most often used to choose such devices.

a. Simplicity of mechanical construction

From this point of view, all the devices we have studied are practically the same. So a better choice can be made using operation simplicity and flexibility. From this point of view, the meander line phase shifters are simpler to operate. They also work in latching operation and it is simple to change the direction of "$\underline{H}_{dc}$," as indicated in Figures 7.7.1 and 7.7.3. In the field displacement phase shifter it is not as simple to change the direction of "$\underline{H}_{dc}$." In fact, this direction must always be orthogonal to the direction of propagation, and for this reason, it is not simple to open a slot inside the ferrite, as we made in Figures 7.7.1 and 7.7.3.

b. Differential phase shifting value

The higher differential phase shifting value belongs to the nonreciprocal meander line phase shifter, with near 140°/cm. Following are the field displacement and meander line reciprocal devices, with near 50°/cm.

c. Figure of merit

The highest figure of merit belongs to the nonreciprocal meander line phase shifter, with near 300°/dB. The meander line reciprocal devices come next with near 80°/dB and finally, field displacement devices come with near 50°/dB.

d. Operating bandwidth

The highest operating bandwidth belongs to field displacement devices, with near 15% of bandwidth. Following are the meander line phase shifters equipped with both reciprocal and nonreciprocal, with near 10% of bandwidth.

REFERENCES

1. L. Lewin, Radiation from discontinuities in stripline, *Proc. of the IEEE*, 163, Feb. 1960.
2. J. M. Drozd, W. T. Joines, Determining Q using S parameter data, *IEEE Trans. on MTT*, 2123, Nov. 1996.
3. En-Yuan Sun, Shuh-Han Chao, Unloaded Q measurement-the critical points method, *IEEE Trans. on MTT*, 1983, Aug. 1995.
4. Chi Wang, Ben-Qing Gao, Ci-Ping Deng, Accurate study of Q-factor of resonator by a finite-difference time-domain method, *IEEE Trans. on MTT*, 1524, July 1995.
5. E. J. Denlinger, Radiation from microstrip resonators, *IEEE Trans. on MTT*, 235, Apr. 1969.
6. L. J. Van der Pauw, The radiation of electromagnetic power by microstrip configurations, *IEEE Trans. on MTT*, 25(9), 719, 1977.
7. E. A. Guillemin, *Synthesis of Passive Networks*, John Wiley & Sons, New York, 1957.
8. L. Weinberg, *Network Analysis and Synthesis*, McGraw Hill, New York, 1962.
9. W. K. Chen, ed., *The Circuits and Filters Handbook*, IEEE and CRC Press, Boca Raton, FL,1995.
10. G. D. Vendelin, A. M. Pavio, U. L. Rhode, *Microwave Circuits Design Using Linear and Nonlinear Techniques*, John Wiley & Sons, New York, 1990.
11. S. A. Maas, *Microwave Mixers*, Artech House, Norwood, MA, 1993.
12. C. Cho, K. C. Gupta, A new design procedure for single layer and two layer three line baluns, *IEEE Trans. on MTT*, 2514, Dec. 1998.
13. R. Schwindt, C. Nguyen, A CAD procedure for the double layer broadside coupled Marchand balun, *Int. Microwave Symp.*, 1, 389, 1994.
14. C. Y. Ho, R. Traynor, New analysis technique builds better baluns, *Microwaves and RF*, 99, Aug. 1985.
15. H. A. Wheeler, Transmission line properties of parallel strips separated by a dielectric sheet, *IEEE Trans. on MTT*, 172, Mar. 1965.
16. B. Bhartia, P. Pramanick, Computer aided design models for broadside coupled striplines and milli-meter wave suspended substrate microstrip lines, *IEEE Trans. on MTT*, 36(11), 1476, 1988. See also, Corrections to Computer aided design models for broadside coupled striplines and millimeter wave suspended substrate microstrip lines, *IEEE Trans. on MTT*, 37(10), 1658, 1989.
17. S. Kumar, C. Tannous, T. Danshhr, A multisection broadband impedance transforming branch-line hybrid, *IEEE Trans. on MTT*, 2517, Nov. 1995.
18. J. L. B. Walker, Improvements to the design of the 0–180° rat race coupler and its application to the design of balanced mixers with high LO to RF isolation, *Microwave Symp.*, 2, 747, 1997.
19. Hee Ran Ahn, I. Wolff, Ik-Soo Chang, Arbitrary termination impedances, arbitrary power divisions and small-sized ring hybrids, *Microwave Symp.*, 1, 285, 1997.
20. K. Lewis, Analysis examines the effects of coupler mismatch, *Microwaves and RF*, 89, Dec. 1998.
21. J. Ho, N. V. Shuley, Wilkinson divider design provides reduced size, *Microwaves and RF*, 104, Oct. 1997.
22. P. Antsos, Modified Wilkinson power dividers for K and Ka band, *Microwave J.*, 98, Nov. 1995.
23. E. J. Wilkinson, An N-way hybrid power divider, *IRE Trans. on MTT*, 116, Jan. 1960.
24. S.B. Cohn, A class of broadband three port TEM mode hybrids, *IEEE Trans. on MTT*, 16(2), 110, 1968.
25. L. I. Parad, R. L. Moynihan, Split-Tee power divider, *IEEE Trans. on MTT*, 91, Jan. 1965.
26. A. A. M. Saleh, Planar electrically symmetric n-way hybrid power dividers-combiners, *IEEE Trans. on MTT*, 6, 555, 1980.
27. Yung-Jinn Chen and Ruey-Beei Wu, A wide-band multiport planar power-divider using matched sectorial components in radial arrangement, *IEEE Trans. on MTT*, 1072, Aug. 1998.
28. E. G. Cristal, L. Young, Theory and tables of optimum symmetrical TEM mode coupled transmission line directional couplers, *IEEE Trans. on MTT*, 13, 544, 1965.
29. E. A. Guillemin, *Synthesis of Passive Networks*, John Wiley & Sons, New York, 1957.
30. L. Weinberg, *Network Analysis and Synthesis*, McGraw Hill, New York, 1962.
31. W. K. Chen, ed., *The Circuits and Filters Handbook*, IEEE and CRC Press, Boca Raton, 1995.
32. L. Young, The analytical equivalence of TEM mode directional couplers and transmission line stepped impedance filters, Proc. IEEE, 110(2), 275, 1963.
33. E. G. Cristal, L. Young, Theory and tables of optimum symmetrical TEM mode coupled transmission line directional couplers, *IEEE Trans. on MTT*, 13, 544, 1965.

34. R. Levy, General synthesis of asymmetric multi element coupled transmission line directional couplers, *IEEE Trans. on MTT*, 226, July 1963.

35. R. Levy, Tables for asymmetric multi element coupled transmission line directional couplers, *IEEE Trans. on MTT*, 275, May 1964.

36. T. J. Russell, A matched line directional divider two way power divider, *Microwave J.*, 92, Nov. 1994.

37. J. Howard, W. C. Lin, Coupler design delivers low ripple performance, *Microwaves and RF*, 72, Sept. 1998.

38. C. B. Sharpe, An equivalence principle for nonuniform transmission line directional couplers, *IEEE Trans. on MTT*, 7, 398, 1967.

39. B. M. Oliver, Directional electromagnetic couplers, *Proc. of the IRE*, 1686, Nov. 1954.

40. C. P. Tresselt, The design and construction of broadband, high directivity, 90 degree couplers using nonuniform line techniques, *IEEE Trans. on MTT*, 12, 647, 1966.

41. C. P. Tresselt, Design and computed theoretical performance of three classes of equal ripple nonuniform line couplers, *IEEE Trans. on MTT*, 17(4), 218, 1969.

42. F. Arndt, High pass transmission line directional coupler, *IEEE Trans. on MTT*, 310, May 1968.

43. F. Arndt, Tables for asymmetric Chebyshev high pass TEM mode directional couplers, *IEEE Trans. on MTT*, 633, Sept. 1970.

44. S. Uysal, H. Aghvami, Synthesis, design and construction of ultra-wide-band nonuniform quadrature directional couplers in inhomogeneous media, *IEEE Trans. on MTT*, 969, June 1989.

45. D. K. Y. Lau, S. P. Marsh, L. E. Davis, R. Sloan, Simplified design technique for high performance microstrip multisection couplers, *IEEE Trans. on MTT*, 2507, Dec. 1998.

46. P. K. Ikailanen, G. L. Matthaei, Wide-band, forward-coupling microstrip hybrids with high directivity, *IEEE Trans. on MTT*, 719, Aug. 1987.

47. J. Lange, Interdigital stripline quadrature hybrid, *IEEE Trans. on MTT*, 1150, Dec. 1969.

48. Y. Tajima, S. Kamihaschi, Multiconductor couplers, *IEEE Trans. on MTT*, 795, Oct. 1978.

49. R. Waugh, D. Lacombe, Unfolding the Lange coupler, *IEEE Trans. on MTT*, 777, Nov. 1972.

50. Wen Pin Ou, Design equations for an interdigitated directional coupler, *IEEE Trans. on MTT*, 253, Feb. 1975.

51. R. M. Osmani, Synthesis of Lange couplers, *IEEE Trans. on MTT*, 168, Feb. 1981.

52. G. Kemp, J. Hodbell, J. W. Biggin, Ultra wideband quadrature coupler, *Electron. Lett.*, 19(6), 197, 1983.

53. J. L. B. Walker, Analysis and design of Kemp type 3 dB quadrature couplers, *IEEE Trans. on MTT*, 1, 88, 1990.

54. S. B. Cohn, F. S. Coale, Directional channel separation filters, *Proc. of the IRE*, 1018, Aug. 1956.

55. F. S. Coale, A traveling wave directional filter, *IRE Trans. on MTT*, 256, Oct. 1956.

56. C. C. Rocha, A. J. M. Soares, H. Abdalla, Jr., A method to design microwave multiplexers using directional filters, *RF Design*, 64, May 1998.

57. M. Abramowitz, I. A. Stegun, *Handbook of Mathematical Functions*, Dover, New York, 1970.

58. G. Matthaei, L. Young, E. M. T. Jones, *Microwave Filters, Impedance Matching Networks and Coupling Structures*, Artech House, Norwood, MA, 1980.

59. F. V. Minnaar, J. C. Coetzee, J. Joubert, A novel ultrawideband microwave differential phase shifter, *IEEE Trans. on MTT*, 1249, Aug. 1997.

60. B. M. Schiffman, A new class of broad band microwave 90 degree phase shifters, *IRE Trans. on MTT*, 232, April 1958.

61. C. E. Free, C. S. Aitchison, Improved analysis and design of coupled line phase shifters, *IEEE Trans. on MTT*, 2126, Sept. 1995.

62. V. P. Meschanov, L V. Metelnikova, V. D. Tupikin, G. G. Chumaevskaya, A new structure of microwave ultrawideband differential phase shifter, *IEEE Trans. on MTT*, 762, May 1994.

63. B. Schiek, J. Köhler, A method for broadband matching of microstrip differential phase shifters, *IEEE Trans. on MTT*, 25(8), 666, 1977.

64. R. B. Wilds, Try $\lambda/8$ stubs for fast fixed phase shifts, *Microwaves*, 67, Dec. 1979.

65. R. J. Garver, D. E. Bergfried, S. J. Raff, B. O. Weinschel, Errors in S_{11} measurements due to the residual standing wave ratio of the measuring equipment, *IEEE Trans. on MTT*, 61, Jan. 1972.

66. F. Di Paolo, A simple high yield 6 to 18 GHz GaAs monolithic phase shifter, *Microwave J.*, 92, April 1997.

67. J. D. Adam, H. Buhay, M. R. Daniel, M. C. Driver, G. W. Eldridge, M. H. Hanes, R. L. Messham, Monolithic integration of an x-band circulator with GaAs MMICS, *Int. Microwave Symp.*, 1, 97, 1995.

68. J. D. Adam, H. Buhay, M. R. Daniel, G, W. Eldridge, M. H. Hanes, R, L. Messham, T. J. Smith, K-band circulators on semiconductor wafers, *Int. Microwave Symp.*, 1, 113, 1996.

69. C. E. Fay, R. L. Comstock, Operation of the ferrite junction circulator, *IEEE Trans. on MTT*, 15, Jan. 1965.

70. H. Bosma, Junction circulators, *Adv. in Microwaves*, 6, 125, 1978.

71. Y. S. Wu, F. J. Rosenbaum, Wide range operation of microstrip circulators, *IEEE Trans. on MTT*, 849, Oct. 1974.

72. Y. Ayasli, Analysis of wide band stripline circulators by integral equation technique, *IEEE Trans. on MTT*, 200, Mar. 1980.

73. J. Lahey, Junction circulator design, *Microwave J.*, 26, Nov. 1989.

74. H. S. Newman, C. M. Krowne, Analysis of ferrite circulators by 2-D finite element and recursive Green's function techniques, *IEEE Trans. on MTT*, 167, Feb. 1998.

75. J. Helszajn, Fabrication of very weakly and weakly magnetized microstrip circulators, *IEEE Trans. on MTT*, 439, May 1998.

76. H. How, C. Vittoria, R. Schmidt, Losses in multiport stripline microstrip circulators, *IEEE Trans. on MTT*, 543, May 1998.

77. H. How, S. A. Oliver, S. W. McKnight, P. M. Zavracky, N. E. McGruer, C. Vittoria, R. Schmidt, Theory and experiment of thin film junction circulator, *IEEE Trans. on MTT*, 1645, Nov. 1998.

78. C. E. Fay, R. L. Comstock, Operation of the ferrite junction circulator, *IEEE Trans. on MTT*, 15, Jan. 1965.

79. G. R. Harrison, G. H. Robinson, B. R. Savage, D. R. Taft, Ferrimagnetic parts for microwave integrated circuits, *IEEE Trans. on MTT*, 577, July 1971.

80. C. E. Fay, R. L. Comstock, Operation of the ferrite junction circulator, *IEEE Trans. on MTT*, 15, Jan. 1965.

81. B. A. Auld, The synthesis of symmetrical waveguide circulators, *IRE Trans. on MTT*, 238, Apr. 1959.

82. U. Milano, J. H. Saunders, L. Davis, Jr., A Y-junction strip line circulator, *IRE Trans. on MTT*, 346, May 1960.

83. J. K. Ackers, A contribution to the theory of stripline junction circulator, *Microwave J.*, 57, July 1967.

84. M. Grace, F. R. Arams, Three port ring circulators, *Proc. of the IRE*, 1497, Aug. 1960.

85. A. M. Borjak, L. E. Davis, Mode 1 and mode 2 designs for 94 GHz microstrip circulators, *IEEE Trans. on MTT*, 310, Sept. 1993.

86. J. L. Allen, D. R. Taft, Ferrite elements for hybrid microwave integrated systems, *IEEE Trans. on MTT*, 405, July 1968.

87. G. R. Harrison, G. H. Robinson, B. R. Savage, D. R. Taft, Ferrimagnetic parts for microwave integrated circuits, *IEEE Trans. on MTT*, 577, July 1971.

88. G. D. Dionne, D. E. Oates, D. H. Temme, Low-loss microwave ferrite phase shifters with superconducting circuits, *Int. Microwave Symp.*, 1, 101, 1994.

89. S. N. Stitzer, Finite element modeling of ferrite phase shifters, *Int. Microwave Symp.*, 1, 125, 1996.

90. D. J. Massé, R. A. Pucel, Microstrip propagation on magnetic substrates. Part II, experiment, *IEEE Trans. on MTT*, 309, May 1972.

91. J. J. Green, F. Sandy, Microwave characterization of partially magnetized ferrites, *IEEE Trans. on MTT*, 641, June 1974.

92. D. J. Massé, R. A. Pucel, Microstrip propagation on magnetic substrates. Part I, design theory, *IEEE Trans. on MTT*, 304, May 1972.

93. G. T. Roome, H. A. Hair, Thin ferrite devices for microwave integrated circuits, *IEEE Trans. on MTT*, 411, July 1968.

94. J. L. Allen, D. R. Taft, Ferrite elements for hybrid microwave integrated systems, *IEEE Trans. on MTT*, 405, July 1968.

95. G. T. Roome, H. A. Hair, Thin ferrite devices for microwave integrated circuits, *IEEE Trans. on MTT*, 411, July 1968.

96. J. A. Weiss, Dispersion and field analysis of a microstrip meander line slow wave structure, *IEEE Trans. on MTT*, 1194, Dec. 1974.

97. M. E. Hines, Reciprocal and nonreciprocal modes of propagation in ferrite stripline and microstrip devices, *IEEE Trans. on MTT*, 442, May 1971.

98. T. M. F Elshafiey, J. T. Aberle, E. B. El Sharawy, Full wave analysis of edge guided mode microstrip isolator, *IEEE Trans. on MTT*, 2661, Dec. 1996.

99. M. E. Hines, Reciprocal and nonreciprocal modes of propagation in ferrite stripline and microstrip devices, *IEEE Trans. on MTT*, 442, May 1971.

100. G. R. Harrison, G. H. Robinson, B. R. Savage, D. R. Taft, Ferrimagnetic parts for microwave integrated circuits, *IEEE Trans. on MTT*, 577, July 1971.

CHAPTER 8

Stripline Devices

8.1 INTRODUCTION

This chapter will cover the most important stripline devices and will complete the study begun in Chapter 7 regarding μstrip devices. All the theory we have developed in that chapter assumed, whenever possible, to approximate the "qTEM" propagation mode in μstrip with a pure "TEM." Since striplines are instead t.l.s that can be assumed to propagate a real "TEM" wave, all the theory developed in Chapter 7 applies to the present case as well.

A lot of devices already studied in Chapter 7 can also be built in stripline technology. For this reason, in this chapter we will refer, whenever possible, to the same devices studied in Chapter 7, but will indicate only the differences between the device in μstrip and stripline. Of course, a general remark can be made for the present case. In fact, due to the equal phase velocity between even and odd waves in "cs" devices, these will have better performances than their "cμ" counterpart.

8.2 TYPICAL TWO PORT NETWORKS

The same typical networks we indicated for the μstrip case can also be built in stripline. The following notes apply:

a. Resonators: Ring resonators, frequently used in μstrip technology, are in our case seldom used, due to the difficulty in holding such circular conductors. Consequently, stripline resonators are practically always made with straight t.l.s.
b. Balun: Although rarely used, a stripline balun is always coplanar with the hot conductor, as indicated in Figure 8.2.1.
c. Stubs: Shortened stubs are more difficult to realize since they require some metallic holder connected to top and bottom conductors at the end of the stub.

8.3 DIRECTIONAL COUPLERS

The same directional couplers studied in Chapter 7 can be equally constructed in stripline. In particular, all the design tables given in Chapter 7 still hold for the stripline counterpart. The following notes apply:

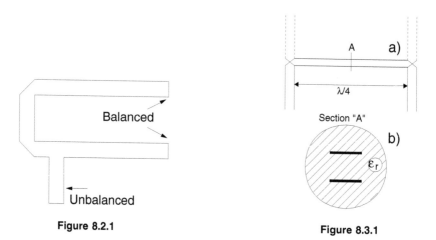

Figure 8.2.1   **Figure 8.3.1**

a. "In-line" or "Wilkinson" — This kind of coupler is quite impractical since the isolation resistor is of course difficult to insert. For this reason, these couplers are usually employed at low frequencies, let us say below 5 GHz.

b. "Wiggly" — These couplers are useless in stripline, since now the phase velocities are naturally equalized.

c. "Interdigital" or "Lange" — It is impractical and it is never used.

d. "Step coupled lines" and "tapered coupling" — a "BCS" or "OBCS" configuration is often used, so that it is possible to have a 3 dB coupling, or a single step coupler. For example, a single step coupler is indicated in Figure 8.3.1, where in part b, a transversal section is depicted to show the "BCS" configuration.

8.4 SIGNAL COMBINERS

The notes for sections 8.2 and 8.3 apply here.

8.5 DIRECTIONAL FILTERS

All the directional filters studied in Chapter 7 can be equally used in striplines. "BCS" can also be used for ring directional couplers of course, and in this case input and output lines are "OSL."

8.6 PHASE SHIFTERS

The following notes apply:

a. Schiffman's phase shifters — The use of "BCS" is theoretically possible, but in practice its use is limited due to the difficulty posed by two short circuits at one extreme the "BCS." A possible solution is indicated in Figure 8.6.1a with a top view and in part b with a lateral view. For example, a type "C" Schiffman's shifter is indicated in Figure 8.6.2.* In this case, input, output, and reference lines are "OSL," while the two coupled lines are "BCS."

    However, a short circuit between "BCS" is good only at low frequencies, let us say below 2 GHz. So, "SCS" are preferred for Schiffman's shifters.

b. Reflection type — Of course, the Lange coupler cannot be used. So, the only divider to be used is the branch line.

* See Chapter 7 for design notes on Schiffman's phase shifters.

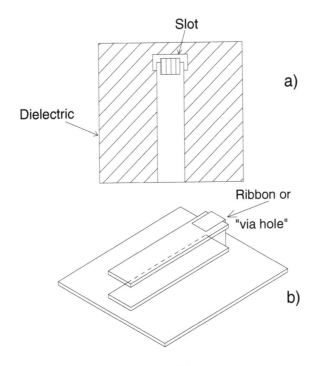

Slot

Dielectric

a)

Ribbon or
"via hole"

b)

Figure 8.6.1

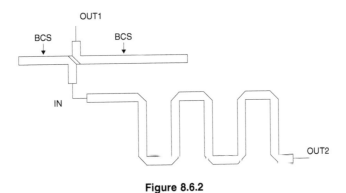

OUT1

BCS

BCS

IN

OUT2

Figure 8.6.2

8.7 THE THREE PORT CIRCULATOR

The theory discussed in Chapter 7 applies to the present case as well.[1,2] A typical construction of a stripline circulator is indicated in Figure 8.7.1. Part a shows a transversal section of the device. Two ferrite cylinders "F1" and "F2" are posed on each side of a circular conductor, with diameter not greater than the diameters of the ferrite cylinders. This circular conductor has three striplines "L_1," "L_2," and "L_3" attached on its border, each one forming an angle of 120° with the nearest stripline. On the other side of the ferrite cylinders ground planes "M1" and "M2" are attached, each for ferrite. These ground planes create the stripline technology of the device. Of course, surrounding the ferrites there is the desired substrate. External to the ground planes, two magnets generate the proper static magnetic field that biases the ferrites. Sometimes it is used as a magnet only.

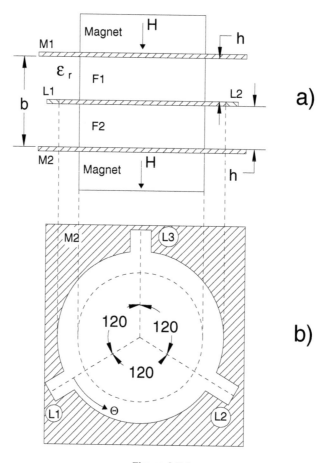

Figure 8.7.1

A top view of the device is drawn in Figure 8.7.1b with the upper magnet, the top ground plane "M1," and the top ferrite "F1" removed. The dashed line represents the ferrite "F2" under the disk conductor. We have indicated the origin of the angles "θ" at the angular center of port lll L_1."

The general theory and design guidelines given in Chapter 7 also apply in this case. The additional dimension to be designed is the height "h" of each ferrite cylinder. In fact, the distance "b" between ground planes is dictated by the required performances of the circulator.* In this case we have:[3]

$$h = \frac{1.48\omega R^2 \varepsilon}{Q_\ell G} \qquad (8.7.1)$$

where:

a. "ε" is the absolute ferrite permittivity
b. "R" is the ferrite cylinder diameter, which practically coincides with the central conductor radius
c. "ω" is the center bandwidth signal angular frequency
d. "Q_ℓ" is the loaded "Q" of the circulator
e. "G" is the circulator conductance at the port disk periphery

* The circulator is usually a "drop in" device, which requires the surrounding electronics to be adapted to its dimensions.

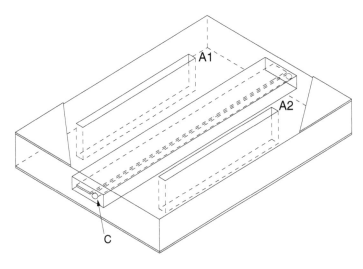

Figure 8.8.1

Electrical performances of stripline circulators are practically the same as their μstrip counterparts, reported in Chapter 7.

8.8 FERRIMAGNETIC PHASE SHIFTERS

The same phase shifter* configurations we studied for μstrip technology, i.e., reciprocal and nonreciprocal, can be constructed in stripline. The quasi-homogeneous structure of the dielectric permits these devices to be analyzed with simple analytical methods.[4] The resulting devices are of course bulky, but the performances are in general better than the μstrip counterpart due the greater interaction with ferrite. The following notes apply:

8.8.1 Reciprocal

The structure of a μstrip reciprocal "ps" studied in Chapter 7 can be also applied to the stripline case. Alternatively, a stripline reciprocal ferrimagnetic phase shifter is indicated in Figure 8.8.1. A ferrite toroid has two apertures "A1" and "A2" where a coil for each aperture is wound. Near the hot conductor, a wire conductor "C" is passed. This conductor could also be avoided and substituted with the hot conductor itself, if proper biasing with high impedance wire is used. When a current "I" feeds the coils in the apertures, a static magnetic "H_{dc}" field is generated as indicated in Figure 8.8.2, i.e., to the magnetic "RF" field "h." When instead a current "I" feeds the wire "C," then a magnetic "H_{dc}" field is generated as indicated in Figure 8.8.3, i.e., parallel to the magnetic "RF" field "h."

In these cases the ferrite, respectively, presents to the wave a permeability "$\mu_{eq\perp}$"** and "1," and the phase shifting "$\Delta\theta$" is:

$$\Delta\theta \propto \left(\beta_z - \beta_{z0}\right)\ell \tag{8.8.1}$$

where "ε_r" is the ferrite permittivity, "ℓ" is approximately the length of the apertures "A1" and "A2," and:

---

* Phase shifters will simply be called "ps."
** This quantity is defined in Chapter 7 and Appendix A7.

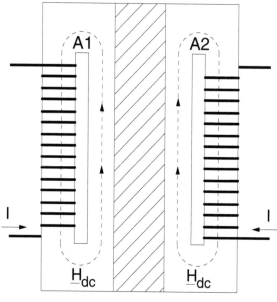

Figure 8.8.2

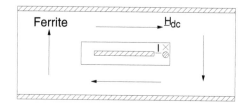

Figure 8.8.3

$$\beta_z = \omega \left(\mu_0 \mu_{eq\perp} \varepsilon_0 \varepsilon_r \right)^{0.5} \tag{8.8.2}$$

$$\beta_{z\parallel} = \omega \left(\mu_0 \varepsilon_0 \varepsilon_r \right)^{0.5} \tag{8.8.3}$$

Typical performances of this device in band 8.2 to 12.4 GHz are[5] a differential phase shift of 30°/cm and a figure of merit near 300°/dB. For latched operation,* these values decrease by approximately 25%. In any case, they are typically two times higher than the values for the μstrip counterpart device.

8.8.2 Nonreciprocal

Similar to the case of the μstrip, a stripline nonreciprocal "ps" can be built using two distinct physical phenomena, i.e.:

1. using the interaction between biased ferrite and "RF" magnetic field with circular polarization
2. impenetrability of ferrite, physical phenomena called "field displacement"

* Latched operation for ferrite devices are explained in Chapter 7.

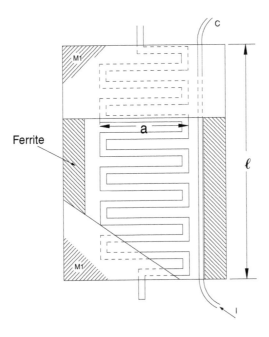

Figure 8.8.4

a. Using Circular Polarized "RF" Magnetic Field

Also in this case, the "ps" employs a meander line. This particular configuration of coupled striplines has been studied in Chapter 7. A nonreciprocal stripline "ps" is indicated in Figure 8.8.4. This is a top view, where the top ground plane "M1" has been partially removed. We see a ferrite toroid and inside it a stripline meander line. A conductor wire "C" travels near the hot conductor, as indicated. In Figure 8.8.5 a cross-sectional view is shown.

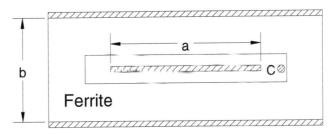

Figure 8.8.5

When a current "I" feeds the wire "C," then a static magnetic "H_{dc}" field is generated as indicated in Figure 8.8.6, i.e., orthogonal to the circularly polarized magnetic "RF" field "h."* We know from the theory of e.m. energy inside ferrimagnetic materials** that when a "TEM" circular polarized wave travels inside an isodirectional magnetized ferrite, it meets two permeabilities, "μ_c" and "μ_d." Associated with the two previous permeabilities there are two phase constants "β_c" and "β_d" given by:

---

* See Chapter 7 to recognize how the magnetic field inside the meander line is circularly polarized.
** See Appendix A7 for energy exchange between wave and ferrimagnetic materials.

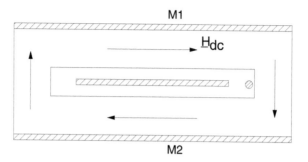

Figure 8.8.6

$$\beta_c \stackrel{\perp}{=} \omega\left(\mu_0\mu_c\varepsilon\right)^{0.5} \tag{8.8.4}$$

$$\beta_d \stackrel{\perp}{=} \omega\left(\mu_0\mu_d\varepsilon\right)^{0.5} \tag{8.8.5}$$

So if we statically magnetize the device as indicated in Figure 8.8.6, then fix a direction of "$\underline{H}_{dc}$," the wave will have a phase constant "β_c" in one direction of propagation and a phase constant "β_d" in the opposite direction of propagation. In such a case the differential phase shift "$\Delta\theta$" will be:

$$\Delta\theta \propto n\left(\beta_c - \beta_d\right)a \tag{8.8.6}$$

where "a" is the length of a complete meander line and "n" is the number of "C-sections" employed in the meander line. Of course, we obtain the same differential phase shift if we fix a direction of propagation and change the direction of "$\underline{H}_{dc}$." Typical performances of this device centered at 5.2 GHz with latched operation are[6] a differential phase shift of 70°/cm and a figure of merit near 80°/dB. The operating bandwidth is near 10%.

A variation of this device employs a dielectric surrounding the hot conductor and the wire as indicated in Figure 8.8.7. Characteristic values for this "ps" are[7] a figure of merit near 310°/dB and a differential phase shift of 40°/cm, in the operating bandwidth of 9 to 10 GHz.

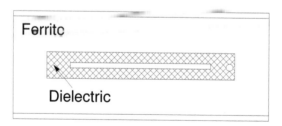

Figure 8.8.7

The coupled meander line stripline phase shifter is a widely used device, which is preferred over its reciprocal counterpart.

b. Using "Field Displacement"

The same theory in Chapter 7 can be applied to the stripline case. So, if the following conditions are properly satisfied:*

---

* See Chapter 7 and Appendix A7 for field displacement theory.

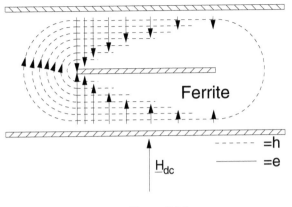

Figure 8.8.8

1. "DC" magnetic field "H_{dc}" strength
2. Direction of "DC" magnetic field
3. Signal frequency
4. Direction of "RF" magnetic field

a "field displacement" effect is established in the propagation. The electromagnetic situation is indicated in Figure 8.8.8.

It is simple to understand that a nonreciprocal phase shifter can be easily realized with an opportune dielectric loading of one edge of the field displacement area. This situation can be observed in Figure 8.8.9. In one side of the field displacement area, a bar of low loss dielectric material is placed, with dielectric constant "$\varepsilon_{r\ell}$." We can realize how, in the direction of propagation where energy is concentrated on the loaded side, the propagation constant "β_ℓ" will be:

$$\beta_\ell \approx \omega\left(\mu_0\mu_p\varepsilon_0\varepsilon_{r\ell}\right)^{0.5} \qquad *(8.8.7)$$

In the opposite direction of propagation, or when "$\underline{H}_{dc}$" changes direction, the propagation constant "β_0" will be:

$$\beta_0 \approx \omega\left(\mu_0\mu_p\varepsilon_0\varepsilon_r\right)^{0.5} \qquad (8.8.8)$$

where "ε_r" is the ferrite dielectric constant.

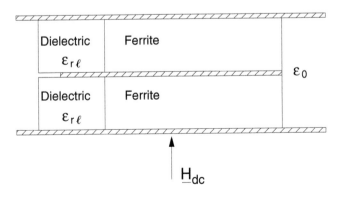

Figure 8.8.9

* "μ_p" is the element of the principal diagonal of the ferrite permeability matrix. See Appendix A7.

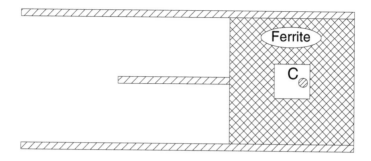

Figure 8.8.10

So the differential phase shift "$\Delta\theta$" will be:

$$\Delta\theta \propto \left(\beta_\ell - \beta_0\right)\ell \tag{8.8.9}$$

where "ℓ" is the dielectric loading bar length. Typical differential phase shift values are[8] near 50°/cm, with figure of merit of 60°/dB, at a frequency of 3.25 GHz and bandwidth near 10%.

To improve these performances, some studies[9] have stated that the configuration indicated in Figure 8.8.10 should have better performance. It is called toroidal field displacement stripline "ps." Typical values of this "ps" are[10] a differential phase shift of 40°/cm and a figure of merit near 500°/dB, centered at a frequency of 5.4 GHz. Of course, the required magnetization is obtained feeding the wire "c" with a current, as indicated in Figure 8.8.11, assuming the current is entering in the plane of this page.

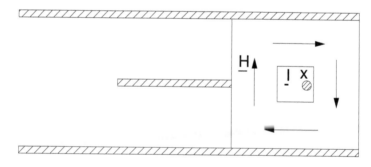

Figure 8.8.11

8.9 FERRIMAGNETIC ISOLATORS

Stripline ferrimagnetic isolators are mostly found as terminated circulators. However, they are also found with other physical phenomena devices where only isolating performances can be built. They are the field displacement isolator and the resonance isolator, which we will now discuss.

8.9.1 Field Displacement

If in the device shown in Figure 8.8.9, we replace the dielectric bar with an absorptive material, we can create an isolator. In the direction of propagation where energy is concentrated on the loaded side, the signal will encounter additional losses introduced by the loading bar, while in the opposite direction of propagation, or when "$\underline{H}_{dc}$" changes direction, the signal will pass practically unattenuated. The performances of this isolator are essentially the same as the µstrip counterpart.

8.9.2 Resonance

A way to cause resonance absorption is to generate circular polarization inside the stripline. As we have studied in the previous section, such polarization can be obtained using a meander line. So, if we bias such a device at the resonance, we can have energy absorption. However, the author is not aware of any results reported in the literature for such a device.

8.10 COMPARISON AMONG FERRIMAGNETIC PHASE SHIFTERS

After having studied the most used ferrimagnetic phase shifters in the previous sections, we think it is useful to do a comparison among all these devices, so that readers can simply choose the phase shifter most suited to their purpose. As discussion topics we will use the phase shifter characteristics most often used to choose such devices.

a. Simplicity of mechanical construction — The meander line phase shifters can be evaluated as the most complicated to build, followed by the field displacement and reciprocal "ps."
b. Differential phase shifting value — The higher differential phase shifting value belongs to the nonreciprocal meander line phase shifter, with near 100°/cm. Then, there are field displacement and reciprocal devices, with near 50°/cm.
c. Figure of merit — The highest figure of merit belongs to the toroidal nonreciprocal field displacement "ps" with near 500°/dB. Then, the nonreciprocal dielectric loaded meander line and the reciprocal devices come, with near 300°/dB. Then, there is the nonreciprocal meander line "ps" with nearly 80°/dB. Finally, there are field displacement nontoroidal devices, with near 60°/dB.
d. Operating bandwidth — The highest operating bandwidth belongs to the toroidal nonreciprocal field displacement "ps" with near 15% of bandwidth. Then the meander line and the reciprocal phase shifters, with near 10% of bandwidth.

We think it is useful to conclude this section with a general comparison among µstrip and stripline "ps"

a. Simplicity of mechanical construction — The most simple "ps" to build is the µstrip device.
b. Differential phase shifting value — The highest differential phase shift value belongs to the nonreciprocal meander line µstrip phase shifter,[11] with near 130°/cm, followed by the stripline counterpart, with near 100°/cm. Then, all the other types of "ps" come, with nearly 50°/cm.
c. Figure of merit — The highest figure of merit belongs to the toroidal nonreciprocal field displacement "ps," with near 500°/dB, followed by the nonreciprocal dielectric loaded stripline meander line, the nonreciprocal µstrip meander line[12] "ps," and the reciprocal stripline devices with near 300°/dB. Then, the nonreciprocal stripline meander line and the µstrip reciprocal "ps" follow, with nearly 80°/dB. Finally, all the other types of "ps" come, with nearly 50°/dB.
d. Operating bandwidth — In practice, all the "ps" have the same bandwidth.

REFERENCES

1. K. M. Gaukel, E. B. El Sharawy, Eigenvalue matrix analysis of segmented stripline junction disk circulator, *IEEE Trans. on MTT*, 1484, July 1995.
2. S. A. Ivanov, Application of the planar model to the analysis and design of the Y junction stripline circulator, *IEEE Trans. on MTT*, 1253, June 1995.

3. C. E. Fay, R. L. Comstock, Operation of the ferrite junction circulator, *IEEE Trans. on MTT*, 15, Jan. 1965.
4. S. N. Stitzer, Modeling a stripline ferrite phase shifter, *Int. Microwave Symp.*, 2, 1117, 1997.
5. T. Nelson, R. A. Moore, E. Wantuch, D. Buck, R. Huber, R. Lee, Small analog stripline X-band ferrite phase shifter, *IEEE Trans. on MTT*, 45, Jan. 1970.
6. R. R. Jones, A slow wave digital ferrite strip transmission line phase shifter, *IEEE Trans. on MTT*, 684, Dec. 1966.
7. E. R. Bertil Hansson, S. Aditya, M. A. Larsson, Planar meanderline ferrite dielectric phase shifter, *IEEE Trans. on MTT*, 209, March 1981.
8. M. E. Hines, Reciprocal and nonreciprocal modes of propagation in ferrite stripline and microstrip devices, *IEEE Trans. on MTT*, 442, May 1971.
9. D. M. Bolle, S. H. Talisa, The edge guided mode nonreciprocal phase shifter, *IEEE Trans. on MTT*, 878, Nov. 1979.
10. L. R. Whicker, R. R. Jones, A digital latching ferrite strip transmission line phase shifter, *IEEE Trans. on MTT*, 781, Nov. 1965.
11. G. T. Roome, H. A. Hair, Thin ferrite devices for microwave integrated circuits, *IEEE Trans. on MTT*, 411, July 1968.
12. J. L. Allen, D. R. Taft, Ferrite elements for hybrid microwave integrated systems, *IEEE Trans. on MTT*, 405, July 1968.

CHAPTER 9

Slot Lines

9.1 GEOMETRICAL CHARACTERISTICS

The physical realization of a slot line is indicated in Figure 9.1.1. This t.l. was first studied by S.B. Cohn,[1] and recently by others.[2,3] It is simple to note that this t.l. is the complement of the microstrip we studied in Chapter 2. In fact, here the conductors are present where they aren't in the microstrip. In contrast to the microstrip case, the slot line is a full planar t.l. since there is no bottom ground conductor. From this point of view, the slot line is very similar to the "CPW" we will study in Chapter 10.

As we can see in Figure 9.1.1, this t.l. is composed of an opening slot of width "s" in a planar conductor so that two conductors are generated. The two conductors of thickness "t" are placed on a dielectric slab of height "h" and dielectric and magnetic constants "ε_r" and "μ_r." The extension "ℓ_1" and "ℓ_2," of the two lateral conductors is assumed to be infinite, and in practice, is many times the length of the signal wavelength. If the lateral conductors cannot satisfy this condition, the new t.l. can be regarded as a coplanar strip "CPS" (studied in Chapter 11).

According to the discussion on propagation modes in Appendix A2, the slot line does not support a "TEM" mode[4,5,6] but it has a zero cutoff frequency. Its fundamental mode is a "qTE,"* with the magnetic field elliptically polarized in longitudinal planes, especially near the air-substrate interface.[7] For this reason the slot line is well suited to be used in ferrimagnetic devices, as we will show later.

Because the number of the electric and magnetic field lines in the air is higher than the number of the same lines for the microstrip case, the slot line effective dielectric constant "ε_{re}" is typically 15% lower than in the microstrip "ε_{re}." Consequently, the maximum reachable characteristic impedance values are higher than in the microstrip case. Of course, the minimum slot line imped-ance** value is higher than in the microstrip case, typically 60% higher.

To avoid e.m. radiation in the air, it is very important to use substrates with high dielectric constants, let us say from a value greater than 10, so that the e.m. field is mainly concentrated inside the dielectric.

Excluding some devices composed only of slot lines, like directional couplers or filters, just to show the possibilities for employing such t.l.s, slot lines present advantages with respect to micros-trips or striplines in balanced mixer circuits and in antenna feeding. This subject will be discussed later in this chapter.

9.2 ELECTRIC AND MAGNETIC FIELD LINES

Some electric "e" and magnetic "h" field lines for the fundamental "qTE" mode in the slot line are indicated in Figure 9.2.1, in a defined cross-section and a defined time. Depending on the

---

* "qTE" means "quasi TE," where quasi (Italian word) means "almost."
** Sometimes with "impedance" we will mean "characteristic impedance," especially when no confusion will arise.

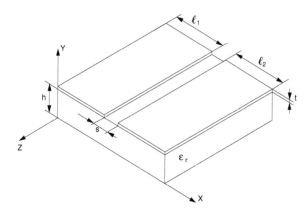

Figure 9.1.1

feeding method, the slot line can be a balanced or an unbalanced t.l. From this point of view, the slot line assumes the characteristics of the feeding line.

The field lines indicated in Figure 9.2.1 are a simplification, especially for the magnetic field. In fact, since the magnetic field lines always try to turn around conductors, they cannot do this for the two lateral conductors of the slot line. So, they are distorted moving on "x" from the center of the slot, and are not in the plane of the figure. For this reason, with a "•" we indicate a vector exiting from the plane of the figure, while with an "x" we indicate a vector entering into this plane. A more defined representation of the magnetic field lines for the fundamental "qTE" mode is indicated in Figure 9.2.2. Note that the slot line's magnetic field lines are quite similar to those in the coplanar waveguide case to be studied in Chapter 10, in which a "qTEM" propagation mode is also not possible. Slot line electric field lines are instead quite dissimilar to those of "CPW," since in this case there is not a central conductor.

There is a very interesting result reported by S.B. Cohn[8] that indicates the intensity of "h_z" and "h_y" fields in the center of the slot vs. the vertical coordinate "y," as depicted in Figure 9.2.3. We see how, near the slots, the fields have their maximum intensity. In addition, since "h_z" and "h_y" are never equal in magnitude, the resulting magnetic field is never perfectly circular polarized, also if near the slot, this can be approximated quite well.

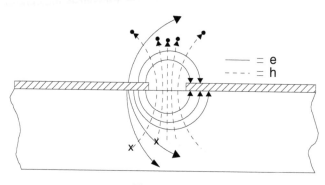

Figure 9.2.1

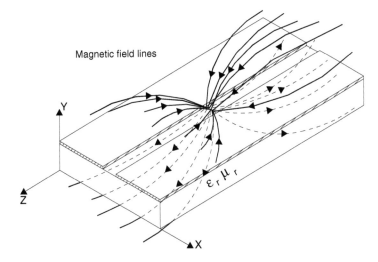

Magnetic field lines

Figure 9.2.2

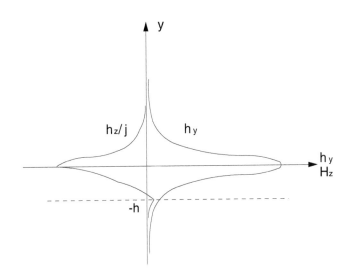

Figure 9.2.3

9.3 SOLUTION TECHNIQUES FOR THE ELECTROMAGNETIC PROBLEM

Some of the first analysis methods for the slot line e.m. problem were developed by S.B. Cohn.[9] We will follow his guidelines, reporting two methods that can be called the "line of magnetic current" and the "transverse resonant" methods. The first method has the advantage of being very simple in its formulation and giving a picture of the slot line as a radiating structure. However, we cannot obtain the characteristic impedance of the line with this method. The second method is instead quite involved in its formulation, but it gives us the expression of the characteristic impedance. Both Cohn's methods assume the conductors have negligible thickness.

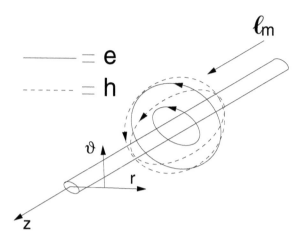

Figure 9.3.1

Full wave slot line analysis has been performed by some researchers[10,11] and the effect of conductor thickness by others.[12]

a. Line of Magnetic Current Method

With this method, the e.m. field produced by the slot line is approximated with that composed of a hypothetical conductor passed through a magnetic current "i_m." Note as the electric field has only the "e_θ" component, while the magnetic field has the components "h_z" and "h_r." Using a cylindrical reference system with axis "r,θ,z," this situation is indicated in Figure 9.3.1.

According to Cohn, as the first approximation we can use the concept of effective relative permittivity, already introduced in our text, and assume that the slot line is surrounded by a homogeneous isotropic dielectric, with:

$$\varepsilon_{re} = (\varepsilon_r + 1)/2 \tag{9.3.1}$$

With the introduction of "ε_{re}," the slot wavelength "λ_s" is;

$$\lambda_s = \lambda_0/\sqrt{\varepsilon_{re}} = \lambda_0\left[2/(\varepsilon_r + 1)\right]^{0.5} \rightarrow (\lambda_0/\lambda_s)^2 \equiv \varepsilon_{re} \tag{9.3.2}$$

The expressions for "h_z," "h_r," and "e_θ" for the situation in Figure 9.3.1 are:

$$h_z = AH_0^{(1)}(k_t r) \tag{9.3.3}$$

where "A" is a generic constant and the quantity "k_t" can be regarded as the coefficient of the transverse wave equation. Supposing a lossless propagation along "z,"* "k_t" is given by:**

$$-k_t^2 = \omega^2\mu_0\varepsilon_0\varepsilon_{re} - \omega^2\mu_0\varepsilon_0$$

---

* Using the notation of Appendix A2, this condition means $k_z \equiv j\beta_z$.
** In Appendix A2 we have shown that "k_t^2" is a negative real number.

so that:

$$k_t = j\omega\left(\mu_0\varepsilon_0\right)\left(\varepsilon_{re} - 1\right)^{0.5} \equiv j\omega\left(\mu_0\varepsilon_0\right)\left[\left(\lambda_0/\lambda_s\right)^2 - 1\right]^{0.5} \stackrel{\perp}{=} j\beta_t \qquad (9.3.4)$$

The other field components are:

$$h_r = -\frac{\beta_z}{k_t^2}\frac{\partial h_z}{\partial r} = \frac{AH_1^{(1)}\left(k_t r\right)}{\left[1 - \left(\lambda_s/\lambda_0\right)^2\right]^{0.5}} \qquad (9.3.5)$$

$$e_\theta = -\frac{j\omega\mu}{k_t^2}\frac{\partial h_z}{\partial r} = \zeta\left(\lambda_s/\lambda_0\right)h_r \qquad (9.3.6)$$

where "$H_0^{(1)}$" and "$H_1^{(1)}$" are the Hankel functions of the first kind, respectively of order zero and one. The functions [13] "$H_n^{(1)}(v)$" and "$H_n^{(2)}(v)$" of the first and second kind of order "n" are defined using the Bessel functions* [14,15,16] according to:

$$H_n^{(1)}\left(v\right) \stackrel{\perp}{=} J_n\left(v\right) + jY_n\left(v\right) \qquad (9.3.7)$$

$$H_n^{(2)}\left(v\right) \stackrel{\perp}{=} J_n\left(v\right) - jY_n\left(v\right) \qquad (9.3.8)$$

For an imaginary argument "jv" with "v" a real variable $>>1$, the Hankel functions "$H_n^{(1)}(v)$" can be approximated as:

$$H_n^{(1)}\left(v\right) \approx 2e^{\left[-v - j(n\pi/2 + \pi/4)\right]}/\left(j\pi v\right)^{0.5} \qquad (9.3.9)$$

If we want to apply this equality to the field expressions, so that the field decay is assured outside the slot, it is necessary that "$k_t r$" be imaginary, i.e., $(\lambda_0/\lambda_s) > 1$ from 9.3.4. From Equation 9.3.2 this corresponds to use of a dielectric with $\varepsilon_r > 1$, to contain radiation. Of course, the higher the "ε_r," the lower the radiation. Cohn has shown that the ratio of the voltage "V(x)" across a half circumference of the electrical field indicated in Figure 9.2.1 and the voltage "V" across the slot is given by:

$$V\left(r\right)/V = 0.5\pi\beta_t r\left|H_1^{(1)}\left(k_t r\right)\right| \qquad (9.3.10)$$

To give some value, for an alumina substrate with $\varepsilon_r = 9.8$ we have 20 dB of attenuation at a distance $r \approx 0.3\lambda_0$.

The result is interesting for the polarization evaluation. In fact, if we do the ratio between "h_z" and "h_r" using 9.3.1 and 9.3.2 we have:

$$\left|h_z/h_r\right| \equiv \left|H_1^{(1)}\left(k_t r\right)/H_0^{(1)}\left(k_t r\right)\right| = \left[1 - \left(\lambda_s/\lambda_0\right)^2\right]^{0.5} \qquad (9.3.11)$$

---

* See Appendix A2 where the Bessel functions are introduced for a practical example. Deep insight into these functions can be found in books indicated in the references at the end of this chapter.

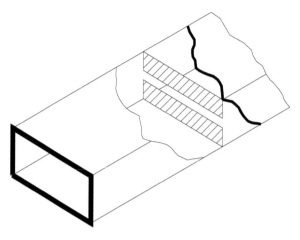

Figure 9.3.2

Since it is possible to show[17] that $|H_1^{(1)}(jv)/H_0^{(1)}(jv)| > 1$ and since the last member is less than one to assure low radiation, and so "k_r" is imaginary, with this hypothesis there is no solution to the previous equation. This means that a perfect circular polarization of the magnetic field does not exist, but only elliptical. Note that the direction of elliptical polarization is dependent on the direction of propagation, i.e., if we change the direction of propagation from "z_0" to "$-z_0$" then the magnetic field changes, the direction of rotation.

b. Transverse Resonance Method

With this method the e.m. study of the slot line is transferred to that of a particular loading used in waveguide, called "iris" and represented in Figure 9.3.2 for a rectangular waveguide. Iris loading is a waveguide technique used to build filters and coupling between resonators.[18,19,20]

The transformation of the slot line in a case representing an iris is performed in the following manner. First, it is assumed that propagation in the slot is in the most general case, i.e., in both directions.* Then, two conducting planes at a distance $z = \lambda_s/2$ and perpendicular to the surface are inserted into the slot line. These two planes do not change the field distributions since for every $\lambda_s/2$ a null of "ε_z" exists. Then, two conducting planes perpendicular to the surface are inserted at a distance "λ_d," symmetrical with respect to the center of the slot, into the slot line. The distance "λ_d" is chosen to be a distance where the intensity of the fields are negligible. These passages are indicated in Figure 9.3.3. Since the slot line is not a TEM t.l. then the characteristic impedance definition is not unique. In these cases it is usual to define the impedance according to the following relationship:

$$\zeta = V^2/2W \tag{9.3.11a}$$

where "V" is the peak voltage across the slot and "W" is the power supported by the wave.

Waveguide iris theory is outside the scope of this text and can be found in the literature. Cohn[21] has customized the general theory to this case, arriving at the expression of the slot line impedance. We will see in the next section how the Cohn expression for slot line impedance has been curve fitted, obtaining closed form expressions.

---

* See Chapter 1 for general theory of transmission line.

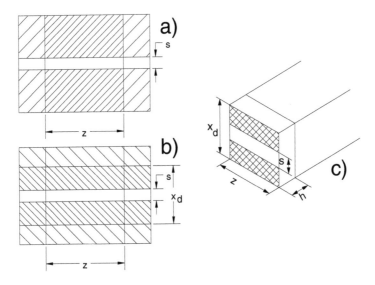

Figure 9.3.3

9.4 CLOSED FORM EQUATIONS FOR SLOT LINE CHARACTERISTIC IMPEDANCE

The closed form we give here has been obtained by curve fitting[22] the results of Cohn's transverse resonant method described in the previous section. These formulas are accurate inside an error of 2% when the slot line physical characteristics are inside these ranges:

$$9.7 \le \varepsilon_r \le 20, \ 0.02 \le s/h \le 1, \ 0.01 \le h/\lambda_0 \le 0.25 / (\varepsilon_r - 1)^{0.5} \tag{9.4.1}$$

So, if $0.02 \le s/h \le 0.2$ we have:

$$\lambda_s / \lambda_0 = 0.923 - 0.448 \log \varepsilon_r + 0.2 s/h \ (0.29 s/h + 0.047) \log (100 h/\lambda_0) \tag{9.4.2}$$

$$Z_1 = 72.62 - 35.19 \log \varepsilon_r + 50 (h/s) (s/h - 0.02) (s/h - 0.1) \tag{9.4.3}$$

$$Z_2 = (44.28 - 19.58 \log \varepsilon_r) \log (100 s/h) \tag{9.4.4}$$

$$\alpha = 0.32 \log \varepsilon_r - 0.11 + (s/h) (1.07 \log \varepsilon_r + 1.44) \tag{9.4.5}$$

$$\beta = (11.4 - 6.07 \log \varepsilon_r - 100 h/\lambda_0)^2 \tag{9.4.6}$$

$$\zeta = Z_1 + Z_2 - \alpha \beta \tag{9.4.7}$$

If $0.2 \le s/h \le 1$ we have:

$$r_1 = 0.987 - 0.483 \log \varepsilon_r + (s/h) (0.111 - 0.0022 \varepsilon_r) \tag{9.4.8}$$

$$r_2 = \left(0.121 + 0.094 s/h - 0.0032\,\varepsilon_r\right)\left[\log\left(100 h/\lambda_0\right)\right] \tag{9.4.9}$$

$$\lambda_s/\lambda_0 = r_1 - r_2 \tag{9.4.10}$$

$$Z_1 = 113.19 - 53.55\log\varepsilon_r + 1.25\left(s/h\right)\left(114.59 - 51.88\log\varepsilon_r\right) \tag{9.4.11}$$

$$Z_2 = 20\left(s/h - 0.2\right)\left(1 - s/h\right) \tag{9.4.12}$$

$$\alpha = 0.15 + 0.23\log\varepsilon_r + \left(s/h\right)\left(2.07\log\varepsilon_r - 0.79\right) \tag{9.4.13}$$

$$\beta = \left[10.25 - 5\log\varepsilon_r + \left(s/h\right)\left(2.1 - 1.42\log\varepsilon_r\right) - 100 h/\lambda_0\right]^2 \tag{9.4.14}$$

$$\zeta = Z_1 + Z_2 - \alpha\beta \tag{9.4.15}$$

In the previous equation with "log" we have indicated the logarithm in base "10." Note that since it is known, the frequency of the signal we apply in the slot, i.e., "λ_0" is known, from 9.4.2 and 9.4.10 we also obtain "ε_{re}," since we know that $\lambda_s = \lambda_0/(\mu_{re}\varepsilon_{re})^{0.5}$ with $\mu_{re} = 1$ for non-ferrimagnetic substrates.*

Formulas for lower permittivity have been obtained by the researchers R. Janaswamy and D. H. Schaubert,[23] through a curve fitting to a full wave analysis. The general condition for the following expression is $0.006 \leq h/\lambda_0 \leq 0.06$. Then, for $2.22 \leq \varepsilon_r \leq 3.8$ we have:

for $0.0015 \leq s/\lambda_0 \leq 0.075$:

$$r_1 = 1.045 - 0.365\,\ell n\varepsilon_r + \frac{6.3\left(s/h\right)\varepsilon_r^{0.945}}{238.64 + 100 s/h} \tag{9.4.16}$$

$$r_2 = \left[0.148 - \frac{8.81\left(\varepsilon_r + 0.95\right)}{100\,\varepsilon_r}\right]\ln\left(h/\lambda_0\right) \tag{9.4.17}$$

$$\lambda_s/\lambda_0 = r_1 - r_2 \tag{9.4.18}$$

$$Z_1 = 60 + 3.69\,\mathrm{sen}\left[\frac{\left(\varepsilon_r - 2.22\right)\pi}{2.36}\right] + 133.5\,\ell n\left(10\varepsilon_r\right)\left(s/\lambda_0\right)^{0.5} \tag{9.4.19}$$

$$Z_2 = 2.81\left[1 - 0.011\varepsilon_r\left(4.48 + \ln\varepsilon_r\right)\right]\left(s/h\right)\ln\left(100 h/\lambda_0\right) \tag{9.4.20}$$

$$Z_3 = 131.1\left(1.028 - \ln\varepsilon_r\right)\left(h/\lambda_0\right)^{0.5} \tag{9.4.21}$$

$$Z_4 = 12.48\left(1 + 0.18\,\ell n\varepsilon_r\right)\frac{s/h}{\left[\varepsilon_r - 2.06 + 0.85\left(s/h\right)^2\right]^{0.5}} \tag{9.4.22}$$

$$\zeta = Z_1 + Z_2 + Z_3 + Z_4 \tag{9.4.23}$$

---

* The case of ferrimagnetic substrates will be studied later in this chapter.

For $0.075 \le s/\lambda_0 \le 1$:

$$r_1 = 1.194 - 0.24 \ell n \, \varepsilon_r - \frac{0.621 \varepsilon_r^{0.835} (s/\lambda_0)^{0.48}}{1.344 + s/h} \tag{9.4.24}$$

$$r_2 = 0.0617 \left[1.91 - \frac{\varepsilon_r + 2}{\varepsilon_r} \right] \ln (h/\lambda_0) \tag{9.4.25}$$

$$\lambda_s / \lambda_0 = r_1 - r_2 \tag{9.4.26}$$

$$Z_1 = 133 + 10.34 (\varepsilon_r - 1.8)^2 \tag{9.4.27}$$

$$\alpha = 2.87 \left[2.96 + (\varepsilon_r - 1.582)^2 \right] \tag{9.4.28}$$

$$\beta = \left\{ [s/h + 2.32 \varepsilon_r - 0.56] \left[(32.5 - 6.67 \varepsilon_r)(100 \, h/\lambda_0)^2 - 1 \right] \right\}^{0.5} \tag{9.4.29}$$

$$Z_3 = 13.23 \left[(\varepsilon_r - 1.722) s/\lambda_0 \right]^2 - (684.45 \, h/\lambda_0)(\varepsilon_r + 1.35)^2 \tag{9.4.30}$$

$$\zeta = Z_1 + \alpha\beta + Z_3 \tag{9.4.31}$$

For $3.8 \le \varepsilon_r \le 9.8$ we have:

for $0.0015 \le s/\lambda_0 \le 0.075$:

$$r_1 = 0.9217 - 0.277 \, \ell n \, \varepsilon_r + 0.0322 (s/h) \left[\varepsilon_r / (s/h + 0.435) \right]^{0.5} \tag{9.4.32}$$

$$r_2 = 0.01 \, \ell n (h/\lambda_0) \left\{ 4.6 - 3.65 / \left[\varepsilon_r^2 (s/\lambda_0)^{0.5} (9.06 - 100 \, s/\lambda_0) \right] \right\} \tag{9.4.33}$$

$$\lambda_s / \lambda_0 = r_1 - r_2 \tag{9.4.34}$$

$$Z_1 = 73.6 - 2.15 \varepsilon_r + (638.9 - 31.37 \varepsilon_r)(s/\lambda_0)^{0.6} \tag{9.4.35}$$

$$Z_2 = \left[36.23 (\varepsilon_r^2 + 41)^{0.5} - 225 \right] \frac{s/h}{s/h + 0.876 \varepsilon_r - 2} \tag{9.4.36}$$

$$Z_3 = 0.51 (\varepsilon_r + 2.12)(s/h) \ln (100 h/\lambda_0) \tag{9.4.37}$$

$$Z_4 = 0.753 \varepsilon_r (h/\lambda_0) / (s/\lambda_0)^{0.5} \tag{9.4.38}$$

$$\zeta = Z_1 + Z_2 + Z_3 - Z_4 \tag{9.4.39}$$

for $0.075 \le s/\lambda_0 \le 1$:

$$r_1 = 1.05 - 0.04\,\varepsilon_r + 0.01411(\varepsilon_r - 1.421)\ln\left[s/h - 2.012(1 - 0.146\,\varepsilon_r)\right] \qquad (9.4.40)$$

$$r_2 = 0.111\,(1 - 0.366\,\varepsilon_r)\,(s/\lambda_0)^{0.5} \qquad (9.4.41)$$

$$r_3 = 0.139\left[1 + 0.52\,\varepsilon_r\,\ln(14.7 - \varepsilon_r)\right](h/\lambda_0)\ln(h/\lambda_0) \qquad (9.4.42)$$

$$\lambda_s/\lambda_0 = r_1 + r_2 + r_3 \qquad (9.4.43)$$

$$Z_1 = 120.75 - 3.74\,\varepsilon_r \qquad (9.4.44)$$

$$\alpha = 50\left[\mathrm{atg}(2\,\varepsilon_r) - 0.8\right](s/h)^{\left[1.11 + 0.132(\varepsilon_r - 27.7)/(100h/\lambda_0 + 5)\right]} \qquad (9.4.45)$$

$$\beta = \ln\left\{100\,h/\lambda_0 + \left[(100\,h/\lambda_0)^2 + 1\right]^{0.5}\right\} \qquad (9.4.46)$$

$$Z_2 = 14.21(1 - 0.458\,\varepsilon_r)(100\,h/\lambda_0 + 5.1\,\ell n\,\varepsilon_r - 13.1)(s/\lambda_0 + 0.33) \qquad (9.4.47)$$

$$\zeta = Z_1 + \alpha\beta + Z_2 \qquad (9.4.48)$$

Note as in all the previous formulas the conductor thickness "t" doesn't appear. Results of conductor thickness "t" analysis[24] have shown that the effects on impedance and phase constant produce a change near some percent, with respect to the case $t = 0$, until $t \le 0.02s$.

Regarding slot line attenuation, experiments performed by some researchers[25] have shown that attenuation is similar to that presented by a microstrip with 50 Ohm of impedance on the same substrate. So, the formulas we gave for microstrips can also be used for slot lines. Some discrepancy can arise when using very narrow slots, below 100 μm.

No closed formulas for slot line attenuation are available. Also, because of the non- "TEM" propagation mode, the Wheeler incremental inductance theory[26] cannot be rigorously applied.

The case of possible magnetic losses, due to the use of ferrimagnetic substrates, will be studied later in this chapter.

9.5 CONNECTIONS BETWEEN SLOT LINES AND OTHER LINES

We think it is important to discuss the most suitable and simple transitions between slot lines and the other most used transmission lines. Every transition always introduces a discontinuity. As a general rule, a transition between t.l.s is practically evaluated as acceptable when the resulting reflection coefficient is typically below 10 dB. Studies on slot line discontinuities[27,28,29] permit evaluation of how one transition can be better than another.

A single slot on an electric plane can be regarded as a particular case of slotline, and in conjunction with a feeding line, this slot becomes an antenna.[30] A real slot line is widely used as a feeding line of a printed antenna called "Vivaldi," indicated in Figure 9.5.1a[31,32] for the single-sided and in Figure 9.5.1 b for the double-sided cases. These antennas have been formerly fed with coaxial cables, but now they are usually fed with microstrips or striplines, with transitions as

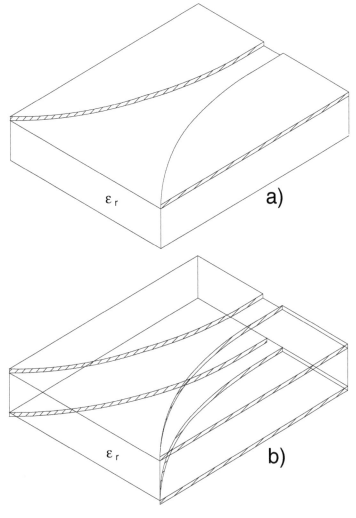

Figure 9.5.1

we will now indicate. A slotline is also used to build a balanced mixer in conjunction with coplanar waveguides.*

a. Connection with Coaxial Cable

This transition is indicated in Figure 9.5.2. For a good transition it is important that the outer conductor of the coaxial cable be short connected to one plane of the slot line. The center conductor of the coaxial cable is of course short connected to the other plane. The transition shown in the figure works quite well until some GHz,[33] but adding some tapering to this transition, an operating bandwidth of 4 to 22 GHz has been reported.[34] Today this transition is seldom used, and microstrips and striplines are employed.

b. Connection with Microstrip

The simplest transition between these two lines is indicated in Figure 9.5.3. With a dashed line we have indicated the bottom conductor while a solid line indicates the upper conductor. As we

* Coplanar waveguides are abbreviated with "CPW."

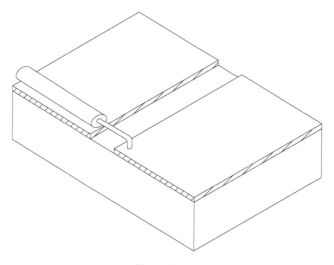

Figure 9.5.2

can see, this transition needs a substrate with both surfaces etched. On one side there is the microstrip, while on the other side there is the slot line. Note that the microstrip is open terminated while the slot line is short terminated, both at a distance approximately a quarter of wavelength long after the crossing. Also note that this transition does not work for DC.

Typical operating bandwidths[35] for the transition indicated in Figure 9.5.3 are near 1.5:1. Using a multistep transformer* for the microstrip line and radial open-ended stubs for both lines, a wider bandwidth near 3 to 18 GHz has been reported.[36]

Another type of transition, quite compact, is shown in Figure 9.5.4. Note as in this case the microstrip is short circuit terminated through the via hole indicated in the drawing, while the slot line is open circuit terminated. In fact, slot line open circuits are usually made with a circle of

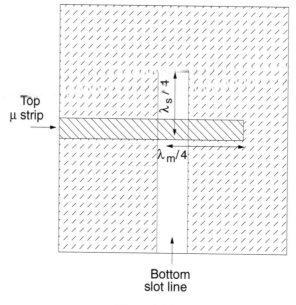

Figure 9.5.3

---

* Microstrip step transformers are studied in Chapter 5.

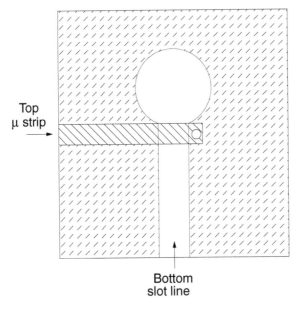

Top
μ strip
→

Bottom
slot line

Figure 9.5.4

radius many times the lowest wavelength of the signal traveling the slot. Slotline-µstrip transition is currently a research topic, with the aim to reach wider operating bandwidths.[37,38]

c. Connection with Stripline

This transition is built with the configuration indicated in Figure 9.5.3, but of course inserted in the middle of the substrate, as the symmetrical stripline geometry requires. This transition is used to feed the double-sided "Vivaldi" antenna indicated in Figure 9.5.1. Performing the same tapering we used for the microstrip case, an operating bandwidth near that of the microstrip case has been reported.[39]

d. Connection with Coplanar Waveguide

Two possible transitions between these two lines are indicated in Figures 9.5.5 and 9.5.6. Since for a complete understanding of these transitions a knowledge of the "CPW" propagation modes

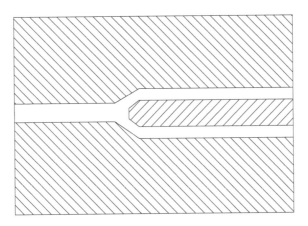

Figure 9.5.5

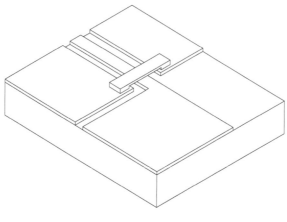

Figure 9.5.6

is required, we refer the reader to Chapter 10 where "CPW" is studied and these transitions are explained in more detail.

9.6 TYPICAL NONFERRIMAGNETIC DEVICES USING SLOT LINES

Some interesting devices can be built employing slot lines, especially when they are used with other t.l.s. The resulting devices are in general smaller with respect to the usual construction. In this section we will discuss such devices.

a. 180° Reciprocal Phase Shifters

Such a device is indicated in Figure 9.6.1. It is built using a slot line and two microstrips, with transitions as indicated in Figure 9.5.3. This device has the characteristic that a signal entering in one microstrip comes out from the other with 180° phase reversal. Of course, to work correctly the bottom conductors must be disconnected from ground, at least near the slot. Experiments performed with short digital pulses[40] have shown an operating bandwidth near one GHz.

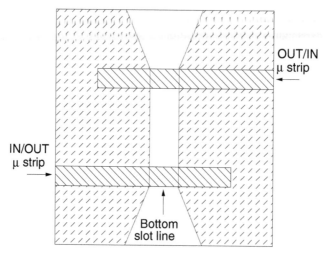

Figure 9.6.1

IN/OUT microstrip

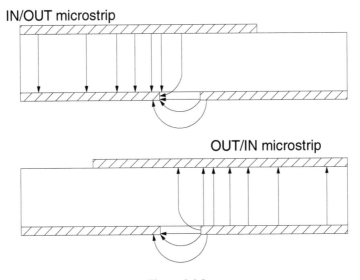

OUT/IN microstrip

Figure 9.6.2

The phase shifting principle is indicated in Figure 9.6.2. Observe that the 180° phase reversal is quite independent from the frequency, so this phase shifter is a high bandwidth device, but in any case with a bandwidth lower than its multistep Shiffman counterpart.* This device can also be realized using the microstrip-slot line transition indicated in Figure 9.5.4.

b. Magic "T"

Using the 180° phase shifter device studied in section "a" above, a new type of magic "T"** can be built. It is indicated in Figure 9.6.3. We can see how the 270° microstrip line has been substituted with a 90° slot line with the transitions indicated in the previous points that perform a

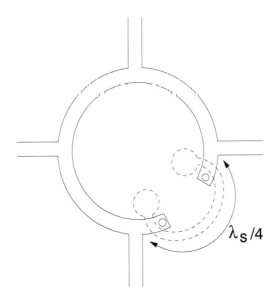

Figure 9.6.3

---

* Shiffman phase shifters are studied in Chapter 5.
** Microstrip magic "T" is studied in Chapter 5.

180°, resulting in a 270° phase shifting. This device is consequently smaller than the typical microstrip magic "T." Measurements performed on this device[41] have shown that the operating bandwidth can reach 100%, which is near four times the operating bandwidth of the usual microstrip magic "T." Another type of magic "T" has been investigated by L. Fan, C. H. Ho, S. Kanamaluru, and K. Chang,[42] which also employs a coplanar waveguide.

c. Mixers

One of the first uses of slot lines in mixers was by the researchers L. E. Dickens and D. W. Maki.[43] They have used a "CPW-slot line" transition, as indicated in Figure 9.5.5, together with a "CPW-microstrip," which we will study in Chapter 10. The characteristics of their mixer are as follows:

RF bandwidth:	8.9 to 9.9 GHz
LO frequency:	7.8 GHz
IF bandwidth:	DC to 1 GHz
Conversion loss:	3.15 dB, max
Image freq. isol:	25 dB, min

Another interesting use of slot lines has been reported by J. A. Eisenberg, J. S. Panelli, and W. Ou.[44] They have employed the transition shown in Figure 9.6.4 to pass from an unbalanced "CPW" to a balanced transmission line called a "coplanar strip" (CPS).* Note they perform such balun operations just using a slot line. Using the balun shown in Figure 9.6.4 they have built a double balanced mixer with the following main characteristics:

RF bandwidth:	6 to 16 GHz
LO bandwidth:	6 to 16 GHz
IF bandwidth:	DC to 1.5 GHz
Conversion loss:	9.1 dB, max
Isol. any port:	18 dB, min

In addition to the good performances of such mixers, these devices have the ability to be fully planar, i.e., the conductors to guide the e.m. energy are on one plane only.

Due to the high interest in mixer devices for communications systems, use of slot lines for these devices is always a research topic.[45,46]

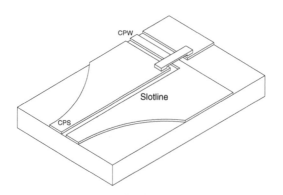

Figure 9.6.4

* Coplanar strips will be studied in Chapter 11.

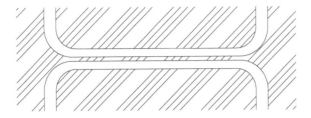

Figure 9.6.5

d. Directional Couplers

One of the first experiments on directional couplers employing slot lines was made by E. A. Mariani and J. P. Agrios.[47] The resulting network is shown in Figure 9.6.5. This is a single $\lambda/4$ step directional coupler. The results of a device working at 2.74 GHz give an operating bandwidth comparable to that of a single $\lambda/4$ step directional coupler in microstrip technology. However, with this configuration it is possible to reach a 3 dB coupling, a value very difficult to achieve using a microstrip.* Other directional couplers can be obtained using a "CPW" in conjunction with a slotline.[48]

e. Filters

The same researchers[49] for the directional coupler discussed in the above section "d" have built a band stop filter using transitions between microstrips and slot lines, as indicated in Figure 9.6.6. The results of a device working at 3.1 GHz have not given particular advantages with respect to the microstrip counterpart. For this reason slot line filters are seldom used in practice, but their study is very interesting from an e.m. point of view.[50]

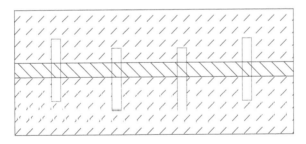

Figure 9.6.6

9.7 MAGNETIZATION OF SLOT LINES ON FERRIMAGNETIC SUBSTRATES

As a result of the discussion in Section 9.2 we know that an elliptic polarization of the magnetic field exists near the region of the slot, with the plane of polarization oriented longitudinally. The existence of this polarization can be used to build nonreciprocal ferrimagnetic devices[51,52,53] like isolators or phase shifters,** as we will study in the next section. The complex argument of energy interaction between e.m. energy and ferrimagnetic materials is discussed in Appendixes A5, A6, and A7. The reader who is not familiar with these topics can read these Appendixes to review the fundamentals of this branch of physics.

---

* See Chapter 5 for practical coupling values of microstrip step couplers.
** Isolators are always nonreciprocal devices, while phase shifters can be, in general, also reciprocal devices.

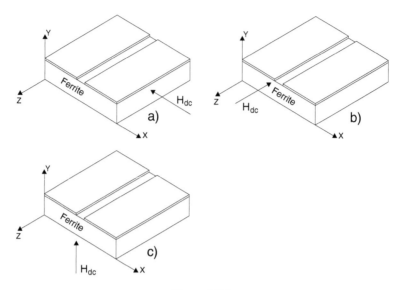

Figure 9.7.1

Indicating with:

"$\underline{H}_{dc}$" a static magnetic field,
"$\underline{h}$" the "RF" magnetic field,
"β"* a direction of propagation

the key points of energy interactions between an "h" elliptically polarized inside a magnetized ferrimagnetic material are:

1. It is "direction sensitive," i.e., if we fixed a direction for "$\underline{H}_{dc}$," for "β" we have the maximum energy exchange while for "$-\beta$" we have the minimum. The same happens if we fix a direction of "β" and change the direction of "$\underline{H}_{dc}$."
2. It is dependent, among other things, on precise relationships between signal frequency, "H_{dc}" intensity, and substrate physical characteristics.

The previous points 1 and 2 are, of course, a high simplification of complex physics interactions, which result in macroscopic effects like "resonance" and "field displacement."**

Once the plane position of elliptical polarization for "h" is known, it is simple to recognize that the direction of "H_{dc}" for maximum energy exchange is as indicated in Figure 9.7.1a. In fact, with these orientations between the "h" polarization plane and "$\underline{H}_{dc}$," the induced spin magnetic vector precession motion can interact with "$\underline{h}$." For the case of microstrips or striplines we have studied how the "$\underline{H}_{dc}$" direction shown in Figure 9.7.1a is used to demagnetize a reciprocal ferrimagnetic phase shifter.***

The second situation for which an energy exchange is possible is indicated in Figure 9.7.1b. In fact, from Section 9.2 we know that an "rf" magnetic field orthogonal to "$\underline{H}_{dc}$" exists. This component is "h_x." Since "h_x" is not elliptical polarized, with this type of magnetization it is possible to build reciprocal phase shifters, as we studied in Chapter 7 or 8 for microstrips or striplines. However, this component is lower than the same component for microstrips or striplines and consequently whenever a reciprocal phase shifter is needed, these last two t.l.s are employed.

---

* With "β" in this case we mean "$\underline{z}_0$" or "$-\underline{z}_0$."
** See Appendix A7 to read about these two phenomenons. The "field displacement" phrase will be abbreviated with "f.d."
*** See Chapters 5 and 7 for ferrimagnetic microstrip or stripline devices.

In the third direction of application for "$\underline{H}_{dc}$," indicated in Figure 9.7.1c, the energy exchange is at a minimum, regardless of the reciprocal directions of "$\underline{H}_{dc}$," "$\underline{\beta}$," or signal frequency, "H_{dc}" intensity and substrate physical characteristics.

It is important to observe that when no external magnetic field is applied, and "RF" signal amplitude is not excessive to self bias the substrate, ferrite is a good dielectric with low attenuation without any particular evident ferrimagnetic effect.

We can conclude these preliminary notes observing that for slotlines the most useful ferrimagnetic devices are always nonreciprocal, and external magnetization is orthogonal to the direction of propagation, just to use the elliptical polarization of "h." In the next sections we will show that it is not necessary to build a slotline directly on ferrimagnetic materials, even if it is preferable. In fact, it is possible to have a nonferrimagnetic material as substrate and add slabs of ferrimagnetic material on the conductors.

In the next section we will assume the reader is familiar with an isolator or phase shifter. Appendix A7 defines such devices.

9.8 SLOT LINE ISOLATORS

Slot line isolators can be built using two energy exchange phenomena, i.e., the "resonance" and the "field displacement" effects.*

9.8.1 Resonance Isolator

A slot line resonance isolator can be built as indicated in Figure 9.8.1. Near the slot of the slot line a cylinder of ferrimagnetic material is positioned. An external static magnetic field "H_{dc}" uniformly magnetizes the cylinder, centered** at the signal frequency where the isolation is desired. Of course, the cylinder can also be substituted with a single slab of ferrite*** positioned above the slot.

The same isolation property can be achieved with the isolator configuration shown in Figure 9.8.2, where the entire slot line is built on ferrimagnetic material.

Of course, some matching is required to adjust the input impedance of these devices to the usual 50 Ohm, since their input impedance is usually lower than this value.

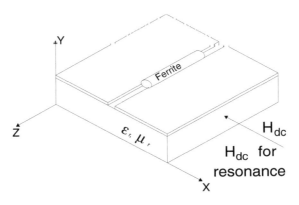

Figure 9.8.1

---

* See Appendix A7 for fundamentals about these two important effects of energy exchange.
** By "centered" we mean that the "H_{dc}" intensity is just what is required to have a precession frequency motion equal to that signal frequency we want to stop. See Appendix A7 for more deep insight into ferrimagnetic energy exchange.
*** "Ferrite" is a particular ferrimagnetic material, like others as "YIG." We will use this term to indicate a generic ferrimagnetic material.

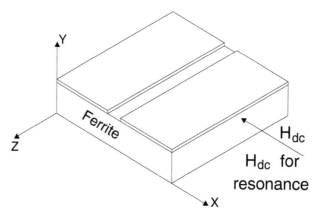

Figure 9.8.2

9.8.2 Field Displacement Isolator

A field displacement (f.d.) slot line isolator can be realized as indicated in Figure 9.8.3. A slot line is built on ferrite, and at the other face, an absorptive material is attached. A static magnetic field "H_{dc}" is applied orthogonally to the direction of propagation "z_0."

To study such a device we need to apply some simplifications to the geometry indicated in Figure 9.8.3 to use the theory we developed in Appendix A7. So we assume:

1. all the e.m. energy is contained inside the ferrite
2. in the slots only the components "h_y" and "h_z" exist, and that they don't vary with the coordinate "x;"* i.e., the fields are uniform
3. only a "TE" mode is supported by the "slot line"

With this hypothesis, we know from Appendix A7 that:
  a. the ferrite presents a permeability "$[\mu_x]$" given by:

$$[\mu_x] = \mu_0 \begin{bmatrix} 1 & 0 & 0 \\ 0 & \mu_p & j\mu_\ell \\ 0 & -j\mu_\ell & \mu_p \end{bmatrix} \qquad (9.8.1)$$

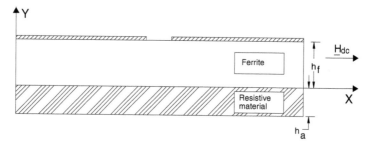

Figure 9.8.3

---

* Using the well-known notation described in Appendix A2 and Chapter 1, this hypothesis means $k_x = 0$.

b. the propagation constants "k_y" and "k_z" are related by:*

$$k_y^2 + k_z^2 = -\omega^2 \mu_0 \varepsilon \mu_{eq\perp} \tag{9.8.2}$$

with the "equivalent permeability ($\mu_{eq\perp}$)" given by:

$$\mu_{eq\perp} \overset{\perp}{=} \frac{\mu_p^2 - \mu_\ell^2}{\mu_p} \tag{9.8.3}$$

For a "TE" mode the field components "h_y" and "h_z" can be written as a function of the transverse electric field "e_x"** i.e.:

$$h_y = \frac{j\mu_p k_z - \mu_\ell k_y}{\omega \mu_0 \left(\mu_\ell^2 - \mu_p^2\right)} e_x \tag{9.8.4}$$

$$h_z = \frac{-j\mu_p k_y - \mu_\ell k_z}{\omega \mu_0 \left(\mu_\ell^2 - \mu_p^2\right)} e_x \tag{9.8.5}$$

Since we assume energy propagation only inside the ferrite, for $y = h_f$ there must be:

$$h_z = 0 \tag{9.8.6}$$

The previous equation can be simply understood forcing, for our hypothesis, the Poynting vector*** to be zero for $y \geq h_f$.

We now force the condition to have a real value "α_y" for "k_y" and an imaginary value "β_z" for "k_z," i.e.:

$$k_y \overset{\perp}{=} \alpha_y \tag{9.8.7}$$

$$k_z \overset{\perp}{=} j\beta_z \tag{9.8.8}$$

Inserting 9.8.3 and the two previous equations into 9.8.4, the condition 9.8.6 is verified if:

$$\alpha_y \overset{\perp}{=} -\left(\mu_\ell / \mu_p\right)\beta_z \tag{9.8.9}$$

Inserting the three previous equations into 9.8.2 we have:

$$\beta_z \overset{\perp}{=} \omega \left(\mu_0 \mu_p \varepsilon\right)^{0.5} \tag{9.8.10}$$

So, if we permit the two previous equations to be verified, then the propagation will be attenuated along "y" and unattenuated along "z."

---

* Remember we have assumed that the e.m. fields have zero dependence on the "x" axis, i.e., $k_x = 0$.
** See Appendices A2 and A7 for field expressions of various modes.
*** See Appendix A2 for the definition of Poynting vector.

In addition, if we do the ratio "e_x/h_z," we have the characteristic impedance given by:

$$\zeta \equiv \left(\mu_0\mu_p/\varepsilon\right)^{0.5} \tag{9.8.11}$$

From the graph (A7.4.1) of "μ_p" reported in Appendix A7, we can observe how for a particular field intensity "H_{dc}"* and a particular frequency it is possible to generate a negative value of "μ_p" and consequently an imaginary impedance. This means that under these circumstances the signal is swept away from the ferrite.

The dependence of the field displacement effect on the directions "$\underline{\beta}_z$" and "$\underline{H}_{dc}$" can be understood observing Table A7.7.1, here reported for simplicity:

Table A7.7.1

Change of sign when **H** changes dir.		Change of sign when prop. changes dir.	
"μ_p"	NO	"μ_p"	NO
"μ_ℓ"	YES	"μ_ℓ"	NO
"$\mu_{eq\perp}$"	NO	"$\mu_{eq\perp}$"	NO
"β_z"	NO	"β_z"	YES

Note that since the sign of "α_y" given by 9.8.9 depends on the product "$\mu_\ell\beta_z$," using the previous table we note how changing directions of both "β_z" and "$\underline{H}_{dc}$" the guiding edge for the e.m. energy doesn't change.

The theoretical concepts we have just reported are graphically represented in Figure 9.8.4. Part a indicates the electrical field lines when the field displacement causes the e.m. energy to be mainly

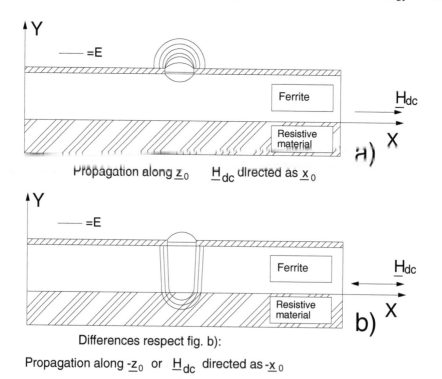

Propagation along $\underline{z}_0$ $\underline{H}_{dc}$ directed as $\underline{x}_0$

Differences respect fig. b):

Propagation along $-\underline{z}_0$ or $\underline{H}_{dc}$ directed as $-\underline{x}_0$

Figure 9.8.4

* "H_{dc}" intensities for field displacement operation are lower than the values for resonance.

on the top surface, i.e., the conductor's surface. This situation is different from the typical nonfer-rimagnetic slot line, where the field lines are mainly concentrated in the dielectric. From the equations we gave before, this situation happens when "$\underline{H}_{dc}$" is directed along "$\underline{x}_0$" and propagation is along "$\underline{z}_0$." Figure 9.8.4b represents what happens when, with respect to part a of the figure, we change the propagation direction or "$\underline{H}_{dc}$" direction. From this figure it is simple to understand that if we attach an "RF" absorptive material on the bottom of the slot line, in the direction that the field travels on this surface a signal attenuation will arise. Consequently, when a change of sign is made to "$\underline{H}_{dc}$" or to the propagation direction, then no loss will be added by this material, since signal propagation is on the opposite surface. Of course, if we move the resistive material on the conductor's surface, we always have a field displacement isolator, but now we have reversed the attenuation direction.

Field displacement isolators are of a wider bandwidth than the resonance isolators and need a lower intensity of "$\underline{H}_{dc}$." They do not reach the isolation values of the resonance counterpart. Experiments[54] on an f.d. isolator similar to that indicated in Figure 9.8.3 have given an isolation of minimum 20 dB and an insertion loss of maximum 3 dB in the 3 to 6 GHz band.

However, in general, devices that only perform isolation are not so widely used, and terminated circulators are preferred due to their high versatility. Slot line circulators have also been investigated in the literature,[55] but no improvements have been given with respect to other planar transmission line circulators.

9.9 SLOT LINE FERRIMAGNETIC PHASE SHIFTERS

Two groups of slot line ferrimagnetic phase shifters* are possible: one that uses the different phase constants "β_c" and "β_d" between the "concordant" and "discordant" waves,** and one that uses the field displacement effect. We will call the first device a "Discon" phase shifter and the second one a "field displacement" phase shifter.

With respect to microstrip or stripline counterparts, slot line phase shifters have the advantage that circular polarization doesn't need to be artificially created. In fact, we know that for microstrip or striplines, circular polarization is obtained with a meander line; but, while all the meander line length is responsible for attenuation, only approximately 65% of its length is responsible for circular polarization. This circumstance results in a higher length for meander line devices with respect to a slot line p.s. A meander line stripline or microstrip p.s. has higher absolute values of figure of merit than the corresponding slot line counterpart, but a slot line device usually has smaller dimensions for the same electrical parameters.

9.9.1 "Discon" Phase Shifter

One of the first experiments performed on this kind of phase shifter was made by G. H. Robinson and J. L. Allen.[56] The common physical principle is the different value of "β_c" and "β_d" and their dependence on "$\underline{H}_{dc}$" and the direction of propagation. These two phase constants are related to the elements of the ferrite permeability matrix by the relationships:

$$\beta_d \stackrel{\pm}{=} \omega\left(\mu_0\varepsilon\right)^{0.5}\left(\mu_p - \mu_\ell\right)^{0.5} \tag{9.9.1}$$

$$\beta_c \stackrel{\pm}{=} \omega\left(\mu_0\varepsilon\right)^{0.5}\left(\mu_p + \mu_\ell\right)^{0.5} \tag{9.9.2}$$

---

* We will use the letters "p.s." to denote a "phase shifter."
** See Appendix A7 for definitions of "concordant" and "discordant" waves.

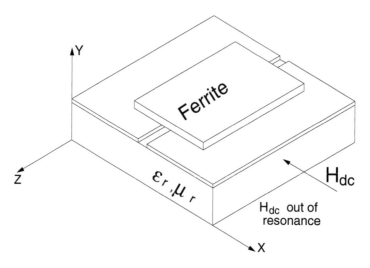

Figure 9.9.1

One possible construction of a "discon" p.s. is indicated in Figure 9.9.1. Note that the situation is practically coincident with that indicated in Figure 9.8.1. The presence of a ferrite slab, as in this case, or ferrite cylinder makes no magnetic difference. The very important difference is that for the isolator case, the "H_{dc}" value is for resonance, while in this case we are not in resonance, obviously. Caution must be used not to bring ferrite into the field displacement region using a wrong value of "H_{dc}".[*] For the "discon" p.s., typical "H_{dc}" values have lower intensity than the required values for a f.d. effect.[**] Of course, a p.s. can also be realized as indicated in Figure 9.8.2, again with "H_{dc}" out of resonance.

An interesting alternative to the planar p.s. structure is the parallelepipedal type, indicated in Figure 9.9.2. This device uses the ferrite residual magnetization.[***] A slot line is built on a surface of a hollow ferrite parallelepiped. A conductor wire "W" is inserted inside the cavity. When a

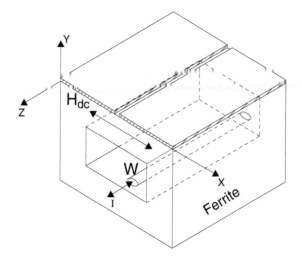

Figure 9.9.2

[*] The proper values of "H_{dc}" depend on the ferrite parameters inserted in the figures of "β_c" and "β_d" we gave in Appendix A7.
[**] Theoretically a "discon" p.s. can also work with "H_{dc}" values above resonance, but such a high value is only employed for high RF power devices.
[***] See appendix A7 for ferrite residual, or permanent, magnetization.

current pulse "I" passes inside the wire, a magnetic field is generated* that, if of proper intensity, can permanently magnetize the ferrite with a residual field "H_{dc}." Changing the direction of "I" is equivalent to changing the direction of "$\underline{H}_{dc}$." Devices working on residual magnetization are called "latching devices." Note that the latching p.s. indicated in Figure 9.9.2 having a closing path for "H_{dc}" does not disperse this static field very much.

Experiments[57] on a "discon" p.s. shown in Figure 9.9.1 working in the frequency range of 8 to 10 GHz have given differential phase shift** "$\Delta\varphi$" values near 12°/cm and figure of merit*** near 100°/dB. Other experiments[58] on the "discon" p.s. composed as shown in Figure 9.8.2, with "$\underline{H}_{dc}$" of course not in resonance, have given differential phase shift values near 30°/cm and figure of merit near 200°/dB for a signal working at 9.5 GHz. A latching[59] p.s. has given differential phase shift values near 20°/cm and figure of merit near 150°/dB, i.e., comparable values with respect to the nonlatching "discon."

9.9.2 Field Displacement Phase Shifter

The operating principles of a field displacement p.s. are exactly the same as those for the f.d. isolator. So, Figure 9.8.4 can represent the construction of an f.d. phase shifter, but for the proper operation it is necessary to change the slab of absorptive material to one with a low loss dielectric slab with permittivity "$\varepsilon_{r\ell}$" as indicated in Figure 9.9.3. In fact, since the phase constant "β_z" along "z" is given by 9.8.10, i.e.:

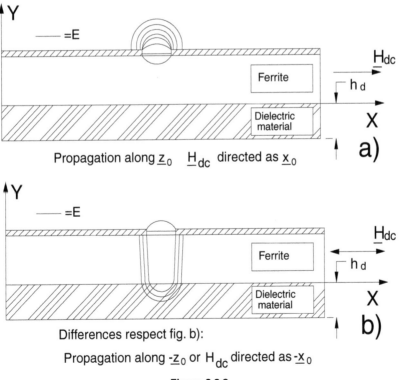

Figure 9.9.3

---

* Fundamental theory of magnetic fields generated by current in conductors is reviewed in Appendix A6.
** "Differential phase shift" is defined in Appendix A7.
*** See Appendix A7 for p.s. quality parameters definitions.

$$\beta_z = \omega \left(\mu_0 \mu_p \varepsilon_0 \varepsilon_{re} \right)^{0.5} \tag{9.9.3}$$

then when the wave is guided by the edge where the dilectric slab is disposed, the value of "ε_{re}" will be:

$$\varepsilon_{re} \approx \left(\varepsilon_f + \varepsilon_{r\ell} \right) / 2 \tag{9.9.4}$$

while when the wave is guided by the opposite edge, the value of "ε_{re}" will be:

$$\varepsilon_{re} \approx \left(\varepsilon_f + 1 \right) / 2 \tag{9.9.5}$$

resulting in a differential phase shifting. In 9.9.4 and 9.9.5 "ε_f" is of course the ferrite permittivity. Note that if we move the added dielectric slab on the conductor surface we always have a field displacement p.s., but now we have reversed the phase shifting sign. Experiments[60] on a f.d.p.s. of the type shown in Figure 9.9.3 have given $\Delta\varphi \approx 20°$/cm in the 6 to 10 GHz bandwidth, and peaks near 40°/cm at 10 GHz.[61]

It is important to conclude this section remembering from the general discussion in Appendix A7, that the distinction between "discon" and f.d. phase shifters is only dependent on the intensity of the applied "H_{dc}" since the two mechanical constructions can be very similar.

9.10 COUPLED SLOT LINES

Coupled slot lines are seldom employed in planar transmission line circuits. In practice, there is no advantage to using such coupled line structures instead of other coupled t.l.s we have studied in this text such as microstrips or striplines. However, we think it is useful to discuss this particular coupled line circuit since it will be recalled in Chapter 10 when we will study coplanar waveguides. Due to the geometric similitude to the "CPW," coupled slot lines are a topic of study.[62,63]

9.10.1 General Characteristics

The geometric structure of side coupled slot lines, simply called "SCSL" is indicated in Figure 9.10.1. In practice, two slot lines share a common conductor as one of the two conductors required for a single slot line. The e.m. field lines can be assumed to be composed by the superposition of the field distribution of the even and odd excitation.* The field distribution for these excitations, also called "modes," is schematically reported in Figure 9.10.2, respectively in parts a and b. Referring to the potential of the separation conductor "w,"** we define as:***

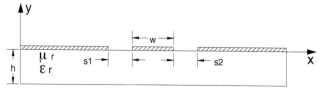

Figure 9.10.1

---

* See Chapter 1 for the even and odd excitation method of studying coupled lines.
** "w" is the width of the separation conductor. For simplicity, with "w" we will also name this conductor.
*** The following definitions a and b are not so general as for the case of Cμ or CS, studied in Chapters 5 and 6, respectively.

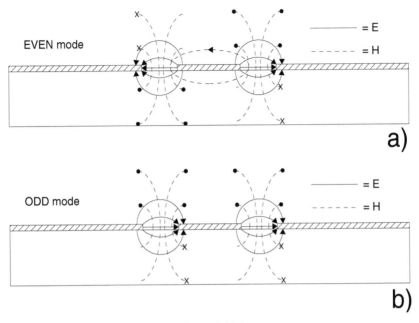

a)

b)

Figure 9.10.2

a. "Even" mode, that one for which "w" can be evaluated as equipotential,
b. "Odd" mode, that one for which "w" can be evaluated as not equipotential.

Note that the even mode also possesses some magnetic field lines that surround the separation conductor, and the e.m. field lines resemble those of the fundamental mode for "CPW."

A structure that can resemble a broad side coupled slot line, simply called "BCSL," is indicated in Figure 9.10.3. This "BCSL" has been studied, using a full wave analysis, by R. Janaswamy [64] but no closed formulas are available. This t.l. has not found applications in practice, since also the Vivaldi antenna, discussed in Section 9.5, cannot be considered a "BCSL" since the distance "s" is not constant.

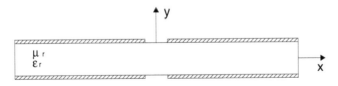

Figure 9.10.3

9.10.2 Analysis

One of the first studies on coupled slot lines was performed by J. B. Knorr and K. D. Kuchler.[65] They used a full wave analysis on this structure. In practice, they have evaluated the Poynting vector* "P_z," along "z," given by:

$$P_z = (\underline{e} \otimes \underline{h}) \cdot \underline{z}_0 \qquad (9.10.1)$$

* See Appendix A2 for Poynting vector definition.

with the assumption:

$$\underline{e} \equiv e_x \underline{x}_0 + e_z \underline{z}_0 \quad \text{and} \quad \underline{h} \equiv h_x \underline{x}_0 + h_z \underline{z}_0 \tag{9.10.2}$$

Then, to define the impedance "ζ"* of this structure they have used the relationship:

$$W = 0.5 v^2 / \zeta \tag{9.10.3}$$

where "v" is the peak voltage along the slot and "W" the mean "rf" power. This procedure is repeated for each mode, giving us the even "ζ_e" and odd "ζ_o" characteristic impedances. The solution of 9.10.3 is not simple since it requires the integration on a surface for "P_z" and along a line for "e." This procedure does not give a closed form expression for the evaluation of impedance. What these researchers have obtained are graphs as indicated in Figures 9.10.4 and 9.10.5, where we have reported "ζ_e" and "ζ_o," and "ε_{re}" and "ε_{ro}," respectively. Note that when "w" increases, the even and odd impedances have a common value as a limit since the two slots become decoupled.

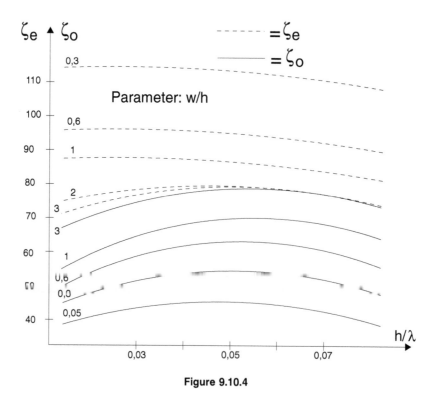

Figure 9.10.4

---

* Since two propagation modes are possible, we will have an even "ζ_e" and odd "ζ_o" impedance.

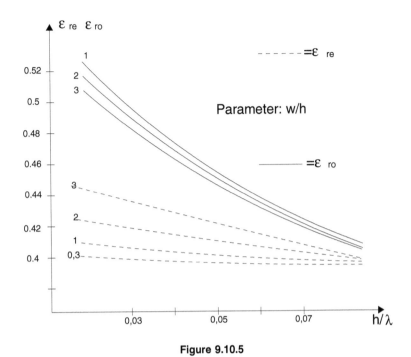

Figure 9.10.5

REFERENCES

1. S.B. Cohn, Slot lines on a dielectric substrate, *IEEE Trans. on MTT*, 768, Oct. 1969.
2. C. Qiang, V.F. Fusco, Three dimensional finite difference time domain slotline analysis on a limited memory personal computer, *IEEE Trans. on MTT*, 358, Feb. 1995.
3. J. Svacina, Dispersion characteristics of multilayered slot lines, a simple approach, *IEEE Trans. on MTT*, 1826, Sept, 1999.
4. J.W. Sheen, Y.D. Lin, Surface wave like mode in slot line, *IEEE MGWL*, 259, July 1996.
5. J.W. Sheen, Y.D. Lin, Propagation characteristics of the slotline first higher order mode, *IEEE Trans. on MTT*, 1774, Nov, 1998.
6. J. Zelchncr, J. Machac, M. Migliozzi, Upper cutoff frequency of the bound wave and new leaky wave on the slotline, *IEEE Trans. on MTT*, 378, Apr. 1998.
7. S.B. Cohn, Slot line field components, *IEEE Trans. on MTT*, 172, Feb. 1972.
8. S.B. Cohn, Slot line field components, *IEEE Trans. on MTT*, 172, Feb. 1972.
9. S.B. Cohn, Slot lines on a dielectric substrate, *IEEE Trans. on MTT*, 172, Feb. 1972.
10. J.B. Knorr, K.D. Kuchler, Analysis of coupled slots and coplanar strips on dielectric substrate, *IEEE Trans. on MTT*, 541, July 1975.
11. J. Citerne, S. Toutain, L. Raczy, Fundamental and higher order modes in microslot lines, Proc. V European Microwave Conference, 273, 1975.
12. T. Kitazawa, Y. Fujiki, Y. Hayashi, M. Suzuki, Slot line with thick metal coating, *IEEE Trans. on MTT*, 580, Sept. 1973.
13. M. Abramowitz, I.A. Stegun, *Handbook of Mathematical Functions*, Dover, New York, 358, 1970.
14. G.N. Watson, *A Treatise on the Theory of Bessel Functions*, Cambridge University Press, London, 1966.
15. A. Ghizetti, L.Marchetti, A. Ossicini, *Lezioni di complementi di matematica*, Editrice Veschi, 373, 1976.
16. M. Abramowitz, I.A. Stegun, *Handbook of Mathematical Functions*, Dover, New York, 358, 1970.
17. E. Jahnke, F. Emde, *Tables of Functions with Formulae and Curves*, Dover, New York, 242, 1945.

18. G. Matthaei, L. Young, E.M.T Jones, *Microwave Filters, Impedance Matching Networks, and Coupling Structures,* Artech House, Norwood, MA, 1980.

19. N. Marcuvitz, *Waveguide Handbook,* MIT Radiation Lab Series, 1950.

20. G. Gerosa, *Appunti di microonde,* Vol. IV, Univ. di Roma, Rome, 1980.

21. S.B. Cohn, Slot line field components, *IEEE Trans. on MTT,* 172, Feb. 1972.

22. R. Garg, K.C. Gupta, Expressions for wavelength and impedance of a slotline, *IEEE Trans. on MTT,* 532, Aug. 1976.

23. R. Janaswamy, D.H. Schaubert, Characteristic impedance of a wide slotline on low permittivity substrates, *IEEE Trans. on MTT,* 900, Aug. 1986.

24. T. Kitazawa, Y.Fujiki, Y. Hayashi, M. Suzuki, Slot line with thick metal coating, *IEEE Trans. on MTT,* 580, Sept. 1973.

25. G.H. Robinson, J.L. Allen, Slot line application to miniature ferrite devices, *IEEE Trans. on MTT,* 1097, Dec. 1969.

26. H.A. Wheeler, Formulas for the skin effect, *Proc. of the IRE,* 30, 412, 1942.

27. J.B. Knorr, J. Saenz, End effect in a shorted slot, *IEEE Trans. on MTT,* 579, Sept. 1973.

28. J.C. Goswami, R. Mittra, An application of FDTD in studying the end effects of slot line and coplanar waveguide with anisotropic substrates, *IEEE Trans. on MTT,* 1653, Sept. 1997.

29. G. Duchamp, L. Casadebaig, S. Gauffre, J. Pistre, An alternative method for end effect characterization in shorted slotlines, *IEEE Trans. on MTT,* 1793, Nov. 1998.

30. B.K. Kormanyos, W. Harokopus Jr., L.P.B. Katehi, GM. Rebeiz, CPW fed active slot antennas, *IEEE Trans. on MTT,* 541, Apr. 1994.

31. P.J. Gibson: The Vivaldi aerial, IX European µwave conference, 102, 1979.

32. R.A. Marino, A novel tapered slot PCS antenna array and model, *Microwave J.,* 90, Jan. 1999.

33. J.B. Knorr, Slot line transitions, *IEEE Trans. on MTT,* 548, May 1974.

34. Y.H. Choung, W.C. Wong, Microwave and millimeter wave slotline transition design, *Microwave J.,* 77, March 1994.

35. J.B. Knorr, Slot line transitions, *IEEE Trans. on MTT,* 548, May 1974.

36. Y.H. Choung, W.C. Wong, Microwave and millimeter wave slotline transition design, *Microwave J.,* 77, March 1994.

37. N.I. Dib, R.N. Simons, L.P.B. Katehi, Broadband uniplanar microstrip to slotline transitions, *Microwave Symp.,* 2, 683, 1995.

38. J. Sercu, N. Fache, F. Libbrecht, P. Lagasse, Mixed potential integral equation technique for hybrid microstrip slotline multilayered circuits using a mixed rectangular triangular mesh, *IEEE Trans. on MTT,* 1162, May 1995.

39. Y.H. Choung, W.C. Wong, Microwave and millimeter wave slotline transition design, *Microwave J.,* 77, March 1994.

40. K.C. Gupta, R. Garg and I.J.Bahl, *Microstrip Lines and Slot lines,* Artech House, Norwood, MA, 253, 1979.

41. L.W. Chua, New broadband matched hybrids for microwave integrated circuits, Proceeding of 1971 European µwave conference, presentation C4/5:1–C4/5:4.

42. L. Fan, C.H. Ho, S. Kanamaluru, K. Chang, Wideband, reduced size uniplanar magic-T, hybrid-ring, and de Ronde's CPW-slot couplers, *IEEE Trans. on MTT,* 2749, Dec. 1995.

43. L.E. Dickens, D.W. Maki, An integrated circuit balanced mixer, image and sum enhanced, *IEEE Trans. on MTT,* 276, March 1975.

44. J.A. Eisenberg, J.S. Panelli and W. Ou, Slotline and coplanar waveguide team to realize a novel MMIC double balanced mixer, *Microwave J.,* 123, Sept. 1992.

45. S.K. Masarweh, T.N. Sherer, K.S. Yngvesson, R.L. Gingras, C. Drubin, A.G. Cardiasmenos, J. Wolverton, Modeling of a monolithic slot ring quasi optical mixer, *IEEE Trans. on MTT,* 1602, Sept. 1994.

46. S.V. Robertson, L.P.B. Katehi, G. M. Rebeiz, A planar quasi optical mixer using a folded slot antenna, *IEEE Trans. on MTT,* 896, Apr. 1995.

47. E.A. Mariani, J.P. Agrios, Slot line filters and couplers, *IEEE Trans. on MTT,* 1089, Dec. 1970.

48. C.H. Ho, Lu Fan, K. Chang, Uniplanar de Ronde's CPW slot directional couplers, *Microwave Symp.,* 3, 1399, 1995.

49. E.A. Mariani, J.P. Agrios, Slot line filters and couplers, *IEEE Trans. on MTT,* 1089, Dec. 1970.

50. G. Duchamp, L. Casadebaig, S. Gauffre, J. Pistre, A new tool for slot microstrip transition simulation, *MGWL*, 276, Sept. 1997.

51. C.K.C. Tzuang, K.F. Fuh, and M. Mrozowski, Complex leaky waves of a partially open nonreciprocal slotline on gyromagnetic substrate, *Microwave Symp.*, 3, 1693, 1994.

52. C. S. Teoh, L.E. Davis, Normal mode analysis of coupled slots with an axially magnetized ferrite substrate, *Microwave Symp.*, 1, 99, 1995.

53. K.F. Fuh, C.K.C. Tzuang, The effects of covering on complex wave propagation in gyromagnetic slotlines, *IEEE Trans. on MTT*, 1100, May 1995.

54. L. Courtois, M. DeVecchis, A new class of non reciprocal components using slot line, *IEEE Trans. on MTT*, 511, June 1975.

55. L. Courtois, M. De Vecchis, A new class of non reciprocal components using slot line, *IEEE Trans. on MTT*, 511, June 1975.

56. G.H. Robinson, J.L. Allen, Slot line application to miniature ferrite devices, *IEEE Trans. on MTT*, 1097, Dec. 1969.

57. G.H. Robinson, J.L. Allen, Slot line application to miniature ferrite devices, *IEEE Trans. on MTT*, 1097, Dec. 1969.

58. G.H. Robinson, J.L. Allen, Slot line application to miniature ferrite devices, *IEEE Trans. on MTT*, 1097, Dec. 1969.

59. G.H. Robinson, J.L. Allen, Slot line application to miniature ferrite devices, *IEEE Trans. on MTT*, 1097, Dec. 1969.

60. G. Bock, New multilayered slot-line structures with high non reciprocity, *Electr. Lett.*, 966, Nov. 1983.

61. El Badawy El Sharawy, R.W. Jackson, Coplanar waveguide and slotline on magnetic substrates, analysis and experiment, *IEEE Trans. on MTT*, 1071, June 1988.

62. Y. D. Lin, Y.B. Tsai, Surface wave leakage phenomena in coupled slot lines, *MGWL*, 338, Oct. 1994.

63. T. Wang, K. Wu, Effects of various suspended mounting schemes on mode characteristics of coupled slotlines considering conductor thickness for wideband MIC applications, *IEEE Trans. on MTT*, May 1995.

64. R. Janaswamy, Even mode characteristics of the bilateral slotline, *IEEE Trans. on MTT*, 760, June 1990.

65. J.B. Knorr, K.D. Kuchler, Analysis of coupled slots and coplanar strips on dielectric substrate, *IEEE Trans. on MTT*, 541, July 1975.

CHAPTER 10

Coplanar Waveguides

10.1 GEOMETRICAL CHARACTERISTICS

The physical realization of a coplanar waveguide* is indicated in Figure 10.1.1. This t.l. was first studied by C.P. Wen.[1] It is important to note that if the word "waveguide" appears in this transmission line name, in this case, there is no similarity to the well-known "waveguides" discussed in Appendix A2. "CPW" is a full planar t.l., since in contrast to the microstrip case, here there is no bottom ground conductor.** From this point of view, the "CPW" is very similar to the slot line studied in Chapter 9. As shown in Figure 10.1.1, this t.l. is composed of a central conductor of width "w," separated from two lateral conductors by a distance "s" called the "slot." All the conductors of thickness "t," are placed on a dielectric slab of height "h" and dielectric and magnetic constant "e_r" and "μ_r." The extension "$w\ell_1$" and "$w\ell_2$" of the two lateral conductors is supposed to be infinite, but in practice is many times the length of the signal wavelength.***

According to the discussion in Appendix A2, the "CPW" has a zero cut-off frequency, but its low order propagation mode is indicated with "qTEM"**** because it is not a real "TEM." However, the error we make in evaluating the fundamental propagation mode as a pure "TEM" is negligible for frequencies up to some tens of GHz.[2,3,4] After this limit, dispersion arises and the propagation mode tends to be nearly a "TE,"***** with the magnetic field elliptically polarized along longitudinal planes. From this point of view, there is a big difference between microstrips and striplines studied in Chapters 2 and 3, while there is some similarity with the slot line propagation mode since this t.l. does not support a real "TEM" mode. Due to the elliptical magnetic field polarization, the "CPW" is a t.l. well suited to have energy exchange with ferrimagnetic materials, as we will show later.

Since the number of the electric and magnetic field lines in the air is higher than the number of the same lines in the microstrip case, the effective dielectric constant "ε_{re}" of "CPW" is typically 15% lower than the "ε_{re}" for microstrips. Consequently, the maximum reachable characteristic impedance values are higher than the microstrip values. In addition, to avoid field radiation in the air, it is very important to use substrates with a high dielectric constant, let us say from a value greater than 10, so that the e.m. field is mainly concentrated inside the dielectric.

A coplanar waveguide, together with a µstrip, is the most studied t.l. due to its "qTEM" propagation mode and its planar structure.[5,6,7,8]

---

* For simplicity the coplanar waveguide transmission line will be simply indicated with "CPW."
** The case of a bottom ground conductor will be studied later.
*** The effect of the limited extension of the lateral ground planes will be studied later.
**** "qTEM" means "quasi TEM."
***** The fact that the propagation mode always tends to be a "TE" mode depends on the natural disposition of the electric field inside the "CPW." Electric field lines are shown in the next section.

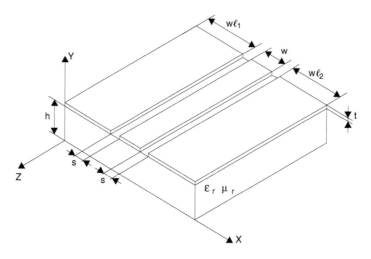

Figure 10.1.1

10.2 ELECTRIC AND MAGNETIC FIELD LINES

Some electric "e" and magnetic "h" field lines for the fundamental "qTEM" mode in "CPW" are indicated in Figure 10.2.1 in a defined cross-section and a defined time. In the fundamental mode, the lateral planes are at signal ground potential and for all the dimensions "w," the central conductor is equipotential. In this case, the "CPW" is an unbalanced t.l. Of course, this is not the only propagation mode for the "CPW" and other modes are possible that are also dependent on the particular feeding line. Later we will show other propagation modes for the "CPW."

The field lines indicated in Figure 10.2.1 are a simplification, especially for the magnetic field. In fact, since where an electric time varying field exists, a magnetic field also exists, then magnetic field lines must exist on the left and right of Figure 10.2.1, in the slots. These magnetic lines are not in the plane of the figure since "h" tries to close its field lines on the two lateral sides where the two ground conductors are placed. For this reason, with a "•" we indicate a vector exiting from the plane of the figure, while with an "x" we indicate a vector entering into this plane.

A more defined representation of the magnetic field lines for the fundamental "qTEM" mode is indicated in Figure 10.2.2. Note the field lines of coplanar waveguide are quite similar to those for the slot line case studied in Chapter 9, and in this case a "qTEM" propagation mode is possible.

Fundamental mode: "e" and "h" field lines.

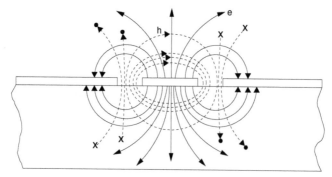

Figure 10.2.1

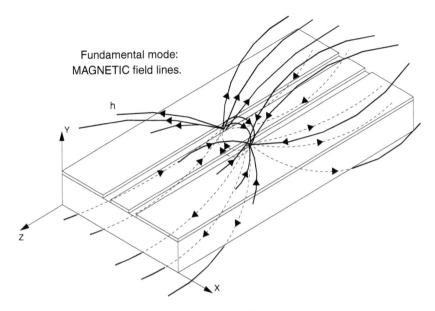

Fundamental mode:
MAGNETIC field lines.

Figure 10.2.2

10.3 SOLUTION TECHNIQUES FOR THE ELECTROMAGNETIC PROBLEM

As we have said in a previous section regarding other transmission lines, in this case the "CPW" can also be studied using quasi static methods*[10,11] or full wave methods: FWM.[12,13,14,15] While full wave methods are at the same time the most accurate tools for obtaining the t.l. characteristics and the most analytically extensive, q.s. methods are quite simple but do not threaten the dispersive nature of a generic t.l. Consequently, the approximation of the q.s. method becomes worse as the t.l. becomes dispersive. It is known that the error in the q.s. method increases if the t.l. does not support a "TEM" or "qTEM" mode. In our case the approximation is very good until 20 GHz and since "CPW" supports a "qTEM" mode. For this reason, we will study this line employing two simple q.s. methods, i.e., the "conformal transformations"** and "finite differences." However, we will mainly concentrate our attention on the the "conformal transformations" method since it is the simplest and most often used q.s. method.

a. Conformal Transformation Method: CTM

As we will study in Appendix A1, with the conformal transformation method the original geometric structure is transformed into a more simple one. With this method, the "CPW" is transformed in a structure that resembles a parallel plate capacitor, for which the capacitance is simply evaluated as $C = \varepsilon S/d$, where "S" is the area of the plate and "d" its distance. Then it is assumed that this parallel plate capacitor represents a "TEM" lossless t.l. with the same capacitance "C," but effective permittivity "ε_{re}" and effective permeability "μ_{re}" given by:

$$\varepsilon_{re} = \left(\varepsilon_r + 1\right)/2 \qquad (10.3.1)$$

$$\mu_{re} = \left(\mu_r + 1\right)/2 \qquad (10.3.2)$$

---

* Quasi static methods will be simply indicated with "q.s." methods. See Appendix A1 for fundamentals on resolution techniques for electrostatic problems.
** The "conformal transformation" is also called "conformal mapping."

i.e., the medium value between the air relative dielectric and magnetic constants and those of the substrate. The reader who is familiar with propagation of e.m. energy through ferrimagnetic materials can recognize that "μ_{re}" given by 10.3.2 is only a simplification. In fact, e.m. propagation inside such materials is, from a theoretical point of view, one of the most formidable branches of physics. We know that every ferrimagnetic material possesses an equivalent permeability that is dependent on many parameters, like reciprocal direction between e.m. energy propagation and direction of the applied magnetic field "H_{dc}," field intensity, signal frequency, ferrimagnetic composition, and more.* We will return to this topic when we discuss ferrimagnetic devices in "CPW" technology. For the moment, the expression of "μ_{re}" in the following formulas will remind us that the presence of any ferrimagnetic material will need to be evaluated.

If "v" is the velocity** of the wave in this "TEM" lossless line, from the value "C" of the capacitance by u.l., the line impedance "ζ" is evaluated by:***

$$\zeta \overset{\perp}{=} \left(L/C\right)^{0.5} \equiv 1/C\,v \tag{10.3.3}$$

where "v" is the light phase speed given by:

$$v \overset{\perp}{=} \left(LC\right)^{-0.5} \equiv v_0 \Big/ \left(\mu_{re}\varepsilon_{re}\right)^{0.5} \tag{10.3.4}$$

where "v_0" is the speed of light in a vacuum.

The application of this procedure to our case is indicated in Figure 10.3.1. The "CPW" is assumed with infinite substrate height "h" and negligible conductor thickness "t." In Figure 10.3.1a, the "CPW" is associated with a complex Cartesian coordinate system in the plane "W." Specializing the Schwarz-Christoffel transformation (see Appendix A1) to the structure in Figure 10.3.1a, we can transform the v < 0 half plane into the parallel plate capacitor indicated in Figure 10.3.1b, which lies in a complex plane "Z." The resulting transformation is given by the following relationship:[16]

$$\frac{dz}{dw} = \frac{A}{\left(w^2 - u_4^2\right)^{0.5}\left(w^2 - u_5^2\right)^{0.5}} \tag{10.3.5}$$

where "A" is a constant. With reference to Figure 10.3.1b, note as the capacitance "C_b"**** per u.l. of this structure is given by:

$$C_b = 2\varepsilon a_5/d \tag{10.3.6}$$

The ratio "a_5/d" is given by:[17]

$$a_5/d = K(p)/K(p') \tag{10.3.7}$$

where "K(p)" is the complete elliptic integral of the first kind, defined in Appendix A8, and the parameters "p" and "p'" are defined as:*****

---

* See Appendix A7 for fundamental information on energy exchange between waves and ferrimagnetic materials.
** In quasi static methods the t.l. are evaluated as nondispersive, so that group and phase velocity coincides.
*** See Chapter 1 for t.l. parameters definitions.
**** The subscript "b" is used to remind we refer to the bottom side of the "CPW" indicated in Figure 10.3.1 a.
***** Sometimes in literature "K(p')" is indicated with "K'(p)." This is only a different symbology, since operatively the integral is evaluated for p'=(1-p²)⁰·⁵.

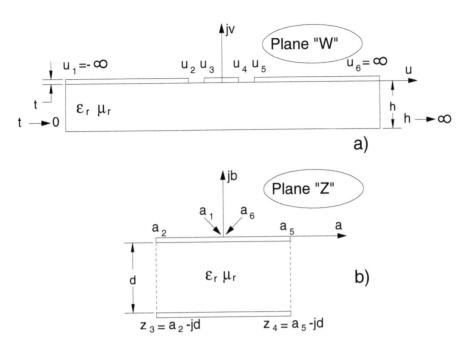

Figure 10.3.1

$$p \overset{\perp}{=} u_4/u_5 \equiv w/(w+2s) \quad \text{and} \quad p' \overset{\perp}{=} \left(1-p^2\right)^{0.5} \tag{10.3.8}$$

Using 10.3.6 and 10.3.7, the capacitance "C_b" per u.l. of the transmission line indicated in Figure 10.3.1b is:

$$C_b = 2\varepsilon_0\varepsilon_r K(p)/K(p') \tag{10.3.9}$$

At this point, the conformal transformation is repeated for the $v > 0$ half plane and, proceeding similarly as we did for the case $v < 0$, the capacitance "C_t"* per u.l. of the transmission line indicated in Figure 10.3.1b with "ε_r" replaced with "ε_0" is:

$$C_t = 2\varepsilon_0 K(p)/K(p') \tag{10.3.10}$$

Once we know "C_b" and "C_t," the "CPW" will have an associated capacitance "C" per u.l. given by:

$$C = C_b + C_t = 2\varepsilon_0\left(1+\varepsilon_r\right) K\,(p)/K(p') \tag{10.3.11}$$

So, using the previous equation in the general definition of the effective relative dielectric constant for the quasi static case,** we have:

$$\varepsilon_{re} \overset{\perp}{=} C/C_0 \equiv \left(\varepsilon_r + 1\right)/2 \tag{10.3.12}$$

* The subscript "t" is used to remind us that we refer to the top side of the "CPW."
** See for example Section 2.4 of Chapter 2.

Inserting 10.3.11 in 10.3.3 and performing simple calculations, we obtain the "CPW" t.l. impedance:

$$\zeta = 30\pi K\left(p'\right)\left(\mu_{re}\right)^{0.5} \Big/ \left(\varepsilon_{re}\right)^{0.5} K\left(p\right) \tag{10.3.13}$$

The effect of finite substrate height "h" has been studied by M. E. Davis, E. W. Williams, and A. C. Celestini,[18] always using conformal mapping. From their studies we have the result that the simple conformal transformation we did disagrees with the values they obtained; within some percent up to h > 2s, while for h = s the error reaches 10%.

The effect of a lower ground plane[19] in the "CPW" was first investigated by G. Ghione and C. Naldi,[20] applying two conformal transformations to the structure indicated in Figure 10.3.2. The characteristic impedance for this t.l. is given by:

$$\zeta = 60\pi\left(\mu_{re}\right)^{0.5} \Big/ \left(\varepsilon_{re}\right)^{0.5} \left[R\left(p\right) + R\left(p_1\right)\right] \tag{10.3.14}$$

where:

$$\varepsilon_{re} = \frac{1 + \varepsilon_r R\left(p_1\right)\big/R\left(p\right)}{1 + R\left(p_1\right)\big/R\left(p\right)} \tag{10.3.15}$$

$$R\left(p\right) \doteq K\left(p\right)\big/K\left(p'\right), \quad R\left(p_1\right) \doteq K\left(p_1\right)\big/K\left(p_1'\right) \tag{10.3.16}$$

$$p_1 \doteq \tanh\left(\pi w/4h\right)\big/\tanh\left[\pi\left(w + 2s\right)/4h\right], \quad p_1' \doteq \left(1 - p_1^2\right)^{0.5} \tag{10.3.17}$$

The effect of the lower ground plane is, among other things,* to decrease the impedance of the original "CPW" indicated in Figure 10.3.1.

Note that the conformal transformations methods always assume the thickness "t" of the strip conductors to be zero. In the next section we will see the effects caused by t > 0 on the "CPW" characteristics.

b. Finite Difference Method: FDM

As we can see in Appendix A1, with the finite difference methods the original geometric structure is evaluated enclosed in a metallic box so that all the e.m. fields can be considered to be inside the box. This geometric situation is indicated in Figure 10.3.3.

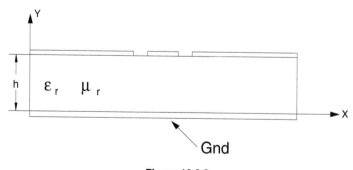

Figure 10.3.2

---

* We will return later to the effects of a bottom ground plane in an "CPW."

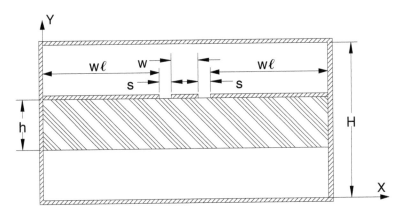

Figure 10.3.3

The general procedure is to obtain the potential "V(x,y,z)" inside this structure and then, with a simple derivative operation, extract the electric field "$\underline{E}$" from the well-known[21,22] relationship* $\underline{E} = -\underline{\nabla}V$. Once the electric field is found, the charge "Q" on the central conductor can be found using the electrostatic Gauss[23] law:

$$Q = \varepsilon \oint_s \underline{E} \bullet \underline{n} dS \qquad (10.3.18)$$

where "S" is a surface surrounding the internal conductor and "$\underline{n}$" is the normal to this surface and directed outside. At this point, the capacity of the t.l. is simply evaluated by the well-known relationship C = Q/V. Then using 10.3.1 through 10.3.4, the t.l. characteristics can be obtained.

Results obtained with this analysis method[24] have produced an error below 3% with respect to the measured values.

We can conclude this section observing that the "CTM" permits closed forms of "CPW," while the "FDM" cannot do this, as we will see later. This last method is more general and accurate than the "CTM."

10.4 CLOSED FORM EQUATIONS FOR "CPW" CHARACTERISTIC IMPEDANCE

The closed forms we give here have been obtained using the "CTM" described in the previous section. For all the "CPW" dimensions, we will refer to Figure 10.4.1, assuming the dimensions "$w\ell_1$" and "$w\ell_2$" to be infinite. Since "CTM" assumes the conductors to be of zero thickness, some correction needs to be made.

To consider the metal thickness "t" we can use the theory employed in Chapter 2 and associate to a metal thickness "t" an extra width "dw" given by:

$$dw = \frac{1.25t}{\pi}\left[1 + \ln\left(\frac{4\pi w}{t}\right)\right] \qquad (10.4.1)$$

So, we define the following new notations:

* See Appendix A8 for delta operator "$\underline{\nabla}$" definition.

$$w_t = w + dw, \quad s_t = s - dw \tag{10.4.2}$$

and the parameter "p" now becomes:

$$p \rightarrow p_t \stackrel{\pm}{=} w_t / (w_t + 2s_t) \quad \text{and} \quad p_t' \stackrel{\pm}{=} (1 - p_t^2)^{0.5} \tag{10.4.3}$$

A better expression of "ε_{re}" with respect to the value $(\varepsilon_r + 1)/2$ has been obtained with a curve fitting procedure performed by I. J. Bahl[25] on the measurement values obtained by M. E. Davis, E. W. Williams, and A. C. Celestini,[26] resulting in:

$$\varepsilon_1 \stackrel{\pm}{=} \tanh\left[1.785\ln(h/s) + 1.75\right] \tag{10.4.4}$$

$$\varepsilon_2 \stackrel{\pm}{=} (ps/h)\left[0.04 - 0.7p + 0.01(1 - 0.1\varepsilon_r)(0.25 + p)\right] \tag{10.4.5}$$

$$\varepsilon_{re} \stackrel{\pm}{=} \left[(\varepsilon_r + 1)/2\right](\varepsilon_1 + \varepsilon_2) \tag{10.4.6}$$

where "p" is defined by 10.3.8. Note how equation 10.4.6 includes the limited extension of "h." The effect of "t" on "ε_{re}" was first evaluated by T. Kitazawa, Y. Hayashi, and M. Suzuki[27,28,29] and were closed form fitted by K. C. Gupta, R. Garg, and I. J. Bahl.[30] We obtain:

$$\varepsilon_{ret} = \left[(\varepsilon_r + 1)/2\right](\varepsilon_1 + \varepsilon_2) - \frac{0.7(\varepsilon_{re} - 1)t/s}{K(p)/K(p') + 0.7t/s} \tag{10.4.7}$$

At this point we have all the relationships required to calculate the characteristic impedance and dielectric constant of our "CPW." In fact, the characteristic impedance is given by 10.3.13 with the insertion of the previous equation and using the parameter "p_t" defined in 10.4.3. Specifically, we have:

$$\zeta = 30\pi K(p_t')(p_{re})^{0.3} / (\varepsilon_{re})^{0.5} K(p_t) \tag{10.4.8}$$

where "ε_{re}" is given by 10.4.7. In a lot of these formulas, the ratio of the complete elliptic integral of the first kind appears. This ratio can be calculated using the approximated expressions given in Appendix A8.

The value of "ζ" and "ε_{re}" produced by the previous formulas has an accuracy better than 2% compared with the measurement results of M. E. Davis, E. W. Williams, and A. C. Celestini,[31] whenever the condition

$$h/s \geq 1 \tag{10.4.9}$$

is satisfied. In practice, equation 10.4.9 is quite often used.

Other closed form expressions, based on "CTM" with the assumption t = 0, have been obtained by T. Q. Deng, M. S. Leong, P. S. Kooi, and T. S. Yeo[32] and compared with "FWM." The result is that when the condition 10.4.9 is verified, the formulas are very accurate up to 20 to 30 GHz, and dispersion can be neglected.

10.5 CLOSED FORM EQUATIONS FOR "CPW" ATTENUATION

As many other t.l.s, the "CPW" losses are due to three causes:

1. Nonperfect conductivity of the conductors, or "conductor loss"
2. Dielectric nonzero conductivity and dumping phenomena
3. Substrate magnetic loss, if the substrate is a ferrimagnetic material
4. Radiation

In this section we will study how to evaluate the first two causes of losses, which are directly related to the geometry of the "CPW" indicated in Figure 10.1.1. Magnetic losses are mainly due to damping phenomena inside ferrimagnetic material and, if the signal frequency is of appropriate value, to resonance absorption.* Radiation losses are strongly depedendent on the surrounding structure near the "CPW" and cannot simply be treated in a general way.

For the present case, if we consider the "CPW" as a t.l. only supporting a "TEM" mode, we can apply the theory developed by H.A. Wheeler.[33,34] The procedure is similar to that used for the μstrip case in Chapter 2, and for this reason, here we will only outline the differences with respect to μstrips. First, consider the attenuation due the nonperfect conductivity of the conductors. In these conditions, we know** that the e.m. energy will penetrate inside the conductors and at the "penetration depth -p-" given by:

$$p \doteq \left(\pi f \mu_c g\right)^{-0.5} \qquad [\text{meters***}] \qquad (10.5.1)$$

the field amplitudes are reduced by "1/e." In Equation 10.5.1, "f" is the signal frequency, and "g" and "μ_c" are the conductor conductivity and absolute permeability, respectively. The effect of each penetration can be regarded as the introduction of an additional series inductance and resistance**** per u.l., indicated respectively with "L_i" and "R_i" and called respectively "incremental inductance" and "incremental resistance." The situation is depicted in Figure 10.5.1. If "L" is the equivalent series inductance per u.l. of the lumped equivalent t.l. for the "CPW," given by:*****

$$L = \zeta / v = \left(\mu_{re}\varepsilon_{re}\right)^{0.5} \Big/ v_0 \qquad [\text{Henry/meter}] \qquad (10.5.2)$$

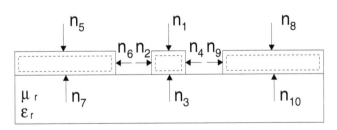

Figure 10.5.1

* See Appendix A7 for energy exchange phenomena between e.m. signal and ferrite.
** See Appendix A2 for e.m. energy penetration inside nonperfect conductors.
*** Remember that unless otherwise stated, in this book we will use the MKSA unit reference system.
**** Since we assume the "CPW" only to support a "TEM" mode, we are referring to the simple low pass equivalent network for a line supporting a "TEM" mode. This topic is covered in Chapter 1.
***** See Chapter 1 for relationships between t.l. characteristics quantities.

each "L_i" and "R_i" are given by:

$$L_i = \frac{p\mu_{cr}}{2} \frac{\partial L}{\partial n} \qquad \text{[Henry/meter]} \qquad (10.5.3)$$

$$R_i \overset{\perp}{=} \omega L_i = \pi f p \mu_{cr} \frac{\partial L}{\partial n} \qquad \text{[}\Omega\text{/meter]} \qquad (10.5.4)$$

where "μ_{cr}" is the conductor's relative permeability and "∂n" is an infinitesimal penetration inside the conductor, positive when the vector "$\underline{n}$" is directed into the conductor. If we define the quantity "R_s," called the "sheet resistance" for the conductor, as:

$$R_s \overset{\perp}{=} \pi f p \mu_c \equiv \left(\pi f \mu_c / g\right)^{0.5} \qquad \text{[}\Omega\text{/square]}* \qquad (10.5.5)$$

Equation 10.5.4 becomes:

$$R_i = \frac{R_s}{\mu_0} \frac{\partial L}{\partial n} \qquad (10.5.6)$$

We now have to take into account all the incremental inductances and resistances, to obtain the whole additional inductance "L_a" and resistance "R_a." With reference to Figure 10.5.1 and using 10.5.3 and 10.5.6, we have:

$$L_a = \sum_{j=1}^{j=10} \left(L_i\right)_j = \frac{1}{2\mu_0} \sum_{j=1}^{j=10} p_j \left(\mu_c\right)_j \frac{\partial L}{\partial n_j} \qquad (10.5.7)$$

$$R_a = \sum_{j=1}^{j=10} \left(R_i\right)_j = \frac{1}{\mu_0} \sum_{j=1}^{j=10} \left(R_s\right)_j \frac{\partial L}{\partial n_j} \qquad (10.5.8)$$

The conductor attenuation coefficient "α_c"** is defined as:

$$\alpha_c = w_c / 2W_t \qquad (10.5.9)$$

where "W_c" and "W_t" are respectively the mean power dissipated in the conductor and the mean transmitted power, given by:

$$W_c = R_a |i|^2, \quad W_t = \zeta |i|^2 \qquad (10.5.10)$$

Using 10.5.8 and 10.5.10, Equation 10.5.9 becomes:

$$\alpha_c = \frac{1}{2\mu_0 \zeta} \sum_{j=1}^{j=10} \left(R_s\right)_j \frac{\partial L}{\partial n_j} \qquad (10.5.11)$$

---

* See Appendix A2 for dimension unit of "conductor resistance."
** We are assuming a longitudinal variation of conductor attenuation with $e^{-\alpha_c \zeta}$. See Chapter 1 for fundamental theory of transmission lines.

From 10.5.2 we note that:

$$\frac{\partial L}{\partial n} = \frac{1}{v_0} \frac{\partial\left[\left(\mu_{re}\varepsilon_{re}\right)^{0.5}\zeta\right]}{\partial n} \perp \frac{1}{v_0} \frac{\partial\zeta_z}{\partial n} \qquad (10.5.12)$$

and so Equation 10.5.11 becomes:

$$\alpha_c = \frac{1}{\zeta 240\pi} \sum_{j=1}^{j=10} \left(R_s\right)_j \frac{\partial\zeta_z}{\partial n_j} \qquad (10.5.13)$$

Since the characteristic impedance "ζ" is also a function of "w," "s," and "t,"* as was shown in the previous paragraph, the derivative "$\partial\zeta_z/\partial n$" is:

$$\frac{\partial\zeta_z}{\partial n} = \frac{1}{dn}\left(\frac{\partial\zeta_z}{\partial s}ds + \frac{\partial\zeta_z}{\partial w}dw + \frac{\partial\zeta_z}{\partial t}dt\right) \qquad (10.5.14)$$

From Figure 10.5.1 we observe that:

$$ds = 2dn, \quad dt = -2dn, \quad dw = -2dn \qquad (10.5.15)$$

and 10.5.13 becomes:

$$\alpha_c = \frac{1}{\zeta 120\pi} R_s\left(\frac{\partial\zeta_z}{\partial s} - \frac{\partial\zeta_z}{\partial w} - \frac{\partial\zeta_z}{\partial t}\right) \qquad (10.5.16)$$

Inserting the expression of "ζ" given in the previous section, I.J. Bahl[35] has obtained the following expression of "α_c" which gives a value in dB per u.l.:

$$\alpha_c = 4.88 * 10^{-4} R_s \varepsilon_{re} \zeta \frac{K^2(p)Q}{K^2(p')\pi s}\left(1 + \frac{w}{s}\right)\frac{N}{D} \quad [db/u.l.] \qquad (10.5.17)$$

where the parameters "p" and "p'" are defined by 10.3.8, and:

$$Q = p\bigg/\left[1 - \left(1 - p^2\right)^{0.5}\right]\left(1 - p^2\right)^{3/4} \quad \text{for} \quad 0 \le p \le 1/\sqrt{2} \qquad (10.5.18)$$

$$Q = \left[K(p')/K(p)\right]^2\big/(1 - p)p^{0.5} \quad \text{for} \quad 1/\sqrt{2} \le p \le 1 \qquad (10.5.19)$$

Of course, in 10.5.17 and 10.5.19 the proper values of the ratio must be inserted between the complete elliptic integrals of the first kind as given in 10.4.10 and 10.4.11. The quantities "N" and "D" are defined as follows:

$$N = \frac{1.25}{\pi}\ln\left(\frac{4\pi w}{t}\right) + 1 + \frac{1.25t}{\pi w} \qquad (10.5.20)$$

* In this theory substrate height "h" is assumed to be theoretically infinite.

$$D = \left\{ 2 + \frac{w}{s} - \frac{1.25t}{\pi s} \left[1 + \ln\left(\frac{4\pi w}{t}\right) \right] \right\}^2$$ (10.5.21)

Note that the "u.l." dimension in 10.5.17 is given by the quantity "s," i.e., the spacing of the lateral ground planes as indicated in Figure 10.1.1.

The incremental inductance rule has been verified to give very accurate results for conductor thicknesses greater than four times "p." This condition is usually verified for every planar transmission line since for the typical conductors, using the value of "p" is lower than some micrometer for frequencies greater than 1 GHz.*

Dielectric loss can be evaluated as has been done for microstrip in Chapter 2, resulting in the same expression:

$$\alpha_d = \frac{20\pi}{\ln 10} \frac{1/\varepsilon_{re} - 1}{1/\varepsilon_r - 1} \frac{\tan\delta}{\lambda_0} \sqrt{\varepsilon_{re}}$$ (10.5.22)

which gives a value of dB/u.l. In the previous equation, "ε_{re}" is given by 10.4.5 through 10.4.7 and "tanδ" is the dielectric "tangent delta."** Of course, the quantities "ε_r" and "ε_{re}" are all relative to the real part of the substrate dielectric constant.*** Also in this case, remember that to evaluate magnetic losses, the "α_d" expression can be formally modified multiplying by $\sqrt{\mu_{re}}$. However, from the discourse in section 10.3 we know that such multiplication is quite often only a notational simplification. In the next section we will study some "CPW" devices that use the RF interaction with ferrimagnetic materials.

Due to the increasing use of "CPW" in microwave circuits, losses in these t.l.s. are currently a topic of study.[16,37]

10.6 CONNECTIONS BETWEEN "CPW" AND OTHER LINES

We think it is important to discuss the most suitable and simple transitions between "CPW" and the other most often used transmission lines. Of course, every transition introduces a discontinuity, and a study of such discontinuities[38,39,40] can help in deciding which transition is better. In general, a transition is evaluated as acceptable when the resulting reflection coefficient is typically below 10 dB.

a. Connection with Coaxial Cable

This transition is indicated in Figure 10.6.1, where we have represented a top view. For a good transition it is important that the outer conductor of the coaxial cable be short connected to the two ground planes of the "CPW." The center conductor of the coaxial cable is of course short connected to the center conductor of the "CPW."

b. Connection with Microstrip

Two transitions are possible between these two lines. The first one is indicated in Figure 10.6.2. This transition is characterized as having the hot conductors of these two lines on the opposite surfaces of the dielectric substrates. The two hot conductors are connected together through a conducting hole, called a "via hole." With this transition, the two lines share the same ground layer.

---

* See Appendix A2 for values of penetration depth inside good conductors.
** See Chapter 1 for "tanδ" definition.
*** See Chapter 1 for complex permittivity definition.

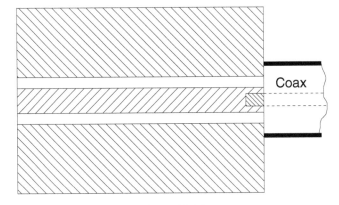

Figure 10.6.1

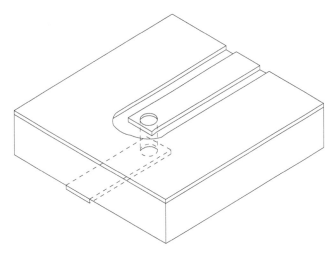

Figure 10.6.2

The second transition is indicated in Figure 10.6.3. This transition is characterized as having the cold conductors of these two lines on the opposite surfaces of the dielectric substrates. It is important that the ground conductor be connected together at the same potential. The two cold conductors are connected together using "via holes." With this transition, the two lines share the same hot conductor. Note the tapering for the cold conductors.

In section 10.3 we discussed the possibility of having a "CPW" with an opposite ground conductor, as indicated in Figure 10.3.2. We will return in the next section to this topic, but for the moment we say that whenever the "CPW" has a bottom ground conductor, the transition in 10.6.3 can be simplified avoiding the tapering in the bottom conductor.

c. Connection with Slot Line

This transition is indicated in Figure 10.6.4, where we have represented a top view. This connection, along with the others previously shown, is the most simple since both the "CPW" and slot line are uniplanar lines. For this reason, "CPW-slotline" transition is currently investigated to improve performances.[41,42,43]

This transition has the characteristic that the signal coming from the slot line doesn't excite the fundamental mode toward the "CPW" while the fundamental mode coming from the "CPW" cannot propagate toward the slot line. This situation is indicated in Figure 10.6.5, respectively in parts a and b. Note that in the case of Figure 10.6.5a the signal generated inside the "CPW" is such that

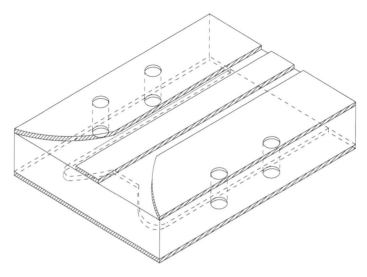

Figure 10.6.3

the center conductor is not equipotential. If we use the "potential" function along the dimension "w," we can recognize how in part a, the potential is an odd function for a reference center in "w/2." For this reason, we will note the electric field lines distribution in the "CPW" indicated in Figure 10.6.5 as corresponding to an "ODD" mode. Conversely, the electric field line distribution corresponding to the fundamental mode will be called the "EVEN" mode since the potential function is even along "w" for a reference center in "w/2." In the next section we will show an application that uses such transition.

Another possible transition between the "CPW" and slot line is indicated in Figure 10.6.6. Comparing this transition with that shown in Figure 10.6.4 we can observe how in Figure 10.6.4 the coupling between the lines can be regarded as an "electrical" one, while for Figure 10.6.6 the coupling can be regarded as "magnetic." Of course, for every transition that involves the "CPW" it is important to assure good continuity to every ground and hot common conductor. This can be accomplished using via holes whenever possible. Note that in the case of Figure 10.6.6, the connection between the two lateral ground planes of the "CPW" to assure the equipotentiality can be performed with an air bridge, a typical technology for "MIC" and "MMIC."

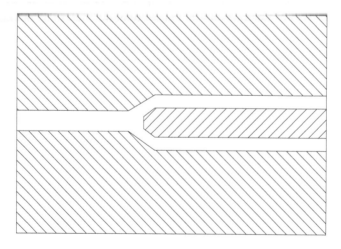

Figure 10.6.4

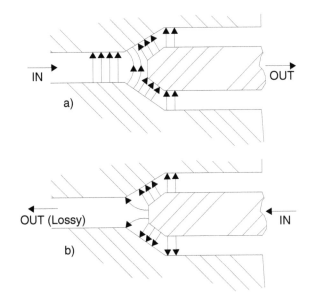

Figure 10.6.5

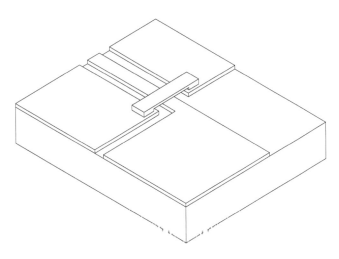

Figure 10.6.6

10.7 TYPICAL NONFERRIMAGNETIC DEVICES USING "CPW"

We think it is important to discuss some devices that can also be built employing "CPWs" quite often giving better results with respect to the use of other t.l.s.

a. Mixers

The use of the "CPW" in mixers was treated in Chapter 9 when introducing the most important slot line networks. For this reason we refer the reader back to Chapter 9. Note that since both these lines are planar, this transition is the most attractive.

Such full planar devices have been proved to work well inside the millimetric wavelength region.[44,45]

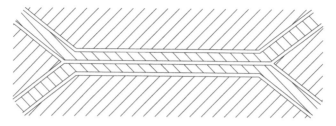

Figure 10.7.1

b. Directional Couplers

One of the first experiments on directional couplers employing "CPWs" was made by C.P. Wen.[46] The resulting network is shown in Figure 10.7.1. Coupled "CWGs" of the type indicated in Figure 10.7.1 are said to be side coupled "CPWs," which we will abbreviate with "SCCPW."

This is a single λ/4 step directional coupler. The operating bandwidth reported by Wen is higher than a single λ/4 step directional coupler in microstrip technology. This can be explained noting that even and odd modes* have phase velocities that are closer with respect to the phase velocities of even and odd modes in microstrip because in the "CPW," the number of field lines in the air is more comparable to the number of field lines in the dielectric. In fact, for microstrips the greatest number of field lines is in the dielectric, since the field is "attracted" by the bottom conductor, while for the "CPW," the field line is "attracted" by the lateral conductor, as shown in Figure 10.2.1. Defining the operating bandwidth as the frequency range where the coupling is below 10% from the designed center frequency value, the performances of the Wen coupler are as indicated below:

RF bandwidth:	1.5 to 2.5 GHz
Mean coupling:	11 dB
Directivity:	>15 dB
Insertion loss:	1 dB max

Of course, more than one section can be used for the coupler as we have shown in the previous chapters. Approximating the propagation mode as a pure "TEM" mode, the tables given in Chapter 7 can be used for this purpose.

Interesting is a work presented by F. Hanna,[47] where the directional coupler indicated with a top view in Figure 10.7.1 is evaluated with a ground plane at the bottom. This structure is depicted with a cross-sectional view in Figure 10.7.2 and is called a "side coupled coplanar waveguide with ground" or simply "SCCPWG." If we define the following parameters:

$$y \overset{\perp}{=} s/(s+2w), \quad p \overset{\perp}{=} (s+2w)/(s+2w+2d), \quad \beta \overset{\perp}{=} \left[\left(1-y^2\right)/\left(1-p^2y^2\right)\right]^{0.5} \qquad (10.7.1)$$

for the odd mode case we have:

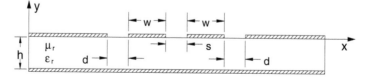

Figure 10.7.2

---

* See Chapter 1 for even and odd mode theory used in coupled lines.

$$a_o \stackrel{\perp}{=} 0.5 \operatorname{senh}^2\left[\left(\pi/2h\right)\left(0.5s+w+d\right)\right] \tag{10.7.2}$$

$$c_o \stackrel{\perp}{=} \operatorname{senh}^2\left[\left(\pi/2h\right)\left(0.5s+w\right)\right]-a_o \tag{10.7.3}$$

$$b_o \stackrel{\perp}{=} \operatorname{senh}^2\left(\pi s/4h\right)-a_o \tag{10.7.4}$$

$$p_o \stackrel{\perp}{=} a_o \frac{\left(a_o^2-b_o^2\right)^{0.5}-\left(a_o^2-c_o^2\right)^{0.5}}{b_o\left(a_o^2-c_o^2\right)^{0.5}+c_o\left(a_o^2-b_o^2\right)^{0.5}} \tag{10.7.5}$$

$$\varepsilon_{reo} = \left[2\varepsilon_r \frac{K(p_o)}{K(p_o')}+\frac{K(\beta)}{K(\beta')}\right] \Big/ \left[2\frac{K(p_o)}{K(p_o')}+\frac{K(\beta)}{K(\beta')}\right] \tag{10.7.6}$$

$$\zeta_o = 120\pi/\varepsilon_{reo}^{0.5}\left[2\frac{K(p_o)}{K(p_o')}+\frac{K(\beta)}{K(\beta')}\right] \tag{10.7.7}$$

For the even mode case we have:

$$a_e \stackrel{\perp}{=} 0.5 \cosh^2\left[\left(\pi/2h\right)\left(0.5s+w+d\right)\right] \tag{10.7.8}$$

$$c_e \stackrel{\perp}{=} \operatorname{senh}^2\left[\left(\pi/2h\right)\left(0.5s+w\right)\right]-a_e+1 \tag{10.7.9}$$

$$b_e \stackrel{\perp}{=} \operatorname{senh}^2\left(\pi s/4h\right)-a_e+1 \tag{10.7.10}$$

$$p_e \stackrel{\perp}{=} a_e \frac{\left(a_e^2-b_e^2\right)^{0.5}-\left(a_e^2-c_e^2\right)^{0.5}}{b_e\left(a_e^2-c_e^2\right)^{0.5}+c_e\left(a_e^2-b_e^2\right)^{0.5}} \tag{10.7.11}$$

$$\varepsilon_{ree} = \left[2\varepsilon_r \frac{K(p_e)}{K(p_e')}+\frac{K(\beta p)}{K\left[(\beta p)'\right]}\right] \Big/ \left[2\frac{K(p_e)}{K(p_e')}+\frac{K(\beta p)}{K\left[(\beta p)'\right]}\right] \tag{10.7.12}$$

$$\zeta_e = 120\pi/\varepsilon_{ree}^{0.5}\left[2\frac{K(p_e)}{K(p_e')}+\frac{K(\beta p)}{K\left[(\beta p)'\right]}\right] \tag{10.7.13}$$

In all the previous formulas the quantity "$K()$" represents the complete elliptic integral of the first kind, already defined in Section 10.5. In addition, any primed quantity "α'" is related to the nonprimed quantity "α" by the known relationship $\alpha' = (1-\alpha^2)^{0.5}$ Using the tables in Chapter 7 for coupled "TEM" t.l.s, the previous formulas give us the ability to synthesize such a coupler. The author suggests that the condition:

$$h \geq s+2(w+d) \tag{10.7.14}$$

must be satisfied to assure a coupled coplanar waveguide mode. CPW directional couplers are very interesting devices, currently being investigated[48,49] which may have better performances.

Modes inside a "SCCPW" will also be discussed in the next section.

Figure 10.7.3

c. Filters

J.K.A. Everard and K.M.K. Cheng[50] have built a "CPW" band pass filter centered at 4.6 GHz. The filter is composed of six sections of the shape indicated in Figure 10.7.3 connected together. The result of their work shows that the radiation of this filter is much lower than a typical band pass filter built with coupled microstrip lines.*

In addition, the "Q" of each resonator indicated in Figure 10.7.3 is higher than the "Q" of microstrip resonators, resulting in a lower loss and sharper edged filter.

Note that the narrow shunt lines connected to the lateral ground planes represent a shunt inductance while the connecting lines perform a impedance transformation.

Also coplanar waveguide filters are currently being theoretically investigated,[51,52,53] especially with the aim of having an accurate synthesis procedure applicable to this t.l.

What we have reported in this section are only some examples of devices where "CPWs" can be used with performances at least comparable to the same devices built with other t.l.s. "CPWs" have also found applications in MMIC amplifiers[54,55,56] and other MIC devices, passive[57] and active,[58] always giving good performances.

10.8 MAGNETIZATION OF "CPW" ON FERRIMAGNETIC SUBSTRATES

As a result of the discussion in Section 10.2 we know that an elliptic polarization of the magnetic field exists near the region of the "CPW" slots, with the plane of polarization oriented longitudinally. The existence of this polarization can be used to build nonreciprocal ferrimagnetic devices like isolators or phase shifters,**[59] as we will study in the next section. The complex topic of energy interaction between e.m. energy and ferrimagnetic materials is discussed in Appendices A5, A6, and A7. The reader who is not familiar with these topics can read these Appendices to review the fundamentals of this branch of physics.

---

* See Chapter 5 for typical microstrip networks.
** Isolators are always nonreciprocal devices, while phase shifters can also be, in general, reciprocal devices.

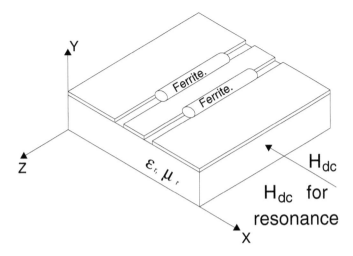

Figure 10.9.1

The applications of those concepts to the "CPW" case can be done as we did in Chapter 9 for the slot line, since both t.l.s have magnetic elliptical polarization in practically the same positions. So, we refer the reader back to Chapter 9 to review the direction in which a "CPW" can be magnetized.

10.9 "CPW" ISOLATORS

The "CPW" isolators can be built using two energy exchange phenomenas, i.e., the "resonance" and the "field displacement" effects.*

10.9.1 Resonance Isolator

A "CPWs" resonance isolator can be built as indicated in Figure 10.9.1. Near the two slots of the "CPW," two cylinders of ferrimagnetic material are set. An external static magnetic field "H_{dc}" uniformly magnetizes the two cylinders, centered** at the signal frequency at which the isolation is desired. The two cylinders can be also substituted with a single slab of ferrite*** positioned above the two slots.

The same isolation property can be achieved with the isolator configuration shown in Figure 10.9.2, where all the "CPW" is built on a ferrimagnetic material.

One experiment[60] on an isolator of the type shown in Figure 10.9.1 has given an attenuation value of 37 dB centered at 6 GHz and a maximum insertion loss of 2 dB.

Of course, some matching is required to adjust the input impedance of these devices to the usual 50Ω since their imput impedance is usually lower than this value.

10.9.2 Field Displacement Isolator

An f.d. "CPW" isolator can be realized as indicated in Figure 10.9.3. A "CPW" is built on ferrite, and at the other face, an absorbtive material is attached. A static magnetic field "$\underline{H}_{dc}$" is applied orthogonally in the direction of propagation "$\underline{z}_0$."

---

* See Appendix A7 for fundamentals on these two important effects of energy exchange.
** By "centered" we mean that the "H_{dc}" intensity is just what is required to have a precession frequency motion equal to that signal frequency we want to stop. See Appendix A7 for more deep insight into ferrimagnetic energy exchange.
*** "Ferrite" is a particular ferrimagnetic material, similar to "YIG." We will use this term to indicate a generic ferrimagnetic material.

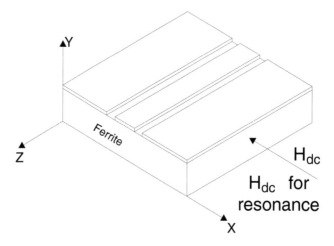

Figure 10.9.2

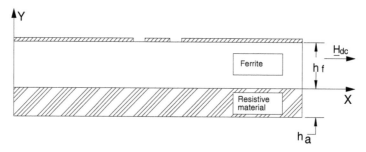

Figure 10.9.3

The f.d. "CPW" phase shifter can be studied in a manner similar to that used in Chapter 9 for slot lines. In this case it is also necessary to assume some simplifications to the geometry indicated in Figure 10.9.3, in order to use the theory we developed in Appendix A7. In particular we assume:

1. All the e.m. energy is contained inside the ferrite
2. In the slots, only the components "h_y" and "h_z" exist, and they don't vary with the coordinate "x,"* i.e., the fields are uniform
3. Only a TE mode is supported by the "CPW"

With these hypotheses, we refer the reader back to Chapter 9 since the same analytical procedure can be used here.

So, an f.d. isolator can be graphically represented as in Figure 10.9.4. In part a we have indicated the electrical field lines when the field displacement causes the e.m. energy to be mainly on the top surface, i.e., the conductor's surface. This situation is different from the typical nonferrimagnetic "CPW," where the field lines are mainly concentrated in the dielectric. From the equations we gave before, this situation happens when "$\underline{H}_{dc}$" is directed along "$\underline{x}_0$" and propagation is along "$\underline{z}_0$." Figure 10.9.4b represents what happens when, with respect to part a of the figure, we change the propagation direction or "$\underline{H}_{dc}$" direction. From this figure it is easy to understand that if we attach an "RF" absorbtive material on the bottom of the "CPW," a signal attenuation will arise in the direction in which the field travels on this surface. Consequently, when a change of sign is made to "$\underline{H}_{dc}$" or to the propagation direction, then no loss will be added by this material since signal propagation is on the opposite surface. Of course, if we move the resistive material on the

* Using the well-known notation described in Appendix A2 and Chapter 1, this hypothesis means $k_x=0$.

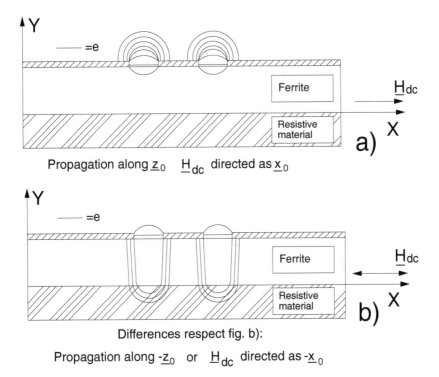

Propagation along $\underline{z}_0$ $\underline{H}_{dc}$ directed as $\underline{x}_0$

Differences respect fig. b):

Propagation along $-\underline{z}_0$ or $\underline{H}_{dc}$ directed as $-\underline{x}_0$

Figure 10.9.4

conductor's surface we always have a field displacement isolator, but now we have reversed the attenuation direction.

Field displacement isolators are wider bandwidth than the resonance isolators, and need a lower intensity of "H_{dc}." They do not reach the isolation values of the resonance counterpart.

However, in general, devices that only perform isolation are not very widely used, regardless of the technology by which they are realized, and terminated circulators are preferred due to their high versatility. Whether "CPW" circulators are theoretically possible is not reported in the literature. This is because the microstrip circulator* is a very simple device with good electrical characteristics and the necessity for another technology is not needed.

10.10 "CPW" FERRIMAGNETIC PHASE SHIFTERS

Two groups of "CPW" ferrimagnetic phase shifters** are possible: one of which uses the different phase constants "β_c" and "β_d" between the "concordant" and "discordant" waves,*** and one that uses the field displacement effect. We will call the first device a "Discon" phase shifter and the second one a "field displacement" phase shifter.

With respect to their microstrip or stripline counterparts, "CPW" phase shifters have the advantage that circular polarization doesn't need to be artificially created. In fact, we know that for microstrip or stripline, circular polarization is obtained with a meander line, but, while all the meander line length is responsible for attenuation, only 65% of its length is responsible for circular polarization. This circumstance results in a higher length for meander line devices with respect to a "CPW" p.s. Meander line stripline or microstrip p.s.s have higher absolute values of figure of

* See Chapter 5 for microstrip circulator technology.
** We will use the letters "p.s." to indicate a "phase shifter."
*** See Appendix A7 for definitions of "concordant" and "discordant" waves.

merit than the corresonding "CPW" counterpart, but "CPW" devices usually have smaller dimensions for the same electrical parameters.

10.10.1 "Discon" Phase Shifter

Many configurations and reported values are available for this kind of phase shifter.[61,62,63] The common physical principle is the different value of "β_c" and "β_d" and their dependence on "$\underline{H}_{dc}$" and direction of propagation. These two phase constants are related to the elements of the ferrite permeability matrix by the relationships:

$$\beta_d \overset{\perp}{=} \omega\left(\mu_0\varepsilon\right)^{0.5}\left(\mu_p - \mu_\ell\right)^{0.5} \tag{10.10.1}$$

$$\beta_c \overset{\perp}{=} \omega\left(\mu_0\varepsilon\right)^{0.5}\left(\mu_p + \mu_\ell\right)^{0.5} \tag{10.10.2}$$

One possible construction of a "discon" p.s. is indicated in Figure 10.10.1. Note that the situation is practically identical to that indicated in Figure 10.9.1. The presence of a ferrite slab, as in this case, or ferrite cylinders, gives no magnetic difference. The very important difference is that for the isolator case, the "H_{dc}" value is for resonance, while in this case we are not in resonance. Caution must be taken to not bring ferrite in the field displacement region using a wrong value of "H_{dc}."* For a "discon" p.s., typical "H_{dc}" values have lower intensity than the required values for f.d. effect.** Of course a p.s. can also be realized as indicated in Figure 10.9.2, again with "H_{dc}" out of resonance.

An interesting alternative to the planar p.s. structure is the parallelepipedal type indicated in Figure 10.10.2. This device uses the ferrite residual magnetization (see Appendix A7). A "CPW" is built on a surface of hollow ferrite parallelepiped. Inside the cavity a conductor wire "W" is inserted. When a current pulse "I" passes inside the wire, a magnetic field is generated*** which, if of proper intensity, can permanently magnetize the ferrite with a residual field "H_{dc}." Changing the direction of "I" is equivalent to changing the direction of "$\underline{H}_{dc}$." Devices working on residual magnetization are called "latching devices." Note that the latching p.s. indicated in Figure 10.10.2 having a closing path for "H_{dc}" does not disperse this static field much.

Experiments[64] on a "discon" p.s. shown in Figure 10.10.1 working in the frequency range 5 to 7 GHz have given differential phase shift**** "$\Delta\varphi$" values near 30°/cm and figures of merit***** near 120°/dB. Other experiments[65] on the "discon" p.s. composed as shown in Figure 10.9.2, with "$\underline{H}_{dc}$" not in resonance, have given differential phase shift values near 25°/cm and figures of merit near 200°/dB for a signal working at 10 GHz. The latching p.s. indicated in Figure 10.10.2 has given differential phase shift values near 8°/cm,[66] quite low with respect to the nonlatching "discon" we have just reported.

10.10.2 Field Displacement Phase Shifter

The operating principles of a field displacement p.s. are exactly the same as those we gave regarding the f.d. isolator. So, Figure 10.9.4 can represent the construction of a f.d. phase shifter, but for the proper operation it is necessary to replace the slab of absorbent material with a low loss

---

* The proper values of "H_{dc}" depend by the ferrite parameters inserted in the figures of "β_c" and "β_d" we gave in Appendix A7.
** Theoretically a "discon" p.s. can also work with "H_{dc}" values above resonance, but such a high value is only employed for high RF power devices.
*** Fundamental theory of magnetic fields generated by current in conductors is reviewed in Appendix A6.
**** "Differential phase shift" is defined in Appendix A7.
***** See Appendix A7 for p.s. quality parameters definitions.

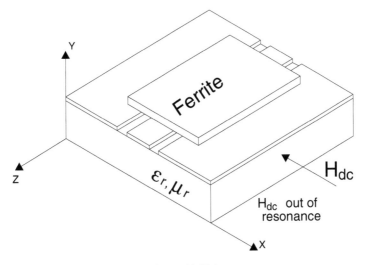

Figure 10.10.1

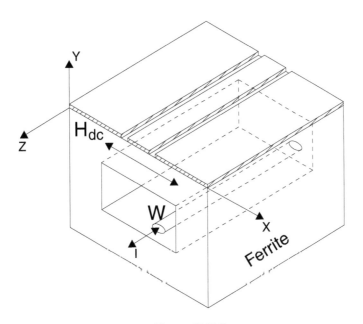

Figure 10.10.2

dielectric slab with permittivity "$\varepsilon_{r\ell}$" as indicated in Figure 10.10.3. In fact, since the phase constant "β_z" along "z" is given by:

$$\beta_z = \omega\left(\mu_0\mu_p\varepsilon_0\varepsilon_{re}\right)^{0.5} \qquad (10.10.3)$$

then when the wave is guided by the edge where the dielectric slab is disposed, the value of "ε_{re}" will be:

$$\varepsilon_{re} \approx \left(\varepsilon_f + \varepsilon_{r\ell}\right)/2 \qquad (10.10.4)$$

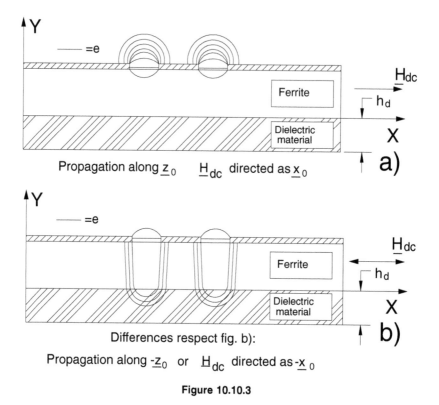

Figure 10.10.3

while when the wave is guided by the opposite edge, the value of "ε_{re}" will be:

$$\varepsilon_{re} \approx \left(\varepsilon_f + 1\right)/2 \qquad (10.10.5)$$

resulting in a differential phase shifting. In 10.10.4 and 10.5.5 "ε_f" is the ferrite permittivity. Note that if we move the added dielectric slab on the conductor's surface, we always have a field displacement p.s., but now we have reversed the phase shifting sign.

Experiments[67] on an f.d.p.s. of the type shown in Figure 10.10.3 has given $\Delta\varphi \approx 30°/cm$ and $\Delta\varphi \approx 45°/cm$ at 10 GHz, respectively, with $\varepsilon_{rf} = 16$ and $\varepsilon_{rd} = 25$ both with $h_d - 1mm$.

10.11 PRACTICAL CONSIDERATIONS

The "CPW" t.l. shown in Figure 10.1.1 is seldom employed as it is indicated. In fact, this t.l. necessarily has to be connected with other t.l.s which requires at least a bottom ground plane* like a microstrip or stripline. In addition, the extension "$w_{\ell1}$" and "$w_{\ell2}$" of the two lateral conductors is seldom as long as required by the ideal definition of "CPW." For these reasons, we think it is important to discuss these two practical aspects of "CPW" use.

10.11.1 The "CPW" with Bottom Ground Conductor

This practical situation is indicated in Figure 10.11.1 with a transverse view, which we will refer to as "CPWG."** Part a represents only the electrical field lines for simplicity, where we

---

* This text discusses planar t.l. For this reason, other transitions are not considered.
** This structure is also called "conductor backed CPW" (CBCPW).

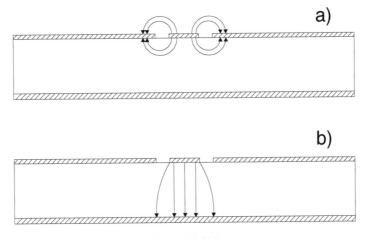

Figure 10.11.1

have indicated the fundamental "CPW" mode, while in part b another possible mode indicates the microstrip mode. Signal propagation through more than one mode is not desired since this causes signal loss and distortion at the receiver end.*[68,69] So if the receiver is built to detect the fundamental "CPW" mode, the other modes must be avoided. This can be performed by setting the bottom ground plane at a distance "h" so that:

$$h > 10(w + 2s) \qquad (10.11.1)$$

We have already studied this case in Section 10.3 where we gave the characteristic impedance formulas.

10.11.2 "CPW" with Bottom Ground Conductor and Lateral Planes with Limited Extension

This situation is represented in Figure 10.11.2. Some researchers[70] have studied this t.l. as the case of three coupled microstrips. The result of their studies is that the coupled microstrip mode can be suppressed if "h" and "w_g" are such that:

$$h > 10(w + 2s) \quad \text{and} \quad w_g > 10(w + 2s) \qquad (10.11.2)$$

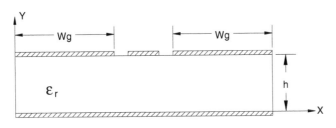

Figure 10.11.2

* See Appendix A2 for this important aspect of information transmission.

Of course, other modes are possible, for example the coupled slot mode.* The rise of this mode can be reduced by connecting the two ground conductors with an air bridge, especially if the t.l. is many wavelengths long. In this last case, more than one connection should be used.

Studies have also been performed on "CPW" without bottom conductors but with lateral ground of limited extension.[71,72]

10.12 COUPLED COPLANAR WAVEGUIDES

Coupled "CPWs" are seldom employed in planar transmission line circuits. In practice, there is no advantage to using such coupled line structures instead of other coupled t.l.s we have studied in this text like microstrips or striplines. In previous sections we have seen some applications of coupled "CPW" like directional couplers.

Since in Chapter 9 we have studied the coupled coplanar slot lines, we think it is useful to discuss some similarities between these two structures.

10.12.1 General Characteristics

The geometric structure of the side coupled "CPW," simply abbreviated with "SCCPW," is depicted in Figure 10.12.1. If w1 = w2 and s1 = s3 this structure is said to be a "symmetrical SCCPW," abbreviated as "SSCCPW," otherwise it is said to be an "asymmetrical SCCPW" sometimes called an "ASCCPW."

Do not confuse this structure with the "SCCPWG" indicated in Figure 10.7.2. Note in fact that in this case no bottom ground conductor is present. In practice, the two center conductors "w_1"** and "w_2" of each otherwise isolated "CPW" are set close together at a distance "s_2." As in many cases of the coupled line theory, the e.m. field lines can be assumed to be composed of the superposition of the field distribution of the even and odd excitation.*** The field distributions for these excitations, also called "modes," are schematically reported in Figure 10.12.2, respectively in parts a and b. Referring to the potential of the internal conductors we define as:

a. "Even" mode that one for which "w_1" and "w_2" have equal potential
b. "Odd" mode that one for which "w_1" and "w_2" have potentials with the opposite sign. Note that the even mode also possesses some magnetic field lines that surround the internal conductors.

A particular structure, which in some way can be regarded as broadside coupled coplanar waveguides, simply called "BCCPW," is indicated in Figure 10.12.3. No practical application of "BCCPW" reported in the literature is known by the author.

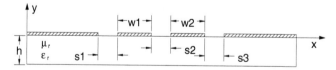

Figure 10.12.1

---

* Coupled slot modes are discussed with more detail in Chapter 9.
** "w_1" and "w_2" is the width of each center conductor. For simplicity, we will also name these conductors "w_1" and "w_2".
*** See Chapter 1 for the even and odd excitation method to study coupled lines.

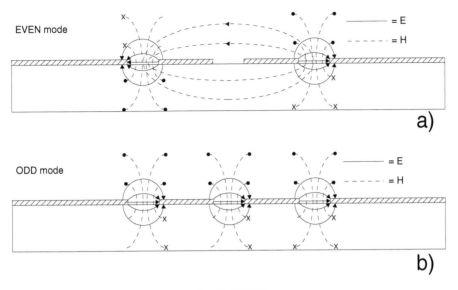

Figure 10.12.2

Figure 10.12.3

10.12.2 Analysis

One of the first studies on "SSCCPW" was done by C.P. Wen.[73] As in the single "CPW," he has applied the "CTM" to the structure indicated in Figure 10.12.1. T. Hatsuda[74] has used the finite difference method to study this structure, while T. Kitazawa, Y. Hayashi, and R. Mittra[75] have also studied the case for nonisotropic dielectrics. In any case, no closed formulas are available for even and odd characteristics.

With reference to Figure 10.12.1, for the case of $w_1 = w_2 \perp w$ and $s_1 = s_3 \perp s$ Hatsuda has given some graphs for even "ζ_e" and odd "ζ_o" characteristic impedances, as indicated in Figure 10.12.4. Losses in "SSCCPW" have been studied by G.Ghione and M.Goano[76] through a "CTM," while "ASCCPWG" has been investigated by K.K.M.Cheng.[77]

S. S. Bedair and I. Wolff[78] have studied an enclosed "BCCPW" structure, using "CTM" indicated in Figure 10.12.5. The closed form equations for the t.l. characterisitics of the even and odd modes, with reference to the symbols indicated in that figure, are as follows:

for the odd mode:

$$C_o \doteq C_{o1} + C_{o2} \tag{10.12.1}$$

$$p_{oi} = \tanh\left(\pi w / 4 h_i\right) / \tanh\left[\pi (w + 2s)/4h_i\right] \quad \text{with } i = 1, 2 \tag{10.12.2}$$

$$C_{oi} = 2\varepsilon_0 \varepsilon_{ri} \frac{K\left(p_{oi}\right)}{K\left(p_{oi}{}'\right)} \quad \text{with } i = 1, 2 \tag{10.12.3}$$

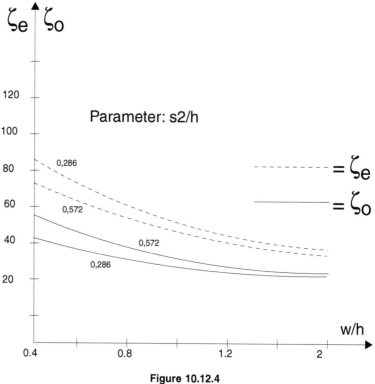

Figure 10.12.4

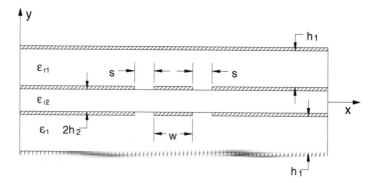

Figure 10.12.5

For the even mode:

$$C_e \stackrel{\pm}{=} C_{o1} + C_{e2} \tag{10.12.4}$$

$$p_e = \operatorname{senh}(\pi w/4h_2)/\operatorname{senh}[\pi(w+2s)/4h_2] \tag{10.12.5}$$

$$C_{e2} = 2\varepsilon_0\varepsilon_{r2}\frac{K(p_e)}{K(p'_e)} \tag{10.12.6}$$

Then, the characteristic impedances and effective dielectric constants are given by:

$$\zeta_m = 1 \Big/ \left[v_0 \big(C_m C_{0m} \big)^{0.5} \right] \qquad (10.12.7)$$

$$\varepsilon_{rem} = C_m / C_{0m} \qquad (10.12.8)$$

where m = e,o, "v_0" is the light speed in the free space, and "C_{0m}" are given by 10.12.1 and 10.12.4 when 10.12.3 and 10.12.6 are evaluated with $\varepsilon_{ri} = 1$.

REFERENCES

1. C. P. Wen, Coplanar waveguide: a surface strip transmission line for nonreciprocal gyromagnetic device application, *IEEE Trans. on MTT*, 1087, Dec. 1969.
2. S. S. Bedair, I. Wolff, Fast, accurate and simple approximate analytic formulas for calculating the parameters of supported coplanar waveguides for MMIC's, *IEEE Trans. on MTT*, 41, Jan. 1992.
3. J. W. Huang, C. K. C. Tzuang, Mode coupling avoidance of shielded conductor backed coplanar waveguide (CBCPW) using dielectric lines compensation, *Microwave Symp.*, 1, 149, 1994.
4. J. Hesselbarth, R. Vahldieck, Leakage suppression in coplanar waveguide circuits by patterned backside metallization, *Microwave Symp.*, 1999, session WE1A.
5. R. Spickermann, N. Dagli, Experimental analysis of millimeter wave coplanar waveguide slow wave structures on GaAs, *IEEE Trans. on MTT*, 1918, Oct. 1994.
6. V. Milanovic, M. Ozgur, D. C. DeGroot, J. A. Jargon, M. Gaitan, and M. E. Zaghloul, Characterization of broad band transmission for coplanar waveguides on CMOS silicon substrates, *IEEE Trans. on MTT*, 632, May 1998.
7. K. J. Herrick, T. A. Schwarz, L. P. B. Katehi, Si-micromachined coplanar waveguides for use in high frequency circuits, *IEEE Trans. on MTT*, 762, June 1998.
8. A. Dehe, H. Klingbeil, C. Weil, H. L. Hartnagel, Membrane supported coplanar waveguides for MMIC and sensor application, *MGWL*, 185, May 1998.
9. M. Hotta, Y. Qian, T. Itoh, Efficient FDTD analysis of conductor backed CPWs with reduced leakage loss, *IEEE Trans. on MTT*, 1585, Aug. 1999.
10. G. Ghione, M. Goano, M. Pirola, Exact, conformal mapping models for the high frequency losses of coplanar waveguides with thick electrodes of rectangular or trapezoidal cross section, *Microwave Symp.*, 1999, session WEF2.
11. Jean Fu Kiang, Quasi TEM analysis of coplanar waveguides with an inhomogeneous semiconductor substrate, *IEEE Trans. on MTT*, 1586, Sept. 1996.
12. J. B. Knorr, K. D. Kuchler, Analysis of coupled slots and coplanar strips on dielectric substrate, *IEEE Trans. on MTT*, 541, July 1975.
13. J. B. Davies, D. Mirshekar-Syahkal, Spectral domain solution of arbitrary coplanar transmission line with multilayer substrate, *IEEE Trans. on MTT*, 143, Feb. 1977.
14. K. K. M. Cheng, and I. D. Robertson, Numerically efficient spectral domain approach to the quasi TEM analysis of supported coplanar waveguide structures, *IEEE Trans. on MTT*, 1958, Oct. 1994.
15. W. Schroeder, I. Wolff, Full wave analysis of the influence of conductor shape and structure details on losses in coplanar waveguide, *Microwave Symp.*, 3, 1273, 1995.
16. C. P. Wen, Coplanar waveguide: a surface strip transmission line for nonreciprocal gyromagnetic device application, *IEEE Trans. on MTT*, 1087, Dec. 1969.
17. C. P. Wen, Coplanar waveguide: a surface strip transmission line for nonreciprocal gyromagnetic device application, *IEEE Trans. on MTT*, 1087, Dec. 1969.
18. M. E. Davis, E. W. Williams, A. C. Celestini, Finite boundary corrections to the coplanar waveguide analysis, *IEEE Trans. on MTT*, 594, Sept. 1973.
19. K. K. M Cheng, Effect of conductor backing on the line to line coupling between parallel coplanar lines, *IEEE Trans. on MTT*, 1132, July 1997.
20. G. Ghione and C. Naldi, Parameters of coplanar waveguides with lower ground plane, *Electr. Lett.*, 734, Sept. 1983.

21. S. Ramo, J. R. Whinnery, T. Van Duzer, *Fields and Waves in Communication Electronics,* John Wiley & Sons New York, 1965.

22. G. Barzilai, Fondamenti di elettromagnetismo, Siderea editor, 75, 1975.

23. S. Ramo, J. R. Whinnery, T. Van Duzer, *Fields and Waves in Communication Electronics,* John Wiley & Sons New York, 1965.

24. T. Hatsuda, Computation of coplanar type strip line characteristics by relaxation method and its application to microwave circuits, *IEEE Trans. on MTT,* 795, Oct. 1975.

25. K. C. Gupta, R. Garg and I. J. Bahl, *Microstrip Lines and Slotlines*, Artech House, Norwood, MA, 275, 1979.

26. M. E. Davis, E. W. Williams, A. C. Celestini, Finite boundary corrections to the coplanar waveguide analysis, *IEEE Trans. on MTT,* 594, Sept. 1973.

27. T. Kitazawa, Y. Hayashi and M. Suzuki, A coplanar waveguide with thick metal coating, *IEEE Trans. on MTT,* 604, Sept. 1976.

28. J. Y. Ke, C. H. Chen, The coplanar waveguides with finite metal thickness and conductivity, *Microwave Symp.,* 3, 1681, 1994.

29. J. Y. Ke, C. H. Chen, Dispersion and attenuation characteristics of coplanar wave guides with finite metalization thickness and conductivity, *IEEE Trans. on MTT,* 1128, May 1995.

30. K. C. Gupta, R. Garg and I. J. Bahl, *Microstrip Lines and Slotlines*, Artech House, Norwood, MA, 275, 1979.

31. M. E. Davis, E. W. Williams, A. C. Celestini, Finite boundary corrections to the coplanar waveguide analysis, *IEEE Trans. on MTT,* 594, Sept. 1973.

32. T. Q. Deng, M. S. Leong, P. S. Kooi and T. S. Yeo, Synthesis formulas simplify coplanar waveguide design, *Microwaves and RF,* 84, March 1997.

33. H. A. Wheeler, Formulas for the skin effect, *Proc. of the IRE,* 30, 412, 1942.

34. R. Sturdivant, Transmission line conductor loss and the incremental inductance rule, *Microwave J.,* 156, Sept. 1995.

35. K. C. Gupta, R. Garg and I. J. Bahl, *Microstrip Lines and Slotlines*, Artech House, Norwood, MA, 275, 1979.

36. L. Vietzorreck, W. Pascher, Modeling of conductor loss in coplanar circuit elements by the method of lines, *IEEE Trans. on MTT,* 2474, Dec. 1997.

37. G. Ghione, M. Goano, The influence of ground plane width on the ohmic losses of coplanar waveguides with finite lateral ground planes, *IEEE Trans. on MTT,* 1640, Sep. 1997.

38. K. Beilenhoff, W. Heinrich, Excitation of the parasitic parallel plate line mode at coplanar discontinuities, *Microwave Symp.,* 3, 1789, 1997.

39. F. L. Lin, R. B. Wu, Analysis of coplanar waveguide discontinuities with finite metallization thickness and nonrectangular edge profile, *IEEE Trans. on MTT,* 2131, Dec. 1997.

40. R. Sturdivant, C. Quan, J. Wooldridge, Transitions and interconnects using coplanar waveguide and other three conductor transmission lines, *Microwave Symp.,* 1, 235, 1996.

41. Y. S. Lin, C. H. Chen, Novel lumped element coplanar waveguide to slotline transitions, *Microwave Symp.,* 1999, session TH1B.

42. K. P. Ma, T. Itoh, A new broadband coplanar waveguide to slotline transition, *Microwave Symp.,* 3, 1627, 1997.

43. K. P. Ma, Y. Qian, T. Itoh, Analysis and applications of a new CPW-Slotline transition, *IEEE Trans. on MTT,* 426, Apr. 1999.

44. M. Schefer, U. Lott, H. Benedickter, Hp. Meier, W. Patrick, W. Bachtold, Active, monolithically integrated coplanar V band mixer, *Microwave Symp.,* 2, 1043, 1997.

45. L. Verweyen, H. Massler, M. Neumann, U. Schaper, W. H. Haydl, Coplanar integrated mixers for 77 GHz automotive applications, *MGWL,* 38, Jan. 1998.

46. C. P. Wen, Coplanar waveguide directional couplers, *IEEE Trans. on MTT,* 318, June 1970.

47. V. F. Hanna, Parameters of coplanar directional couplers, Proc. 15 Europ. μwave Conf., 820, 1985. An evident typographic error in expression (8) of this article has been corrected here.

48. C. H. Ho, Lu Fan, Kai Chang, New uniplanar coplanar waveguide couplers, *Microwave Symp.,* 1, 285, 1994.

49. P. Pieters, S. Brebels, W. De Raedt, E. Beyne, Broadband coplanar couplers in multilayer thin film MCMD technology, *Microwave Symp.,* 1999, session TH4B.

50. J. K. A. Everard and K. M. K. Cheng, High performance direct coupled bandpass filters on coplanar waveguide, *IEEE Trans. on MTT,* 1568, Sept. 1993.

51. F. L. Lin, Chien Wen Chiu, Ruey Beei Wu, Coplanar waveguide bandpass filter: a ribbon-of-brick-wall design, *IEEE Trans. on MTT,* 1589, July 1995.

52. R. Kulke, I. Wolff, Design of passive coplanar filters in V-band, *Microwave Symp.,* 3, 1647, 1996.

53. A. Vogt, W. Jutzi, An HTS narrow bandwidth coplanar shunt inductively coupled microwave bandpass filter on $LaAlO_3$, *IEEE Trans. on MTT,* 492, Apr. 1997.

54. M. Riaziat, S. Bandy, G. Zdasiuk, Coplanar waveguides for MMICS, *Microwave J.,* 125, June 1987.

55. K. Minot, B. Nelson, W. Jones, A low noise phase linear distributed coplanar waveguide amplifier, *IEEE Trans. on MTT,* 1650, Sept. 1993.

56. D. L. Edgar, H. McLelland, S. Ferguson, N. I. Cameron, M. Holland, I. G. Thayne, M. R. S. Taylor, C. R. Stanley, S. P. Beaumont, 94 and 150 GHz coplanar waveguide MMIC amplifiers realized using InP technology, *Microwave Symp.,* 1999, session MO3C.

57. C. H. Ho, L. Fan, and K. Chang, New uniplanar coplanar waveguide hybrid ring couplers and magicT, *IEEE Trans. on MTT,* 2440, Dec. 1994.

58. M. Houdart, Coplanar lines: applications to broadband microwave integrated circuits, Proc. of the 6th European Micr. Conf., Roma, 49, 1976.

59. L. Zhou, L. H. Davis, FEM analysis of nonreciprocity of a coplanar waveguide with a transversely magnetised ferrite layer, *Microwave Symp.,* 2, 1109, 1997.

60. C. P. Wen, Coplanar waveguide: a surface strip transmission line for nonreciprocal gyromagnetic device application, *IEEE Trans. on MTT,* 1087, Dec. 1969.

61. C. P. Wen, Coplanar waveguide: a surface strip transmission line for nonreciprocal gyromagnetic device application, *IEEE Trans. on MTT,* 1087, Dec. 1969.

62. El Badawy El Sharawy, R. W. Jackson, Coplanar waveguide and slotline on magnetic substrates: analysis and experiment, *IEEE Trans. on MTT,* 1071, June 1988.

63. S. K. Koul, B. Bhat, *Microwave and Millimeter Wave Phase Shifters,* vol. I, Artech House, Norwood, MA, 1991.

64. C. P. Wen, Coplanar waveguide: a surface strip transmission line for nonreciprocal gyromagnetic device application, *IEEE Trans. on MTT,* 1087, Dec. 1969.

65. El Badawy El Sharawy, R. W. Jackson, Coplanar waveguide and slotline on magnetic substrates: analysis and experiment, *IEEE Trans. on MTT,* 1071, June 1988.

66. El Badawy El Sharawy, R. W. Jackson, Coplanar waveguide and slotline on magnetic substrates: analysis and experiment, *IEEE Trans. on MTT,* 1071, June 1988.

67. El Badawy El Sharawy, R. W. Jackson, Coplanar waveguide and slotline on magnetic substrates: analysis and experiment, *IEEE Trans. on MTT,* 1071, June 1988.

68. Yaozhong Liu, Kimin Cha, T. Itoh, Non leaky coplanar (NLC) waveguides with conductor backing, *IEEE Trans. on MTT,* 1067, May 1995.

69. Yu De Lin, Jyh Wen Sheen, Surface wave leakage of coplanar wave guide with nearby back conductor plane, *Microwave Symp.,* 3, 1701, 1994.

70. M. Riaziat, I. J. Feng, R. Majidi-Ahy, B. A. Auld, Single mode operation of coplanar waveguides, *Electr. Lett.,* 1281, Nov. 1987.

71. F. Brauchler, S. Robertson, J. East. L. P. B. Katehi, W-band finite ground coplanar (FGC) line circuit elements, *Microwave Symp.,* 3, 1845, 1996.

72. Lei Zhu, Ke Wu, Unified CAD oriented circuit model of finite ground coplanar waveguide gap structure for uniplanar M(H)MICS, *Microwave Symp.* 1999, session MO1D.

73. C. P. Wen, Coplanar waveguide directional couplers, *IEEE Trans. on MTT,* 318, June 1970.

74. T. Hatsuda, Computation of coplanar type strip line characteristics by relaxation method and its application to microwave circuits, *IEEE Trans. on MTT,* 795, Oct. 1975.

75. T. Kitazawa, Y. Hayashi, R. Mittra, Asymmetrical coupled coplanar type transmission lines with anisotropic substrates, IEE Proc., 265, Aug. 1986.

76. G. Ghione, M. Goano, A closed form CAD oriented model for the high frequency conductor attenuation of symmetrical coupled coplanar waveguides, *IEEE Trans. on MTT,* 1065, July 1997.

77. K. K. M. Cheng, Characteristics of asymmetrical coupled lines of a conductor backed coplanar type, *IEEE Trans. on MTT,* 460, Mar. 1997.

78. S. S. Bedair, I. Wolff, Fast and accurate analytic formulas for calculating the parameters of a general broadside coupled coplanar waveguide for (M)MIC applications, *IEEE Trans. on MTT,* 843, May 1989.

CHAPTER 11

Coplanar Strips

11.1 GEOMETRICAL CHARACTERISTICS

The physical realization of a coplanar strip* is indicated in Figure 11.1.1. This t.l. was first studied by C.P. Wen.[1] It is realized by setting two conductor strips of width "w_1" and "w_2" in close proximity supported by a dielectric of thickness "h." Note that on the other side of the dielectric there is no ground plane. If $w_1 \neq w_2$ the structure is said to be an "asymmetric CPS," simply called "ACPS." Unless otherwise stated we will assume $w_1 = w_2$, i.e., the symmetric case, still simply called "CPS."

"CPS" can be regarded as the complement of the "CPW" since conductors are present where they are absent in "CPW." Do not confuse "CPS" with two coupled lines. In this case, each conductor is not a single t.l., but both conductors are a single t.l. instead. There will always be a ground conductor, for example, that corresponds to the enclosed box, but "CPS" is theoretically assumed to be suspended on an infinite, thick dielectric.** This condition is in practice approximated for h > 5(s+2w).

"CPS" is a full planar t.l., similar to slot lines or "CPW." According to the discussion in Appendix A2, the "CPS" has a zero cutoff frequency, but its low order propagation mode is not a real "TEM" due to the bottom and top dielectric discontinuity. The fundamental mode is then indicated with "qTEM"*** because it resembles a "TEM" mode since the longitudinal field components are smaller than the transverse ones, and quite often they can be neglected. This is true until frequencies of some GHz. After this limit, dispersion arises and the propagation mode tends to be nearly a "TE."**** Similar to the "CPW" case, the magnetic field can be considered elliptically polarized along longitudinal planes. Due to the elliptical magnetic field polarization, the "CPS" should be a t.l. suited to energy exchange with ferrimagnetic materials. Nevertheless, the minimal use of this t.l. is due mainly to the difficulty with μstrips***** and consequently there are no reports of the use of ferrimagnetic materials.

Because the number of the electric and magnetic field lines in the air is higher than the number of the same lines in the microstrip case, the effective dielectric constant "ε_{re}" of "CPS" is typically 20% lower than the "ε_{re}" for microstrips. Consequently, the maximum reachable characteristic impedance values are higher than the microstrip values. In addition, to avoid field radiation in the air, it is very important to use substrates with high dielectric constants, let us say from a value greater than 10, so that the e.m. field is mainly concentrated inside the dielectric.

---

* For simplicity the coplanar strips transmission line will be simply indicated with "CPS."
** The effect of finite dielectric thickness will be studied later.
*** "qTEM" means "quasi TEM" where quasi = almost.
**** The fact that the propagation mode always tends to be a "TE" mode depends on the natural disposition of the electric field inside the "CPS." Electric field lines are shown in the next section.
***** Transitions among "CPS" and other lines are studied later in this chapter.

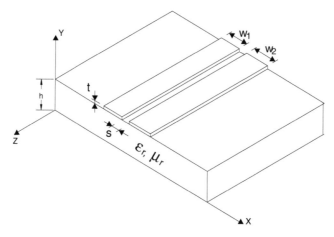

Figure 11.1.1

11.2 ELECTRIC AND MAGNETIC FIELD LINES

Some electric "e" and magnetic "h" field lines for the fundamental "qTEM" mode in "CPS" are indicated in Figure 11.2.1, in a defined cross-section and a defined time. In the fundamental mode, both conductors are equipotential. Different from the case of "CPW," this t.l. is naturally "balanced." So, the feeding and the loading of this t.l. need to be balanced, as indicated in Figure 11.2.1. We have already seen* how to balance µstrips, but in that case the resulting balanced t.l. is not planar. In this case, instead, we have a planar balanced line.

If a ground conductor is present, some µstrip mode could arise. We have already encountered this phenomenon in "CPW." Applying the results for "CPW" of some researchers,[2,3,4] these unwanted modes can be limited setting the bottom ground plane at a distance "h" so that:

$$h > 10(w + s) \tag{11.2.1}$$

The previous equation is quite conservative, but in practice it is often used.

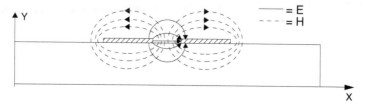

Figure 11.2.1

11.3 SOLUTION TECHNIQUES FOR THE ELECTROMAGNETIC PROBLEM

As we said in the previous section regarding other transmission lines, the "CPS" can be studied using quasi static methods[5,6]** or full wave methods: "FWM."[7,8]

---

* See Chapter 7 for µstrip devices.
** See Appendix A1 for quasi static analysis methods.

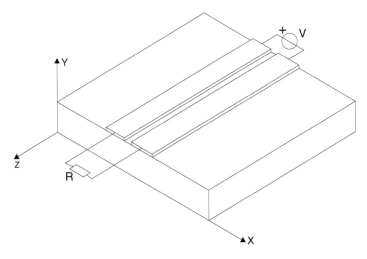

Figure 11.2.2

In this section we will study the "CPS" with a quasi static method mainly because the "CPS" is the geometric complement of "CPW." So, we can adapt the quasi static study of "CPW" to our problem, using the same conformal transformation.

With this method the original geometric structure is transformed into a more simple one. In particular, the "CPS" is transformed in a structure that resembles a parallel plate capacitor, for which the capacitance is simply evaluated as $C = \varepsilon S/d$, where "S" is the area of the plate and "d" the distance. The required analytical transformation is the same we used in Chapter 10 for "CPW." The transformation of the plane for "CPS" case is reported in Figure 11.3.1 parts b and c, together with the "CPW" case indicated in part a.

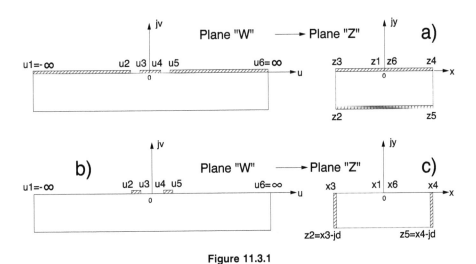

Figure 11.3.1

We can repeat all the analytical steps given for "CPW" case, where the substrate is supposed to be of infinite extent below the strips and $w1 \equiv w2$. The most important difference is that now the capacitance "C" of the parallel plate structure in Figure 11.3.1c is:

$$C = \varepsilon_0 \varepsilon_r \, d/2x_4 \tag{11.3.1}$$

where:

$$d/x_4 = K(p)/K(p') \tag{11.3.2}$$

$$p = s/(s + 2w) \quad \text{and} \quad p' \overset{\perp}{=} \left(1 - p^2\right)^{0.5} \tag{11.3.3}$$

So, applying the transformation for both half planes of Figure 11.3.1b, the resulting capacitance "C" by unit length of the associated "TEM" t.l. for our "CPS" is:

$$C = \varepsilon_0 \left(1 + \varepsilon_r\right) K\left(p'\right)/2K(p) \tag{11.3.4}$$

In addition, the equivalent "TEM" t.l. will have an effective permittivity "ε_{re}" and effective permeability "μ_{re}" given by:

$$\varepsilon_{re} = \left(\varepsilon_r + 1\right)/2 \tag{11.3.5}$$

$$\mu_{re} = \left(\mu_r + 1\right)/2 \tag{11.3.6}$$

i.e., the medium value between the air relative dielectric and magnetic constants and those of the substrate. Expression 11.3.5 for "ε_{re}" can be obtained as we did in Chapter 10 for the "CPW" case.* Equation 11.3.4 is only a simplification of the real "μ_{re}" since this quantity is dependent on many parameters like reciprocal direction between e.m. energy propagation and direction of the applied magnetic field "H_{dc}," field intensity, signal frequency, ferrimagnetic composition, and more.** For the moment, the expression of "μ_{re}" in the following formulas will remind us that the presence of any ferrimagnetic material will need to be evaluated.

Now, from the well-known relation for "TEM" t.l***:

$$\zeta = 1/Cv \tag{11.3.7}$$

where "v" is the light phase speed given by:

$$v \overset{\perp}{=} \left(LC\right)^{-0.5} \equiv v_0 / \left(\mu_{re}\varepsilon_{re}\right)^{0.5} \tag{11.3.8}$$

we can simply evaluate the "CPS" characteristic impedance. We have

$$\zeta = 120\pi K(p) \left(\mu_{re}\right)^{0.5} / \left(\varepsilon_{re}\right)^{0.5} K\left(p'\right) \tag{11.3.9}$$

The effect of finite substrate height "h" has been studied by some researchers[9,10] always using a conformal mapping. From this analysis we have the same impedance expression of the previous equation, but with a value of "ε_{re}" and a new parameter "p_1" given by:

---

* The expression for "μ_{re}" can also be obtained for duality, if it is a simplification.
** See Appendix A7 for fundamental on energy exchange between waves and ferrimagnetic materials.
*** See Chapter 1 for t.l. parameters definitions.

$$\varepsilon_{re} = 1 + \frac{\varepsilon_r - 1}{2} \frac{K(p')}{K(p)} \frac{K(p_1)}{K(p_1')} \tag{11.3.10}$$

$$p_1 = senh(\pi s/4h)/senh\left[(\pi/2h)(w+s/2)\right] \tag{11.3.11}$$

Comparisons among impedance values with or without considering the finite value of substrate height have shown that the effective dielectric constant and impedance variation are negligible if the following condition is verified:

$$h \geq 2(s+2w) \tag{11.3.12}$$

The expressions in this section do not contain any approximations, excluding the assumption of negligible strip thickness and "TEM" propagation mode. Regarding the ratio of elliptic integrals, approximated closed form expressions are given in Appendix A8.

Results of full wave analysis suggest that when increasing the operating frequency, the impedance and the ratio $\lambda_g/\lambda_0 = 1/\sqrt{\varepsilon_{re}}$ both decrease.

In the next section we will give simple design formulas that also take into account the "conductor thickness "t.""

11.4 DESIGN EQUATIONS

The closed form we give here has been obtained using the "CTM" described in the previous section. For all the "CPS" dimensions we will refer to Figure 11.1.1.

To consider the metal thickness "t," we can use the theory employed in Chapter 2 and associate an extra width "dw" to a metal thickness "t" given by:

$$dw = \frac{1.25t}{\pi}\left[1 + \ln\left(\frac{4\pi w}{t}\right)\right] \tag{11.4.1}$$

So, we define the following new notations:

$$w_t = w + dw, \quad s_t = s - dw \tag{11.4.2}$$

and the parameter "p" now becomes:

$$p \rightarrow p_t \stackrel{\perp}{=} s_t/(s_t + 2w_t) \quad \text{and} \quad p_t' \stackrel{\perp}{=} \left(1 - p_t^2\right)^{0.5} \tag{11.4.3}$$

A better value of "ε_{re}" with respect to the value $(\varepsilon_r + 1)/2$ can be obtained using the same expression we got for "CPW" with proper variable substitution $s \leftrightarrow w$, resulting in:

$$\varepsilon_1 \stackrel{\perp}{=} \tanh\left[1.785\ln(h/w) + 1.75\right] \tag{11.4.4}$$

$$\varepsilon_2 \stackrel{\perp}{=} (pw/h)\left[0.04 - 0.7p + 0.01(1 - 0.1\varepsilon_r)(0.25 + p)\right] \tag{11.4.5}$$

$$\varepsilon_{re} \overset{\perp}{=} \left(\varepsilon_r + 1\right)\left(\varepsilon_1 + \varepsilon_2\right)/2 \tag{11.4.6}$$

where "p" is defined by 11.3.3. Note how Equation 10.4.6 includes the limited extension of "h." So, to evaluate "ζ" we can use expression 11.3.9, where "p" and "ε_{re}" are respectively substituted by "p_t" and the previous equation.

The effect of "t" on "ε_{re}" can be evaluated using an expression similar to that used for the "CPW." We obtain[11]:

$$\varepsilon_{re} = \left[\left(\varepsilon_r + 1\right)/2\right]\left(\varepsilon_1 + \varepsilon_2\right) - \frac{1.4\left(\varepsilon_{re} - 1\right)t/s}{K(p')/K(p) + 1.4\,t/s} \tag{11.4.7}$$

At this point we have all the relationships required to calculate the "ζ" and "ε_{re}" of our "CPS" as function of all geometric parameters. In fact, the characteristic impedance is given by 11.3.13 with the insertion of the previous equation and using the parameter "p_t" defined in 11.4.3. Specifically, we have:

$$\zeta = 30\pi K\left(p_t'\right)\left(\mu_{re}\right)^{0.5} \Big/ \left(\varepsilon_{re}\right)^{0.5} K\left(p_t\right) \tag{11.4.8}$$

where "ε_{re}" is given by 11.4.7.

In this case, "p_t" and "p," which respectively appear in the "ζ" and "ε_{re}" expressions, are the parameters to be evaluated for choosing the best approximation for the elliptic integral ratio to use in the "ζ" and "ε_{re}" expressions.

The value of "ζ" and "ε_{re}" produced by the previous formulas has an accuracy of better than 5% compared with the measurement results of Knorr and Kuchler,[12] which assume t = 0.

11.5 ATTENUATION

As with many other t.l.s, the "CPS" losses are due to three causes:

1. Imperfect conductivity of the conductors, or "conductor loss"
2. Dielectric nonzero conductivity and dumping phenomena
3. Substrate magnetic loss, if the substrate is a ferrimagnetic material
4. Radiation

This section will present how to evaluate the first two causes of losses, which are directly related to the geometry of the "CPS" indicated in Figure 11.1.1. Magnetic losses are mainly due to damping phenomena inside ferrimagnetic material and, if the signal frequency is of appropriate value, to resonance absorption.* Radiation losses are strongly dependent on the surrounding structure near the "CPS" and cannot be simply treated in a general way.

For the present case, if we consider the "CPS" as a t.l. only supporting a "TEM" mode we can apply the theory developed by H.A. Wheeler.[13,14] The procedure is similar to that for the µstrip case in Chapter 2, and for this reason here we will only outline the differences with respect to "CPW." For the other concepts and common formulas it is possible, for example, to see the loss evaluation for the "CPW" case in Chapter 2.

The additional inductance "L_a" and resistance "R_a" with reference to Figure 11.5.1, are given by:

* See Appendix A7 for energy exchange phenomena between e.m. signal and ferrite.

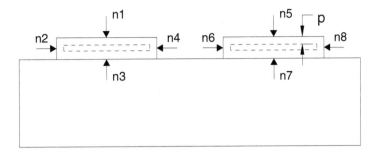

Figure 11.5.1

$$L_a = \sum_{j=1}^{j=8}\left(L_i\right)_j = \frac{1}{2\mu_0}\sum_{j=1}^{j=8}p_j(\mu_c)_j\frac{\partial L}{\partial n_j} \qquad (11.5.1)$$

$$R_a = \sum_{j=1}^{j=8}\left(R_i\right)_j = \frac{1}{\mu_0}\sum_{j=1}^{j=8}\left(R_s\right)_j\frac{\partial L}{\partial n_j} \qquad (11.5.2)$$

where:

  a. "L_i" is the "incremental inductance" per u.l.
  b. "R_i" is the "incremental resistance" per u.l.
  c. "p" is the "penetration depth," [u.l.]
  d. "μ_c" is the conductor absolute permeability
  e. "R_s" is the conductor "sheet resistance," [Ω/square]*
The conductor attenuation coefficient "α_c"** is defined as:

$$\alpha_c = W_c/2W_t \qquad (11.5.3)$$

where "W_c" and "W_t" are respectively the mean power dissipated in the conductor and the mean transmitted power, given by:

$$W_c = R_a|i|^2, \quad W_t = \zeta|i|^2 \qquad (11.5.4)$$

Consequently, the conductor attenuation constant does not depend on the additional inductance "L_a." Using 11.5.2 and 11.5.4, Equation 11.5.3 becomes:

$$\alpha_c = \frac{1}{2\mu_0\zeta}\sum_{j=1}^{j=8}\left(R_s\right)_j\frac{\partial L}{\partial n_j} \qquad (11.5.5)$$

From 3.4.1 and 3.4.2 it follows that:

$$L = \zeta/v = \zeta\left(\mu_{re}\varepsilon_{re}\right)^{0.5}/v_0 \qquad (11.5.6)$$

* See Appendix A2 for measurement unit of "conductor resistance."
** We are assuming a longitudinal variation of conductor attenuation with $e^{-\alpha_c z}$. See Chapter 1 for fundamental theory of transmission lines.

and so:

$$\frac{\partial L}{\partial n} = \frac{1}{v_0} \frac{\partial \left[\left(\mu_{re} \varepsilon_{re} \right)^{0.5} \zeta \right]}{\partial n} \perp \frac{1}{v_0} \frac{\partial \zeta_z}{\partial n}$$

(11.5.7)

where $\zeta_z \perp (\mu_{re} \varepsilon_{re})^{0.5} \zeta$. Observing that $\mu_0 v_0 = \zeta_v \equiv 120\pi$ and using Equation 11.5.7, Equation 11.5.5 becomes:

$$\alpha_c = \frac{1}{\zeta 240\pi} \sum_{j=1}^{j=8} (R_s)_j \frac{\partial \zeta_z}{\partial n_j}$$

(11.5.8)

Since the "ζ" is a function of "w," "t," and "s,"* as was shown in the previous section, the derivative "$\partial \zeta_z / \partial n$" is:

$$\frac{\partial \zeta_z}{\partial n} = \frac{1}{dn} \left(\frac{\partial \zeta_z}{\partial w} dw + \frac{\partial \zeta_z}{\partial s} ds + \frac{\partial \zeta_z}{\partial t} dt \right)$$

(11.5.9)

From Figure 11.5.1 we observe that:

$$dw = -2dn, \quad dt = -2dn, \quad ds = 2dn$$

(11.5.10)

and 11.5.8 becomes:

$$\alpha_c = \frac{R_s}{\zeta 120\pi} \left(-\frac{\partial \zeta_z}{\partial w} + \frac{\partial \zeta_z}{\partial s} - \frac{\partial \zeta_z}{\partial t} \right)$$

(11.5.11)

Of course, the value given by "α_c" is in neper/meter.**

Now using in 11.5.11 the expression "ζ" given in the previous section, Bahl and others[15] have obtained the following expression for "α_c" which provides the value of the attenuation in dB for unit of length:

$$\alpha_c = 17.34 \frac{R_s}{\zeta} \left[\frac{K(p)}{K(p')} \right]^2 \frac{Q}{\pi s} \left(1 + \frac{w}{s} \right) \frac{N}{D} \quad [dB/u.1.]$$

(11.5.12)

where:

$$N = \frac{1.25}{\pi} \ln \frac{4\pi w}{t} + 1 + \frac{1.25t}{\pi w}$$

(11.5.13)

$$D = \left\{ 1 + \frac{2w}{s} + \frac{1.25t}{\pi s} \left[1 + \ln \left(\frac{4\pi w}{t} \right) \right] \right\}^2$$

(11.5.14)

and "Q" is given by:

* In this theory substrate height "h" is assumed to be theoretically infinite.
** See Chapter 1 for attenuation constants dimensions.

$$Q = \frac{p}{\left[1 - \left(1 - p^2\right)^{0.5}\right]\left[1 - p^2\right]^{3/4}} \quad \text{for } 0 \leq p \leq 0.707 \tag{11.5.15}$$

$$Q = \frac{\left[K(p')/K(p)\right]^2}{(1 - p)\sqrt{p}} \quad \text{for } 0.707 \leq p \leq 1 \tag{11.5.16}$$

The incremental inductance rule has been verified to give very accurate results for conductor thickness greater than four times the penetration "p." This condition is usually verified for every planar transmission line since for the typical conductors used, the value of the penetration "p" is lower than some micrometer for frequencies greater than 1 GHz.*

Dielectric loss can be evaluated as already done for the microstrip in Chapter 2, resulting in the same expression:

$$\alpha_d = \frac{20\pi}{\ln 10} \frac{1/\varepsilon_{re} - 1}{1/\varepsilon_r - 1} \frac{\tan\delta}{\lambda_0} \sqrt{\varepsilon_{re}} \tag{11.5.17}$$

which gives a value of dB/u.l. In the previous equation "ε_{re}" is given by 11.4.4 through 11.4.7 and "$\tan\delta$" is the dielectric "tangent delta."** Of course, the quantities "ε_r" and "ε_{re}" are all relative to the real part of the substrate dielectric constant.*** Remember that magnetic losses could be present in the dielectric, the previous expression for "α_d" can be formally modified multiplying by $\sqrt{\mu_{re}}$. We know that it is a simplification, as we explained in Chapter 10 for "CPW."

11.6 CONNECTIONS BETWEEN "CPS" AND OTHER LINES

Excluding some "artistical" connection with t.l. composed of more than one layer, like µstrip or stripline, "CPS" is suited to be connected with full planar t.l.s. These are slot line, essentially a balanced t.l., and "CPW," which is in origin a unbalanced t.l. The study of discontinuities[16,17] places an important role on realizing a proper transition between t.l.s. In this section we will indicate the most simple interconnections between "CPW," slot line, and "CPS." In general, a transition between t.l.s. is practically evaluated as acceptable when the resulting reflection coefficient is typically below 10 dB in the operating bandwidth.

"CPS-slotline" and "CPW-CPS"[18] transitions are indicated respectively in Figures 11.6.1 and 11.6.2. The proper tapering for both transitions is often a practical work, since no closed formulas for such transition designs are available.

11.7 USE OF "CPS"

There are not many applications of "CPS" that cannot be achieved with other t.l.s. A useful application is in the "mixer" devices[19] in conjunction with other planar t.l.s, as we have already introduced in Chapter 9.

Another use of "CPS" is like a guiding structure for balanced signals, like the "ECL" devices. These have complementary outputs that can be transmitted and received in a balanced way. In these cases, "CPS" are well suited to the purpose.[20,21]

---

* See Appendix A2 for values of penetration depth inside good conductors.
** See Chapter 1 for "$\tan\delta$" definition.
*** See Chapter 1 for complex permittivity definition.

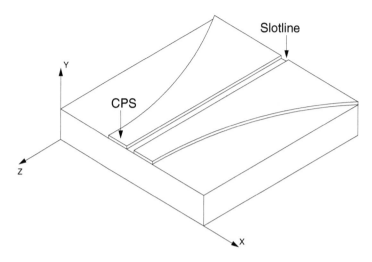

Figure 11.6.1

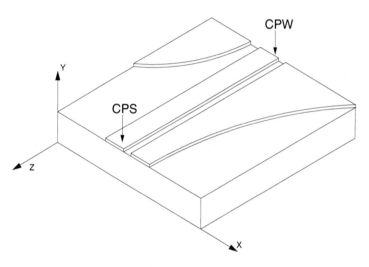

Figure 11.6.2

Use of "CPS" has also been made in filters[22,23] and MMIC.[24]

As we noticed at the beginning of this section, another potential use is in building gyromagnetic devices. However, no experiment in literature is known to the author of this text. This lack is due to the little use of this t.l., mainly because "CPS" is not well-suited to transitions with μstrips.

REFERENCES

1. C. P. Wen, Coplanar waveguide: a surface strip transmission line for nonreciprocal gyromagnetic device application., *IEEE Trans. on MTT*, 1087, Dec. 1969.
2. M. Riaziat, I. J. Feng, R. Majidi-Ahy, B. A. Auld, Single mode operation of coplanar waveguides, *Electron. Lett.*, 1281, Nov. 1987.
3. Y. D. Lin, J. W. Sheen, C. Y. Chang, Surface wave leakage properties of coplanar strips., *Microwave Symp.*, 229, 1995.
4. A. B. Yakovlev, G. W. Hanson, On the nature of critical points in leakage regimes of a conductor backed coplanar strip line., *IEEE Trans. on MTT*, 87, Jan. 1997.

5. C. P. Wen, Coplanar waveguide: a surface strip transmission line for nonreciprocal gyromagnetic device application., *IEEE Trans. on MTT*, 1087, Dec. 1969.

6. V. F. Hanna, Finite boundary corrections to coplanar stripline analysis, *Electron. Lett.*, 604, July 1980.

7. J. B. Knorr, K. D. Kuchler, Analysis of coupled slots and coplanar strips on dielectric substrate, *IEEE Trans. on MTT*, 541, July 1975.

8. S. G. Pintzos, Full wave spectral domain analysis of coplanar strips, *IEEE Trans. on MTT*, 239, Feb. 1991.

9. V. F. Hanna, Finite boundary corrections to coplanar stripline analysis, *Electron. Lett.*, 604, July 1980.

10. G. Ghione, C. Naldi, Analytical formulas for coplanar lines in hybrid and monolithic MICs., *Electron. Lett.*, 179, Feb. 1984. These researchers have varied Hanna's result, in Reference 6, to have a better approximation to a spectral domain analysis for the structure.

11. K. C. Gupta, R. Garg, I. J. Bahl, *Microstrip Lines and Slotlines*. Artech House, 280, 1979.

12. J. B. Knorr, K. D. Kuchler, Analysis of coupled slots and coplanar strips on dielectric substrate, *IEEE Trans. on MTT*, 541, July 1975.

13. H. A. Wheeler, Formulas for the skin effect, Proc. of the IRE, 30, 412, 1942.

14. R. Sturdivant, Transmission line conductor loss and the incremental inductance rule, *Microwave J.*, 156, Sept. 1995.

15. K. C. Gupta, R. Garg, I. J. Bahl, *Microstrip Lines and Slotlines*. Artech House, 280, 1979.

16. K. Goverdhanam, R. N. Simons, N. Dib, L. P. B. Katehi, Coplanar stripline components for high frequency applications, *Microwave Symp.*, 2, 1193, 1996.

17. R. N. Simons, N. L. Dib, L. P. B. Katehi, Modeling of coplanar stripline discontinuities, *IEEE Trans. on MTT*, 711, May 1996.

18. L. Zhu, K. Wu, Hybrid FGCPW/CPS scheme in the building block design of low cost uniplanar and multilayer circuit and antenna, *Microwave Symp.*, 1999, session WE1A.

19. J. A. Eisenberg, J. S. Panelli and W. Ou, Slotline and coplanar waveguide team to realize a novel MMIC double balanced mixer, *Microwave J.*, 123, Sept. 1992.

20. M. Y. Frankel, R. H. Voelker, J. N. Hilfiker, Coplanar transmission lines on thin substrates for high speed low loss propagation, *IEEE Trans. on MTT*, 396, Mar. 1994.

21. H. Cheng, J. F. Whitaker, T. M. Weller, L. P. B. Katehi, Terahertz bandwidth pulse propagation on a coplanar stripline fabricated on a thin membrane, *MGWL*, 89, Mar. 1994.

22. J. E. Oswald. P. H. Siegel, The application of the FDTD method to millimeter wave filter circuits including the design and analysis of a compact coplanar strip filter for THz frequencies, *Microwave Symp.*, 1, 309, 1994.

23. K. Goverdhanam, R. N. Simons, L. P. B. Katehi, Coplanar stripline propagation characteristics and bandpass filter, *MGWL*, 214, Aug. 1997.

24. W. N. Manng, D. P. Butler, W. Xiong, W. Kula, R. Sobolewski, Propagation characteristics of monolithic YBaCuO coplanar strip transmission lines fabricated by laser writing patterning technique, *MGWL*, 132, May 1994.

APPENDIX **A1**

Solution Methods for Electrostatic Problems

A1.1 THE FUNDAMENTAL EQUATIONS OF ELECTROSTATICS

Electrostatics is that branch of physics that studies the forces among electric charges that are fixed in the reference system. A lot of formulas are used as foundations of this branch. As in other Appendixes of this text, here we will discuss just those relationships that are necessary to better understand our text. The reader interested in a broader knowledge of these topics can refer to the texts reported in the bibliography.[1,2,3]

a. Equations for Electric "E" and Electric Flux Density "D" Fields

The two fundamental equations for such fields are:

$$\underline{\nabla} \otimes \underline{E} = 0 \qquad\qquad (A1.1.1)$$

$$\underline{\nabla} \bullet \underline{D} = \rho \qquad\qquad (A1.1.2)$$

where symbols "$\underline{\nabla}\otimes$" and "$\underline{\nabla}\bullet$" are respectively the "curl" and the "divergence," defined in Appendix A8, while "ρ" is the volumetric charge density. The vector "$\underline{D}$" is related to "$\underline{E}$" by the well-known formula:

$$\underline{D} = \varepsilon\underline{E} \qquad\qquad (A1.1.3)$$

In the most general case, the medium permittivity "ε" is a function of coordinates and of the applied electric field "$\underline{E}$" strength. The vector "D" is also called the "electric displacement vector."

b. Poisson* and Laplace** Equations

Remembering the well-known result that:***

$$\underline{\nabla}\otimes\left(\underline{\nabla}s\right) = 0 \qquad\qquad (A1.1.4)$$

where "s" is a scalar function, for A1.1.1 we can set:

---

* Denis Poisson, French mathematician, born in Pithiviers in 1781 and died in Paris in 1840.
** Pier Simon de Laplace, French mathematician, born in Beaumont an Auge in 1749 and died in Paris in 1827.
*** The "gradient operation -$\underline{\nabla}$-" is defined in Appendix A8.

$$\underline{E} \overset{\perp}{=} -\underline{\nabla} V \tag{A1.1.5}$$

where "V" is a scalar function called "potential."

Inserting A1.1.3 into A1.1.2 and applying the relationship:

$$\underline{\nabla} \bullet s\underline{v} = \underline{v} \bullet \underline{\nabla}s + s\underline{\nabla} \bullet \underline{v} \tag{A1.1.6}$$

where "s" is a scalar function and "$\underline{v}$" a vector, we have:

$$\underline{\nabla} \bullet \varepsilon\underline{E} = \underline{E} \bullet \underline{\nabla}\varepsilon - \varepsilon\nabla^2 V \tag{A1.1.7}$$

where $\nabla^2 \overset{\perp}{=} \underline{\nabla} \bullet \underline{\nabla}$. Now, let us assume a medium where "ε" is not a function of coordinates. Then, from A1.1.7 we have:

$$\underline{\nabla} \bullet \varepsilon\underline{E} \equiv \underline{\nabla} \bullet \underline{D} = -\varepsilon\nabla^2 V \tag{A1.1.8}$$

and using A1.1.2 we have:

$$\nabla^2 V = -\rho/\varepsilon \tag{A1.1.9}$$

which is called "Poisson's equation." If the previous equation is evaluated in a region where no charges are placed, then we have:

$$\nabla^2 V = 0 \tag{A1.1.10}$$

which is called "Laplace's equation."

c. Boundary Conditions

If some discontinuities exist in the medium where "E" and "D" are present, they must satisfy the following conditions:

$$\underline{n} \otimes \left(\underline{E}_2 - \underline{E}_1\right) \overset{\perp}{=} 0 \tag{A1.1.11}$$

$$\underline{n} \bullet \left(\underline{D}_2 - \underline{D}_1\right) \overset{\perp}{=} q_s \tag{A1.1.12}$$

where "q_s" is the charge surface density, measured in Coulomb/m^2 in MKSA reference system unit,* and "$\underline{n}$" is the normal to the discontinuity surface. The two previous equations are explained saying that the tangential component of the "$\underline{E}$" is continuous crossing the discontinuity, while the normal component of "$\underline{D}$" changes by a quantity "q_s."

Equations A1.1.11 and A1.1.12 can be rewritten where only the potential is envolved, just using Equations A1.1.3 and A1.1.5. In fact, we have:

$$\underline{n} \otimes \left(\underline{E}_2 - \underline{E}_1\right) \equiv \underline{n} \otimes \left(\underline{\nabla}V_1 - \underline{\nabla}V_2\right) \tag{A1.1.13}$$

---

* Unless otherwise stated, we will use the MKSA reference system.

$$\underline{n} \bullet (\underline{D}_2 - \underline{D}_1) \equiv \underline{n} \bullet (\varepsilon_1 \underline{\nabla} V_1 - \varepsilon_2 \underline{\nabla} V_2) \qquad (A1.1.14)$$

In some electrostatic problems it is more convenient to write the boundary conditions as in the two previous equations.

d. Green's* Function

Green's function is the name of the solution function of the Poisson's equation when the charge volumetric density is theoretically a pulse, called as "Dirac's** function" and indicated with "δ." This ideal situation is represented in practice when charges are of negligible dimension with respect to the region under study and the observation distance from the charge. In the case of a single Dirac charge density we will show that it is easy to extract the Green's function. In fact, in this case we have:

$$\nabla^2 V = -\delta / \varepsilon \doteq \nabla^2 G \qquad (A1.1.15)$$

where "δ" is the Dirac pulse function to which we give the dimension of 1 Coulomb/m³. Let us set a spherical coordinate system with the origin in the charge position, as indicated in Figure A1.1.1. For reasons of geometric symmetry we can write:

$$\partial / \partial \theta = \partial / \partial \varphi = 0 \qquad (A1.1.16)$$

Using the expression of "∇^2" in spherical coordinates given in Appendix A8 and applying the previous condition to A1.1.15 we have:

$$\frac{1}{r^2} \frac{d}{dr} \left(r^2 \frac{dG}{dr} \right) = -\delta / \varepsilon \qquad (A1.1.17)$$

where "r" is the distance of the observation point from the charge. Since the elementary space volume "dv" is:

$$dv = r^2 \operatorname{sen}\theta d\varphi d\theta dr \qquad (A1.1.18)$$

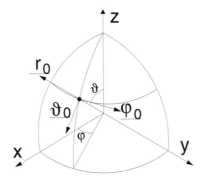

Figure A1.1.1

* George Green, English mathematician, born in Sneinton in 1793 and died there in 1841.
** Paul Dirac, English physicist, born in Bristol in 1902 and died in Tallahassee, FL, U.S.A., in 1984.

integrating Equation A1.1.17 in the whole volume we have:

$$\int_0^r \int_0^\pi \int_0^{2\pi} \frac{1}{r^2} \frac{d}{dr}\left(r^2 \frac{dG}{dr}\right) r^2 \text{ sen } \theta \, dr \, d\theta \, d\varphi = -\frac{1}{\varepsilon} \qquad (A1.1.19)$$

The second member of the previous equation comes from the fact that:

$$\int_V \delta dv = 1 \qquad (A1.1.20)$$

From A1.1.19 we have:

$$4\pi r^2 dG/dr = -1/\varepsilon \qquad (A1.1.21)$$

which, integrated in "r" and setting G = !0 at infinity, gives:

$$G(r) = 1/4\pi\varepsilon r \qquad (A1.1.22)$$

which is Green's function for our free space system. Green's function is assumed as the potential distribution in the space due to a unitary charge positioned in any desired place. Since Poisson's equation is linear, if the charge has a value "q," it is sufficient to multiply the second member of the previous equation times "q."

Applying Equation A1.1.5 at A1.1.22 we obtain the electric field distribution given by:

$$E(r) = q/4\pi\varepsilon r^2 \qquad (A1.1.23)$$

Another situation where it is easy to solve the Poisson's equation is where the charge distribution can be evaluated as a line of charge. In this case "δ" is the two-dimensional Dirac pulse function, to which we give the dimension of 1 Coulomb/m. A practical representation is indicated in Figure A1.1.2, where a generic structure with cylindrical symmetry is represented. Let us assume the line of charge to be parallel to the longitudinal symmetry axis, so that "P" is the intercept of the line charge with the cross-sectional view of the geometry. Also in this case, for reasons of geometric symmetry we can write:

$$\partial/\partial\theta = \partial/\partial z = 0 \qquad (A1.1.24)$$

Using the expression of "∇^2" in cylindrical coordinates given in Appendix A8 and applying the previous condition to A1.1.15 we have:

$$\frac{1}{r} \frac{d}{dr}\left(r \frac{dG}{dr}\right) = -\delta/\varepsilon \qquad (A1.1.25)$$

where "r" is the distance of the observation point from the line of charge. Since the elementary surface "dS" of the cross-sectional surface indicated in Figure A1.1.2 is:

$$dS = rd\theta dr \qquad (A1.1.26)$$

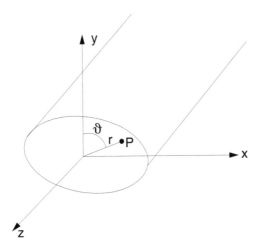

Figure A1.1.2

integrating Equation A1.1.25 on the whole surface we have:

$$\int_0^r \int_0^{2\pi} \frac{1}{r} \frac{d}{dr}\left(r\frac{dG}{dr}\right) r dr d\theta = -\frac{1}{\varepsilon} \qquad (A1.1.27)$$

The second member of the previous equation comes from the fact that:

$$\int_s \delta dS = 1 \qquad (A1.1.28)$$

From A1.1.27 we have:

$$2\pi r d G/dr = -1/\varepsilon \qquad (A1.1.29)$$

where integrated in "r" and setting G = !0 at a distance "r_a" we have:

$$G(r) = -(1/2\pi\varepsilon)\left[\ln(r) - \ln(r_a)\right] = -(1/2\pi\varepsilon)\left[\ln(r/r_a)\right] \qquad (A1.1.30)$$

Since the Poisson's equation is linear, if the line of charge has a value "q_ℓ"* it is sufficient to multiply the second member of the previous equation times "q_ℓ."

Of course, in this case we can also apply Equation A1.1.5 to the previous equation to extract the electric field.

e. Gauss's Law

Applying the well-known divergence, or Gauss's** theorem, to A1.1.2 we have the famous Gauss' law, i.e.:

---

* $[q_\ell]$ = Coulomb/meter.
** Carl Friedrich Gauss, German physicist, born in Brunswick in 1777 and died in Gottingen in 1855.

$$\oint_s \underline{D} \bullet \underline{n} dS = q \qquad (A1.1.31)$$

where "S" is a surface that contains a volume "V," "$\underline{n}$" the "normal" to this surface and directed outside the region under study, and "q" the whole charge contained in the volume "V." Gauss' law is widely used to evaluate the impedance of the t.l. in a lot of transmission line problems supporting a "TEM" mode.

A1.2 GENERALITIES ON SOLUTION METHODS FOR ELECTROSTATIC PROBLEMS

In practice it is possible to specify some electrostatic problem whose geometric aspect and charge distribution suggest to us that it is best studied with one method rather than another. In this section we will try to define such electrostatic problems and will briefly discuss what type of solution methods exist, while in the next section we will study each method more deeply.

These methods give an exact solution for an electrostatic problem. Sometimes, due to their relative simplicity, these methods are applied to time varying problems. In these cases it is usually said that the problem is studied using a "quasi* static" method.** This extension is not theoretically exact, but their use can give approximated results, with a little effort, for the real situation that should be studied with Maxwell's equations*** and involving a lot of mathematics. In our text we have often used these methods to evaluate transmission line characteristics, like characteristic impedance and effective dielectric constant.

a. Finite Difference Method

The power of the finite difference method**** is that it gives an approximate solution to problems where the exact solution appears to be difficult. This method is particularly suited to problems where the geometric structure is limited with known boundary electrical conditions, and the equation to be resolved is a differential equation. This type of equation is very common in our problems and for this reason this method is the most frequently used.

The solution is simpler to reach if the limiting structure has a simple geometric shape, like a rectangle.

b. Image Charge Method

Contrary to the finite difference method, this is an exact method. It is particularly suited in all problems where the electrostatic situation is composed of charges and planes.

c. Conformal Transformation Method*****

This method is also an exact one. It consists of transforming the original geometric structure under study into a simpler one, where the electric field in all desired points can be obtained immediately. This geometrical transformation is usually performed through differential equations. Once the problem for the simpler structure is solved, with some simple change of variable, the solution for the original structure can be found.

---

*   "Quasi" means "almost."
**   The word "quasi static" is simply abbreviated with "qstatic."
***   Maxwell's equations are reviewed in Appendix A2.
****   The finite difference method is simply abbreviated with "FDM."
*****   Conformal transformation method is simply abbreviated with "CTM."

We want to conclude this introduction noting that the finite difference method is the simplest to execute with an electronic computer.

A1.3 FINITE DIFFERENCE METHOD

As an example of this method, let us suppose to want to obtain the potential inside the structure indicated in Figure A1.3.1. Since no charges are assigned in this structure we have to solve the Laplace equation $\nabla^2 V = 0$. Let us start with a rectangular Cartesian reference system and divide the internal structure into small rectangular areas called "meshes," given by the intersections of "r" horizontal rows, and "c" vertical columns. The higher the number of these meshes the better the approximation, but the longer the time required to reach the solution.

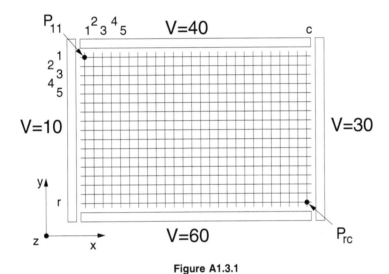

Figure A1.3.1

To simplify the explanation of this method we will assume that the potential does not vary with "z" as the geometrical symmetry suggests to us, i.e.:

$$\partial^2/\partial z^2 \doteq 0 \qquad\qquad (A1.3.1)$$

and consequently the equation to solve is:

$$\partial^2/\partial x^2 + \partial^2/\partial y^2 = 0 \qquad\qquad (A1.3.2)$$

However, the finite difference method can also be applied to full three-dimensional problems. Now, let us assume we know the potential in a generic point "$P_{x,y}$" with coordinates "x,y" in the grid, and evaluate the potential in a point at a distance "$\pm x'$" and "$\pm y'$." "x'" and "y'" are respectively the horizontal and vertical dimension of the mesh. Using Taylor's expansion, we have:

$$V(x + x', y) = V(x, y) + \sum_{n=1}^{n=\infty} \frac{d^n V}{dx^n} \frac{x'^n}{n!} \qquad\qquad (A1.3.3)$$

$$V(x - x', y) = V(x, y) + \sum_{n=1}^{n=\infty} \frac{d^n V}{dx^n} \frac{(-x')^n}{n!} \tag{A1.3.4}$$

$$V(x, y + y') = V(x, y) + \sum_{n=1}^{n=\infty} \frac{d^n V}{dx^n} \frac{y'^n}{n!} \tag{A1.3.5}$$

$$V(x, y - y') = V(x, y) + \sum_{n=1}^{n=\infty} \frac{d^n V}{dx^n} \frac{(-y')^n}{n!} \tag{A1.3.6}$$

Now we do the approximation that the potential does not vary so abruptly to consider all the infinite derivatives in the previous equation. Then we stop the derivatives to the second order, so that, for example, A1.3.4 and A1.3.5 can be rewritten as:

$$V(x, y + y') = V(x, y) + (dV/dy) \, y' + (d^2V/dy^2)(y'^2/2) \tag{A1.3.7}$$

$$V(x, y - y') = V(x, y) - (dV/dy) \, y' + (d^2V/dy^2)(y'^2/2) \tag{A1.3.8}$$

Summing the two previous equations we have:

$$d^2V/dy^2 = \frac{V(x, y + y') + V(x, y - y') - 2V(x, y)}{y'^2} \tag{A1.3.9}$$

Similarly, we can have a derivative with respect to "x," i.e.:

$$d^2V/dx^2 = \frac{V(x + x', y) + V(x - x', y) - 2V(x, y)}{x'^2} \tag{A1.3.10}$$

Using the two previous equations, A1.3.2 can be rewritten as:

$$V(x, y) = \left[V(x + x', y) + V(x - x', y) + V(x, y + y') + V(x, y - y') \right]/4 \tag{A1.3.11}$$

which is, of course, an approximation of A1.3.2, but well suited to be implemented on a computer. Just to show how Equation A1.3.11 can be used, let us evaluate the potential "V_{11}" at the first intersection point "P_{11}." We have:

$$V(x - x', y) = 10, \quad V(x + x', y) = 0, \quad V(x, y + y') = 40, \quad V(x, y - y') = 0 \tag{A1.3.12}$$

from which, using A1.3.11, it follows that $V_{11} = 12.5$. For example, for the point "P_{12}" we can write:

$$V(x - x', y) = 12.5, \quad V(x + x', y) = 0, \quad V(x, y + y') = 40, \quad V(x, y - y') = 0 \tag{A1.3.13}$$

from which, using A1.3.12, it follows that $V_{12} = 13.125$.

So, proceeding as we have indicated, all the potentials inside the structure in Figure A1.3.1 can be evaluated. This procedure is repeated many times, increasing the accuracy of the potential values obtained. Of course, a procedure is needed to terminate the iteration. It is common practice to use the "relaxation method" to achieve such termination. With this method a residual "$R_{ij,k}$" for the position "i,j" at the "k-th" iteration is defined as:

$$R_{ij,k} \doteq V_{ij,k} - V_{ij,k-1} \tag{A1.3.14}$$

Once the residual is obtained, the potential at the "k-th" iteration is evaluated as:

$$V_{ij,k} \doteq V_{ij,k-1} + \Gamma R_{ij,k-1} \tag{A1.3.15}$$

where "Γ" is a constant called the "relaxation constant." The situation when values of "Γ" increase or decrease the residual are called, respectively, "super relaxation" or "under relaxation." So, the procedure is obviously stopped when the residual is under a minimum value and is almost equal for any point of the grid.

A1.4 IMAGE CHARGE METHOD

With this method, fictitious charges are inserted in particular points of the geometry under study so that together with the assigned charges, the boundary conditions are verified but the problem can be simplified. To do an example, let us study the situation represented in Figure A4.1.1. A charge "q" is positioned at a distance "d" from an infinite conducting plane, and we want to evaluate the potential at any points of the half semiplane where the charge "q" is positioned.

The boundary condition on the perfect conducting wall forces the potential to be null. We can observe that the same condition can be verified if we set a charge of value "–q" at a distance "– d" from the wall, as indicated in Figure A1.4.2. We know, as we have studied in Section A1.1, that the electric potential "V(r)" produced by a charge is:

$$V(r) = q/4\pi\varepsilon r \tag{A1.4.1}$$

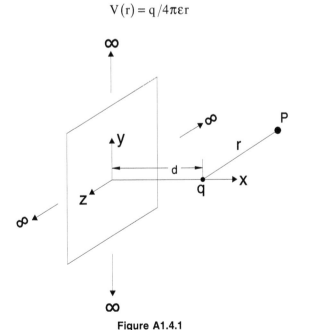

Figure A1.4.1

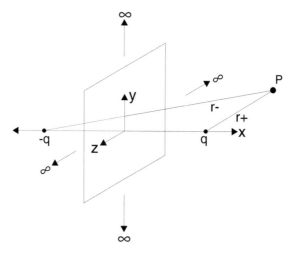

Figure A1.4.2

Using this result, in our case the solution is very simple, since applying the superposition principle we can write:

$$V(r) = (q/4\pi\varepsilon)\left(1/r^+ - 1/r^-\right) \tag{A1.4.2}$$

where:

$$r^+ = \left[(x-d)^2 + y^2 + z^2\right]^{0.5} \tag{A1.4.3}$$

$$r^- = \left[(x+d)^2 + y^2 + z^2\right]^{0.5} \tag{A1.4.4}$$

Equation A1.4.2, with the insertion of A1.4.3 and A1.4.4, is the solution of our problem. However, note that for x < 0 the two electrostatic problems indicated in Figures A1.4.1 and A1.4.2 do not give the same result. So, with the image charge method, caution must be used in evaluating the field in regions not involved in the original problem and/or where boundary conditions are not satisfied.

Another problem where the application of this method is profitable is indicated in Figure A1.4.3. Here two media exist, indicated with "1" and "2," the first with dielectric constant "ε_1" and the other with dielectric constant "ε_2." The separation interface between the two media is evaluated as a perfect plane, which different from the previous example is not a conducting one. A charge "q" is placed in media "2" at a distance "d" from the interface layer. We want to evaluate the potential in any portion of this structure. Indicating with "V_1" and "V_2" the potentials respectively in regions "1" and "2," since on the interface layer there are no charges, the contour conditions on this plane are:

$$V_1 = V_2 \tag{A1.4.5}$$

$$\varepsilon_1\left(dV_1/dr\right)\underline{r} \bullet \underline{n}_1 + \varepsilon_2\left(dV_2/dr\right)\underline{r} \bullet \underline{n}_2 \overset{\perp}{=} 0 \tag{A1.4.6}$$

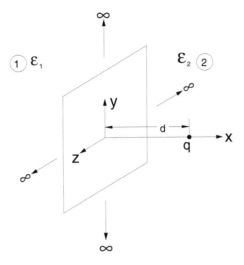

Figure A1.4.3

The representation of vectors is indicated in Figure A1.4.4. Here "P" is a generic point on the interface layer, "P_r" and "P_ℓ" are the positions of the right and left charges and "α" is the angle between the normal "$\underline{n}_1$" in "P" and the versor "$\underline{r}_{01}$." To resolve our problem we use the superposition principle. So, we will apply the following procedure:

1. fill the space with only the media "1" plus a ficticious charge "q_r" in the same position "x_0" of "q."
2. fill the space with only the media "2" plus a ficticious charge "q_ℓ" as image of "q" i.e., at the position "$-x_0$" from the origin of the reference system.

The potential "V_1" immediately at the left of the separation plane for this case is:

$$V_1(r) = q_r / 4\pi\varepsilon_1 r \qquad (A1.4.7)$$

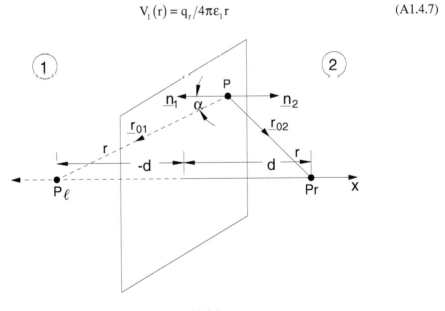

Figure A1.4.4

The potential "V_2" immediately at the right of the separation plane is:

$$V_2(r) = q_\ell/4\pi\varepsilon_2\, r + q/4\pi\varepsilon_2\, r \tag{A1.4.8}$$

Equations A1.4.5 and A1.4.6 are then rewritten as:

$$q_\ell/4\pi\varepsilon_2\, r + q/4\pi\varepsilon_2\, r =! q_r/4\pi\varepsilon_1\, r \tag{A1.4.9}$$

$$\varepsilon_2\cos\theta\left[q/4\pi\varepsilon_2 r^2 - q_\ell/4\pi\varepsilon_2 r^2\right] =! \varepsilon_1\cos\theta q_r\big/4\pi\varepsilon_1 r^2 \tag{A1.4.10}$$

From these last two equations we have:

$$q_\ell = \frac{\varepsilon_2 - \varepsilon_1}{\varepsilon_2 + \varepsilon_1}\, q \tag{A1.4.11}$$

$$q_r = \frac{2\varepsilon_1}{\varepsilon_2 + \varepsilon_1}\, q \tag{A1.4.12}$$

So, once these two charges are defined, the electric field in any position can be evaluated applying the superposition principle.

A1.5 FUNDAMENTALS ON FUNCTIONS WITH COMPLEX VARIABLES

We know that a complex number "z" can be represented in two ways:

1. with real "x" and imaginary "y," using the notation:

$$z \overset{\perp}{=} x + jy \tag{A1.5.1}$$

2. with modulus "r" and phase angle "θ," using the notation:

$$z = re^{j\theta} \tag{A1.5.2}$$

So, a generic function "f(z)" working on a complex number "z" can in general be represented as a transformation from a plane "z" to a plane "w" obtained by w = f(z). In general, the number "w" is also complex, i.e., we can set:

$$w \overset{\perp}{=} u + jv \tag{A1.5.3}$$

where both "u" and "v" are functions of "x" and "y," i.e., we can set:

$$u = u(x,y) \quad v = v(x,y) \tag{A1.5.4}$$

Among the infinite functions in electromagnetism, the analytic functions* are a very important class. It is possible to show that a function $w = f(z)$ is analytic if the transforming functions A1.5.4 satisfy the following Cauchy**-Riemann*** conditions:

$$\partial u/\partial x = \partial v/\partial y, \quad \partial u/\partial y = -\partial v/\partial x \tag{A1.5.5}$$

Deriving the first of A1.5.5 with respect to "x," the second with respect to "y," and combining them we have:

$$\partial^2 u/\partial x^2 + \partial^2 u/\partial y^2 = 0 \tag{A1.5.6}$$

Similarly, if we derive the first of A1.5.5 with respect to "y," the second with respect to "x" and combine we have:

$$\partial^2 v/\partial x^2 + \partial^2 v/\partial y^2 = 0 \tag{A1.5.7}$$

From A1.5.6 and A1.5.7 we can deduce that the real and imaginary parts of an analytic function are a solution of the two-dimensional Laplace equation. From an electromagnetic point of view, it means that these functions can represent a potential function. It is simple to verify that the electric fields obtained from the real and imaginary parts of an analytic function are orthogonal. In fact, using the functions "u" and "v" in Equation A1.1.5 we can write:

$$\underline{E}_u = -\underline{\nabla}u = -(\partial u/\partial x)\,\underline{x}_0 - (\partial u/\partial y)\underline{y}_0 \tag{A1.5.8}$$

$$\underline{E}_v = -\underline{\nabla}v = -(\partial v/\partial x)\,\underline{x}_0 - (\partial v/\partial y)\underline{y}_0 \tag{A1.5.9}$$

If now we do the scalar product "$\underline{E}_u \bullet \underline{E}_v$" and apply the Cauchy-Riemann conditions, we have $\underline{E}_u \bullet \underline{E}_v = 0$, i.e., these electric fields are orthogonal.

Another important characteristic of "u" and "v" functions is that once defined, one of these is to represent the potential while the other is proportional to the flux "φ_u" of the electric flux density "D_u" through a line. In fact, with reference the Figure A1.5.1 and using the function "u" to represent the potential, we can write:

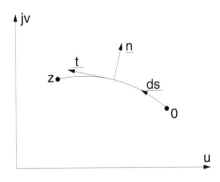

Figure A1.5.1

---

* An analytic function is a complex one whose derivative is unique and independent of the phase angle "θ" of which we arrive at the derivative point.

** Augustine Cauchy, French mathematician, born in Paris in 1789 and died in Sceaux in 1857.

*** Bernhard Riemann, German mathematician, born in Breselenz in 1826 and died in Selasca in 1866.

$$\varphi_u = \int_0^z \underline{D}_u \bullet \underline{n} \, ds \equiv \varepsilon \int_0^z \underline{E}_u \bullet \underline{n} \, ds \qquad (A1.5.10)$$

Observing that:

$$\underline{n} = -j\underline{t}, \quad \underline{t} \, ds = \underline{x}_0 \, dx + \underline{y}_0 \, dy \qquad (A1.5.11)$$

Equation A1.5.10 becomes:

$$\varphi_u = -j\varepsilon \int_0^z \underline{E}_u \bullet \left(\underline{x}_0 \, dx + \underline{y}_0 \, dy \right) \qquad (A1.5.12)$$

Inserting in the previous equation, the expression A1.5.8 and the Cauchy-Riemann conditions, from Equation A1.5.12 we have:*

$$\varphi_u = -\varepsilon \int_0^z \left(\frac{\partial v}{\partial y} \, dy + \frac{\partial v}{\partial x} \, dx \right) = -\varepsilon \left[v(z) - v(0) \right] \qquad (A1.5.13)$$

from which we see how the flux "φ_u" of "D_u" is just proportional to the "v" function.

Another important characteristic of the analytic functions is that they preserve, in the transformation, the value of the original angle. To show that, let us refer to Figure A1.5.2. Here in the "Z" plane two generic curves "C_{1z}" and "C_{2z}" are represented, transformed in the "W" plane in the curves "C_{1w}" and "C_{2w}." Due to the condition on the derivative of an analytic function, for the intersection points "z" and "w" it must be:

$$\lim_{\Delta_{1z} \to 0} \frac{\Delta_{1w}}{\Delta_{1z}} = ! \lim_{\Delta_{2z} \to 0} \frac{\Delta_{2w}}{\Delta_{2z}} \qquad (A1.5.14)$$

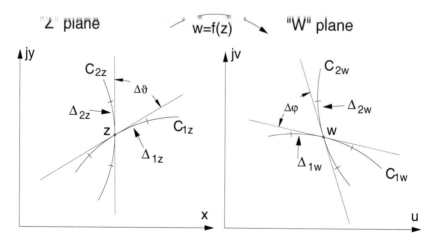

Figure A1.5.2

---

* Note that $j\underline{x}_0 = \underline{y}_0$ and $j\underline{y}_0 = -\underline{x}_0$.

Now, let us write the complex numbers in the "Z" and "W" planes, respectively, with modulus "r" and angle "θ" and with modulus "γ" and angle "φ." So, the previous equation becomes:

$$\lim_{\Delta_{1z} \to 0} \frac{\Delta_{1\gamma}}{\Delta_{1r}} e^{(\varphi 1 - \theta 1)} =! \lim_{\Delta_{2z} \to 0} \frac{\Delta_{2\gamma}}{\Delta_{2z}} e^{(\varphi 2 - \theta 2)} \tag{A1.5.15}$$

This equation is surely verified if:

$$\Delta_{1\gamma} \backslash \Delta_{1r} \equiv \Delta_{2\gamma} \backslash \Delta_{2r} \quad \text{and} \quad \varphi_1 - \theta_1 \equiv \varphi_2 - \theta_2 \tag{A1.5.16}$$

or, just rewriting the previous equation in a different way,

$$\Delta_{1\gamma} \equiv \Delta_{1r} \left(\Delta_{2\gamma} \backslash \Delta_{2r} \right) \quad \text{or} \quad \Delta_{2\gamma} \equiv \Delta_{2r} \left(\Delta_{1\gamma} \backslash \Delta_{1r} \right) \tag{A1.5.17}$$

and

$$\varphi_1 - \varphi_2 \equiv \theta_1 - \theta_2 \tag{A1.5.18}$$

Equation A1.5.17 says that in the "W" plane the modulus increments are obtained from those of the "Z" plane just after a scaling factor $A \doteq \Delta_{1\gamma} \backslash \Delta_{1r} \equiv \Delta_{2\gamma} \backslash \Delta_{2r}$. Equ A1.5.18 says that the angles between locally intersecting lines are preserved between the two planes. Due to the validity of Equations A1.5.17 and A1.5.18 the analytic functions are also called "conformal transformations." In the next two sections we will see some applications for the conformal transformation method, using these analytic functions.

A1.6 CONFORMAL TRANSFORMATION METHOD

Conformal transformation techniques are a very powerful tool to resolve many electrostatic problems. Under some particular approximations, many electrodynamic problems can be studied with this method also. For example, in our text we have applied the conformal transformation method to study coplanar waveguides. Here, we will see how a typical electrostatic problem can be studied with this method. In the next section we will study a particular transformation, the Schwarz*-Christoffel** transformation.

Let us examine the situation indicated in Figure A1.6.1. Here, an empty cylinder with circular section of radius "r" is held to a zero potential. Inside it a longitudinal distribution of charge exists, located at the point "z_ℓ." We want to evaluate the potential inside the cylinder.

Let us apply the following functional transformation to the structure indicated in Figure A1.6.1a:

$$w = -j \frac{z - r}{z + r} \doteq f(z) \tag{A1.6.1}$$

so that we transform the original geometry indicated in Figure A1.6.1b. The previous equation is called "bilinear transformation." In particular, the circumference in part a of the figure is transformed in the real axis "u" in part b, and the internal part of the circumference is transformed in the upper half plane of the "W" plane. The point "z_ℓ" is transformed in the imaginary point "jv_ℓ." So, the

* Karl Schwarz, German mathematician, born in Hermsdorf in 1843 and died in Berlin in 1921.
** Elwin Christoffel, German mathematician, born in Montjoie in 1829 and died in Strasburg in 1900.

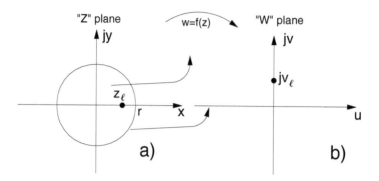

Figure A1.6.1

original problem of finding the potential inside the cylinder is transformed in the problem to finding the potential in the upper plane of Figure A1.6.1b. In this case, the solution can be found by applying the "image charge" method, studied in section A1.4. So, we insert a line of charge in the position "$-jv_\ell$" so that the boundary condition of zero potential on the "u" axis is respected. Then, we use the superposition principle, employing the potential expression we got for the structures with cylindric symmetry, given in Equation A1.1.30. So, in our case we have:

$$v(w) = -\frac{q_\ell}{2\pi\varepsilon}\ln\frac{w - jv_\ell}{w + jv_\ell} \tag{A1.6.2}$$

from which the potential in the original structure indicated in Figure A1.6.1a is obtained by substituting for "w," resulting in:

$$v(z) = -\frac{q_\ell}{2\pi\varepsilon}\ln\frac{w - f(z)}{w - f(z)^*} \tag{A1.6.3}$$

The real part of the previous equation is the potential inside the structure indicated in Figure A1.6.1a. This is true applying the boundary conditions to the conclusions obtained in section A1.5 regarding real and imaginary parts of an analytical function. In fact, in that section we showed that the real and imaginary parts of an analytic function are solutions of Laplace's equation.

A1.7 THE SCHWARZ-CHRISTOFFEL TRANSFORMATION

The Schwarz-Christoffel transformation is a powerful tool to study electrostatic problems whose geometry can be regarded as a polygon. Is is also assumed that the polygon can be opened at some side. The internal region of the original structure is transformed into the upper complex half "W" plane, while the bounding perimeter is transformed in the real axis of the "W" plane. This situation is represented in Figure A1.7.1. The function that permits the transformation between the two planes is:

$$z = A\int_w (w - w_1)^{-\frac{\alpha_1}{\pi}}(w - w_2)^{-\frac{\alpha_2}{\pi}}\ldots(w - w_n)^{-\frac{\alpha_n}{\pi}}dw + C \tag{A1.7.1}$$

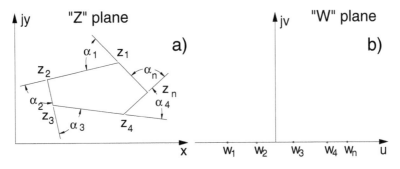

Figure A1.7.1

where "A" and "C" are two complex constants, the first proportional to the polygon dimension and orientation in the "Z" plane, the second proportional to the position of such polygon. The points "w_i" can theoretically be arbitrarily chosen to lie on the "u" axis; a proper choice of their position can simplify the solution of the problem.

Moving counterclockwise on the polygon corresponds to moving from left to right on the "u" axis. Particular attention is required to measure the angles. Note in fact that the angles are measured, let us say, in a counterclockwise manner, i.e., from the hypothetical line that extends the segment just covered and the nearest segment at left. As we have said, the transformation is also applicable to open polygons, and also if some point "w_i" is chosen to be at infinity on the "u" axis; in this last case it does not appear in Equation A1.7.1.

To do an example of the Schwarz-Christoffel transformation, let us evaluate the equipotential lines at the end of a parallel plate capacitor, indicated in Figure A1.7.2a. To simplify the analytical representation, let us suppose the plates to be of infinite extent for $x < 0$. We want to transform the capacitor upper and lower plates respectively into the $u < 0$ and $u > 0$ half axis. In this case, the following positions hold:

$$\alpha_2 = -\pi, \ z_2 = jd \rightarrow w_2 \overset{\perp}{=} -1 \tag{A1.7.2}$$

$$\alpha_3 = \pi, \ z_3 = -\infty \rightarrow w_3 \overset{\perp}{=} 0 \tag{A1.7.3}$$

$$\alpha_4 = -\pi, \ z_4 = 0 \rightarrow w_4 \overset{\perp}{=} 1 \tag{A1.7.4}$$

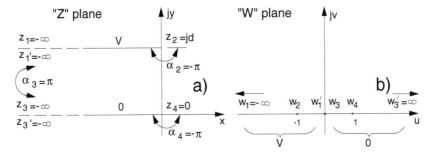

Figure A1.7.2

Inserting the previous points into equ. A1.7.1 we have:

$$z = A \int_{w} (w+1)\, w^{-1} (w-1)\, dw + C \qquad (A1.7.5)$$

which, integrated, gives:

$$z = (A/2)w^2 - A \ln(w) + c \qquad (A1.7.6)$$

If we insert in Equation A1.7.6 the points correspondence indicated in Equation A1.7.4 we have:

$$c \overset{\perp}{=} - A/2 \qquad (A1.7.7)$$

Now, let us set $w = |w|e^{j\varphi}$, so that Equation A1.7.6 becomes:

$$z = (A/2)w^2 - A\left(\ln|w| + j\varphi\right) + C \qquad (A1.7.8)$$

If we insert in the previous equation the points correspondence indicated in Equation A1.7.2, observing that $w_2 = -1$, the phase angle is $\varphi = \pi$, we have:

$$A \overset{\perp}{=} -d/\pi \qquad (A1.7.9)$$

and so Equation A.1.7.6 becomes:

$$z = (d/\pi)\left[\left(1 - w^2\right)/2 + \ln(w)\right] \qquad (A1.7.10)$$

This equation permits the transformation between parts a and b of Figure A1.7.2, but unfortunatly it is neither simple to invert nor simple to relate "x" and "y" with "u" and "v." So, it is convenient to apply another transformation to the previous equation, which permits these tasks. To this purpose, we use the following equation:

$$s \overset{\perp}{=} p + jq = (V/j\pi)\ln(w) \qquad (A1.7.11)$$

which transforms Figure A1.7.2b as indicated in Figure A1.7.3b. For simplicity, in Figure A1.7.3a we have reported Figure A1.7.2b. Note as Equation A1.7.11 transforms the negative and positive half axis of the "W" plane respectively in the $p = 0$ and $p = V$ axis of the "S" plane. As a consequence, the resulting structure is still a parallel plate capacitor. Equation A1.7.11 is simply invertible, obtaining:

$$w = e^{j\pi s/V} \qquad (A1.7.12)$$

and inserted into A1.7.10 gives:

$$z = (d/\pi)\left[0.5\left(1 - e^{j2\pi s/V}\right) + j\pi s/V\right] \qquad (A1.7.13)$$

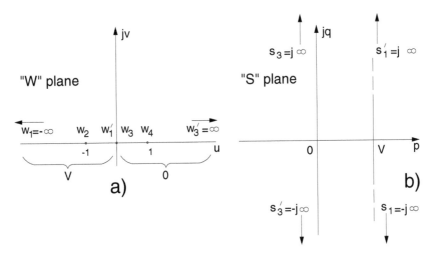

Figure A1.7.3

Now, since $z \doteq x + jy$ and $s \doteq p + jq$, equating real with real and imaginary with imaginary parts at both members of previous equation, we have:

$$x = (d/2\pi)\left[1 - e^{j2\pi s/V} \cos(2\pi p/V)\right] - dg/V \qquad (A1.7.14)$$

$$y = dp/V - (d/2\pi)\left[1 - e^{j2\pi s/V} \operatorname{sen}(2\pi p/V)\right] \qquad (A1.7.15)$$

The two previous equations represent the coordinates in the "Z" plane of potential and electric flux density, as a function of coordinates "p" and "q" in the "S" plane. From Figure A1.7.3b we recognize that if we fix a value for "p" it means to fix a potential; so, in these conditions if we vary "q" we have the potential lines for Figure A1.7.2a. Similarly, we recognize that if we fix a value for "q" it means to fix an electric flux. So, in these conditions if we vary "p" we have the electric flux lines for Figure A1.7.2a.

In Figure A1.7.4 we have reported, (drawn as explained) some of these lines near the end of the parallel plate capacitor, for a normalized d = 1 and V = 1. Here, equipotential lines and equal

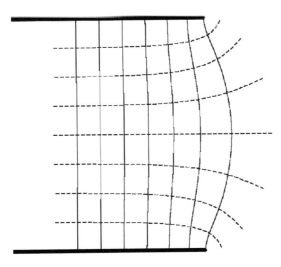

Figure A1.7.4

electric flux lines are respectively indicated with dashed and continuous lines. Note that because the capacitor is not closed, field lines go beyond the plates. This is the well-known effect called the "end effect," which is applicable for any structure that realizes a capacitor, for example an open end microstrip. As it is known, the end effect causes an increase of capacitance.

REFERENCES

1. S. Ramo, J.R. Whinnery, T. Van Duzer, *Fields and Waves in Communication Electronics*, John Wiley & Sons, New York, 1965.
2. G. Barzilai, *Fundamentals of Electromagnetism*, Siderea, Roma, 1975.
3. P. Bernardi, *Exercises in Electromagnetism*, Siderea, Roma, 1969.

Wave Equations, Waves, and Dispersion

A2.1 INTRODUCTION

In this Appendix we will review the most important analytical electromagnetic relationships, which are fundamentals of this text. Of course, we do not want to treat all the topics that are the source of these relationships because we do not want to make a text on electromagnetism. A complete treatment of these topics would require a book of at least the same length as this text. For this reason, it is required that the reader know the following arguments for a proper reading of this appendix:

1. Fundamental theory of electromagnetism, particularly regarding the following point 2
2. Theory of Maxwell's* equations
3. Matrix and vectorial algebra

Concerning points 1, 2, and 3, we will indicate, in the following sections, good texts where the reader can find background information. Since we think that the reader of this text surely has the background required for points 1 and 2, we have inserted in Appendix A8 all the necessary definitions on matrix and vector algebra required for a proper reading of this appendix.

A2.2 MAXWELL'S EQUATIONS AND BOUNDARY CONDITIONS

For an isotropic system not in motion, the Maxwell's equations are:

$$\underline{\nabla} \bullet (\varepsilon \underline{e}) = \rho \tag{A2.2.1}$$

$$\underline{\nabla} \bullet (\underline{b}) = 0 \tag{A2.2.2}$$

$$\underline{\nabla} \otimes \underline{e} = -\partial \underline{b}/\partial t \tag{A2.2.3}$$

$$\underline{\nabla} \otimes (\underline{b}/\mu) = \partial(\varepsilon \underline{e})/\partial t + \underline{J} \tag{A2.2.4}$$

where "J" is the current surface density impressed by the generators, measured in Amp/m². If the medium is also homogeneous and isotropic, then "ε" and "μ" can be set outside the operators; for example, Equation A2.2.1 becomes:

* James Clark Maxwell, English physicist, born in Edinburgh in 1831, died in Cambridge in 1879.

$$\underline{\nabla} \bullet \underline{e} = \rho/\varepsilon \qquad\qquad (A2.2.5)$$

The e.m. field must satisfy some conditions at dielectric interfaces and on conductors. These conditions are called "boundary conditions" and once satisfied, assure that the solution of the Maxwell equation is unique. We will give such boundary conditions without proof, sending the reader to the voluminous literature on the subject.[1,2,3,4,5] In the following equations, "n" is the normal at the interface surface under study, while subscripts "1" and "2" individuate the fields inside regions "1" and "2."

a. Condition for the electrical displacement field vector "$\underline{d}$:"

$$\underline{n} \bullet \left(\underline{d}_2 - \underline{d}_1 \right) = q_s \qquad\qquad (A2.2.6)$$

   i.e., the difference between the normal components to the local interface surface is equal to the surface charge "q_s."*

b. Condition for the electrical field vector "$\underline{e}$:"

$$\underline{n} \otimes \left(\underline{e}_2 - \underline{e}_1 \right) = 0 \qquad\qquad (A2.2.7)$$

   i.e., the difference between the tangential components to the local interface surface is continuous. The case where one of the two media is a perfect conductor is interesting; in this case, the electric field is zero inside the conductor and from the previous equation we have $\underline{n} \otimes \underline{e}_2 = 0$, i.e., the tangential electric field is zero on a perfect conductor.

c. Condition for the magnetic induction field vector "$\underline{b}$:"

$$\underline{n} \bullet \left(\underline{b}_2 - \underline{b}_1 \right) = 0 \qquad\qquad (A2.2.8)$$

   i.e., the difference between the normal components to the local interface surface is zero. This condition can also be obtained using the "duality principle" applied to Equation A2.2.6. This principle states that when a solution of the Maxwell's equation for a field is known, it is possible to have the solution for the other field if we make the following substitutions:

$$e \to h, \ h \to -e, \ \mu \leftrightarrow \varepsilon, \ J \to J_m, \ J_m \to -J, \ \rho \to \rho_m, \ \rho_m \to -\rho \qquad (A2.2.9)$$

   where "J," "ρ," "J_m," and "ρ_m" are, respectively, the general surface current and volumetric charge, respectively, for the electric and magnetic case. Also, if the solution obtained with the duality principle is surely a solution of the Maxwell's equations, this solution could not correspond to a physical case. For this reason, it is always necessary to verify that the new solution is applicable to the original physical problem. In this case, observing that magnetic free charges are not known in practice, applying the "duality principle" to Equation A2.2.6 we have Equation A2.2.8.

d. Condition for the magnetic field vector "$\underline{h}$:"

$$\underline{n} \otimes \left(\underline{h}_2 - \underline{h}_1 \right) = \underline{i}_\ell \qquad\qquad (A2.2.10)$$

---

* Since we use MKSA reference system unless otherwise stated, $[q_s]$ = Coulomb/m².

i.e., the difference between the tangential components to the local interface surface is equal to the linear current "i_ℓ."* The case where one of the two media is a perfect conductor is interesting; in this case, the magnetic field is zero inside the conductor and from the previous equation we have $\underline{n} \otimes \underline{h}_2 = i_\ell$, i.e., the tangential magnetic field is zero on a perfect conductor.

A2.3 WAVE EQUATIONS IN HARMONIC TIME DEPENDENCE

In homogeneous, isotropic, and linear media where relative permeability "μ_r," relative permittivity "ε_r." and conductivity "g" are space, time, and field intensity invariant, if field sources are at the infinite, then magnetic "$\underline{h}$" and electric "$\underline{e}$" field vectors satisfy the following equations:

$$\nabla^2(\underline{e}) = k^2 \underline{e} \qquad (A2.3.1)$$

$$\nabla^2(\underline{h}) = k^2 \underline{h} \qquad (A2.3.2)$$

where:

$$k^2 \overset{\perp}{=} -\omega^2 \mu \varepsilon_c \quad \mu \overset{\perp}{=} \mu_0 \mu_r \quad \varepsilon_c \overset{\perp}{=} \varepsilon - jg/\omega \quad \varepsilon \overset{\perp}{=} \varepsilon_0 \varepsilon_r \overset{\perp}{=} \varepsilon_{ar} - j\varepsilon_{aj} \quad \varepsilon_r \overset{\perp}{=} \varepsilon_{rr} - j\varepsilon_{rj} \quad \omega \overset{\perp}{=} 2\pi f \quad (A2.3.3)$$

and "f" is the signal frequency. Note that "k" is, in general, a complex quantity. A2.3.1 and A2.3.2 are called "homogeneous wave equations in harmonic time dependence" or simply "homogeneous wave equations." Sometimes, Equations A2.3.1 and A2.3.2 are also called "Helmholtz** equations." In these equations, with "∇^2" we mean the "Laplacian*** operator" or simply "Laplacian," which is defined in Appendix A8. Note that Equations A2.3.1 and A2.3.2 are mathematically equivalent, so to find the solutions of wave equations we can concentrate only on one equation, without losing the generality of the solution. We will use A2.3.2. So, representing "$\underline{h}$" in Cartesian coordinates, i.e.:

$$\underline{h} \overset{\perp}{=} h_x \underline{x}_0 + h_y \underline{y}_0 + h_z \underline{z}_0 \qquad (A2.3.4)$$

and applying Equation A2.3.4 into A2.3.2 we have that this equation is satisfied if:

$$\nabla^2 h_x = k^2 h_x \qquad (A2.3.5)$$

$$\nabla^2 h_y = k^2 h_y \qquad (A2.3.6)$$

$$\nabla^2 h_z = k^2 h_z \qquad (A2.3.7)$$

Also in this case equations A2.3.5 through A2.3.7 are mathematically equivalent, and we will concentrate only on A2.3.7. To solve such an equation, the "variable separation method, VSM" is generally used, which consists of finding a function "$h_z(x,y,z)$" given by the product of functions of only a variable, i.e.:

$$h_z(x, y, z) \overset{\perp}{=} h_{0z} X(x) Y(y) Z(z) \qquad (A2.3.8)$$

---

* $[i_\ell] = \text{Amp/m}$.
** H.L.F. Von Helmholtz, German physicist, born in 1821 in Potsdam and died in Charlottensburg in 1894.
*** Pier Simon de Laplace, French mathematician, born in Beaumont en Auge in 1749 and died in Paris in 1827.

Now, to simplify the notation, we will omit the coordinate dependence every time so we will not generate confusion. So, inserting A2.3.8 into A2.3.7 we have:

$$h_{0z}\left[\frac{d^2X}{dx^2}YZ + \frac{d^2Y}{dy^2}XZ + \frac{d^2Z}{dz^2}XY\right] = k^2 h_{0z}XYZ \qquad (A2.3.9)$$

from which, with simple passages:

$$\frac{d^2X}{Xdx^2} + \frac{d^2Y}{Ydy^2} + \frac{d^2Z}{Zdz^2} = k^2 \qquad (A2.3.10)$$

Equation A2.3.10 is surely satisfied if:

$$\frac{d^2X}{Xdx^2} \doteq k_x^2 \qquad \frac{d^2Y}{Ydy^2} \doteq k_y^2 \qquad \frac{d^2Z}{Zdz^2} \doteq k_z^2 \qquad (A2.3.11)$$

with the condition:

$$k_x^2 + k_y^2 + k_z^2 =! k^2 \qquad (A2.3.12)$$

Since "k" is in general a complex number, the same holds for "k_x," "k_y," and "k_z." Also in this case all equations in A2.3.11 are mathematically equivalent, and we will concentrate only on the last one. In mechanics theory, equations of this type are called "harmonic motion equations." The solution of this equation is simple and can be found by setting $Z(z) \doteq M_z e^{k_z z}$, where "$M_z$" is a scalar and "$k_z$" is called the "propagation constant" with dimensions "1/m" in MKSA system unit.*
So, in our case the general solution is a linear combination of exponentials:

$$Z(z) = M_z^+ e^{-k_z z} + M_z^- e^{k_z z} \qquad (A2.3.13)$$

or a linear combination of hyperbolic sinus and cosinus:

$$Z(z) = A\cosh(k_z z) + B\mathrm{senh}(k_z z) \qquad (A2.3.14)$$

since hyperbolic sinus and cosinus are defined as:

$$\cosh(z) \doteq \frac{e^z + e^{-z}}{2} \qquad \mathrm{senh}(z) \doteq \frac{e^z - e^{-z}}{2}$$

that is, like a linear combination of exponentials. Also for Equation A2.3.14, the quantities "A" and "B" are scalars.
The case where the quantity "k_z" is imaginary, i.e., $k_z \equiv jk_{zj}$ is interesting. In this case the solution of equations like A2.3.11 is a linear combination of sinus and cosinus, i.e.:

---

* Unless otherwise stated we will use MKSA system unit.

$$Z(z) = A' \cos\left(k_{zj} z\right) + B' \sin\left(k_{zj} z\right) \quad \text{with } k_z \overset{\perp}{=} jk_{zj}. \tag{A2.3.15}$$

The choice of which kind of solution between A2.3.13 and A2.3.14 it is better to employ depends on the known boundary conditions of the electromagnetic problem. Exponential solution is useful when one extreme of the sources is theoretically to the infinite, while hyperbolic solutions are useful when considering a limited region of space. The term that contains the negative exponential is called "progressive," since it decreases in amplitude in the positive direction of "z" while the other is called "regressive," which decreases in amplitude when "z" decreases in amplitude in the negative direction of "z."

Using only the progressive terms for the exponential form for "X(x)," "Y(y)," and "Z(z)," Equation A2.3.8 can be rewritten as:

$$h_z(x, y, z) \overset{\perp}{=} h_{0z} e^{-k_x x - k_y y - k_z z} \tag{A2.3.16}$$

and the associated vector "$\underline{h}_z(x,y,z)$" will be:

$$\underline{h}_z(x, y, z) \overset{\perp}{=} h_{0z} e^{-k_x x - k_y y - k_z z} \underline{z}_0 \tag{A2.3.17}$$

So, the magnetic field vector "$\underline{h}(x,y,z)$" in the Cartesian system unit will be:

$$\underline{h}(x, y, z) \equiv \underline{h}_x(x, y, z) + \underline{h}_y(x, y, z) + \underline{h}_z(x, y, z) \overset{\perp}{=} h_0 e^{-k_x x - k_y y - k_z z} \tag{A2.3.18}$$

and similarly, for the electric field:

$$\underline{e}(x, y, z) \equiv \underline{e}_x(x, y, z) + \underline{e}_y(x, y, z) + \underline{e}_z(x, y, z) \overset{\perp}{=} e_0 e^{-k_x x - k_y y - k_z z} \tag{A2.3.19}$$

The vectors "$\underline{h}_0$" and "$\underline{e}_0$" are, in general, complex and contain the time dependence. This topic will be treated next.

It is useful at this point to indicate another simple and compact representation for Equations A2.3.18 and A2.3.19. If we define "$\underline{k}$" and position "$\underline{r}$" vectors as:

$$\underline{k} \overset{\perp}{=} k_x \underline{x}_0 + k_y \underline{y}_0 + k_z \underline{z}_0 \quad \text{and} \quad \underline{r} \overset{\perp}{=} r_x \underline{x}_0 + r_y \underline{y}_0 + r_z \underline{z}_0 \tag{A2.3.20}$$

Equations A2.3.18 and A2.3.19 become:

$$\underline{h}(x, y, z) \equiv \underline{h}_0 e^{-\underline{k} \cdot \underline{r}} \quad \text{and} \quad \underline{e}(x, y, z) \equiv \underline{e}_0 e^{-\underline{k} \cdot \underline{r}} \tag{A2.3.21}$$

where with "$\underline{k} \bullet \underline{r}$" we indicate the product vector between "$\underline{k}$" and "$\underline{r}$." The vector "$\underline{h}_0$" and "$\underline{e}_0$" are, in general, complex, and we can write:

$$\underline{h}_0 \overset{\perp}{=} \underline{h}_{0r} + j\underline{h}_{0j} \quad \text{and} \quad \underline{e}_0 \overset{\perp}{=} \underline{e}_{0r} + j\underline{e}_{0j} \tag{A2.3.22}$$

We can also associate the responsibility of time dependence to "$\underline{h}_0$" and "$\underline{e}_0$."

In the next section we will make a closer study of the propagation vectors and their relationships.

A2.4 THE PROPAGATION VECTORS AND THEIR RELATIONSHIPS WITH ELECTRIC AND MAGNETIC FIELDS

The most known propagation vector "$\underline{k}$" is composed of the propagation constants we gave in the previous section. In the Cartesian coordinate system it is defined as:

$$k_x \underline{x}_0 + k_y \underline{y}_0 + k_z \underline{z}_0 \overset{\perp}{=} \underline{k} \tag{A2.4.1}$$

being "$\underline{x}_0$," "$\underline{y}_0$," and "$\underline{z}_0$" the axis versors of our reference system. By definition of vector product, we have:

$$\underline{k} \cdot \underline{k} \equiv k_x^2 + k_y^2 + k_z^2 \equiv k^2 \tag{A2.4.2}$$

According to A2.3.3, "k" is in general a complex quantity, and we set:

$$k = j\omega\left(\mu\varepsilon_c\right)^{0.5} \equiv j\omega\left[\mu\left(\varepsilon - jg/\omega\right)\right]^{0.5} \overset{\perp}{=} k_r + jk_j \tag{A2.4.3}$$

In general, "k" is dependent on the physical characteristics of the propagation medium. In Chapter 1, where we studied the general theory of transmission lines, we gave the expression of "k" as:

$$k = \left[\left(R + j\omega L\right)\left(G_p + j\omega C\right)\right]^{0.5} \overset{\perp}{=} k_r + jk_j \tag{1.4.1}$$

where "R," "L," "G_p," and "C" are, respectively, the series resistance, series inductance, parallel conductance, and parallel capacitance for the unit length of the line, so that the dimensions of "k" are 1/m. But also in our case, the dimensions of "k" are 1/m, since the following dimensions apply for "μ," "ε," and "ω:"*

$$[\mu] = \Omega.\sec/m \quad [\varepsilon] = \sec/m\Omega \quad [\omega] = 1/\sec \tag{A2.4.4}$$

Expressions A2.4.3 and 1.4.1 can always be set equal to each other, but the procedure can be quite tedious. For example, in a lossless ideal coaxial cable with dielectric with parameters "ε" and "μ," the capacitance and inductance per unit length are:[6,7]

$$C = 2\pi\varepsilon/\log\left(r_e/r_i\right) \quad L = \mu\log\left(r_e/r_i\right)/2\mu \tag{A2.4.5}$$

where "log" is the logarithm in base "10." If we insert Equations A2.4.5 into 1.4.1 we have the same expression A2.4.3 evaluated with "g = 0," i.e., with zero losses.

Squaring A2.4.3 and equating imaginary and real parts at both members, we have:

$$k_r = \omega\left\{0.5\mu\varepsilon_{ar}\left[\left(1 + \left(d/\varepsilon_{ar}\right)^2\right)^{0.5} - 1\right]\right\}^{0.5} \tag{A2.4.6}$$

---

* Whenever confusion will not arise, we will indicate the variable dimensions with square brackets.

$$k_j = \omega \left\{ 0.5\mu\varepsilon_{ar} \left[\left(1 + \left(d/\varepsilon_{ar} \right)^2 \right)^{0.5} + 1 \right] \right\}^{0.5} \qquad (A2.4.7)$$

where $d \doteq \varepsilon_{aj} + g/\omega$. For an ideal lossless dielectric, i.e., with $\varepsilon_{rj} = 0$ and $g = 0$, we have:

$$k_r = 0 \quad \text{and} \quad k_j = \omega \left(\mu\varepsilon \right)^{0.5} \equiv 2\pi \left(\mu_r \right)^{0.5} / \lambda_g \rightarrow k \equiv jk_j \qquad (A2.4.8)$$

For a good dielectric, i.e., those used for guided propagation, the product "$\omega\varepsilon_{aj}$" is much higher than "g" and $d/\varepsilon_{ar} \approx \tan\delta*$ so that from A2.4.6 we have:

$$k_r = \omega \left\{ 0.5\mu\varepsilon_{ar} \left[\left(1 + \left(\tan\delta \right)^2 \right)^{0.5} - 1 \right] \right\}^{0.5} \qquad (A2.4.9)$$

Then, supposing $\varepsilon_{rj} \ll \varepsilon_{rr}$, we can expand with McLaurin** series the previous equation, and terminating the expansion to the third term we have:

$$k_r = \frac{\omega \tan\delta \sqrt{\varepsilon_{rr}}}{2v_0} = \frac{\pi \tan\delta \sqrt{\varepsilon_{rr}}}{\lambda_0} \qquad (A2.4.10)$$

The previous expression, representing only the dielectric losses, can be applied to any t.l. supporting a "TEM" mode, for example a stripline or a coaxial cable. Conductor losses are strongly dependent on the cross-section of the t.l., and for this reason, "TEM" t.l.s with different cross-sections have in general different expressions and values of conductor loss.

Another expression for "$\underline{k}$" is the so called "alphabeta $\underline{k}$ vector," given by:

$$\underline{k} \doteq \underline{\alpha} + j\underline{\beta} \qquad (A2.4.11)$$

where "$\underline{\alpha}$" and "$\underline{\beta}$" are called "attenuation vector" and "phase vector," respectively, both real numbers. The "alphabeta $\underline{k}$ vector" can be easily set in relationship to "k_r" and "k_j." We can write:

$$\underline{k} \cdot \underline{k} = k^2 = \left(k_r + jk_j \right)^2 \equiv \left(\underline{\alpha} + j\underline{\beta} \right) \cdot \left(\underline{\alpha} + j\underline{\beta} \right) \qquad (A2.4.12)$$

Then, explicating and equating imaginary and real parts at both members of the previous equation we have:

$$\alpha^2 - \beta^2 = k_r^2 - k_j^2 \qquad (A2.4.13)$$

$$\underline{\alpha} \cdot \underline{\beta} = k_r k_j \qquad (A2.4.14)$$

Note that if "$\underline{\alpha}$" and "$\underline{\beta}$" vectors are parallel, then $\underline{\alpha} \bullet \underline{\beta} \equiv \alpha\beta$ that with A2.4.13 results in:

$$\alpha \equiv k_r \quad \text{and} \quad \beta \equiv k_j \qquad (A2.4.15)$$

* The quantity "$\tan\delta$" was defined in Chapter 1.
** Colin McLaurin, English mathematician, born in Kilmodan in 1698 and died in Edinburgh in 1746.

In the next section we will define various types of waves according to the value of the vector product "$\underline{\alpha} \bullet \underline{\beta}$."

With the introduction of the "alphabeta $\underline{k}$ vector," Equation A2.3.21 becomes:

$$\underline{h}(x, y, z) \equiv \underline{h}_0 e^{-(\underline{\alpha}+j\underline{\beta})\bullet \underline{r}} \quad \text{and} \quad \underline{e}(x, y, z) \equiv \underline{e}_0 e^{-(\underline{\alpha}+j\underline{\beta})\bullet \underline{r}} \tag{A2.4.16}$$

Note the term "$e^{-\underline{\alpha} \bullet \underline{r}}$" is responsible for the field amplitude variation, while the term "$e^{-j\underline{\beta} \bullet \underline{r}}$" is responsible for the field phase variation. Considering "$\underline{\alpha}$" and "$\underline{\beta}$" as reference vectors, the condition:

$$\underline{\alpha} \bullet \underline{r} = \text{constant} \tag{A2.4.17}$$

represents a plane orthogonal to the vector "$\underline{\alpha}$," and it is called an "equiamplitude plane." Similarly, the condition:

$$\underline{\beta} \bullet \underline{r} = \text{constant} \tag{A2.4.18}$$

represents a plane orthogonal to the vector "$\underline{\beta}$" and it is called an "equiphase plane." It is for this reason that fields given by expressions A2.4.16 are said to represent a "plane wave." If the projection of vector "$\underline{r}$" along "$\underline{\alpha}$" or "$\underline{\beta}$" has to be constant, then the extreme of "$\underline{r}$" must be on a plane orthogonal to "$\underline{\alpha}$" or "$\underline{\beta}$."

Important relationships hold between vectors "$\underline{k}$," "$\underline{h}$," and "$\underline{e}$." To show them, we need to start from the two Maxwell's equations in harmonic time dependence, i.e.:

$$\underline{\nabla} \otimes \underline{e} = -j\omega\mu\underline{h} \quad \text{and} \quad \underline{\nabla} \otimes \underline{h} = j\omega\varepsilon_c\underline{e} \tag{A2.4.19}$$

Then, applying the "curl operator" definition given in Appendix A8, the first of A2.4.19 becomes:

$$\underline{\nabla} \otimes \underline{e} \equiv \begin{pmatrix} \underline{x}_0 & \underline{y}_0 & \underline{z}_0 \\ -k_x & -k_y & -k_z \\ e_x & e_y & e_z \end{pmatrix} \equiv -\underline{k} \otimes \underline{e} = -j\omega\mu\underline{h} \tag{A2.4.20}$$

and so, using the expression A2.3.21, we have:

$$\underline{k} \otimes \underline{e}_0 \overset{\perp}{=} j\omega\mu\underline{h}_0 \tag{A2.4.21}$$

Similarly proceeding with the second of A2.4.19 we have:

$$\underline{k} \otimes \underline{h}_0 \overset{\perp}{=} -j\omega\varepsilon\underline{e}_0 \tag{A2.4.22}$$

Doing the scalar product between "$\underline{k}$" and the first members of A2.4.21 and A2.4.22 we can write:

$$\underline{k} \bullet \underline{h}_0 = 0 \tag{A2.4.23}$$

$$\underline{k} \bullet \underline{e}_0 = 0 \tag{A2.4.24}$$

Equations A2.4.21 through A2.4.24 are very important relationships for plane waves. We will see later some examples where their use will give us important results.

A2.5 THE TIME DEPENDENCE

In all this text, as already stated, we will assume that time dependence of our electromagnetic field is sinusoidal, i.e., of the form:

$$z(t) \stackrel{\perp}{=} M \operatorname{sen}(\omega t + \varphi) \tag{A2.5.1}$$

with "ω" the signal angular frequency, "φ" the initial phase, and "M" a generic real constant. For simplicity, we will set the origin of the time in a point where the initial phase is zero. Since the sinus function is proportional to the imaginary part of the exponential, it is easy to see that to extract the time dependence from any wave expression it is enough to multiply the function time "$e^{j\omega t}$" and extract the imaginary part.

For example, to have the time dependence explicated for the magnetic field given in A2.4.16 we can multiply time "$e^{j\omega t}$" and extract the imaginary part. So:

$$\underline{h}(x,y,z,t) \equiv \operatorname{Im}\left(\underline{h}_0 e^{-(\alpha+j\beta)\cdot\underline{r}} e^{j\omega t}\right) \tag{A2.5.2}$$

or:

$$\underline{h}(x,y,z,t) \equiv e^{-\alpha\cdot\underline{r}}\left\{\underline{h}_{0r}\operatorname{sen}(\omega t - \beta\cdot\underline{r}) + \underline{h}_{0j}\cos(\omega t - \beta\cdot\underline{r})\right\} \tag{A2.5.3}$$

A similar expression can be obtained for the electric field in A2.4.16. However, anytime we do not consider it important, the time dependence will be omitted.

A2.6 PLANE WAVE DEFINITIONS

Plane wave definitions are related to the result of the scalar product of "$\underline{\alpha}$" and "$\underline{\beta}$" given in Section A2.4. Of course, the existence of these plane waves can be rigorously shown starting from the general Maxwell's equation and can be found in many texts.[8,9,10]

a. Uniform Plane Wave This is a plane wave for which equiamplitude and equiphase planes are parallel. Consequently $\underline{\alpha} \bullet \underline{b} \neq !0$ since, as was shown in Section A2.4, $\underline{\alpha} \equiv k_r$ and $\underline{\beta} \equiv k_j$ when "$\underline{\alpha}$" and "$\underline{\beta}$" vectors are parallel. Examples of uniform plane waves are the electromagnetic field at great distance from a dipole.

Uniform plane waves are usually simply called "UPW."

b. Unattenuated Uniform Plane Wave This is a uniform plane wave that propagates in a lossless media, i.e., a media with g = 0. Since in this case $k_r \equiv 0$, then $\underline{\alpha} \bullet \underline{\beta} = !0$ because, according to point a, now a $\equiv 0$.

c. Nonuniform Plane Wave This is a plane wave for which equiamplitude and equiphase planes are not parallel. The possible case of "$\underline{\alpha} \perp \underline{\beta}$" is given next in item d.

Nonuniform plane waves are usually simply called as "NUPW."

d. Unattenuated Nonuniform Plane Wave This is a nonuniform plane wave that propagates in a lossless media, i.e., a media with g = 0. Since $k_r \equiv 0$ for g = 0, then $\underline{\alpha} \bullet \underline{\beta} = !0$, as in item b. From the wave theory used in electromagnetism, "β" represents the phase constant and it cannot be zero, otherwise we do not have propagation through waves. So, the condition $\underline{\alpha} \bullet \underline{\beta} = !0$ can be satisfied for:

1. $\alpha = 0$, and $\beta \neq 0$, which results in a particular case of uniform plane wave discussed in item b. Since $\alpha = 0$, the constant phase plane is also surely a constant amplitude plane, which is the condition for a uniform plane wave.
2. $\underline{\alpha} \perp \underline{\beta}$, with $\alpha \neq 0$ and $\beta \neq 0$, which represents our case, d.

Note that condition 2 results in a wave whose amplitude decreases exponentially in some direction, with constant "α" if it propagates in a lossless media.

e. "TEM" Wave The "TEM" wave is a plane wave with the characteristic that electric and magnetic fields have only transverse components, with respect to the direction of propagation. "TEM" is the acronym for "Transverse Electric and Magnetic" wave.

f. "TE" Wave The "TE" wave is a plane wave with the characteristic that the electric field is the only field with transverse components, with respect to the direction of propagation. "TE" is the acronym for "Transverse Electric" wave.

g. "TM" Wave The "TM" wave is a plane wave with the characteristic that the magnetic field is the only field with transverse components, with respect to the direction of propagation. "TM" is the acronym for "Transverse Magnetic" wave.

h. Standing Waves All the previous waves defined above can be progressive, regressive, or a combination of both. When both progressive and regressive waves exist, it is common practice to say that a standing wave phenomenon exists. From the general line theory we discussed in Chapter 1, we know that standing waves exist every time a mismatching exists in a transmission line. In the case of waves, the same holds, with the addition that since wave propagation can also be sustained in the vacuum, it is possible to have standing waves if a progressive wave is reflected from a surface. As an example, let us suppose that the electric field "$\underline{e}$" of an unattenuated "UPW" coincides on a perfect plane conductor parallel to the electric field. The Cartesian coordinate system is aligned with the direction "x" coincident to the linear polarized electric field, and the direction "z" is coincident with the direction of propagation. The "y" axis is aligned to the direction of the magnetic field, and the origin of the reference system is on the conductor plane. We can write:

$$e_x = e^+ e^{-j\beta z} + e^- e^{j\beta z} \tag{A2.6.1}$$

If we indicate with "ζ" the characteristic impedance of the "UPW,"* for the magnetic field we can write:**

$$h_y = \left[e^+ e^{-j\beta z} - e^- e^{j\beta z} \right] / \zeta \tag{A2.6.2}$$

In this case, we know that the electric field must be zero on the conductor surface, and so for $z = 0$ we have:

$$e^+ = ! - e^- \stackrel{\perp}{=} A \tag{A2.6.3}$$

Inserting A2.6.3 into A2.6.1 and A2.6.2 we have:

---

* We will see in Section A2.11 that for the case of an "UPW" we have $\zeta = (\mu/\varepsilon_0)^{0.5}$.
** See Section 1.3 of Chapter 1, since the representation A2.6.1 is the same as that developed for voltage in a generic transmission line.

$$e_x = A\left[e^{-j\beta z} - e^{j\beta z}\right] \equiv -2\,j\,A\,\text{sen}\,(\beta z) \tag{A2.6.4}$$

$$h_y = A\left[e^{-j\beta z} + e^{j\beta z}\right]\big/\zeta \equiv 2A\cos(\beta z)/\zeta \tag{A2.6.5}$$

Note as in this case of stationary waves, the amplitude of the electric field has a coordinate shape proportional to "sen(βz)," while the amplitude of the magnetic field has a coordinate shape proportional to "cos(βz)." This means that for these fields, zeros exist in a precise coordinate of the propagation axis, which do not change positions. For example, "e_x" is zero every $\beta z = 2n\pi$, with "n" an integer number. It is for this reason that a situation like this is said to be a "stationary wave phenomenon." In addition, "e_x" and "h_y" are in space quadrature, since one moves with "sinus" and the other with "cosinus" and in time quadrature due to the presence of the term "j."

A2.7 EVALUATION OF ELECTROMAGNETIC ENERGY

Every electromagnetic wave brings energy with itself. To make a practical example of how important it is to evaluate the e.m. energy, we can simply think of a microwave oven, where an electromagnetic field near some GHz is able to cook foods. The determination of e.m. energy is also very important to evaluate possible risks for electromagnetic compatibility, i.e., how much our device can radiate energy outside the area where the energy should be confined.

A very important relationship in electromagnetism is the "Poynting* theorem." This theorem represents in electromagnetism the general principle of energy conservation. To introduce such an important result, we have to begin our study with the general Maxwell's equations:

$$\underline{\nabla} \otimes \underline{e} = -\frac{\partial \underline{b}}{\partial t} - \underline{J}_{im} \tag{A2.7.1}$$

$$\underline{\nabla} \otimes \underline{h} = -\frac{\partial \underline{d}}{\partial t} + \underline{J}_{ie} + g\underline{e} \tag{A2.7.2}$$

The terms "$\underline{J}_{ie}$" and "$\underline{J}_{im}$" represent the surface current density impressed by the generators, respectively of "electric" and "magnetic" types. Of course, the term "$\underline{J}_{im}$" is used to satisfy the "duality principle."

Multiplying scalarly time "$\underline{h}$" A2.7.1 and time "$\underline{e}$" A2.7.2, and subtracting, we have:

$$\underline{h} \cdot \underline{\nabla} \otimes \underline{e} - \underline{e} \cdot \underline{\nabla} \otimes \underline{h} = -\underline{h} \cdot \frac{\partial \underline{B}}{\partial t} - \underline{e} \cdot \frac{\partial \underline{D}}{\partial t} - \underline{h} \cdot \underline{J}_{im} - \underline{e} \cdot \underline{J}_{ie} - g\underline{e} \cdot \underline{e} \tag{A2.7.3}$$

Since we know that:

$$\underline{h} \cdot \underline{\nabla} \otimes \underline{e} - \underline{e} \cdot \underline{\nabla} \otimes \underline{h} \equiv \underline{\nabla} \cdot (\underline{e} \otimes \underline{h}) \tag{A2.7.4}$$

Equation A2.7.3 can be rewritten as:

$$\underline{\nabla} \cdot (\underline{e} \otimes \underline{h}) + \underline{h} \cdot \frac{\partial \underline{B}}{\partial t} + \underline{e} \cdot \frac{\partial \underline{D}}{\partial t} + g e^2 = -\underline{h} \cdot \underline{J}_{im} - \underline{e} \cdot \underline{J}_{ie} \tag{A2.7.5}$$

* J.H. Poynting, English physicist, born in Monton in 1852 and died in Birmingham in 1914.

Since the dimensions of "$\underline{\nabla} \cdot (\underline{e} \otimes \underline{h})$" are so that:

$$[\underline{\nabla} \cdot (\underline{e} \otimes \underline{h})] = W/m^3 \qquad (A2.7.6)$$

if we integrate Equation A2.7.5 in the volume "V," which contains the region under study, we have a power as result. To simplify the calculations, we can apply the "divergence theorem,"* also called "Gauss' theorem," which states:

$$\int_V \underline{\nabla} \cdot \underline{v}\,dV = \oint_S \underline{v} \cdot \underline{n}\,dS \qquad (A2.7.7)$$

where "$\underline{v}$" is a generic vector, "S" is the surface that contains the volume "V," and "$\underline{n}$" is the "normal" to this surface and directed outside the region under study. Using A2.7.7, A2.7.5 becomes:

$$\int_S (\underline{e} \otimes \underline{h}) \cdot \underline{n}\,dS + \int_V \left(\underline{h} \cdot \frac{\partial \underline{B}}{\partial t} + \underline{e} \cdot \frac{\partial \underline{D}}{\partial t} \right) dV + \int_V g e^2 dV = -\int_V \left(\underline{h} \cdot \underline{J}_{im} + \underline{e} \cdot \underline{J}_{ie} \right) dV \qquad (A2.7.8)$$

Equation A2.7.8 is the "Poynting theorem." Taking in analysis the first member of this equation, we have that:

1. The first integral represents the e.m. power leaving the surface "S," and it is called the "flux of Poynting vector"
2. The second integral represents the e.m. power that is stored inside the region "S"
3. The third integral represents the power lost in dissipation

At the second member we have the power that is impressed by the generators. So, from this analysis it is simple to understand how the Poynting theorem is an application to the electromagnetic case of the general energy conservation principle.

The vector:

$$\underline{P} \overset{\perp}{=} \underline{e} \otimes \underline{h} \qquad (A2.7.9)$$

is called the "Poynting vector," with dimensions W/m^2. Sometimes, this vector is associated with the dimension of W/unit surface, but this is not theoretically correct, since the application of A2.7.8 has to be performed on closed surfaces "S." In other words, there could be situations in which the Poynting vector is different from zero, while its integral on a closed surface is zero. A practical situation like this is a region where electrostatic charges and magnetostatic fields exist. Here, surely subregions exist where static electric and magnetic fields exist and are not parallel. Consequently the Poynting vector here is different than zero.** But if we integrate on a closed surface, the result is zero. Since inside this surface there does not exist currents for our electrostatic problem, consequently:

$$\underline{\nabla} \otimes \underline{e} \equiv 0 \quad \text{and} \quad \underline{\nabla} \otimes \underline{h} \equiv 0 \qquad (A2.7.10)$$

So, remembering A2.7.4 we have:

---

* The divergence, or Gauss theorem is defined in Appendix A8 together other important vectorial relationships.
** Of course, in this electrostatic example, the power flux has little meaning.

$$\int_S (\underline{e} \otimes \underline{h}) \cdot \underline{n} dS \equiv 0 \qquad (A2.7.11)$$

i.e., the flux of the Poynting vector is zero, and locally the Poynting vector is different from zero. So, caution must be taken when evaluating the flux of the Poynting vector through an open surface.

A2.8 WAVES IN GUIDING STRUCTURES WITH CURVILINEAR ORTHOGONAL COORDINATE REFERENCE SYSTEM

We will call a "guiding structure" every structure that is able to guide e.m. energy, and where some energy can be lost during the path inside the guiding structure. Of course, this is another definition of the "transmission line" we gave in Chapter 1. In our text we have studied a lot of these guiding structures, like microstrips, striplines, slotlines, etc.

A general definition of a guiding structure where it is useful to apply a "curvilinear orthogonal coordinate reference system" is where this structure is obtained as the translation along an axis, called the "longitudinal axis," of a conductor ring of any shape. This reference system is composed of a longitudinal coordinate with versor "ℓ_0" and two transversal coordinates with versors "t_{10}" and "t_{20}," these last being on a plane that is orthogonal to the direction "ℓ." Practical examples of such a reference system are the cylindrical or Cartesian reference system. A representation of a generic guiding structure and its associated coordinate system is indicated in Figure A2.8.1. With "S" we indicate the area of a transversal surface, whose perimeter is indicated with "p." In addition, "t" is an axis that is always locally tangent to "p" and always orthogonal to "ℓ," and "n" is an axis orthogonal to "t." So, "n" and "t" form a Cartesian coordinate reference system, while in the most general case "ℓ," "t_1," and "t_2" are curvilinear coordinates and do not necessarily represent a Cartesian coordinate system.

Solving the Maxwell's equations for a structure as indicated in Figure A2.8.1, it is possible to show that only "TE" and "TM" modes can exist inside this structure. The word "mode" intends the natural possible configurations of the e.m. field inside a region without sources, i.e., a region like our guiding structure indicated in Figure A2.8.1 where the sources are assumed to be only on the extreme terminals. The fields associated with these modes can be obtained using a potential

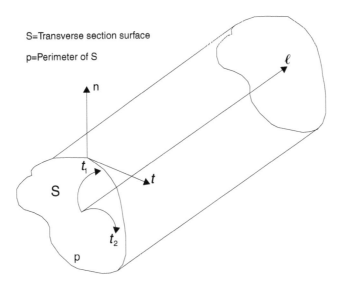

Figure A2.8.1

function.* In a system with such a coordinate system, every potential function "$P(t_1,t_2,\ell)$" which can generate an e.m. field, for instance the "vector potential function -A-" for a "TM" wave, is defined as being composed of the product of a function that only depends on a transverse coordinate with a function that only depends on the longitudinal coordinate. Indicating with "$T(t_1,t_2)$" and "$L(\ell)$" respectively, the first and second function, we will write:

$$P(t_1,t_2,\ell) \stackrel{\perp}{=} T(t_1,t_2)L(\ell) \qquad (A2.8.1)$$

Functions "T" and "L" can be in general complex. Note Equation A2.8.1 is an aspect of the VSM we have previously introduced.

In an e.m. ambient defined for the homogeneous wave equation in Section A2.4, every "$P(t_1,t_2,\ell)$" must satisfy the homogeneous wave equation, i.e.:

$$\nabla^2(P) = k^2 P \qquad (A2.8.2)$$

In our reference coordinate system it is possible to write the Laplacian as:

$$\nabla^2(\ ) \equiv \nabla_t^2(\ ) + \frac{\partial^2(\ )}{\partial \ell^2} \qquad (A2.8.3)$$

where "$\nabla_t^2(\ )$," called the "transverse Laplacian," is defined when the cross-section of the guiding structure is specified. Some expressions for Laplacian operator in the most used coordinate systems can be found in Appendix A8. We will see in the next sections some solutions of this equation.

Combining together A2.8.1 through A2.8.3 we have:

$$L\nabla_t^2 T + T\frac{\partial^2 L}{\partial \ell^2} = k^2 TL \rightarrow \frac{\nabla_t^2}{T} + \frac{\partial^2 L}{L\partial \ell^2} = k^2 \qquad (A2.8.4)$$

Proceeding as we did regarding the wave equation, we can set:

$$\frac{\nabla_t^2 T}{T} \stackrel{\perp}{=} k_t^2 \quad \text{and} \quad \frac{\partial^2 L}{L\partial \ell^2} \stackrel{\perp}{=} k_\ell^2 \qquad (A2.8.5)$$

with the condition:

$$k_t^2 + k_\ell^2 =! \ k^2 \equiv -\omega^2 \mu \varepsilon_c \qquad (A2.8.6)$$

The first equation in A2.8.5, i.e.:

$$\nabla_t^2 T = k_t^2 T \qquad (A2.8.7)$$

---

* The use of electrodynamic "potential functions" is an e.m. theory method to develop the e.m. fields in a region. "Potential functions" theory can be found in the following texts:
R.E. Collin, *Foundations for Microwave Engineering*. McGraw Hill, 56, 1992.
G. Barzilai, *Fondamenti di elettromagnetismo*, Siderea editore, 292, 1975.
G. Gerosa, *Appunti di microonde*, Università di Roma, "La Sapienza," I-28, 1980.

is called the "transverse wave equation," and in general its solution is not very simple, since it is strongly dependent on the cross-section of the guiding structure. In addition, due to the different condition that the function "T" has to verify on the boundary of the region where the field is contained, the values of "k_t" are in general different for the various propagation modes inside the region. For this reason, also "k_ℓ" values obtained from A2.8.6 are in general different among modes. The parameter "k_t" is called the "cut-off wave number" or "transverse wave number."

Contrarily, the solution of the second equation of A2.8.5 is very simple, since it is the so-called "harmonic motion equation" we have already encountered in this appendix. So, the general solution is a linear combination of exponentials:

$$L(\ell) = Pe^{-k_\ell \ell} + Re^{k_\ell \ell} \tag{A2.8.8}$$

Using the potential function procedure, the equations for a "TM" mode inside the structure indicated in Figure A2.8.1 are:

$$\underline{h}_\ell = 0 \tag{A2.8.9}$$

$$\underline{h}_t = L\underline{\nabla}_t T_{(m)} \otimes \underline{\ell}_0 \tag{A2.8.10}$$

$$\underline{e}_\ell = \frac{-k_{t(m)}^2}{j\omega\varepsilon_c} LT_{(m)}\underline{\ell}_0 \tag{A2.8.11}$$

$$\underline{e}_t = \frac{1}{j\omega\varepsilon_c} \frac{\partial L}{\partial \ell} \underline{\nabla}_t T_{(m)} \tag{A2.8.12}$$

where the subscript "(m)" denotes a generic "TM" mode. The operator "$\underline{\nabla}_t()$" is called the "transverse gradient" and is defined when the cross-section of the guiding structure is specified. Some expressions for gradient operator in the most used coordinate systems can be found in Appendix A8. If we are in the case where only the progressive wave exists, then in the previous equation the term $k_{\ell(m)}/j\omega\varepsilon_c \doteq \zeta_{TM}$ appears, which is called "TM mode impedance."

The equations for "TE" modes can be found directly from A2.8.9 through A2.8.12, applying the duality principle expressed in A2.2.9. So, we can write:

$$\underline{e}_\ell = 0 \tag{A2.8.13}$$

$$\underline{e}_t = -L\underline{\nabla}_t T_{[m]} \otimes \underline{\ell}_0 \tag{A2.8.14}$$

$$\underline{h}_\ell = \frac{-k_t^2}{j\omega\mu} LT_{[m]}\underline{\ell}_0 \tag{A2.8.15}$$

$$\underline{h}_t = \frac{1}{j\omega\mu} \frac{\partial L}{\partial \ell} \underline{\nabla}_t T_{[m]} \tag{A2.8.16}$$

where the subscript "[m]" denotes a generic "TE" mode. If we are in the case where only the progressive wave exists, then in the previous equation the term $j\omega\mu/k_{\ell[m]} \doteq \zeta_{TE}$ appears, which is called "TE mode impedance."

Of course, the boundary conditions that "T" must verify on the guiding structure contour are different from a "TE" or "TM" mode. In fact, the general boundary condition is that the electric field tangential component on the contour "s" of the guiding structure must be zero. So, for a:

1. "TM" mode, it means that:

$$\underline{e}_\ell = \frac{-k_{t(m)}^2}{j\omega\varepsilon_c} LT_{(m)}\underline{\ell}_0 =! \, 0 \quad \text{and} \quad \underline{e}_t \cdot \underline{t} = \frac{1}{j\omega\varepsilon_c}\frac{\partial L}{\partial\ell}\underline{\nabla}_t T_{(m)} \cdot \underline{t} =! \, 0 \qquad (A2.8.17)$$

Since $L \neq ! \, 0$ to have propagation, and $k_{t(m)}^2 \neq ! \, 0$ inside a guiding structure for a not null mode, from the first of A2.8.17 it follows that:

$$T_{(m)} =! \, 0 \quad \text{on "p"} \quad \text{for a "TM" mode}. \qquad (A2.8.18)$$

From the second of A2.8.17 we have:

$$\underline{e}_t \cdot \underline{t} = \frac{1}{j\omega\varepsilon_c}\frac{\partial L}{\partial\ell}\underline{\nabla}_t T_{(m)} \cdot \underline{t} \equiv \frac{1}{j\omega\varepsilon_c}\frac{\partial L}{\partial\ell}\frac{\partial T_{(m)}}{\partial t} =! \, 0 \qquad (A2.8.19)$$

but $(\partial T_{(m)}/\partial t)=0$ if $T_{(m)} = 0$, as stated in A2.8.18. So, $T_{(m)} = 0$ on "p" is the necessary and sufficient contour condition that "$T_{(m)}$" must satisfy for "TM" modes inside a guiding structure.

2. "TE" mode, it means that:

$$\underline{e}_t \cdot \underline{t} = -L\left(\underline{\nabla}_t T_{[m]} \otimes \underline{\ell}_0\right) \cdot \underline{t} \equiv -L\underline{\nabla}_t T_{[m]} \cdot \underline{\ell}_0 \otimes \underline{t} \equiv -L\underline{\nabla}_t T_{[m]} \cdot \underline{n} \equiv -L\left(\partial T_{[m]}/\partial n\right) =! \, 0 \quad (A2.8.20)$$

And so:

$$\partial T_{[m]}/\partial n =! \, 0 \quad \text{on "p"} \quad \text{for a "TE" mode}. \qquad (A2.8.21)$$

It is very interesting to show that with the conditions A2.8.18 and A2.8.19 for "T," the parameter "k_t^2" is a negative real number. Applying the Green two dimensions identity given in Appendix A8 to the functions "T" and "T*" we have:

$$\oint_p T\frac{\partial T^*}{\partial n}\,dp = \int_S\left(\underline{\nabla}_t T \cdot \underline{\nabla}_t T^* + T\nabla_t^2 T^*\right)dS \qquad (A2.8.22)$$

The first member is always zero, since for a "TM" mode $T = 0$ while for a "TE" mode $\partial T/\partial n \equiv \partial T^*/\partial n = 0$. So, the previous equation becomes:

$$\int_S\underline{\nabla}_t T \cdot \underline{\nabla}_t T^*\,dS = -\int_S T\nabla_t^2 T^*\,dS \qquad (A2.8.23)$$

From A2.8.7 it follows that "T*" must satisfy:

$$\nabla_t^2 T^* = k_t^{2^*} T^* \qquad (A2.8.24)$$

A2.8.23 with the use of A2.8.24 becomes:

$$k_t^{2*} = -\frac{\int_S \underline{\nabla}_t T \cdot \underline{\nabla}_t T^* dS}{\int_S TT^* dS} \qquad (A2.8.25)$$

Since the functions that appear in the above integrals are positive, the same is true for the result of the integral. So "k_t^{2*}" is a real negative number and, being real we have that $k_t^2 \equiv k_t^{2*}$; consequently, also "k_t^2" is a real negative number, for every mode.

Using A2.8.25 it is simple to show that a "TEM" mode cannot exist inside a guiding structure as indicated in Figure A2.8.1, unless it is null. A "TEM" mode can be obtained for example starting with a "TM" mode and setting $e_\ell = !0$. So, from A2.8.11 we have:

$$\underline{e}_\ell = \frac{-k_{t(m)}^2}{j\omega\varepsilon_c} LT_{(m)}\ell_0 = !0 \qquad (A2.8.26)$$

The functions "L" and "$T_{(m)}$" cannot be zero in any point of "S," otherwise no wave could exist. Consequently, $k_t^2 = !0$. But, from A2.8.25 it would follow that "$\underline{\nabla}_t T_{(m)}$" must be zero on the section "S," and from A2.8.10 and A2.8.12 it should follow that all the field is zero. So, a non null "TEM" mode cannot exist inside the guiding structure indicated in Figure A2.8.1. In general, a "TEM" mode can exist inside guiding structures where the field is contained inside two separate conductors and at least one of these conductors has limited extension. For example, the coplanar waveguide "CPW"* can support a "TEM" wave, at least in a limited bandwidth, while slot lines cannot have a "TEM" mode. "CPW" has the central conductor with limited extension while the slot line also having the field guided between two conductors, has theoretically infinite extensions.

In the next sections we will show the two most used guiding structures, which belong to the general discussion made here. They are the rectangular and circular waveguide.

We want to conclude this section giving some properties relative to function "T" for the generic structure indicated in Figure A2.8.1. Note from Figure A2.8.1 that "S" is the transverse section of the guiding structure. The properties are:

a. orthogonality of functions "T"

$$\int_S T_{(m)} T_{(n)} dS = 0 \qquad (A2.8.27)$$

$$\int_S T_{[m]} T_{[n]} dS = 0 \qquad (A2.8.28)$$

for $k_{t(m)}^2 \neq k_{t(n)}^2$ and $k_{t[m]}^2 \neq k_{t[n]}^2$.

b. orthogonality of vectors "$\nabla_t T$"

$$\int_S \underline{\nabla}_t T_{(m)} \cdot \underline{\nabla}_t T_{(n)} dS = 0 \qquad (A2.8.29)$$

$$\int_S \underline{\nabla}_t T_{[m]} \cdot \underline{\nabla}_t T_{[n]} dS = 0 \qquad (A2.8.30)$$

for $k_{t(m)}^2 \neq k_{t(n)}^2$ and $k_{t[m]}^2 \neq k_{t[n]}^2$.

* Coplanar waveguide transmission line is studied in Chapter 10.

c. generic properties

$$\int_S \underline{\nabla}_t T_{(m)} \cdot \underline{\nabla}_t T_{(n)} \otimes \underline{\ell}_0 \, dS = 0 \qquad\qquad (A2.8.31)$$

$$\int_S \underline{\nabla}_t T_{(m)} \cdot \underline{\nabla}_t T_{[n]} \otimes \underline{\ell}_0 \, dS = 0 \qquad\qquad (A2.8.32)$$

$$\int_S \underline{\nabla}_t T_{[m]} \cdot \underline{\nabla}_t T_{(n)} \otimes \underline{\ell}_0 \, dS = 0 \qquad\qquad (A2.8.33)$$

Note that Equations A2.8.27 through A2.8.30 are only true when we use different values of "k_t" for each mode, while Equations A2.8.31 through A2.8.33 are always true. In the next section we will use such relationships.

A2.9 "TE" AND "TM" MODES IN RECTANGULAR WAVEGUIDE

A rectangular waveguide is an e.m. guiding structure whose transverse section is a rectangle. Its representation is indicated in Figure A2.9.1. The most useful coordinate system is the Cartesian one.

The waveguide is usually made of aluminium, copper, or brass, and sometimes the internal faces are covered with a thin layer of gold or silver to increase their conductivity.

According to the introduction of this appendix, we will not give here all the passages to reach the solution, leaving this job to dedicated texts.[11,12]

To have the expressions of "TE" and "TM" modes in rectangular waveguide, we have to find the function "T" for the structure in Figure A2.9.1. The function "T" has to satisfy the transverse wave equation A2.8.7. So, applying the VSM to "T" and writing:

$$T(x,y) \stackrel{\perp}{=} X(x)Y(y) \qquad\qquad (A2.9.1)$$

and expressing the "∇_t^2" in Cartesian coordinate system, A2.8.7 with simple passages becomes:

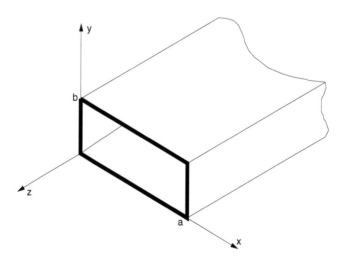

Figure A2.9.1

$$\frac{d^2X}{Xdx^2} + \frac{d^2Y}{Ydy^2} = k_t^2 \qquad (A2.9.2)$$

where for notational simplicity we have omitted the variable dependence for "X" and "Y." Equation A2.9.2 is surely satisfied if:

$$\frac{d^2X}{Xdx^2} \doteq k_{xx}^2 \quad \text{and} \quad \frac{d^2Y}{Ydy^2} \doteq k_{yy}^2 \qquad (A2.9.3)$$

with the condition:

$$k_{xx}^2 + k_{yy}^2 =! \, k_t^2 \qquad (A2.9.4)$$

Since "k_t^2" is a negative real number, then "k_{xx}" and "k_{yy}" are imaginary, and we set:

$$k_{xx} \doteq jk_x \quad \text{and} \quad k_{yy} \doteq jk_y \qquad (A2.9.5)$$

with "k_x" and "k_y" as real numbers. Then, Equation A2.9.2 becomes:

$$\frac{d^2X}{Xdx^2} \doteq -k_x^2 \quad \text{and} \quad \frac{d^2Y}{Ydy^2} \doteq -k_y^2 \qquad (A2.9.6)$$

with the condition:

$$k_x^2 + k_y^2 =! -k_t^2 \qquad (A2.9.7)$$

Equations A2.9.6 are the "harmonic motion equations," and its solution is a linear combination of sinus and cosinus, i.e.:

$$X(x) = C_1 \cos(k_x x) + C_2 \text{sen}(k_x x) \qquad (A2.9.8)$$

$$Y(y) = D_1 \cos(k_y y) + D_2 \text{sen}(k_y y) \qquad (A2.9.9)$$

The longitudinal function "$L(\ell)$" in the Cartesian coordinate system indicated in Figure A2.9.1 becomes:

$$L(z) = Pe^{-k_z z} + R^{k_z z} \qquad (A2.9.10)$$

where "k_z," using condition A2.9.7, must satisfy:

$$k_t^2 + k_z^2 =! \, k^2 \doteq -\omega^2 \mu \varepsilon_c \qquad (A2.9.11)$$

So, applying the contour conditions A2.8.18 and A2.8.21 to the structure indicated in Figure A2.9.1 we can specify the function "T" for each mode and finally the expressions for the fields. We have:

1. TE Mode

The function "T" and "k_t^2" are:

$$T(x,y) = q_1 \cos(m\pi x/a)\cos(n\pi y/b) \tag{A2.9.12}$$

$$k_t^2 \to k_{t,mn}^2 = -(m\pi/a)^2 - (n\pi/b)^2 \tag{A2.9.13}$$

where "m" and "n" are two real, integer and positive numbers never both equal to zero, and "q_1" is a generic constant. Inserting A2.9.13 into A2.9.11 we have:

$$k_{z,mn} = \left[(m\pi/a)^2 + (n\pi/b)^2 - \omega^2\mu\varepsilon_c\right]^{0.5} \tag{A2.9.14}$$

So, "TE" modes in rectangular waveguides are indicated as "TE_{mn}." Inserting A2.9.12 into A2.8.13 through A2.8.16 we have the fields for a "TE_{mn}" mode:

$$e_z = 0 \tag{A2.9.15}$$

$$\underline{e}_x = C(n\pi/b)\cos(m\pi x/a)\,\text{sen}\,(n\pi y/b)\,\underline{x}_0 \tag{A2.9.16}$$

$$\underline{e}_y = -C(m\pi/a)\,\text{sen}\,(m\pi x/a)\cos(n\pi y/b)\,\underline{y}_0 \tag{A2.9.17}$$

$$\underline{h}_x = -\left(e_y/\zeta_{TEmn}\right)\underline{x}_0 \tag{A2.9.18}$$

$$\underline{h}_y = -\left(e_x/\zeta_{TEmn}\right)\underline{y}_0 \tag{A2.9.19}$$

$$\underline{h}_z = C\left(-k_t^2/j\omega\mu\right)\cos(m\pi x/a)\cos(n\pi y/b)\,\underline{z}_0 \tag{A2.9.20}$$

where the term "ζ_{TEmn}" is the "TE_{mn}" wave impedance, given by:

$$\zeta_{TEmn} \overset{\perp}{=} j\omega\mu/k_{z,mn} \tag{A2.9.21}$$

and "C" is a generic constant. All the field expressions, considering for simplicity to be only in a progressive propagation, need to be multiplied by "$e^{-k_{z,mn}z}$."

From the "TE_{mn}" field expressions it is simple to recognize how, for the "TE_{10}," the only non-zero fields "e_y," "h_x," and "h_z" have the shape indicated in Figure A2.9.2. Here we have adopted the convention that with lines of different lengths, we indicate the different strength of the field. For example, the electric field component "e_y" is maximum at the center of the waveguide and null at x = 0 and x = a. Note two coordinates $x = x_{c1}$ and $x = x_{c2}$ exist where the transverse magnetic components "h_x" and "h_z" have equal amplitude and in time quadrature. This means that in these two longitudinal planes, the magnetic field is circularly polarized, and this characteristic is used to build ferrimagnetic devices, as isolators or phase shifters, which are studied in Appendix A7. Also note how at x = a/2 the magnetic field is instead linearly polarized, since only the component "h_x" exists, while the electric field is always linearly polarized.

It is interesting to show an important characteristic of rectangular waveguides, i.e., their high-pass filter property. Assuming the medium inside the waveguide to be lossless, we see from A2.9.14 that "k_z" can be real, null, or imaginary if:

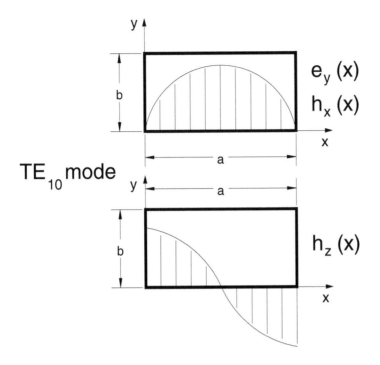

Figure A2.9.2

$$(m\pi/a)^2 + (n\pi/b)^2 - \omega^2\mu\varepsilon >, =, < 0$$

The propagation along "z" without attenuation corresponds to an imaginary "k_z," and from the previous equation, we can see that it happens above a frequency "f_c," called the "cut-off frequency," given by:

$$f_{c,mn} = \left(-k_t^2/\mu\varepsilon\right)^{0.5}/2\pi \equiv \left[(m/a)^2 + (n/b)^2\right]^{0.5}/2(\mu\varepsilon)^{0.5} \qquad (A2.9.22)$$

From A2.9.22 we see that every mode has its own cut-off frequency, which means that the frequency of the signal we send into the waveguide must be higher than the cut-off frequency of the mode we want to use. Assuming we feed the waveguide using its widest side, from the previous equation we see that the mode with the lowest "$f_{c,mn}$" is the "TE_{10}," and for this reason it is called the "dominant mode." It is simple to recognize that if we increase the frequency of the signal that travels inside the waveguide, then it changes its mode of propagation, because the frequency overcomes the cut-off frequencies of higher order modes. This is not a desirable phenomenon since every mode of propagation has its own polarization and the receiver is usually optimized to detect only one polarization, to increase the sensitivity. So, caution must be used in choosing the proper waveguide for the signal frequency that travels inside it. To this purpose, a lot of waveguides exist with dimensions "a" and "b" just optimized for a dedicated bandwidth.

Since the "TE_{10}" is quite often the most used mode in a rectangular waveguide, we think it is useful to indicate the fields of this mode, obtained from expressions A2.9.15 through A2.9.20 with m = 1 and n = 0. Indicating only the nonnull fields, we have:

$$\underline{e}_y = -C(\pi/a)\operatorname{sen}(\pi x/a)\underline{y}_0 \qquad (A2.9.23)$$

$$\underline{h}_x = -\left(e_y/\zeta_{TE10}\right)\underline{x}_0 \tag{A2.9.24}$$

$$\underline{h}_z = C\left(\pi^2/j\omega\mu a^2\right)\cos\left(\pi x/a\right)\underline{z}_0 \tag{A2.9.25}$$

Note that from A2.9.22 we have:

$$f_{c,10} = v/2a \rightarrow \lambda_{c,10} = 2a \tag{A2.9.26}$$

with $v \pm 1/(\mu\varepsilon)^{0.5}$, i.e., the light speed in the dielectric that fills the waveguide.

With the introduction of A2.9.22, "$k_{z,mn}$" and "ζ_{TEmn}" are expressed as functions of "$f_{c,mn}$." Assuming the case of lossless dielectric, i.e., $\varepsilon_c \equiv \varepsilon$, inserting A2.9.22 into A2.9.14, with $\varepsilon_c \equiv \varepsilon$, we have:

$$k_{z,mn} = j\omega\left(\mu\varepsilon\right)^{0.5}\left(1 - \omega_{c,mn}^2/\omega^2\right)^{0.5} \tag{A2.9.27}$$

and inserting this equation into A2.9.21:

$$\zeta_{TEmn} \pm \zeta\left(1 - \omega_{c,mn}^2/\omega^2\right)^{-0.5} \tag{A2.9.28}$$

where "ζ" is the impedance presented to a plane wave by the dielectric that fills the waveguide, i.e., $\zeta \pm (\mu/\varepsilon)^{0.5}$.

Using A2.9.26 through A2.9.28 and introducing a constant "A" defined as $A \pm -C\omega_{c,10}/v$ with dimensions V/m, Equations A2.9.23 through A2.9.25 can be set as functions of cut-off wavelength, i.e.:

$$\underline{e}_y = A\text{sen}\left(\pi x/a\right)\underline{y}_0 \tag{A2.9.29}$$

$$\underline{h}_x = \left[-e_y\left(1 - \omega_{c,10}^2/\omega^2\right)^{0.5}/\zeta\right]\underline{x}_0 \tag{A2.9.30}$$

$$h_z = j\left(A\omega_{c,10}/\omega\zeta\right)\cos\left(\pi x/a\right)z_0 \tag{A2.9.31}$$

All the field expressions A2.9.29 through A2.9.31, considering for simplicity as only being in a progressive propagation, need to be multiplied by "$e^{-k_{z,10}z}$."

2. "TM" Mode

The function "T" is:

$$T(x,y) = q_2\text{sen}\left(m\pi x/a\right)\text{sen}\left(n\pi y/b\right) \tag{A2.9.32}$$

where "m" and "n" are two real, integer and positive numbers never equal to zero, and "q_2" is a generic constant. "k_t^2" has the same expression we gave for "TE" modes, and consequently also "$k_{z,mn}$." Since "m" and "n" are never equal to zero, the lower order "TM" mode in a rectangular waveguide is a "TM_{11}."

Inserting A2.9.32 into A2.8.9 through A2.8.12 we have the fields for a "TM_{mn}" mode:

$$h_z = 0 \tag{A2.9.33}$$

$$\underline{h}_x = D(n\pi/b)\operatorname{sen}(m\pi x/a)\cos(n\pi y/b)\underline{x}_0 \tag{A2.9.34}$$

$$\underline{h}_y = -D(m\pi/a)\cos(m\pi x/a)\operatorname{sen}(n\pi y/b)\underline{y}_0 \tag{A2.9.35}$$

$$\underline{e}_x = h_y \zeta_{TMmn} \underline{x}_0 \tag{A2.9.36}$$

$$\underline{e}_y = h_x \zeta_{TMmn} \underline{y}_0 \tag{A2.9.37}$$

$$\underline{e}_z = D\left(-k_t^2/j\omega\varepsilon_c\right)\operatorname{sen}(m\pi x/a)\operatorname{sen}(n\pi y/b)\underline{z}_0 \tag{A2.9.38}$$

where the term "ζ_{TMmn}" is the "TM_{mn}" wave impedance, given by:

$$\zeta_{TMmn} \overset{\perp}{=} k_{z,mn}/j\omega\varepsilon_c \tag{A2.9.39}$$

and "D" is a generic constant. All the field expressions, considering for simplicity as only being in a progressive propagation, need to be multiplied by "$e^{-k_{z,mn}z}$."

From the "TM_{mn}" field expression it is simple to recognize how, for the "TM_{11}," the field components "h_x" and "h_y" have the shape indicated in Figure A2.9.3. Note that in this case these components also have dependence along the coordinate "y."

Excluding the field component expressions given in A2.9.33 through A2.9.39, for modes "TM" we can repeat the same discussions we made above for modes "TE." In particular, we want to observe that modes "TE" and "TM" with equal subscripts "mn" have equal cut-off frequency, but, since for "TE" modes one subscript can be zero, then the lowest cut-off frequency belongs to "TE_{10}" mode, i.e., the dominant mode.

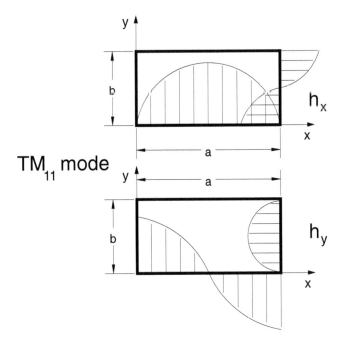

Figure A2.9.3

A2.10 "TE" AND "TM" MODES IN CIRCULAR WAVEGUIDE

Circular waveguides are another guiding structure often used in microwave technology, especially to feed antennas for radar systems. Appendix A7 also shows some applications of these waveguides to build isolators, phase shifters, and circulators. As for the rectangular waveguide, here the metals used to build this waveguide are usually aluminium, copper, or brass, sometimes with the internal faces covered with a thin layer of gold or silver to increase their conductivity.

A mechanical representation of this waveguide is indicated in Figure A2.10.1a, while in part b we have indicated the most useful reference system for this structure, i.e., the cylindrical system. This reference system is composed of a longitudinal axis "z," a radial axis "r," and an angular axis "θ," so that "z," "r," and "θ" are orthogonal to each other. Of course, the versors are "z_0," "r_0," and "θ_0," respectively.

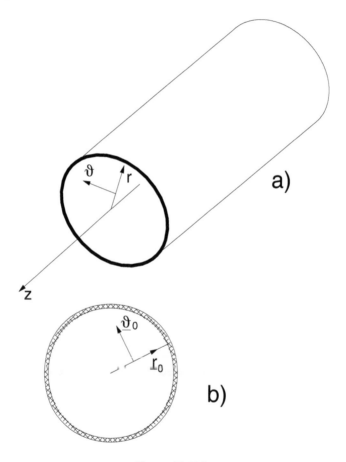

Figure A2.10.1

To have the expressions of "TE" and "TM" modes in circular waveguide we have to find the function "T" (defined in Section A2.8) for the structure in Figure A2.10.1. The function "T" has to satisfy the transverse wave equation A2.8.7. So, applying the VSM to "T" and writing:

$$T(r,\theta) \stackrel{\perp}{=} R(r)\,\Theta(\theta) \tag{A2.10.1}$$

and expressing the "∇_t^2" in cylindrical coordinate system,* A2.8.7 with simple passages becomes:

* See Appendix A8 for some representation of "∇_t^2."

$$r^2 \frac{d^2R}{Rdr^2} + r \frac{dR}{Rdr} + \frac{d^2\Theta}{\Theta d\theta^2} = r^2 k_t^2 \qquad (A2.10.2)$$

where for notational simplicity we have omitted the variable dependence for "R" and "Θ." Different from the case we studied for the rectangular waveguides, now the second member is not a constant since it depends on the product "$r^2 k_t^2$." So, let us rewrite the previous equation moving the second member to first one, i.e.:

$$r^2 \frac{d^2R}{Rdr^2} + r \frac{dR}{Rdr} - r^2 k_t^2 + \frac{d^2\Theta}{\Theta d\theta^2} = 0 \qquad (A2.10.3)$$

Note that since:

a. The first three terms are only dependent on "r" while...
b. The fourth term is only dependent on "θ," and...
c. The sum of the terms must be zero

then these two groups of terms must be equal to a constant. So let us write:

$$\frac{d^2\Theta}{\Theta d\theta^2} \overset{\perp}{=} -n^2 \qquad (A2.10.4)$$

with "n" a real number. This equation is again the "harmonic motion equation," and in this case the solution is a linear combination of sinus and cosinus, i.e.:

$$\Theta(\theta) = C_1 \operatorname{sen}(n\theta) + C_2 \cos(n\theta) \qquad (A2.10.5)$$

where "C_1" and "C_2" are generic constants. From Figure A2.10.1 we recognize that function "$\Theta(\theta)$" is periodic in "2π," and consequently "n" in addition to being a real number, must also be an integer. It is convenient to rewrite Equation A2.10.5 as:

$$\Theta(\theta) = P \cos(n\theta + \varphi) \qquad (A2.10.6)$$

where "P" and "φ" are generic constants. Then, inserting A2.10.4 into A2.10.3, for what we have said there must be:

$$r^2 \frac{d^2R}{Rdr^2} + r \frac{dR}{Rdr} - r^2 k_t^2 - n^2 = 0 \qquad (A2.10.7)$$

Since in Section A2.8 we said that "k_t^2" is a real and negative number, we set:

$$-k_t^2 \overset{\perp}{=} K^2 \qquad (A2.10.8)$$

with "K" a real number. Equation A2.10.7 with simple passages can be written as:

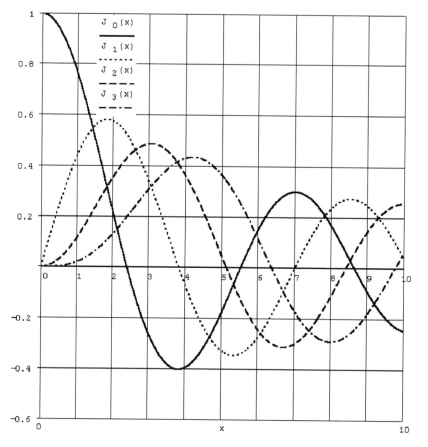

Figure A2.10.2

$$\frac{d^2R}{K^2dr^2} + \frac{dR}{K^2rdr} + \left[1 - \left(n/Kr\right)^2\right]R = 0 \qquad (A2.10.9)$$

which is the Bessel differential equation of the first kind and order "n." The solution of A2.10.9 can be set as a linear combination of Bessel functions of first "$J_n(Kr)$" and second "$Y_n(Kr)$" kind of order "n." Bessel equations and functions are very important and used in many electromagnetic problems. They can be found in a lot of mathematical books.[13,14,15] So, the general solution of A2.10.9 is:

$$R(Kr) = D_1 J_n(Kr) + D_2 Y_n(Kr) \qquad (A2.10.10)$$

where "D_1" and "D_2" are generic constants. In Figures A2.10.2 and A2.10.3 we have indicated the shapes of the first four Bessel functions of the first and second kind. We see how all the "$Y_n(x)$" goes to infinity for x = 0. For this reason, this function cannot represent our physical case, where all the fields have finite value. It means that in our case the constant $D_2 = !0$, and consequently the solution of A2.10.9 is:

$$R(Kr) = D_1 J_n(Kr) \qquad (A2.10.11)$$

and the resulting transverse function is:

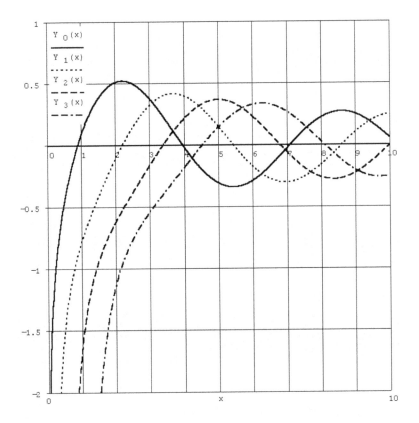

Figure A2.10.3

$$T(r,\theta) \overset{\perp}{=} CJ_n(Kr)\cos(n\theta + \varphi) \tag{A2.10.12}$$

Note that the value of the subscript "n" that appears in the Bessel function "$J_n(Kr)$" is the same as the number "n" that appears inside "$\cos(n\theta + \varphi)$."

Once we have found the function "T" for our circular waveguide, to find the field expressions we can apply the boundary conditions A2.8.18 and A2.8.21 to the structure indicated in Figure A2.10.1. We have:

1. TE Mode

The "k_t^2" is given by:

$$k_t^2 \overset{\perp}{=} -K^2 \rightarrow k_{t,mn}^2 \overset{\perp}{=} -K_{mn}^2 = -(z'_{mn}/a)^2 \tag{A2.10.13}$$

where "m" and "n" are two real, integer and positive numbers, "a" is the waveguide radius, and "z_{mn}'" is the m-th zero of the derivative of "$J_n(Kr)$." So, "TE" modes in circular waveguides are indicated as "TE_{mn}." Since, conventionally, the roots "z_{mn}'" of the derivative of "$J_n(Kr)$" are indicated starting with "1," then "m" cannot be equal to "0," while "n" can be.* Some values of "z_{mn}'" are listed in Table A2.10.1**.

---

* In some literature "TE" modes in circular waveguides are indicated as "TE_{nm}." Since the m-th zero always starts conventionally with "1," with this notation it is the second subscript that cannot be equal to zero.
** It is common practice to set comma as superscript to indicate a derivative. So, "$J_0'(x)$" means the derivative respect "x" of "$J_0(x)$." Of course, the comma as superscript to "z_{mn}" reminds one that these zeros are relative to "$J_0'(x)$." This notation has been adapted into Table A2.10.1.

Table A2.10.1

Derivative of Bessel function:	First three zeros		
$J_0'(x)$	$z_{10}' = 3.832$	$z_{20}' = 7.016$	$z_{30}' = 10.174$
$J_1'(x)$	$z_{11}' = 1.841$	$z_{21}' = 5.331$	$z_{31}' = 8.536$
$J_2'(x)$	$z_{12}' = 3.054$	$z_{22}' = 6.706$	$z_{32}' = 9.970$

So, inserting A2.10.13 into Equation A2.8.6 it is possible to evaluate "k_z," given by:

$$k_{z,mn} = \left[\left(z_{mn}'/a \right)^2 - \omega^2 \mu \varepsilon_c \right]^{0.5} \tag{A2.10.14}$$

and the cut-off frequency, with a lossless dielectric filling the waveguide, will be:

$$f_{c,mn} = \left(-k_t^2/\mu\varepsilon \right)^{0.5} \big/ 2\pi \equiv z_{mn}' \big/ 2\pi a (\mu\varepsilon)^{0.5} \tag{A2.10.15}$$

From the previous equation and the values given in Table A2.10.1 we recognize that the lower cut-off "TE" mode for circular waveguides is the "TE_{11}." When we discuss the "TM" modes for our waveguide, we will verify that the "TE_{11}" is the dominant mode, i.e., it has the lowest cut-off frequency of any mode in the circular waveguide.

Inserting A2.10.12 into A2.8.13 through A2.8.16 we have the fields for a "TE_{mn}" mode, given by:

$$e_z = 0 \tag{A2.10.16}$$

$$\underline{e}_r = C\left(n/r\right) J_n\left(K_{mn}r\right) \text{sen}\left(n\theta + \varphi\right)\underline{r}_0 \tag{A2.10.17}$$

$$\underline{e}_\theta = CKJ_n'\left(K_{mn}r\right)\cos\left(n\theta + \varphi\right)\underline{\theta}_0 \tag{A2.10.18}$$

$$\underline{h}_r = -\left(e_\theta/\zeta_{TEmn}\right)\underline{r}_0 \tag{A2.10.19}$$

$$\underline{h}_\theta = \left(e_r/\zeta_{TEmn}\right)\underline{y}_0 \tag{A2.10.20}$$

$$\underline{h}_z = C\left(K_{mn}^2/j\omega\mu\right)J_n\left(K_{mn}r\right)\cos\left(n\theta + \varphi\right)\underline{z}_0 \tag{A2.10.21}$$

where the term "ζ_{TEmn}" is the "TE_{mn}" wave impedance given by A2.9.21:

$$\zeta_{TEmn} \overset{\perp}{=} j\omega\mu/k_{z,mn} \tag{A2.9.21}$$

and "C" is a generic constant. All the field expressions, considered for simplicity as being only in a progressive propagation, need to be multiplied by "$e^{-k_{z,mn}z}$."

Since the "TE_{11}" is quite often the most used mode in a circular waveguide, we think it is useful to indicate the fields of this mode, obtained from the expressions A2.10.16 through A2.10.21 with m = n = 1. Indicating only the nonnull fields, we have:

$$\underline{e}_r = (C/r)J_1(K_{11}r)\operatorname{sen}(\theta + \varphi)\underline{r}_0 \tag{A2.10.22}$$

$$\underline{e}_\theta = CKJ_1'(K_{11}r)\cos(\theta + \varphi)\underline{\theta}_0 \tag{A2.10.23}$$

$$\underline{h}_r = -(e_\theta/\zeta_{TE11})\underline{r}_0 \tag{A2.10.24}$$

$$\underline{h}_\theta = (e_r/\zeta_{TE11})\underline{y}_0 \tag{A2.10.25}$$

$$\underline{h}_z = C(K_{11}^2/j\omega\mu)J_1(K_{11}r)\cos(\theta + \varphi)\underline{z}_0 \tag{A2.10.26}$$

Note that from A2.10.15 we have:

$$f_{c,11} = 1.841v/2\pi a \rightarrow \lambda_{c,11} \approx 3.41a \tag{A2.10.27}$$

with $v \doteq 1/(\mu\varepsilon)^{0.5}$, i.e., the light speed in the dielectric that fills the waveguide. From the "TE$_{11}$" field expressions it is simple to recognize how the electric and transverse magnetic field lines, for $\varphi = 0$, have the shape indicated in Figure A2.10.4, parts a and b, respectively. For the electric field lines, note how for the $\theta = 0$ or $\theta = \pi$ line, i.e., the "horizontal"* line, the electric field has only the "e_θ" component, while for the $\theta = \pi/2$ or $\theta = -\pi/2$ line, i.e., the "vertical"** line, the electric field has only the "e_r" component. In any other position, both components exist. Of course, the

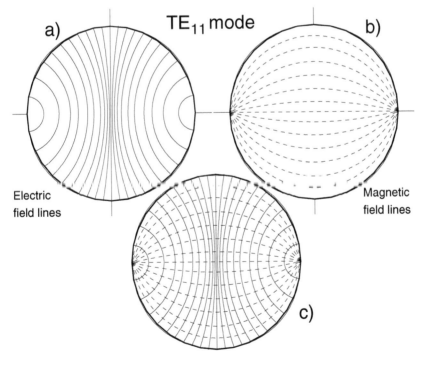

TE$_{11}$ mode

a)

b)

Electric
field lines

Magnetic
field lines

c)

Figure A2.10.4

* Of course, for a circular cross-section it is not a well-defined "horizontal" axis. However, here we mean one of the two Cartesian axes with the origin centered in the center of the circular cross-section.
** We can repeat what we said for the "horizontal" axis. In this case we refer to the remaining axis. Of course, vertical and horizontal axes need to be related to the feeding axis of the circular waveguide.

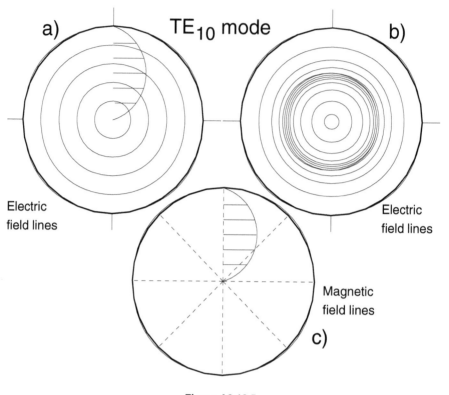

Figure A2.10.5

"e_θ" component for the $\theta = 0$ and $\theta = \pi$ line is zero for $r = 1$ since we are considering that the waveguide is built with a perfect conductor. For electric field lines, note also that if the tangential component is always zero on the waveguide internal surface, the remaining radial component changes its intensity with "θ" according to "$\text{sen}\theta$;" so for $\theta = 0$ and $\theta = \pi$ we also have $e_r = 0$, together with the already known $e_\theta = 0$. The magnetic fields lines can be drawn reminding one that it is always locally orthogonal to the electric field. In Figure A2.10.4c we have indicated all the transverse field lines. Note that the shape of the field lines for $\varphi \neq 0$ can be obtained from that indicated in Figure A2.10.4 by applying an angular rotation of "φ" to that shape.

Modes with $n = 0$ are called "electrical circular modes" because the electric field has "e_θ" as the only nonzero component. These modes have "e_θ," "h_r," and "h_z" as the only nonzero field components. In Figure A2.10.5 we have represented the components "e_θ" and "h_r" for "TE_{10}" mode. We see in Figure A2.10.5a how the electric field is just circumferential, but its intensity is zero at $r = 0$ and $r = a$, if "a" is the waveguide radius. The intensity dependence with "r" is indicated with the curve and horizontal lines in part "a." This representation follows the same format used for rectangular waveguides. However, if we use the notation to condense more lines where the intensity of the field is higher, then we have the representation indicated in Figure A2.10.5b, where we can see how the lines are condensed at $r = a/2$. The magnetic field lines are indicated in Figure A2.10.5c, where we see how they are only radial, i.e., always orthogonal to "e_θ," with the intensity along "r" as that of "e_θ." Electrical circular modes are very attractive for energy propagation along this waveguide. Note that theoretically, longitudinal currents are zero since $h_\theta = 0$, and the same happens for circumferential currents due to "h_r" since $h_r = 0$ for $r = a$. The last magnetic component "h_z" gives a circumferential current that has the ability to be zero for $f \to \infty$. So these modes are attractive to transmit signals with low attenuation, provided they use very high frequencies. In practice, a small amount of current on the internal surface of the waveguide will always exist due to waveguide

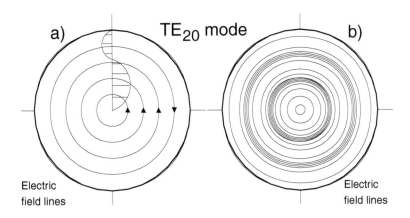

Figure A2.10.6

finite conductivity, but in any case these modes encounter the lowest attenuation inside the waveguide. However, note that "TE_{m0}" modes are nonfundamentals, so they always propagate together with the fundamental mode "TE_{11}." In Figure A2.10.6 we have indicated the electric field lines for "TE_{20}" mode, for convenience.

2. TM Mode

The "k_t^2" is given by:

$$k_t^2 \overset{\pm}{=} -K^2 \rightarrow k_{t,mn}^2 \overset{\pm}{=} -K_{mn}^2 = -\left(z_{mn}/a\right)^2 \tag{A2.10.28}$$

where "m" and "n" are two real, integer and positive numbers, "a" is the waveguide radius and "z_{mn}" is the m-th zero of "$J_n(Kr)$." So, also the "TM" modes in circular waveguides are indicated as "TM_{mn}," and the main difference with "TE" modes is in the different values of "k_t^2" given by A2.10.28. Since, conventionally, also the roots "z_{mn}" of "$J_n(Kr)$" are indicated starting with "1," then "m" cannot be equal to "0," while "n" can be.* Some values of "z_{mn}" are listed in Table A2.10.2.

Table A2.10.2

Bessel function:	First three zeros		
$J_0(x)$	$z_{10} = 2.405$	$z_{20} = 5.520$	$z_{30} = 8.654$
$J_1(x)$	$z_{11} = 3.832$	$z_{21} = 7.016$	$z_{31} = 10.174$
$J_2(x)$	$z_{12} = 5.135$	$z_{22} = 8.417$	$z_{32} = 11.620$

So, inserting A2.10.28 into Equation A2.8.6, it is possible to evaluate "k_z," given by:

$$k_{z,mn} = \left[\left(z_{mn}/a\right)^2 - \omega^2\mu\varepsilon_c\right]^{0.5} \tag{A2.10.29}$$

and the cut-off frequency, with a lossless dielectric filling the waveguide, will be:

$$f_{c,mn} = \left(-k_t^2/\mu\varepsilon\right)^{0.5}/2\pi \equiv z_{mn}/2\pi a\left(\mu\varepsilon\right)^{0.5} \tag{A2.10.30}$$

* In some literature "TM" modes in circular waveguides are indicated as "TM_{nm}." Since the m-th zero always starts conventionally with "1," with this notation it is the second subscript that cannot be equal to zero.

Using the previous equation and the values given in Table A2.10.2 we confirm that the dominant mode for circular waveguides is the "TE_{11}."

Using A2.10.12 into A2.8.9 through A2.8.12 we have the fields for a "TM_{mn}" mode, given by:

$$h_z = 0 \tag{A2.10.31}$$

$$\underline{h}_r = -C(n/r)J_n(K_{mn}r)\operatorname{sen}(n\theta + \varphi)\underline{r}_0 \tag{A2.10.32}$$

$$\underline{h}_\theta = CKJ_n'(K_{mn}r)\cos(n\theta + \varphi)\underline{\theta}_0 \tag{A2.10.33}$$

$$\underline{e}_r = \zeta_{TMmn}\underline{h}_\theta\underline{r}_0 \tag{A2.10.34}$$

$$\underline{e}_\theta = -\zeta_{TMmn}\underline{h}_r\underline{y}_0 \tag{A2.10.35}$$

$$\underline{h}_z = C(K_{mn}^2/j\omega\varepsilon_c)J_n(K_{mn}r)\cos(n\theta + \varphi)\underline{z}_0 \tag{A2.10.36}$$

where the term "ζ_{TMmn}" is the "TM_{mn}" wave impedance given by A2.9.39:

$$\zeta_{TMmn} \stackrel{\perp}{=} k_{z,mn}/j\omega\varepsilon_c \tag{A2.9.39}$$

and "C" is a generic constant. All the field expressions, considered for simplicity as being only in a progressive propagation, need to be multiplied by "$e^{-k_{z,mn}z}$."

The "TM" counterpart of electric circular modes are the "TM_{m0}" modes, called "magnetic circular modes." The field lines for "TM_{10}" are indicated in Figure A2.10.7, whose explanation follows what we said regarding Figure A2.10.5.

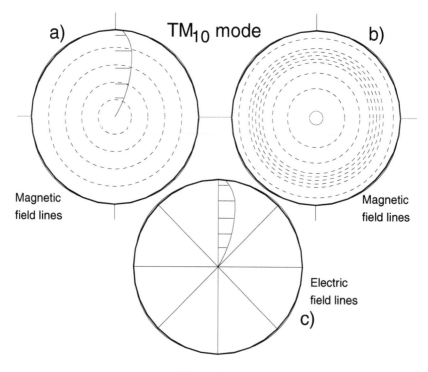

Figure A2.10.7

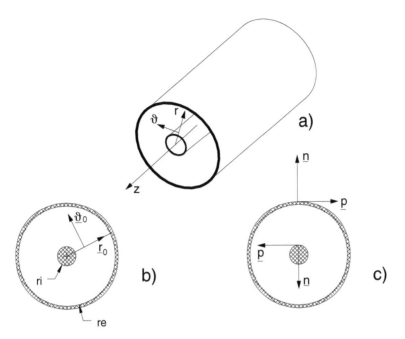

Figure A2.11.1

A2.11 UNIFORM PLANE WAVES AND "TEM" EQUATIONS

Signal transmission is made in two ways, i.e., "guided" or "nonguided." The most used propagation mode in the guided case is the "TEM" mode in some suitable transmission line, while for the nonguided case it is the uniform plane wave "UPW" through air or vacuum. For the "UPW" transmission mode, we saw that this wave is what we receive, at a distance much higher than the signal wavelength, from an antenna transmitting the signal. It is simple to recognize how "UPW" is diffused, thinking of how many transmissions are made using antennas. Also for the "TEM" case it is simple to recognize how it is diffused, noting for instance how many televisions are in the world, each one connected to the antenna using a coaxial cable, and the coaxial cable is a transmission line that can support "TEM" mode, as we will show later.

In this section we will study the related equations of these transmission modes and how they can be generated.

A2.11.1 Modes Inside a Coaxial Cable

The coaxial cable is represented in Figure A2.11.1. As the name suggests, it is realized using two concentric circular conductors, the internal with radius "r_i" and the external with radius "r_e." The internal conductor does not need to be solid, since it is employed at frequencies where it cannot become a circular waveguide. Sometimes, this conductor is not really a cylinder, but instead it is made of a conductor plait. The same can happen for the outer conductor. The e.m. field is completely contained inside the medium between the two concentric conductors, which also support the internal conductor. This medium is made of low loss dielectric.

The study of the propagation modes inside the coaxial cable follows what we did in Sections A2.8 and A2.10. So, for the transverse function "T" we can use Equation A2.10.1, i.e.:

$$T(r,\theta) \stackrel{\perp}{=} R(r)\,\Theta\,(\theta) \tag{A2.10.1}$$

with:

$$\Theta(\theta) = P\cos(n\theta + \varphi) \tag{A2.10.6}$$

$$R(Kr) = D_1 J_n(Kr) + D_2 Y_n(Kr) \tag{A2.10.10}$$

$$-k_t^2 \overset{!}{=} K^2 \tag{A2.10.8}$$

In this case, we cannot impose "D_2" equal to zero since the value $r = 0$ where $Y_n(Kr) = \infty$ is outside the region where the fields can be.* We will now study which modes can exist inside a coaxial cable.

a. "TM" Mode

The contour condition for "T" is:

$$T_{(m)} \overset{!}{=} 0 \quad \text{on "p" for a "TM" mode.} \tag{A2.8.18}$$

The perimeter of the structure is, in our case, composed of two parts and so the previous condition becomes:

$$D_1 J_n(Kr_i) + D_2 Y_n(Kr_i) \overset{!}{=} 0 \tag{A2.11.1}$$

$$D_1 J_n(Kr_e) + D_2 Y_n(Kr_e) \overset{!}{=} 0 \tag{A2.11.2}$$

that combined together give:

$$\frac{J_n(Kr_e)}{Y_n(Kr_e)} \overset{!}{=} \frac{J_n(Kr_i)}{Y_n(Kr_i)} \tag{A2.11.3}$$

The solution in "K" of A2.11.3 is not simple analytically, but it can be obtained simply and with good approximation using a graphic method. An example of this method is indicated in Figure A2.11.2, where we have used n = 0 and assumed $r_e/r_i = 3$. From these graphs we have to read the values of "k" given by the intersections of the curves $J_n(Kr_i)/Y_n(Kr_i)$ with the curves $J_n(Kr_e)/Y_n(Kr_e)$. Note that we have infinite "TM" modes, which we will indicate as "TM$_{mn}$," where "m" represents the index of that "K" value that verifies Equation A2.11.3 and "n" is the order of the Bessel functions.

Once we obtain "K," using Equation A2.10.8 we have the "k_t^2" for every "TM" mode, and consequently we can extract all its field components.

b. "TE" Mode

The contour condition for "T" is:

$$\partial T_{[m]}/\partial n \overset{!}{=} 0 \quad \text{on "p" for a "TE" mode.} \tag{A2.8.21}$$

---

* Unless otherwise stated we are evaluating the case of perfect conductors. So, from the contour conditions it is known that an e.m. field cannot exist inside a perfect conductor.

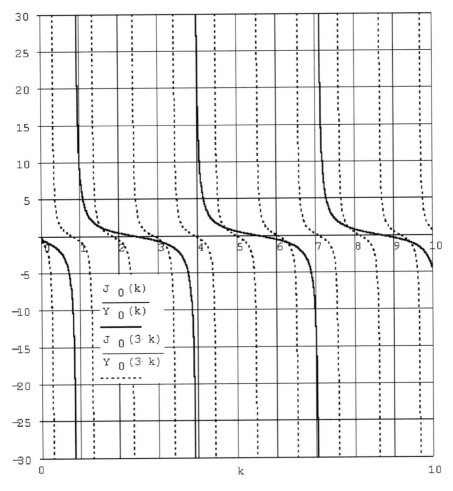

$$\frac{J_0(k)}{Y_0(k)}$$ ———

$$\frac{J_0(3 \cdot k)}{Y_0(3 \cdot k)}$$ - - - - - - -

Figure A2.11.2

which in our case becomes:

$$\partial R/\partial r = ! \, 0 \quad \text{for } r = r_i \text{ and } r = r_e. \tag{A2.11.4}$$

Since:

$$\partial R/\partial r = K\left[D_1 J_n'(Kr) + D_2 Y_n'(Kr)\right] \tag{A2.11.5}$$

A2.8.21 becomes:

$$K\left[D_1 J_n'(Kr_i) + D_2 Y_n'(Kr_i)\right] = ! \, 0 \tag{A2.11.6}$$

$$K\left[D_1 J_n'(Kr_e) + D_2 Y_n'(Kr_e)\right] = ! \, 0 \tag{A2.11.7}$$

which, combined, give:

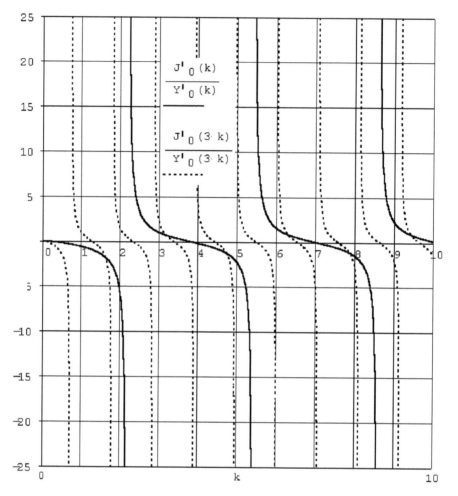

Figure A2.11.3

$$\frac{J'_n(Kr_e)}{Y'_n(Kr_e)} = !\; \frac{J'_n(Kr_i)}{Y'_n(Kr_i)}$$ (A2.11.8)

The previous equation can be solved with good approximation using a graphic method, as we did for "TM" modes. The resulting graph of A2.11.8 is indicated in Figure A2.11.3 where we have used n = 0 and supposed r_e/r_i = 3. Note that now we also have infinite "TE" modes, which we will indicate as "TE_{mn}."

Also in this case, once we obtain "K," using Equation A2.10.8 we have the "k_t^2" for every "TE" mode, and consequently we can extract all its field components.

c. "TEM" Mode

In general, a "TEM" mode can be thought to be generated by a:

1. "TE" mode forcing $\underline{h}_z$ = !0
2. "TM" mode forcing $\underline{e}_z$ = !0

Remembering the field Equation A2.8.11 for "e_z" in a "TM" mode and A2.8.15 for "h_z" in a "TE" mode, the previous conditions 1 and 2 bring $k_t = !\,0$ for a "TEM" mode. This condition gives us an important result. Since in this case $k_t = 0$, then for a lossless cable:

$$k_t^2 + k_z^2 \equiv k_z^2 = !\,k^2 = -\omega^2 \mu \varepsilon$$

which means that for a "TEM" mode the cut-off frequency is zero, since "k_z" is always imaginary.

The condition $k_t = 0$ does not assure the real existence of a "TEM" mode, and to this purpose it is necessary to find a "T" function that satisfies the contour conditions for such a mode. Using the previous point 2 as the beginning condition to search a "TEM" mode,* we know that for the electric "e" field of a "TM" mode in general, we have:

$$\underline{e} = \underline{e}_t + \underline{e}_z \tag{A2.11.9}$$

Now, since:

3. $e_z = !\,0$ to have a "TEM" mode
4. the tangential component "e_τ" of the electric field must be zero on the contour

then from A2.11.9 we have:

$$e_\tau = \underline{e}_t \cdot \underline{p} = !\,0 \tag{A2.11.10}$$

The field "e_t" can be obtained from Equation A2.8.12 applied to our coordinate system, and so:

$$\underline{e}_t = \frac{1}{j\omega\varepsilon_c} \frac{\partial L}{\partial z} \underline{\nabla}_t T \tag{A2.11.11}$$

Inserting this Equation into A2.11.10 we have the condition that function "T" must have a "TEM" mode inside the structure in Figure A2.11.1, i.e.:

$$\nabla_t T \cdot \underline{p} \equiv \partial T / \partial p = !\,0 \quad \text{on "p" for a "TEM" mode,} \tag{A2.11.12}$$

This equation means that "T" must be constant on our contour, i.e., $T = T_i$ and $T = T_e$ on internal and external "p."** Noting that $\partial p = r\partial\theta$ and using Equation A2.10.1, the application of A2.11.12 to our case results in:

$$R(r_i)\,\Theta'(\theta) = !\,0 \quad \text{and} \quad R(r_e)\,\Theta'(\theta) = !\,0 \tag{A2.11.13}$$

The condition:

$$R(r_i) = R(r_e) = 0$$

cannot be accepted,*** then necessarily:

---

* It is possible to prove that a "TEM" mode cannot be generated by a "TE" mode in a structure limited by electric walls.
** Condition "$\partial T/\partial p = !\,0$ on 'p' " theoretically does not mean that "T" must assume two different values "T_i" and "T_e" on $r = r_i$ and $r = r_e$, but it is possible to prove that if $T_i = T_e$ then the "TEM" field must be zero.
*** We said that it is possible to prove that if $T_i = T_e$ then the "TEM" field must be zero.

$$\Theta'(\theta) = !\,0 \rightarrow \Theta(\theta) = Q \qquad\qquad\qquad (A2.11.14)$$

where "Q" is a constant. Then from A2.10.6 it follows that n = ! 0. Equation. A2.11.14 means that "T" does not vary with "θ." Since "$\Theta(\theta)$" is known,* it remains to evaluate "R(r)." From Equation A2.10.7 evaluated with $k_t^2 = 0$ and n = 0 we have:

$$r^2 \frac{d^2R}{dr^2} + r\frac{dR}{dr} = 0 \rightarrow \frac{d}{dr}\left(r\frac{dR}{dr}\right) = 0 \rightarrow rR' = C \qquad (A2.11.15)$$

where "C" is a generic constant. From the expression of "$\underline{\nabla}_t T$" in cylindric coordinate given in Appendix A8 and noting that from A2.11.12 it follows** that $\partial T/\partial\theta = 0$; we have $\underline{\nabla}_t T \equiv (\partial T/\partial r)\underline{r}_0$. Since as a result of this discussion we have T = ΘR(r); then using A2.11.15 we have:

$$\frac{dR}{dr} \equiv \frac{1}{Q}\frac{dT}{dr} = \frac{C}{r} \rightarrow \underline{\nabla}_t T \equiv (D/r)\underline{r}_0 \qquad (A2.11.16)$$

where D = CQ is a constant.

At this point, since we have all the functions to evaluate the "TEM" mode from "TM" Equations A2.8.9 through A2.8.12 inside the coaxial cable, we have that the field components are:

$$\underline{e}_z = 0 \qquad\qquad\qquad\qquad (A2.11.17)$$

$$\underline{h}_z = 0 \qquad\qquad\qquad\qquad (A2.11.18)$$

$$\underline{h}_t(r,z) = -D\frac{L(z)}{r}\,\underline{\theta}_0 \qquad\qquad\qquad (A2.11.19)$$

$$\underline{e}_t(r,z) = \frac{D}{r j\omega\varepsilon_c}\frac{dL(z)}{dz}\,\underline{r}_0 \qquad\qquad (A2.11.20)$$

If we assume a case of progressive propagation, we have $L(z) = e^{-jk_z z}$. In this case it is simple to define the "TEM" impedance "ζ_{TEM}" given by:

$$\zeta_{TEM} \doteq k_z/j\omega\varepsilon_c \equiv (\mu/\varepsilon_c)^{0.5} \qquad\qquad (A2.11.21)$$

where the last quantity comes from the fact that since $k_t = 0$ then:

$$k_z \equiv j\omega(\mu\varepsilon_c)^{0.5} \qquad\qquad\qquad (A2.11.22)$$

It is interesting to conclude this study on coaxial cable saying that the first non- "TEM" mode to start at high frequency, as we are going to specify, is the "TE_{11}." An approximate relationship that gives the limit wavelength "λ_c" below which this mode starts to propagate is:

$$\lambda_c \approx \pi(r_i + r_e) \qquad\qquad\qquad (A2.11.23)$$

---

* We said that "$\Theta(\theta)$" must be a constant, whose value will be determined by the contour conditions.
** From A2.11.14 we recognize that "T" is constant with "θ."

whose error is inside some percent up to $r_e \leq 5r_i$. To do an example, let us assume a coaxial cable with $r_i = 0.5$mm, $r_e = 1.8$mm filled with a lossless dielectric with $\varepsilon_r = 2$. From the previous equation we have $\lambda_c \approx 7.2$mm, which corresponds to a frequency "f_c" of $f_c \approx 29.4$ GHz. We see how higher order modes usually start at frequencies where coaxial cables are seldom used. However, if a coaxial cable should be used, its dimension could be adjusted to avoid the growth of non- "TEM" modes.

A2.11.2 Uniform Plane Wave

A pure uniform plane wave can be thought to propagate inside the space between two infinite parallel conductors. This wave is also what we have at great distance from a dipole antenna, typically many times the antenna length.

A simple analytical way to generate a uniform plane wave is to solve the Maxwell's equations for the field produced in the space by a time-varying current "K" on a conductor plane, as indicated in Figure A2.11.4. To show that, let us start applying the boundary condition for the magnetic field on the conductor surface. Indicating with "$\underline{h}_t^+$" and "$\underline{h}_t^-$" the tangential components on the plane of the magnetic field, respectively, for $z = 0^+$ and $z = 0^-$ we have:

$$\underline{z}_0 \otimes \left(\underline{h}_t^+ - \underline{h}_t^- \right) =! - K\underline{x}_0 \tag{A2.11.24}$$

where the negative sign of the second member comes out from the direction of "$\underline{K}$." Note that $[K] = $ Amp/m. Since the plane with current is symmetrical with respect to $z = 0$, we have $h_t^+ \equiv -h_t^-$ and the previous equation becomes:

$$2\underline{z}_0 \otimes \underline{h}_t^+ =! - K\underline{x}_0 \tag{A2.11.25}$$

from which we see that:

$$\underline{h}_t^+ = \left(K/2 \right) \underline{y}_0 \tag{A2.11.26}$$

This is the result of the boundary condition for the structure indicated in Figure A2.11.4, and represents the source of the field. Now we want to solve the Maxwell's equation for the semispace extending from $z = 0$ to $z = \infty$, forcing the fields to be null at this infinite distance. The equations to be solved are the A2.4.19, i.e.:

$$\underline{\nabla} \otimes \underline{e} = -j\omega\mu\underline{h} \quad \text{and} \quad \underline{\nabla} \otimes \underline{h} = j\omega\varepsilon_c\underline{e} \tag{A2.4.19}$$

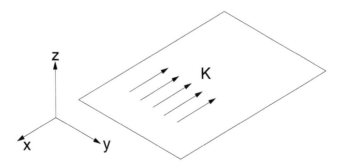

Figure A2.11.4

since we are supposing that "K" moves with time in a sinusoidal shape. Permeability "μ" and complex permittivity "ε_c" are relative to the medium that surrounds the plane. So, for symmetry we see that:

$$\frac{\partial}{\partial x} = \frac{\partial}{\partial y} = 0$$

and the first and second equations in A2.4.19 become:*

$$\frac{\partial e_y}{\partial z} = j\omega\mu h_x, \quad \frac{\partial e_x}{\partial z} = -j\omega\mu h_y, \quad h_z = 0 \qquad \text{(A2.11.27)}$$

$$\frac{\partial h_y}{\partial z} = -j\omega\varepsilon_c e_x, \quad \frac{\partial h_x}{\partial z} = j\omega\varepsilon_c e_y, \quad e_z = 0 \qquad \text{(A2.11.28)}$$

Note that among Equations A2.11.27 and A2.11.28 it is possible to find two pairs of equations, one formed with "e_x" and "h_y" and the other with "e_y" and "h_x." In any case, these two pairs represent the same wave, the only difference being the different polarization. For this reason, we can use only one pair and say that a uniform plane wave is represented by:

$$\frac{\partial e_x}{\partial z} = -j\omega\mu h_y \quad \text{and} \quad \frac{\partial h_y}{\partial z} = -j\omega\varepsilon_c e_x \qquad \text{(A2.11.29)}$$

Another very important thing to observe is that since the plane wave at distance from a dipole antenna is locally a "TEM," similar to what we have obtained with Equations A2.11.27 and A2.11.28 where $h_z = e_z = 0$, we can say that a "UPW" is also a "TEM" wave.

From Equation A2.11.27 we can note an analytical analogy with the generic transmission line equations we studied in Chapter 1. Note that if we substitute to "e_x," "h_y," "$j\omega\mu$," and "$j\omega\varepsilon_c$" respectively, "v," "i," "Z_s," and "Y_p" we have exactly the transmission line Equations 1.2.16 and 1.2.18, with the only notational difference that in the case of a transmission line the direction of propagation has been indicated with "x" and here with "z."** From this analogy, to obtain the solutions of these equations we can proceed as we did in Chapter 1. Proceeding speedily, from Equation A2.11.27 we can obtain:

$$\frac{d^2 e_x}{dz^2} = -\omega^2 \mu\varepsilon_c e_x \qquad \text{(A2.11.30)}$$

whose general solution is:

$$e_x(z) = e^+ e^{(-k_z z)} + e^- e^{(k_z z)} \qquad \text{(A2.11.31)}$$

with:

$$k_z \overset{\perp}{=} j\omega(\mu\varepsilon_c)^{0.5} \qquad \text{(A2.11.32)}$$

---

* See Appendix A8 for "$\underline{\nabla} \otimes (\ )$" in Cartesian coordinate.
** Usually, in wave theory the direction of propagation is indicated with "z." By analogy, the same "z" could be used in Chapter 1 when speaking about transmission lines. Anyway, there we have preferred to use "x" instead of "z," to not create confusion with "Z_s" which represents the line series impedance.

Inserting Equation A2.11.31 into the second of A2.11.29 we have:

$$h_y(z) = -\frac{1}{j\omega\mu}\left[-k_z e^+ e^{(-k_z z)} + k_z e^- e^{(k_z z)}\right] \tag{A2.11.33}$$

or:

$$h_y(z) = h^+ e^{(-k_z z)} + h^- e^{(k_z z)} \tag{A2.11.34}$$

with:

$$h^+ \overset{\perp}{=} e^+ (\varepsilon_c/\mu)^{0.5} \quad \text{and} \quad h^- \overset{\perp}{=} -e^- (\varepsilon_c/\mu)^{0.5} \tag{A2.11.35}$$

To the quantity "ζ_{UPW}" given by:

$$\zeta_{UPW} \overset{\perp}{=} (\mu/\varepsilon_c)^{0.5} \tag{A2.11.36}$$

is given the name of "uniform plane wave impedance," which coincides with the ratio "e^+/h^+" as we can recognize from the first of A2.11.35. Note as A2.11.36 is the same expression we have for a "TEM" mode, indicated in A2.11.21.

A2.12 DISPERSION

To describe the dispersion phenomena it is necessary to define the phase and group velocity of a wave. To this purpose, let us evaluate a wave with a phase dependence, as a function of time and coordinate, of the type:

$$\text{sen}(\omega t - \beta z) \tag{A2.12.1}$$

"z" being the direction of propagation, and the other variables now well known.

They are defined:

1. "phase velocity" — the quantity "v_p" given by:

$$v_p \overset{\perp}{=} \omega/\beta \tag{A2.12.2}$$

2. "group velocity" — the quantity "v_g" given by:

$$v_q \overset{\perp}{=} d\omega/d\beta \tag{A2.12.3}$$

A lot of transmission lines, and electrical networks,* have a linear relationship between phase constant and frequency, i.e., "ω." In this case $v_p \equiv v_g$. But also a lot of transmission lines and networks exist where the relationship between phase constant and frequency is not linear. In this

---

* Dispersion phenomena is not only a transmission line characteristic, but it also appears in a lot of electrical networks, for example filters.

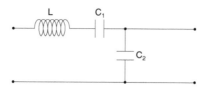

Figure A2.12.1

last case the device is said to be "dispersive," or to cause dispersion. The reason for this name lies in one of the greatest effects of dispersion. To explain it, let us consider the transmission of an impulse in a dispersive transmission line. As we know from the signal analysis theory, such an impulse can be represented by a Fourier series, where each term is a sinusoidal factor with its amplitude and frequency. Each one of these factors can be evaluated by its "v_p" and, since the line is dispersive, the term will have different "v_p." These components will also travel with different speeds inside the line, and at the receiving side the resulting impulse will be distorted. This is the most undesirable effect of the dispersive network, that is the distortion of the impulses that travel inside them. This effect is more evident as the bandwidth of the transformed impulse is higher, i.e., as the rise and fall time of the impulses are shorter. Dispersion is said to be "normal" if the variation of "v_p" with frequency is of opposite sign with respect to the variation of frequency, i.e., if $dv_p/d\omega < 0$, otherwise the dispersion is said to be "anomalous."

An interesting diagram to represent the dispersion is the "$\omega\beta$" diagram. The analytical procedure is generally simple for filters, while for transmission lines it becomes simple after we have its equivalent filter circuit.* To do an example, let us evaluate the "$\omega\beta$" diagram for the transmission line indicated in Figure A2.12.1. Remembering the theory developed in Chapter 1, we have:

$$Z_s = j\omega L + 1/j\omega C_1 \quad \text{and} \quad Y_p = j\omega C_2 \tag{A2.12.4}$$

Setting $r_c \triangleq C_2/C_1$ we can write:

$$k = \alpha + j\beta \triangleq \left(Z_s Y_p\right)^{0.5} = \left(r_c - \omega^2 LC_2\right)^{0.5} \tag{A2.12.5}$$

and we can have the following three cases:

a. $r_c = \omega^2 LC_2$
   In this case $k \equiv 0$ and $\alpha = \beta = 0$. This situation is obtained for an angular frequency "ω_c" given by:

$$\omega_c = \left(r_c/LC_2\right)^{0.5} \equiv \left(LC_1\right)^{-0.5} \tag{A2.12.6}$$

b. $r_c < \omega^2 LC_2$ or, using A2.12.6, $\omega > \omega_c$.
   In this case $k \equiv j\beta$, with:

$$\beta = \left(\omega^2 LC_2 - r_c\right)^{0.5} = \omega\left[LC_2\left(1 - \omega_c^2/\omega^2\right)\right]^{0.5} \tag{A2.12.7}$$

c. $r_c > \omega^2 LC_2$ or, using A2.12.6, $\omega < \omega_c$.
   In this case $k \equiv \alpha$, with:

---

* We will study in the next section the networks representative of transmission lines.

$$\alpha = \left(r_c - \omega^2 LC_2\right)^{0.5} = \omega\left[LC_2\left(\omega_c^2/\omega^2 - 1\right)\right]^{0.5} \qquad (A2.12.8)$$

Note that the line indicated in Figure A2.12.1 also introduces attenuation for $\omega < \omega_c$ if it is lossless.

Connecting in one graph the solutions A2.12.6 through A2.12.8 we have the result indicated in Figure A2.12.2.

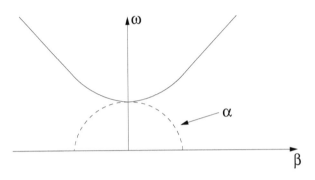

Figure A2.12.2

To make another example, it is simple to recognize that for a simple "LC" low pass lossless filter, i.e., represented by the network in Figure A2.12.1 without the capacitor "C_1," the "$\omega\beta$" diagram is as represented in Figure A2.12.3.

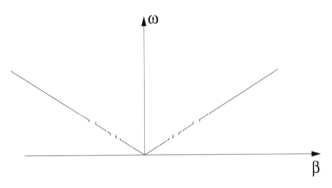

Figure A2.12.3

Using the "$\omega\beta$" diagram we can give a graphical representation of "v_p" and "v_g." From the definitions of these quantities given in A2.12.2 and A2.12.3, it is simple to recognize that:

I. "v_p" is given by the tangent of the angle formed between the "β" axis and the line that starts from the axis origin and the desired point "Q" on the "$\omega\beta$" diagram. In Figure A2.12.4 we have indicated this angle with "θ_p"

II. "v_g" is given by the tangent at the desired point "Q" on the "$\omega\beta$" diagram. The angle formed between the "β" axis and the tangent in "Q" has been indicated with "θ_g."

Note that for the point "I" in Figure A2.12.4 phase velocity is infinite and $\theta_p = 90°$, while group velocity is zero and $\theta_g = 0$. Waves that posses $v_p > (\mu_0\varepsilon_0)^{0.5}$ are said to be "fast," otherwise are said

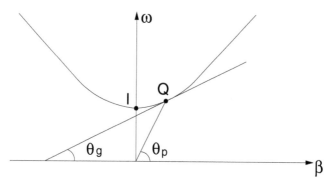

Figure A2.12.4

to be "slow." Note that an infinite value for "v_p" does not have a physical meaning since the phase is not a physical quantity, and consequently this result does not contradict the relativistic result that the maximum speed for a physical quantity is equal to the speed of light. Dispersive lines are characterized by $v_p \neq v_g$, at least for some values of "ω." Dispersive lines can have frequency ranges where "v_p" and "v_g" are equal or very similar. For example, the network indicated in Figure A2.12.2 for $\omega \gg \omega_c$ has $v_p \approx v_g$, i.e., the "$\omega\beta$" diagram becomes a line.

A2.13 ELECTRICAL NETWORKS ASSOCIATED WITH PROPAGATION MODES

It is interesting to explain how the transmission modes studied previously can be associated with lumped networks, through some manipulation of their characteristic analytical formulas. The common procedure is to transform in some way the field equations in the general transmission line equations we studied in Chapter 1, i.e.:

$$\frac{dv(x)}{dx} = - z_s i(x) \tag{1.2.16}$$

$$\frac{di(x)}{dx} = y_p v(x) \tag{1.2.10}$$

The starting points are the propagation modes inside structures with curvilinear orthogonal coordinate reference systems, i.e., the "TM" and "TE" modes of Section A2.8 and the "TEM" mode of Section A2.11. In particular, only the transverse component of each mode can be used to obtain the equivalent transmission line equations. The longitudinal axis will be indicated with "z." We will also see how the "UPW" can be associated with its lumped network.

1. Associated Network for a "TM" Mode

To extract the equivalent network for such a mode it is necessary to rewrite the Maxwell's equations in a way that only the transverse field components are involved. To do that, insert into A2.4.19 the expression of "delta" operator specifying its transverse and longitudinal components. We have:

$$\underline{\nabla} \otimes \underline{e} = \left(\underline{\nabla}_t + \frac{\partial}{\partial z} \underline{z}_0 \right) \otimes \underline{e} = \underline{\nabla}_t \otimes \underline{e} + \underline{z}_0 \otimes \frac{\partial \underline{e}}{\partial z} = -j\omega\mu\underline{h} \qquad (A2.13.1)$$

$$\underline{\nabla} \otimes \underline{h} = \left(\underline{\nabla}_t + \frac{\partial}{\partial z} \underline{z}_0 \right) \otimes \underline{h} = \underline{\nabla}_t \otimes \underline{h} + \underline{z}_0 \otimes \frac{\partial \underline{h}}{\partial z} = j\omega\varepsilon_c\underline{e} \qquad (A2.13.2)$$

Now, let us do for A2.13.1 a left vector product with "$\underline{z}_0$" and for A2.13.2 a scalar product with "$\underline{z}_0$." We have:

$$\underline{z}_0 \otimes \left(\underline{\nabla}_t \otimes \underline{e} \right) + \underline{z}_0 \otimes \left(\underline{z}_0 \otimes \frac{\partial \underline{e}}{\partial z} \right) = j\omega\mu \left(\underline{h} \otimes \underline{z}_0 \right) \qquad (A2.13.3)$$

$$\underline{z}_0 \cdot \left(\underline{\nabla}_t \otimes \underline{h} \right) + \underline{z}_0 \cdot \left(\underline{z}_0 \otimes \frac{\partial \underline{h}}{\partial z} \right) = j\omega\varepsilon_c\underline{e} \cdot \underline{z}_0 \qquad (A2.13.4)$$

Executing the calculations,* the previous two equations become:

$$\underline{\nabla}_t e_z - \frac{\partial \underline{e}_t}{\partial z} = j\omega\mu \left(\underline{h}_t \otimes \underline{z}_0 \right) \qquad (A2.13.5)$$

$$\underline{\nabla}_t \cdot \underline{h} \otimes \underline{z}_0 = j\omega\varepsilon_c e_z \qquad (A2.13.6)$$

Extracting "e_z" from A2.13.6 and inserting into A2.13.5 we have:

$$\frac{\partial \underline{e}_t}{\partial z} = -j\omega\mu\left(\underline{h}_t \otimes \underline{z}_0\right) + \frac{1}{j\omega\varepsilon_c}\underline{\nabla}_t\underline{\nabla}_t \cdot \left(\underline{h}_t \otimes \underline{z}_0\right) \qquad (A2.13.7)$$

from which, for duality, we can have the equivalent equation for the magnetic field, i.e.:

$$\frac{\partial \underline{h}_t}{\partial z} = j\omega\varepsilon_c\left(\underline{e}_t \otimes \underline{z}_0\right) - \frac{1}{j\omega\mu}\underline{\nabla}_t\underline{\nabla}_t \cdot \left(\underline{e}_t \otimes \underline{z}_0\right) \qquad (A2.13.8)$$

Equations A2.13.7 and A2.13.8 are the desired expressions of Maxwell's equations where only the transverse fields appear.

Now, for Equations A2.8.10 and A2.8.12 of "TM" modes let us set:

$$L(z) \overset{\pm}{=} i(z) \quad \text{and} \quad \frac{1}{j\omega\varepsilon_c}\frac{dL}{dz} \overset{\pm}{=} -v(z) \ ** \qquad (A2.13.9)$$

With these positions, Equations A2.8.10 and A2.8.12 can be rewritten as:***

$$\underline{h}_t = i\underline{\nabla}_t T \otimes \underline{z}_0 \qquad (A2.13.10)$$

$$\underline{e}_t = -v\underline{\nabla}_t T \qquad (A2.13.11)$$

* See Appendix A8 for the operation $\underline{a} \otimes \underline{b} \otimes \underline{c}$.
** The partial derivative has been substituted with absolute derivative, since for our geometry "L" is only dependent on "z."
*** For notation simplicity we will omit for "i" and "v" their dependence with "z."

For Equations A2.8.14 and A2.8.16 of "TE" modes let us set:

$$L(z) \stackrel{\perp}{=} v(z) \quad \text{and} \quad \frac{1}{j\omega\mu}\frac{dL}{dz} \stackrel{\perp}{=} -i(z) \tag{A2.13.12}$$

With these positions, Equations A2.8.14 and A2.8.16 for "TE" modes can be rewritten as:

$$\underline{e}_t = -v\underline{\nabla}_t T \otimes \underline{z}_0 \tag{A2.13.13}$$

$$\underline{h}_t = -i\underline{\nabla}_t T \tag{A2.13.14}$$

Now, let us suppose that inside our guiding structure the electric and magnetic transverse components are given as the sum of the whole "TE" and "TM" fields. With the notation used in Section A2.8, we have:

$$\underline{h}_t = \sum_{(n)} i_{(n)}\underline{\nabla}_t T_{(n)} \otimes \underline{z}_0 - \sum_{[n]} i_{[n]}\underline{\nabla}_t T_{[n]} \tag{A2.13.15}$$

$$\underline{e}_t = -\sum_{(n)} v_{(n)}\underline{\nabla}_t T_{(n)} - \sum_{[n]} v_{[n]}\underline{\nabla}_t T_{[n]} \otimes \underline{z}_{(} \tag{A2.13.16}$$

Inserting the two previous equations into A2.13.7, we have:*

$$-\sum_{(n)} \frac{dv_{(n)}}{dz}\underline{\nabla}_t T_{(n)} - \sum_{[n]} \frac{dv_{(n)}}{dz}\underline{\nabla}_t T_{[n]} \otimes \underline{z}_0 = \cdots$$

$$\cdots = -j\omega\mu\left(-\sum_{(n)} i_{(n)}\underline{\nabla}_t T_{(n)} - \sum_{[n]} i_{[n]}\underline{\nabla}_t T_{[n]} \otimes \underline{z}_0\right) - \frac{\sum_{(n)} k_{t(n)}^2 i_{(n)}\underline{\nabla}_t T_{(n)}}{j\omega\varepsilon_c} \tag{A2.13.17}$$

where we have used the following simplifications:

a. $(\underline{\nabla}_t T \otimes \underline{z}_0) \otimes \underline{z}_0 \equiv -\underline{\nabla}_t T$
b. $\underline{\nabla}_t\underline{\nabla}_t \bullet \sum_{(n)} i_{(n)}\underline{\nabla}_t T_{(n)} = \underline{\nabla}_t\sum_{(n)} i_{(n)}\underline{\nabla}_t \bullet \underline{\nabla}_t T_{(n)} = \underline{\nabla}_t \sum_{(n)} i_{(n)}\underline{\nabla}_t^2 T_{(n)} = \sum_{(n)} i_{(n)} k_{t(n)}^2 \underline{\nabla}_t T_{(n)}$
c. $\underline{\nabla}_t\underline{\nabla}_t \bullet \sum_{[n]} i_{[n]} (\underline{\nabla}_t T_{[n]} \otimes \underline{z}_0) = \underline{\nabla}_t\sum_{[n]} i_{[n]}\underline{\nabla}_t \bullet (\underline{\nabla}_t T_{[n]} \otimes \underline{z}_0) = 0$

Now, for A2.13.17 let us do a scalar product with "$\underline{\nabla}_t T_{(m)}$"** and integrate on a transverse section "S" of our guiding structure. We have:

$$-\sum_{(n)} \frac{dv_{(n)}}{dz}\int_s \underline{\nabla}_t T_{(m)} \bullet \underline{\nabla}_t T_{(n)} dS - \sum_{[n]} \frac{dv_{[n]}}{dz}\int_s \underline{\nabla}_t T_{(m)} \bullet \underline{\nabla}_t T_{[n]} \otimes \underline{z}_0 dS$$

$$= j\omega\mu\left(\sum_{(n)} i_{(n)}\int_s \underline{\nabla}_t T_{(m)} \bullet \underline{\nabla}_t T_{(n)} dS - \sum_{[n]} i_{[n]}\int_s \underline{\nabla}_t T_{(m)} \bullet \underline{\nabla}_t T_{[n]} \otimes \underline{z}_0 dS\right) \tag{A2.13.18}$$

$$+ -\frac{\sum_{(n)} k_{t(n)}^2 i_{(n)}\int_s \underline{\nabla}_t T_{(m)} \bullet \underline{\nabla}_t T_{(n)} dS}{j\omega\varepsilon_c}$$

---

* The partial derivatives can now be substituted with absolute derivatives, since "v" and "i" are only dependent to "z."
** The subscript "(m)" represents a generic "TM" mode.

Using the relationships given in Section A2.8 it is simple to recognize that the only nonzero elements are:

a. The first term at first member, if $m \equiv n$
b. The first and third term at second member, if $m \equiv n$.

To simplify the notation, we can set:

$$\int_{s} \underline{\nabla}_{t} T \cdot \underline{\nabla}_{t} T dS \overset{\perp}{=} 1 \qquad (A2.13.19)$$

which does not limit the generality of the solution, since the previous equation set to "1" the constant that appears in "T," but we can insert the constants generality to the constant that appears in "L(z)." So, using A2.13.19, the orthogonality properties given in Section A2.8, A2.13.18 becomes:

$$\frac{dv_{(m)}}{dz} = -\left[j\omega\mu - \frac{k^2_{t(m)}}{j\omega\varepsilon_c} \right] i_{(m)} \qquad (A2.13.20)$$

To obtain an equivalent relationship for the current "$i_{(m)}$" we have to insert Equations A2.13.15 and A2.13.16 into A2.13.8. Then we do a scalar product with "$\underline{\nabla}_t T_{(m)} \otimes \underline{z}_0$" and integrate on a transverse section "S" of our guiding structure. We have:

$$\frac{di_{(m)}}{dz} = -j\omega\varepsilon_c v_{(m)} \qquad (A2.13.21)$$

From Equations A2.13.20 and A2.13.21 we can recognize that the associated network for a "TM" mode inside our guiding structure with curvilinear orthogonal coordinates reference system is as indicated in Figure A2.13.1. The element's values are:

$$L = \mu, \quad C_1 = \varepsilon/-k_t^2, \quad R_1 = -k_t^2/g, \quad C_2 = \varepsilon, \quad R_2 = 1/g \qquad (A2.13.22)$$

It is interesting to observe that if we insert into the definition of propagation constant given in Chapter 1, i.e..

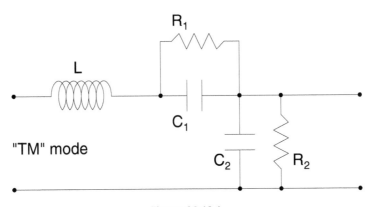

Figure A2.13.1

$$k = \left(Z_s Y_p \right)^{0.5} \tag{1.3.4}$$

the expression of "Z_s" and "Y_p" given by A2.13.20 and A2.13.21, i.e.:

$$Z_s = j\omega\mu - \frac{k_{t(m)}^2}{j\omega\varepsilon_c} \quad \text{and} \quad Y_p = j\omega\varepsilon_c \tag{A2.13.23}$$

we have:

$$k^2 = -\omega^2\mu\varepsilon_c - k_{t(m)}^2 \tag{A2.13.24}$$

So, the value of "k" for the network indicated in Figure A2.13.1 coincides with the value of "k_z" for the "TM" mode inside the guiding structure, obtained by A2.8.6 when evaluated for $\ell_0 \equiv z_0$. Similarly, inserting A2.13.23 into the definition of transmission line impedance we gave in Chapter 1, we have:

$$\zeta \overset{\perp}{=} \left(Z_s / Y_p \right)^{0.5} = k_z / j\omega\varepsilon_c \tag{A2.13.25}$$

which is equal to the "TM" mode impedance "ζ_{TM}" we gave in Section A2.8.

2. Associated Network for a "TE" Mode

For the "TE" mode in the previous point 1 we have set:

$$L(z) \overset{\perp}{=} v(z) \quad \text{and} \quad \frac{1}{j\omega\mu}\frac{dL}{dz} \overset{\perp}{=} -i(z) \tag{A2.13.12}$$

For A2.13.17 let us do a scalar product with "$\underline{\nabla}_t T_{[m]} \otimes \underline{z}_0$"* and integrate on a transverse section "S" of our guiding structure. Then, using A2.13.19, the orthogonality properties given in Section A2.8, m = n, we have.**

$$\frac{dv_{[m]}}{dz} = -j\omega\mu i_{[m]} \tag{A2.13.26}$$

To obtain an equivalent relationship for the current "$i_{[m]}$" we have to insert Equations A2.13.15 and A2.13.16 into A2.13.8. Then we do a scalar product with "$\underline{\nabla}_t T_{[m]}$" and integrate on a transverse section "S" of our guiding structure. We have:

$$\frac{di_{[m]}}{dz} = -\left[j\omega\varepsilon_c - \frac{k_{t[m]}^2}{j\omega\mu} \right] v_{[m]} \tag{A2.13.27}$$

---

* The subscript "[m]" represents a generic "TE" mode.
** The procedure is similar to what we did in previous point 1 for "TM" mode.

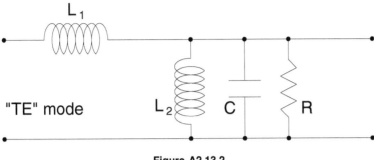

Figure A2.13.2

From Equations A2.13.26 and A2.13.27 we can recognize that the associated network for a "TE" mode inside our guiding structure with curvilinear orthogonal coordinates reference system is as indicated in Figure A2.13.2. The element's values are:

$$L_1 = \mu, \quad L_2 = \mu/-k_t^2, \quad C = \varepsilon, \quad R = 1/g \tag{A2.13.28}$$

Similar to the case of "TM" mode, it is possible to verify that also in this case using the values of "Z_s" and "Y_p" obtained from A2.13.26 and A2.13.27, the values of "k" and "ζ" for the network indicated in Figure A2.13.2 are coincident with "k_z" and "ζ_{TE}," when evaluated for $\ell_0 \equiv z_0$, we gave in Section A2.8.

3. Associated Network for a "TEM" Mode

For a "TEM" mode we set:

$$L(z) \stackrel{\perp}{=} i(z) \quad \text{and} \quad \frac{1}{j\omega\varepsilon_c} \frac{dL}{dz} \stackrel{\perp}{=} -v(z) \tag{A2.13.29}$$

In this case "i(z)" and "v(z)" satisfy the transmission line equations, and so it is very simple to associate an equivalent network. Inserting the first equation into the second equation of A2.13.26 we have:

$$\frac{di}{dz} = -j\omega\varepsilon_c v \tag{A2.13.30}$$

Deriving in "z" assuming this equation and supposing to have only the progressive wave we have:

$$\frac{d^2 i}{dz^2} = k_z^2 i = -\omega^2 \mu\varepsilon_c i = -j\omega\varepsilon_c \frac{dv}{dz} \tag{A2.13.31}$$

from which:

$$\frac{dv}{dz} = -j\omega\mu i \tag{A2.13.32}$$

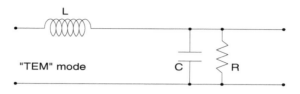

Figure A2.13.3

From Equations A2.13.30 and A2.13.32 we can recognize that the associated network for a "TEM" mode inside our guiding structure with curvilinear orthogonal coordinates reference system is as indicated in Figure A2.13.3. The element's values are:

$$L = \mu, \quad C = \varepsilon, \quad R = 1/g \tag{A2.13.33}$$

For the "TEM" case it is simple to verify that using the values of "Y_p" and "Z_s" obtained from A2.13.30 and A2.13.32, the values of "k" and "ζ" for the network indicated in Figure A2.13.3 are:

$$k = j\omega \left(\mu \varepsilon_c\right)^{0.5} \tag{A2.13.34}$$

$$\zeta = \left(\mu/\varepsilon_c\right)^{0.5} \tag{A2.13.35}$$

coincident with the "k_z" and "ζ_{TEM}" values we gave in A2.11.22 and A2.11.21 for the "TEM" mode inside a coaxial cable.

4. Associated Network for a "UPW"

For a "UPW" it is easy to extract the associated network. From Equations A2.11.29 setting:

$$e_x(z) = v(z), \quad Z_s = j\omega\mu, \quad Y_p = j\omega\varepsilon_c, \quad h_y(z) = i(z) \tag{A2.13.36}$$

we have:

$$\frac{dv}{dz} = Z_s i \quad \text{and} \quad \frac{di}{dz} = -Y_p v \tag{A2.13.37}$$

From Equation A2.13.37 we can recognize that the associated network for a "UPW" inside a medium with permeability "μ" and permittivity "ε_c" with a Cartesian coordinate system is as indicated in Figure A2.13.4. The element's values are:

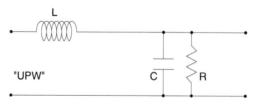

Figure A2.13.4

$$L = \mu, \quad C = \varepsilon, \quad R = 1/g \tag{A2.13.38}$$

Confronting Figure A2.13.3 and A2.13.4 we recognize how a "TEM" mode and an "UPW" have the same type of associated network. Also in this case it is simple to verify that using the values of "Y_p" and "Z_s" obtained from A2.13.37, the values of "k" and "ζ" for the network indicated in Figure A2.13.4 are again given by A2.13.34 and A2.13.35, i.e., coincident with the "k_z" and "ζ_{UPW}" values we gave in A2.11.32 and A2.11.36 for an "UPW."

A2.14 FIELD PENETRATION INSIDE NONIDEAL CONDUCTORS

In all the previous paragraphs we have assumed that all the conductors guiding the e.m. fields are perfect, i.e., they have infinite conductivity, i.e., $g = \infty$. In this section we will study how the e.m. fields penetrate inside a real conductor that is characterized by a finite value of "g." To do that, let us write the Maxwell homogeneous equations:

$$\underline{\nabla} \otimes \underline{e} = -j\omega\mu\underline{h} \quad \text{and} \quad \underline{\nabla} \otimes \underline{h} = -j\omega\varepsilon\underline{e} + g\underline{e} \tag{A2.14.1}$$

Doing the "curl" to both members and applying the identity:

$$\underline{\nabla} \otimes \underline{\nabla} \otimes \underline{e} = \underline{\nabla}\underline{\nabla} \cdot \underline{e} - \nabla^2\underline{e} \tag{A2.14.2}$$

A2.14.1 becomes:

$$\underline{\nabla}\underline{\nabla} \cdot \underline{e} - \nabla^2\underline{e} = -j\omega\mu\underline{\nabla} \otimes \underline{h} \tag{A2.14.3}$$

Since also inside a nonideal conductor there is no volumetric free charge* then $\underline{\nabla} \bullet \underline{e} = 0$, In addition, in a good conductor $g \gg \omega\varepsilon$, and so we can neglect the first term in the second equation of A2.14.1. With these introductions, the previous equation becomes:

$$\nabla^2\underline{e} = j\omega\mu\,\underline{\nabla} \otimes \underline{h} \tag{A2.14.4}$$

Let us resolve A2.14.4 assuming the electric field is dependent only on a coordinate, and to be in the practical situation reported in Figure A2.14.1. The Cartesian coordinate system is as indicated, and the electric field is only dependent on "x." In this case, Equation A2.14.4 becomes:

$$\frac{d^2e_z}{dx^2} = j\omega\mu g e_z \underline{\perp} k^2 e_z, \text{with } k^2 = j\omega\mu g \tag{A2.14.5}$$

Note that the previous equation is similar to what we have resolved in Section A2.3. Then, the general solution is:

$$e_z = C_1 e^{-kx} + C_2 e^{kx} \tag{A2.14.6}$$

The constants "C_1" and "C_2" can be determined observing that for:

---

* Of course, also if we are considering a nonperfect conductor we would always treat with a good conductor, for which the volumetric free charge may be neglected.

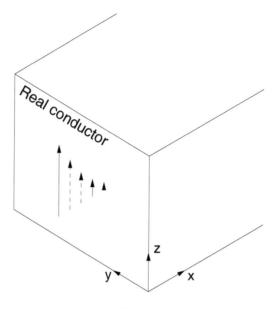

Figure A2.14.1

a. $x = \infty$ the field must be zero. So: $C_2 = ! \; 0$.
b. $x = 0$ the field must be equal to the incident value, that we will indicate as "e_0." So: $C_1 = e_0$.

Consequently, A2.14.6 becomes:

$$e_z = e_0 e^{-kx} \tag{A2.14.7}$$

Applying the identity $\sqrt{j} = (1+j)/\sqrt{2}$ to the definition of "k" given in A2.14.5 we have:

$$k = k_r + jk_j = (1+j)/p \tag{A2.14.8}$$

where the quantity "p," called "penetration depth," is given by:

$$p \overset{\perp}{=} (\pi f \mu g)^{-0.5} \tag{A2.14.9}$$

with dimensions [meter]. Note that the quantity "p" gives the amplitude decay rate of the electric field inside the good conductor. Note that the smaller the "p" value is, the smaller is the penetration depth, and the higher the "k" value is, the higher is the decay rate.

It is interesting to note that for the magnetic field "h" and the surface current "J" an equation similar A2.14.4 holds. In fact, for the magnetic field we have:

$$\underline{\nabla} \otimes \underline{\nabla} \otimes \underline{h} = \underline{\nabla}\underline{\nabla} \cdot \underline{h} - \nabla^2 \underline{h} = j\omega\varepsilon\underline{\nabla} \otimes \underline{e} + g\underline{\nabla} \otimes \underline{e} \tag{A2.14.10}$$

Since:

1. from the Maxwell's equations $\underline{\nabla} \bullet \underline{h} = 0$
2. in a good conductor $g \gg \omega\varepsilon$

using the first of A2.14.1 the previous equation becomes:

$$\nabla^2 \underline{h} = j\omega\mu g \underline{h} \tag{A2.14.11}$$

For the surface current density "J" using the first of A2.14.1 we have:

$$\underline{J} = g\underline{e} \rightarrow \underline{\nabla} \otimes \underline{J} = g\underline{\nabla} \otimes \underline{e} = -j\omega\mu g\underline{h} \tag{A2.14.12}$$

Then

$$\underline{\nabla} \otimes \underline{\nabla} \otimes \underline{J} = \underline{\nabla}\underline{\nabla} \cdot \underline{J} - \nabla^2 \underline{J} = -j\omega\mu g\underline{\nabla} \otimes \underline{h} \tag{A2.14.13}$$

Since in a good conductor:

3. $\underline{\nabla} \cdot \underline{J} = 0$
4. $g \gg \omega\varepsilon$

using the second of A2.14.1 the previous equation becomes:

$$\nabla^2 \underline{J} = j\omega\mu g\underline{J} \tag{A2.14.14}$$

So, solutions of the form A2.14.6 and A2.14.7 are also applicable for magnetic field and surface current density.

In our case of nonideal conductors, but still "good" ones, an internal impedance "ζ_i" is defined as the ratio between the electric field on the surface and the total current in the conductor. With reference to Figure A2.14.1, the electric field on the surface, i.e., x = 0, is $e_{z0} = J_0/g$. Consequently, the total current "I_z" per unit width of the conductor is:

$$I_z = \int_0^\infty J_z(x)dx = \int_0^\infty J_0 e^{\frac{1+j}{p}x}dx = J_0\frac{p}{1+j} \tag{A2.14.15}$$

with dimensions [A/meter]. So, the internal impedance "ζ_i" is:

$$\zeta_i \overset{\perp}{=} e_{z0}/I_z \overset{\perp}{=} R_s + j\omega L_i = (1+j)/pg \quad [\Omega] \tag{A2.14.16}$$

From this equation we note "ζ_i" has the real part is equal to the imaginary part, which means that:

$$R_s \equiv \omega L_i = 1/pg \qquad [\Omega] \tag{A2.14.17}$$

The quantity "R_s" is called the "sheet resistance" of the conductor. In literature it is possible to find "R_s" called "surface resistance." We think that this name can generate confusion since "R_s" has [Ω] as physical dimension and not [Ω/unit surface]. In practice, "R_s" is measured as [Ω/square]. The word "square" means that the value of "R_s" is a function of the ratio between the longitudinal "ℓ" and transverse "t" absolute length of the conductor. For example, if one conductor has $\ell = 2t$ its "R_s" along the direction "ℓ_0" is twice that of the same conductor with $\ell = t$ along the direction "ℓ_0."

Inserting A2.14.9 into A2.14.17 we have:

$$R_s = \left(\pi\mu f/g\right)^{0.5} \qquad \left[\Omega/square\right] \qquad (A2.14.18)$$

Alternatively, if we extract from A2.14.9 the value of "g" and we insert it into A2.14.17 we have:

$$R_s = \pi\mu fp \qquad \left[\Omega/square\right] \qquad (A2.14.19)$$

In Table A2.14.1 we report the conductivity, penetration depth, and "R_s" for typical conductors used in transmission lines:

Table A2.14.1

Metal	g(Sie/m)	p (m)	R_s (W)
Silver	$6.17*10^7$	$0.0642/\sqrt{f}$	$2.52*10^{-7}\sqrt{f}$
Copper	$5.8*10^7$	$0.066/\sqrt{f}$	$2.61*10^{-7}\sqrt{f}$
Gold	$4.35*10^7$	$0.0763/\sqrt{f}$	$3.01*10^{-7}\sqrt{f}$
Aluminum	$3.72*10^7$	$0.0826/\sqrt{f}$	$3.26*10^{-7}\sqrt{f}$
Brass	$1.57*10^7$	$0.127/\sqrt{f}$	$5.01*10^{-7}\sqrt{f}$

It is interesting to observe how copper, which is readily available, has a conductivity second only to silver. However copper oxidizes, and for this reason it is covered with gold when good transmission lines are needed.

Figures A2.14.2 and A2.14.3 report the penetration depth "p" in meters as a function of frequency in MHz, for Silver (Ag), Copper (Cu), and Gold (Au). Note that starting from the frequency of 1 GHz, the penetration depth inside such conductors are only some micrometers and decrease with frequency.

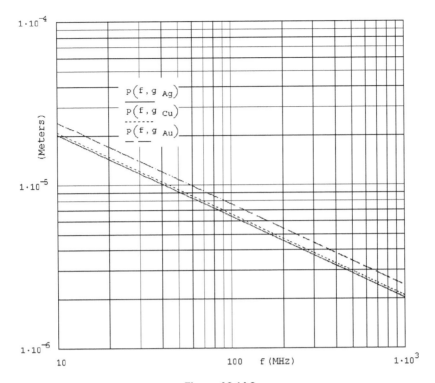

Figure A2.14.2

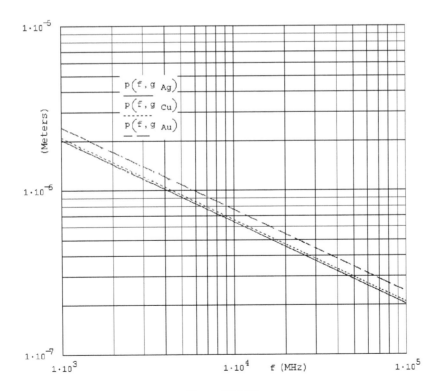

Figure A2.14.3

REFERENCES

1. S. Ramo, J. R. Whinnery, T. Van Duzer, *Fields and Waves in Communication Electronics*, John Wiley & Sons, New York, 568, 1980.
2. G. Barzilai, Fondamenti di elettromagnetismo, Siderea editore, Roma, 125, 1975.
3. R. E. Collin, *Foundations for Microwave Engineering*, McGraw Hill, New York, 1992.
4. G. Gerosa, Appunti di microonde, Università degli studi di Roma La Sapienza, 1980.
5. G. Franceschett, Campi elettromagnetici, Boringhieri editore, 1982.
6. S. Ramo, J. R. Whinnery, T. Van Duzer, *Fields and Waves in Communication Electronics*, John Wiley & Sons, New York, 568, 1980.
7. G. Barzilai, Fondamenti di elettromagnetismo, Siderea editore, Roma, 125, 1975.
8. R. E. Collin, *Foundations for Microwave Engineering*, McGraw Hill, New York, 1992.
9. G. Gerosa, Appunti di microonde, Università degli studi di Roma "La Sapienza," 1980.
10. G. Franceschetti, Campi elettromagnetici, Boringhieri editore, 1982.
11. R. E. Collin, *Foundations for Microwave Engineering*, McGraw Hill, New York, 1992.
12. G. Gerosa, Appunti di microonde, Università degli studi di Roma "La Sapienza," 1980.
13. G. N. Watson, *A Treatise on the Theory of Bessel Functions*, Cambridge University Press, London, 1966.
14. A. Ghizetti, L. Marchetti, A. Ossicini, Lezioni di complementi di matematica, Editrice Veschi, 373, 1976.
15. M. Abramowitz, I. A. Stegun, *Handbook of Mathematical Functions*, Dover, New York, 358, 1970.

Using the previous equation and the values given in Table A2.10.2 we confirm that the dominant mode for circular waveguides is the "TE_{11}."

Using A2.10.12 into A2.8.9 through A2.8.12 we have the fields for a "TM_{mn}" mode, given by:

$$h_z = 0 \tag{A2.10.31}$$

$$\underline{h}_r = -C(n/r)J_n(K_{mn}r)\operatorname{sen}(n\theta + \varphi)\underline{r}_0 \tag{A2.10.32}$$

$$\underline{h}_\theta = CKJ'_n(K_{mn}r)\cos(n\theta + \varphi)\underline{\theta}_0 \tag{A2.10.33}$$

$$\underline{e}_r = \zeta_{TMmn}h_\theta\underline{r}_0 \tag{A2.10.34}$$

$$\underline{e}_\theta = -\zeta_{TMmn}h_r\underline{y}_0 \tag{A2.10.35}$$

$$\underline{h}_z = C(K_{mn}^2/j\omega\varepsilon_c)J_n(K_{mn}r)\cos(n\theta + \varphi)\underline{z}_0 \tag{A2.10.36}$$

where the term "ζ_{TMmn}" is the "TM_{mn}" wave impedance given by A2.9.39:

$$\zeta_{TMmn} \stackrel{\perp}{=} k_{z,mn}/j\omega\varepsilon_c \tag{A2.9.39}$$

and "C" is a generic constant. All the field expressions, considered for simplicity as being only in a progressive propagation, need to be multiplied by "$e^{-k_{z,mn}z}$."

The "TM" counterpart of electric circular modes are the "TM_{m0}" modes, called "magnetic circular modes." The field lines for "TM_{10}" are indicated in Figure A2.10.7, whose explanation follows what we said regarding Figure A2.10.5.

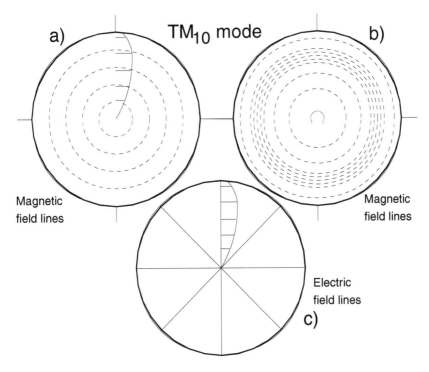

Figure A2.10.7

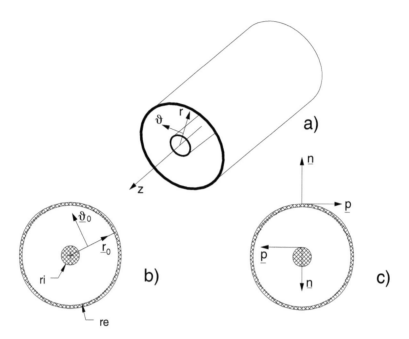

Figure A2.11.1

A2.11 UNIFORM PLANE WAVES AND "TEM" EQUATIONS

Signal transmission is made in two ways, i.e., "guided" or "nonguided." The most used propagation mode in the guided case is the "TEM" mode in some suitable transmission line, while for the nonguided case it is the uniform plane wave "UPW" through air or vacuum. For the "UPW" transmission mode, we saw that this wave is what we receive, at a distance much higher than the signal wavelength, from an antenna transmitting the signal. It is simple to recognize how "UPW" is diffused, thinking of how many transmissions are made using antennas. Also for the "TEM" case it is simple to recognize how it is diffused, noting for instance how many televisions are in the world, each one connected to the antenna using a coaxial cable, and the coaxial cable is a transmission line that can support "TEM" mode, as we will show later.

In this section we will study the related equations of these transmission modes and how they can be generated.

A2.11.1 Modes Inside a Coaxial Cable

The coaxial cable is represented in Figure A2.11.1. As the name suggests, it is realized using two concentric circular conductors, the internal with radius "r_i" and the external with radius "r_e." The internal conductor does not need to be solid, since it is employed at frequencies where it cannot become a circular waveguide. Sometimes, this conductor is not really a cylinder, but instead it is made of a conductor plait. The same can happen for the outer conductor. The e.m. field is completely contained inside the medium between the two concentric conductors, which also support the internal conductor. This medium is made of low loss dielectric.

The study of the propagation modes inside the coaxial cable follows what we did in Sections A2.8 and A2.10. So, for the transverse function "T" we can use Equation A2.10.1, i.e.:

$$T(r,\theta) \stackrel{.}{=} R(r)\,\Theta(\theta) \tag{A2.10.1}$$

with:

$$\Theta(\theta) = P\cos(n\theta + \varphi) \qquad\qquad (A2.10.6)$$

$$R(Kr) = D_1 J_n(Kr) + D_2 Y_n(Kr) \qquad\qquad (A2.10.10)$$

$$-k_t^2 \overset{!}{=} K^2 \qquad\qquad (A2.10.8)$$

In this case, we cannot impose "D_2" equal to zero since the value $r = 0$ where $Y_n(Kr) = \infty$ is outside the region where the fields can be.* We will now study which modes can exist inside a coaxial cable.

a. "TM" Mode

The contour condition for "T" is:

$$T_{(m)} =! \, 0 \quad \text{on "p" for a "TM" mode.} \qquad\qquad (A2.8.18)$$

The perimeter of the structure is, in our case, composed of two parts and so the previous condition becomes:

$$D_1 J_n(Kr_i) + D_2 Y_n(Kr_i) =! \, 0 \qquad\qquad (A2.11.1)$$

$$D_1 J_n(Kr_e) + D_2 Y_n(Kr_e) =! \, 0 \qquad\qquad (A2.11.2)$$

that combined together give:

$$\frac{J_n(Kr_e)}{Y_n(Kr_e)} =! \, \frac{J_n(Kr_i)}{Y_n(Kr_i)} \qquad\qquad (A2.11.3)$$

The solution in "K" of A2.11.3 is not simple analytically, but it can be obtained simply and with good approximation using a graphic method. An example of this method is indicated in Figure A2.11.2, where we have used $n - 0$ and assumed $r_e/r_i = 3$. From these graphs we have to read the values of "k" given by the intersections of the curves $J_n(Kr_i)/Y_n(Kr_i)$ with the curves $J_n(Kr_e)/Y_n(Kr_e)$. Note that we have infinite "TM" modes, which we will indicate as "TM_{mn}," where "m" represents the index of that "K" value that verifies Equation A2.11.3 and "n" is the order of the Bessel functions.

Once we obtain "K," using Equation A2.10.8 we have the "k_t^2" for every "TM" mode, and consequently we can extract all its field components.

b. "TE" Mode

The contour condition for "T" is:

$$\partial T_{[m]}/\partial n =! \, 0 \quad \text{on "p" for a "TE" mode.} \qquad\qquad (A2.8.21)$$

---

* Unless otherwise stated we are evaluating the case of perfect conductors. So, from the contour conditions it is known that an e.m. field cannot exist inside a perfect conductor.

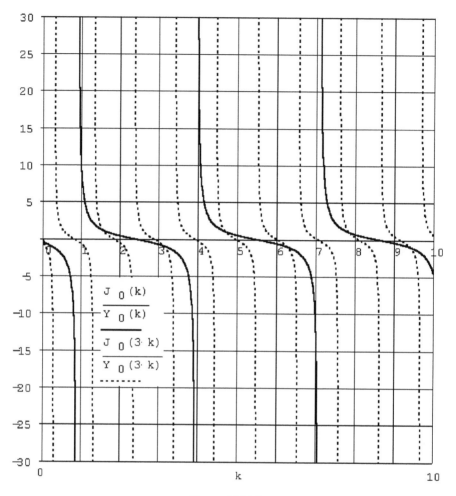

Figure A2.11.2

which in our case becomes:

$$\partial R/\partial r = ! \, 0 \quad \text{for } r = r_i \text{ and } r = r_e. \tag{A2.11.4}$$

Since:

$$\partial R/\partial r = K\left[D_1 J_n'(Kr) + D_2 Y_n'(Kr)\right] \tag{A2.11.5}$$

A2.8.21 becomes:

$$K\left[D_1 J_n'(Kr_i) + D_2 Y_n'(Kr_i)\right] = ! \, 0 \tag{A2.11.6}$$

$$K\left[D_1 J_n'(Kr_e) + D_2 Y_n'(Kr_e)\right] = ! \, 0 \tag{A2.11.7}$$

which, combined, give:

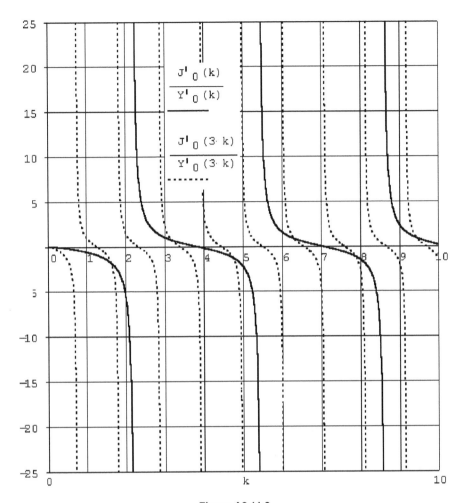

Figure A2.11.3

$$\frac{J'_n\left(Kr_e\right)}{Y'_n\left(Kr_e\right)} - !\frac{J'_n\left(Kr_i\right)}{Y'_n\left(Kr_i\right)} \qquad (\Lambda2.11.8)$$

The previous equation can be solved with good approximation using a graphic method, as we did for "TM" modes. The resulting graph of A2.11.8 is indicated in Figure A2.11.3 where we have used n = 0 and supposed r_e/r_i = 3. Note that now we also have infinite "TE" modes, which we will indicate as "TE$_{mn}$."

Also in this case, once we obtain "K," using Equation A2.10.8 we have the "k_t^2" for every "TE" mode, and consequently we can extract all its field components.

c. "TEM" Mode

In general, a "TEM" mode can be thought to be generated by a:

1. "TE" mode forcing $\underline{h}_z$ = !0
2. "TM" mode forcing $\underline{e}_z$ = !0

Remembering the field Equation A2.8.11 for "e_z" in a "TM" mode and A2.8.15 for "h_z" in a "TE" mode, the previous conditions 1 and 2 bring $k_t = !\,0$ for a "TEM" mode. This condition gives us an important result. Since in this case $k_t = 0$, then for a lossless cable:

$$k_t^2 + k_z^2 \equiv k_z^2 = !\,k^2 = -\omega^2\mu\varepsilon$$

which means that for a "TEM" mode the cut-off frequency is zero, since "k_z" is always imaginary.

The condition $k_t = 0$ does not assure the real existence of a "TEM" mode, and to this purpose it is necessary to find a "T" function that satisfies the contour conditions for such a mode. Using the previous point 2 as the beginning condition to search a "TEM" mode,* we know that for the electric "e" field of a "TM" mode in general, we have:

$$\underline{e} = \underline{e}_t + \underline{e}_z \tag{A2.11.9}$$

Now, since:

3. $e_z = !\,0$ to have a "TEM" mode
4. the tangential component "e_τ" of the electric field must be zero on the contour

then from A2.11.9 we have:

$$e_\tau = \underline{e}_t \cdot \underline{p} = !\,0 \tag{A2.11.10}$$

The field "e_t" can be obtained from Equation A2.8.12 applied to our coordinate system, and so:

$$\underline{e}_t = \frac{1}{j\omega\varepsilon_c} \frac{\partial L}{\partial z} \underline{\nabla}_t T \tag{A2.11.11}$$

Inserting this Equation into A2.11.10 we have the condition that function "T" must have a "TEM" mode inside the structure in Figure A2.11.1, i.e.:

$$\underline{\nabla}_t T \cdot \underline{p} \equiv \partial T/\partial p = !\,0 \quad \text{on "p" for a "TEM" mode.} \tag{A2.11.12}$$

This equation means that "T" must be constant on our contour, i.e., $T = T_i$ and $T = T_e$ on internal and external "p."** Noting that $\partial p = r\partial\theta$ and using Equation A2.10.1, the application of A2.11.12 to our case results in:

$$R(r_i)\,\Theta'(\theta) = !\,0 \quad \text{and} \quad R(r_e)\,\Theta'(\theta) = !\,0 \tag{A2.11.13}$$

The condition:

$$R(r_i) = R(r_e) = 0$$

cannot be accepted,*** then necessarily:

* It is possible to prove that a "TEM" mode cannot be generated by a "TE" mode in a structure limited by electric walls.
** Condition "$\partial T/\partial p = !\,0$ on 'p'" theoretically does not mean that "T" must assume two different values "T_i" and "T_e" on $r = r_i$ and $r = r_e$, but it is possible to prove that if $T_i = T_e$ then the "TEM" field must be zero.
*** We said that it is possible to prove that if $T_i = T_e$ then the "TEM" field must be zero.

$$\Theta'(\theta) = !\, 0 \rightarrow \Theta(\theta) = Q \tag{A2.11.14}$$

where "Q" is a constant. Then from A2.10.6 it follows that n = ! 0. Equation. A2.11.14 means that "T" does not vary with "θ." Since "$\Theta(\theta)$" is known,* it remains to evaluate "R(r)." From Equation A2.10.7 evaluated with $k_t^2 = 0$ and n = 0 we have:

$$r^2 \frac{d^2R}{dr^2} + r \frac{dR}{dr} = 0 \rightarrow \frac{d}{dr}\left(r \frac{dR}{dr}\right) = 0 \rightarrow rR' = C \tag{A2.11.15}$$

where "C" is a generic constant. From the expression of "$\underline{\nabla}_t T$" in cylindric coordinate given in Appendix A8 and noting that from A2.11.12 it follows** that $\partial T/\partial \theta = 0$; we have $\underline{\nabla}_t T \equiv (\partial T/\partial r)\underline{r}_0$. Since as a result of this discussion we have T = ΘR(r); then using A2.11.15 we have:

$$\frac{dR}{dr} \equiv \frac{1}{Q}\frac{dT}{dr} = \frac{C}{r} \rightarrow \underline{\nabla}_t T \equiv (D/r)\underline{r}_0 \tag{A2.11.16}$$

where D = CQ is a constant.

At this point, since we have all the functions to evaluate the "TEM" mode from "TM" Equations A2.8.9 through A2.8.12 inside the coaxial cable, we have that the field components are:

$$\underline{e}_z = 0 \tag{A2.11.17}$$

$$\underline{h}_z = 0 \tag{A2.11.18}$$

$$\underline{h}_t(r,z) = -D\frac{L(z)}{r}\underline{\theta}_0 \tag{A2.11.19}$$

$$\underline{e}_t(r,z) = \frac{D}{rj\omega\varepsilon_c}\frac{dL(z)}{dz}\underline{r}_0 \tag{A2.11.20}$$

If we assume a case of progressive propagation, we have L(z) = $e^{-jk_z z}$. In this case it is simple to define the "TEM" impedance "ζ_{TEM}" given by:

$$\zeta_{TEM} \overset{!}{-} k_z/j\omega\varepsilon_c = (\mu/\varepsilon_c)^{0.5} \tag{A2.11.21}$$

where the last quantity comes from the fact that since $k_t = 0$ then:

$$k_z \equiv j\omega(\mu\varepsilon_c)^{0.5} \tag{A2.11.22}$$

It is interesting to conclude this study on coaxial cable saying that the first non- "TEM" mode to start at high frequency, as we are going to specify, is the "TE_{11}." An approximate relationship that gives the limit wavelength "λ_c" below which this mode starts to propagate is:

$$\lambda_c \approx \pi(r_i + r_e) \tag{A2.11.23}$$

---

* We said that "$\Theta(\theta)$" must be a constant, whose value will be determined by the contour conditions.
** From A2.11.14 we recognize that "T" is constant with "θ."

whose error is inside some percent up to $r_e \leq 5r_i$. To do an example, let us assume a coaxial cable with $r_i = 0.5$mm, $r_e = 1.8$mm filled with a lossless dielectric with $\varepsilon_r = 2$. From the previous equation we have $\lambda_c \approx 7.2$mm, which corresponds to a frequency "f_c" of $f_c \approx 29.4$ GHz. We see how higher order modes usually start at frequencies where coaxial cables are seldom used. However, if a coaxial cable should be used, its dimension could be adjusted to avoid the growth of non- "TEM" modes.

A2.11.2 Uniform Plane Wave

A pure uniform plane wave can be thought to propagate inside the space between two infinite parallel conductors. This wave is also what we have at great distance from a dipole antenna, typically many times the antenna length.

A simple analytical way to generate a uniform plane wave is to solve the Maxwell's equations for the field produced in the space by a time-varying current "K" on a conductor plane, as indicated in Figure A2.11.4. To show that, let us start applying the boundary condition for the magnetic field on the conductor surface. Indicating with "$\underline{h}_t^+$" and "$\underline{h}_t^-$" the tangential components on the plane of the magnetic field, respectively, for $z = 0^+$ and $z = 0^-$ we have:

$$\underline{z}_0 \otimes \left(\underline{h}_t^+ - \underline{h}_t^- \right) = ! -K\underline{x}_0 \qquad (A2.11.24)$$

where the negative sign of the second member comes out from the direction of "$\underline{K}$." Note that $[K] = $ Amp/m. Since the plane with current is symmetrical with respect to $z = 0$, we have $h_t^+ \equiv -h_t^-$ and the previous equation becomes:

$$2\underline{z}_0 \otimes \underline{h}_t^+ = ! -K\underline{x}_0 \qquad (A2.11.25)$$

from which we see that:

$$\underline{h}_t^+ = (K/2)\underline{y}_0 \qquad (A2.11.26)$$

This is the result of the boundary condition for the structure indicated in Figure A2.11.4, and represents the source of the field. Now we want to solve the Maxwell's equation for the semispace extending from $z = 0$ to $z = \infty$, forcing the fields to be null at this infinite distance. The equations to be solved are the A2.4.19, i.e.:

$$\underline{\nabla} \otimes \underline{e} = -j\omega\mu\underline{h} \quad \text{and} \quad \underline{\nabla} \otimes \underline{h} = j\omega\varepsilon_c\underline{e} \qquad (A2.4.19)$$

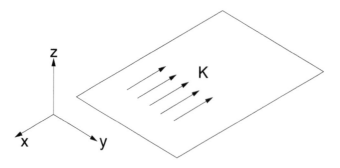

Figure A2.11.4

since we are supposing that "K" moves with time in a sinusoidal shape. Permeability "μ" and complex permittivity "ε_c" are relative to the medium that surrounds the plane. So, for symmetry we see that:

$$\frac{\partial}{\partial x} = \frac{\partial}{\partial y} = 0$$

and the first and second equations in A2.4.19 become:*

$$\frac{\partial e_y}{\partial z} = j\omega\mu h_x, \quad \frac{\partial e_x}{\partial z} = -j\omega\mu h_y, \quad h_z = 0 \qquad (A2.11.27)$$

$$\frac{\partial h_y}{\partial z} = -j\omega\varepsilon_c e_x, \quad \frac{\partial h_x}{\partial z} = j\omega\varepsilon_c e_y, \quad e_z = 0 \qquad (A2.11.28)$$

Note that among Equations A2.11.27 and A2.11.28 it is possible to find two pairs of equations, one formed with "e_x" and "h_y" and the other with "e_y" and "h_x." In any case, these two pairs represent the same wave, the only difference being the different polarization. For this reason, we can use only one pair and say that a uniform plane wave is represented by:

$$\frac{\partial e_x}{\partial z} = -j\omega\mu h_y \quad \text{and} \quad \frac{\partial h_y}{\partial z} = -j\omega\varepsilon_c e_x \qquad (A2.11.29)$$

Another very important thing to observe is that since the plane wave at distance from a dipole antenna is locally a "TEM," similar to what we have obtained with Equations A2.11.27 and A2.11.28 where $h_z = e_z = 0$, we can say that a "UPW" is also a "TEM" wave.

From Equation A2.11.27 we can note an analytical analogy with the generic transmission line equations we studied in Chapter 1. Note that if we substitute to "e_x," "h_y," "$j\omega\mu$," and "$j\omega\varepsilon_c$" respectively, "v," "i," "Z_s," and "Y_p" we have exactly the transmission line Equations 1.2.16 and 1.2.18, with the only notational difference that in the case of a transmission line the direction of propagation has been indicated with "x" and here with "z."** From this analogy, to obtain the solutions of these equations we can proceed as we did in Chapter 1. Proceeding speedily, from Equation A2.11.27 we can obtain:

$$\frac{d^2 e_x}{dz^2} = -\omega^2 \mu\varepsilon_c e_x \qquad (A2.11.30)$$

whose general solution is:

$$e_x(z) = e^+ e^{(-k_z z)} + e^- e^{(k_z z)} \qquad (A2.11.31)$$

with:

$$k_z \stackrel{\perp}{=} j\omega(\mu\varepsilon_c)^{0.5} \qquad (A2.11.32)$$

---

* See Appendix A8 for "$\underline{\nabla} \otimes ()$" in Cartesian coordinate.
** Usually, in wave theory the direction of propagation is indicated with "z." By analogy, the same "z" could be used in Chapter 1 when speaking about transmission lines. Anyway, there we have preferred to use "x" instead of "z," to not create confusion with "Z_s" which represents the line series impedance.

Inserting Equation A2.11.31 into the second of A2.11.29 we have:

$$h_y(z) = -\frac{1}{j\omega\mu}\left[-k_z e^+ e^{(-k_z z)} + k_z e^- e^{(k_z z)}\right] \qquad (A2.11.33)$$

or:

$$h_y(z) = h^+ e^{(-k_z z)} + h^- e^{(k_z z)} \qquad (A2.11.34)$$

with:

$$h^+ \overset{\perp}{=} e^+\left(\varepsilon_c/\mu\right)^{0.5} \quad \text{and} \quad h^- \overset{\perp}{=} -e^-\left(\varepsilon_c/\mu\right)^{0.5} \qquad (A2.11.35)$$

To the quantity "ζ_{UPW}" given by:

$$\zeta_{UPW} \overset{\perp}{=} \left(\mu/\varepsilon_c\right)^{0.5} \qquad (A2.11.36)$$

is given the name of "uniform plane wave impedance," which coincides with the ratio "e^+/h^+" as we can recognize from the first of A2.11.35. Note as A2.11.36 is the same expression we have for a "TEM" mode, indicated in A2.11.21.

A2.12 DISPERSION

To describe the dispersion phenomena it is necessary to define the phase and group velocity of a wave. To this purpose, let us evaluate a wave with a phase dependence, as a function of time and coordinate, of the type:

$$sen(\omega t - \beta z) \qquad (A2.12.1)$$

"z" being the direction of propagation, and the other variables now well known.
They are defined:

1. "phase velocity" — the quantity "v_p" given by:

$$v_p \overset{\perp}{=} \omega/\beta \qquad (A2.12.2)$$

2. "group velocity" — the quantity "v_g" given by:

$$v_q \overset{\perp}{=} d\omega/d\beta \qquad (A2.12.3)$$

A lot of transmission lines, and electrical networks,* have a linear relationship between phase constant and frequency, i.e., "ω." In this case $v_p \equiv v_g$. But also a lot of transmission lines and networks exist where the relationship between phase constant and frequency is not linear. In this

---

* Dispersion phenomena is not only a transmission line characteristic, but it also appears in a lot of electrical networks, for example filters.

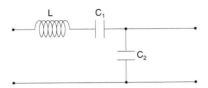

Figure A2.12.1

last case the device is said to be "dispersive," or to cause dispersion. The reason for this name lies in one of the greatest effects of dispersion. To explain it, let us consider the transmission of an impulse in a dispersive transmission line. As we know from the signal analysis theory, such an impulse can be represented by a Fourier series, where each term is a sinusoidal factor with its amplitude and frequency. Each one of these factors can be evaluated by its "v_p" and, since the line is dispersive, the term will have different "v_p." These components will also travel with different speeds inside the line, and at the receiving side the resulting impulse will be distorted. This is the most undesirable effect of the dispersive network, that is the distortion of the impulses that travel inside them. This effect is more evident as the bandwidth of the transformed impulse is higher, i.e., as the rise and fall time of the impulses are shorter. Dispersion is said to be "normal" if the variation of "v_p" with frequency is of opposite sign with respect to the variation of frequency, i.e., if $dv_p/d\omega < 0$, otherwise the dispersion is said to be "anomalous."

An interesting diagram to represent the dispersion is the "$\omega\beta$" diagram. The analytical procedure is generally simple for filters, while for transmission lines it becomes simple after we have its equivalent filter circuit.* To do an example, let us evaluate the "$\omega\beta$" diagram for the transmission line indicated in Figure A2.12.1. Remembering the theory developed in Chapter 1, we have:

$$Z_s = j\omega L + 1/j\omega C_1 \quad \text{and} \quad Y_p = j\omega C_2 \tag{A2.12.4}$$

Setting $r_c \doteq C_2/C_1$ we can write:

$$k = \alpha + j\beta \doteq \left(Z_s Y_p\right)^{0.5} = \left(r_c - \omega^2 L C_2\right)^{0.5} \tag{A2.12.5}$$

and we can have the following three cases:

a. $r_c = \omega^2 L C_2$
   In this case $k \equiv 0$ and $\alpha = \beta = 0$. This situation is obtained for an angular frequency "ω_c" given by:

$$\omega_c = \left(r_c/LC_2\right)^{0.5} \equiv \left(LC_1\right)^{-0.5} \tag{A2.12.6}$$

b. $r_c < \omega^2 LC_2$ or, using A2.12.6, $\omega > \omega_c$.
   In this case $k \equiv j\beta$, with:

$$\beta = \left(\omega^2 LC_2 - r_c\right)^{0.5} = \omega\left[LC_2\left(1 - \omega_c^2/\omega^2\right)\right]^{0.5} \tag{A2.12.7}$$

c. $r_c > \omega^2 LC_2$ or, using A2.12.6, $\omega < \omega_c$.
   In this case $k \equiv \alpha$, with:

---

* We will study in the next section the networks representative of transmission lines.

$$\alpha = \left(r_c - \omega^2 LC_2\right)^{0.5} = \omega\left[LC_2\left(\omega_c^2/\omega^2 - 1\right)\right]^{0.5} \tag{A2.12.8}$$

Note that the line indicated in Figure A2.12.1 also introduces attenuation for $\omega < \omega_c$ if it is lossless.

Connecting in one graph the solutions A2.12.6 through A2.12.8 we have the result indicated in Figure A2.12.2.

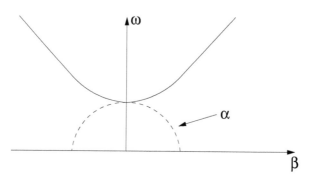

Figure A2.12.2

To make another example, it is simple to recognize that for a simple "LC" low pass lossless filter, i.e., represented by the network in Figure A2.12.1 without the capacitor "C_1," the "$\omega\beta$" diagram is as represented in Figure A2.12.3.

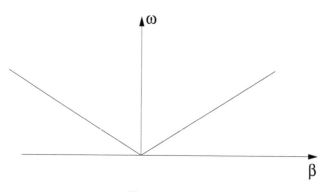

Figure A2.12.3

Using the "$\omega\beta$" diagram we can give a graphical representation of "v_p" and "v_g." From the definitions of these quantities given in A2.12.2 and A2.12.3, it is simple to recognize that:

    I. "v_p" is given by the tangent of the angle formed between the "β" axis and the line that starts from the axis origin and the desired point "Q" on the "$\omega\beta$" diagram. In Figure A2.12.4 we have indicated this angle with "θ_p"

    II. "v_g" is given by the tangent at the desired point "Q" on the "$\omega\beta$" diagram. The angle formed between the "β" axis and the tangent in "Q" has been indicated with "θ_g."

Note that for the point "I" in Figure A2.12.4 phase velocity is infinite and $\theta_p = 90°$, while group velocity is zero and $\theta_g = 0$. Waves that posses $v_p > (\mu_0\varepsilon_0)^{0.5}$ are said to be "fast," otherwise are said

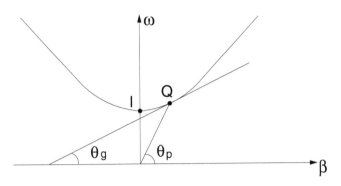

Figure A2.12.4

to be "slow." Note that an infinite value for "v_p" does not have a physical meaning since the phase is not a physical quantity, and consequently this result does not contradict the relativistic result that the maximum speed for a physical quantity is equal to the speed of light. Dispersive lines are characterized by $v_p \neq v_g$, at least for some values of "ω." Dispersive lines can have frequency ranges where "v_p" and "v_g" are equal or very similar. For example, the network indicated in Figure A2.12.2 for $\omega \gg \omega_c$ has $v_p \approx v_g$, i.e., the "$\omega\beta$" diagram becomes a line.

A2.13 ELECTRICAL NETWORKS ASSOCIATED
WITH PROPAGATION MODES

It is interesting to explain how the transmission modes studied previously can be associated with lumped networks, through some manipulation of their characteristic analytical formulas. The common procedure is to transform in some way the field equations in the general transmission line equations we studied in Chapter 1, i.e.:

$$\frac{dv(x)}{dx} = - z_s i(x) \tag{1.2.16}$$

$$\frac{di(x)}{dx} = - y_p v(x) \tag{1.2.18}$$

The starting points are the propagation modes inside structures with curvilinear orthogonal coordinate reference systems, i.e., the "TM" and "TE" modes of Section A2.8 and the "TEM" mode of Section A2.11. In particular, only the transverse component of each mode can be used to obtain the equivalent transmission line equations. The longitudinal axis will be indicated with "z." We will also see how the "UPW" can be associated with its lumped network.

1. Associated Network for a "TM" Mode

To extract the equivalent network for such a mode it is necessary to rewrite the Maxwell's equations in a way that only the transverse field components are involved. To do that, insert into A2.4.19 the expression of "delta" operator specifying its transverse and longitudinal components. We have:

$$\underline{\nabla} \otimes \underline{e} = \left(\underline{\nabla}_t + \frac{\partial}{\partial z} \underline{z}_0 \right) \otimes \underline{e} = \underline{\nabla}_t \otimes \underline{e} + \underline{z}_0 \otimes \frac{\partial \underline{e}}{\partial z} = -j\omega\mu\underline{h} \qquad (A2.13.1)$$

$$\underline{\nabla} \otimes \underline{h} = \left(\underline{\nabla}_t + \frac{\partial}{\partial z} \underline{z}_0 \right) \otimes \underline{h} = \underline{\nabla}_t \otimes \underline{h} + \underline{z}_0 \otimes \frac{\partial \underline{h}}{\partial z} = j\omega\varepsilon_c\underline{e} \qquad (A2.13.2)$$

Now, let us do for A2.13.1 a left vector product with "$\underline{z}_0$" and for A2.13.2 a scalar product with "$\underline{z}_0$." We have:

$$\underline{z}_0 \otimes \left(\underline{\nabla}_t \otimes \underline{e} \right) + \underline{z}_0 \otimes \left(\underline{z}_0 \otimes \frac{\partial \underline{e}}{\partial z} \right) = j\omega\mu \left(\underline{h} \otimes \underline{z}_0 \right) \qquad (A2.13.3)$$

$$\underline{z}_0 \cdot \left(\underline{\nabla}_t \otimes \underline{h} \right) + \underline{z}_0 \cdot \left(\underline{z}_0 \otimes \frac{\partial \underline{h}}{\partial z} \right) = j\omega\varepsilon_c\underline{e} \cdot \underline{z}_0 \qquad (A2.13.4)$$

Executing the calculations,* the previous two equations become:

$$\underline{\nabla}_t e_z - \frac{\partial \underline{e}_t}{\partial z} = j\omega\mu \left(\underline{h}_t \otimes \underline{z}_0 \right) \qquad (A2.13.5)$$

$$\underline{\nabla}_t \cdot \underline{h} \otimes \underline{z}_0 = j\omega\varepsilon_c e_z \qquad (A2.13.6)$$

Extracting "e_z" from A2.13.6 and inserting into A2.13.5 we have:

$$\frac{\partial \underline{e}_t}{\partial z} = -j\omega\mu \left(\underline{h}_t \otimes \underline{z}_0 \right) + \frac{1}{j\omega\varepsilon_c} \underline{\nabla}_t \underline{\nabla}_t \cdot \left(\underline{h}_t \otimes \underline{z}_0 \right) \qquad (A2.13.7)$$

from which, for duality, we can have the equivalent equation for the magnetic field, i.e.:

$$\frac{\partial \underline{h}_t}{\partial z} = j\omega\varepsilon_c \left(\underline{e}_t \otimes \underline{z}_0 \right) - \frac{1}{j\omega\mu} \underline{\nabla}_t \underline{\nabla}_t \cdot \left(\underline{e}_t \otimes \underline{z}_0 \right) \qquad (A2.13.8)$$

Equations A2.13.7 and A2.13.8 are the desired expressions of Maxwell's equations where only the transverse fields appear.

Now, for Equations A2.8.10 and A2.8.12 of "TM" modes let us set:

$$L(z) \stackrel{\perp}{=} i(z) \quad \text{and} \quad \frac{1}{j\omega\varepsilon_c} \frac{dL}{dz} \stackrel{\perp}{=} -v(z) ** \qquad (A2.13.9)$$

With these positions, Equations A2.8.10 and A2.8.12 can be rewritten as:***

$$\underline{h}_t = i\underline{\nabla}_t T \otimes \underline{z}_0 \qquad (A2.13.10)$$

$$\underline{e}_t = -v\underline{\nabla}_t T \qquad (A2.13.11)$$

* See Appendix A8 for the operation $\underline{a} \otimes \underline{b} \otimes \underline{c}$.
** The partial derivative has been substituted with absolute derivative, since for our geometry "L" is only dependent on "z."
*** For notation simplicity we will omit for "i" and "v" their dependence with "z."

For Equations A2.8.14 and A2.8.16 of "TE" modes let us set:

$$L(z) \overset{\perp}{=} v(z) \quad \text{and} \quad \frac{1}{j\omega\mu} \frac{dL}{dz} \overset{\perp}{=} -i(z) \tag{A2.13.12}$$

With these positions, Equations A2.8.14 and A2.8.16 for "TE" modes can be rewritten as:

$$\underline{e}_t = -v\underline{\nabla}_t T \otimes \underline{z}_0 \tag{A2.13.13}$$

$$\underline{h}_t = -i\underline{\nabla}_t T \tag{A2.13.14}$$

Now, let us suppose that inside our guiding structure the electric and magnetic transverse components are given as the sum of the whole "TE" and "TM" fields. With the notation used in Section A2.8, we have:

$$\underline{h}_t = \sum_{(n)} i_{(n)} \underline{\nabla}_t T_{(n)} \otimes \underline{z}_0 - \sum_{[n]} i_{[n]} \underline{\nabla}_t T_{[n]} \tag{A2.13.15}$$

$$\underline{e}_t = -\sum_{(n)} v_{(n)} \underline{\nabla}_t T_{(n)} - \sum_{[n]} v_{[n]} \underline{\nabla}_t T_{[n]} \otimes \underline{z}_(\tag{A2.13.16}$$

Inserting the two previous equations into A2.13.7, we have:*

$$-\sum_{(n)} \frac{dv_{(n)}}{dz} \underline{\nabla}_t T_{(n)} - \sum_{[n]} \frac{dv_{(n)}}{dz} \underline{\nabla}_t T_{[n]} \otimes \underline{z}_0 = \cdots$$

$$\cdots = -j\omega\mu \left(-\sum_{(n)} i_{(n)} \underline{\nabla}_t T_{(n)} - \sum_{[n]} i_{[n]} \underline{\nabla}_t T_{[n]} \otimes \underline{z}_0 \right) - \frac{\sum_{(n)} k_{t(n)}^2 i_{(n)} \underline{\nabla}_t T_{(n)}}{j\omega\varepsilon_c} \tag{A2.13.17}$$

where we have used the following simplifications:

a. $(\underline{\nabla}_t T \otimes \underline{z}_0) \otimes \underline{z}_0 \equiv -\underline{\nabla}_t T$
b. $\underline{\nabla}_t \underline{\nabla}_t \bullet \sum_{(n)} i_{(n)} \underline{\nabla}_t T_{(n)} = \sum \sum_{(u)} i_{(u)} \underline{\nabla}_t \bullet \underline{\nabla}_t T_{(n)} = \underline{\nabla}_t \sum_{(n)} i_{(n)} \underline{\nabla}_t{}^2 T_{(n)} = \sum_{(n)} i_{(n)} k_{t(n)}^2 \underline{\nabla}_t T_{(n)}$
c. $\underline{\nabla}_t \underline{\nabla}_t \bullet \sum_{[u]} i_{[u]} (\underline{\nabla}_t T_{[u]} \otimes \underline{z}_0) = \sum \sum_{[u]} i_{[u]} \underline{\nabla}_t \bullet (\underline{\nabla}_t T_{[u]} \otimes \underline{z}_0) = 0$

Now, for A2.13.17 let us do a scalar product with "$\underline{\nabla}_t T_{(m)}$"** and integrate on a transverse section "S" of our guiding structure. We have:

$$-\sum_{(n)} \frac{dv_{(n)}}{dz} \int_s \underline{\nabla}_t T_{(m)} \bullet \underline{\nabla}_t T_{(n)} dS - \sum_{[n]} \frac{dv_{[n]}}{dz} \int_s \underline{\nabla}_t T_{(m)} \bullet \underline{\nabla}_t T_{[n]} \otimes \underline{z}_0 dS$$

$$= j\omega\mu \left(\sum_{(n)} i_{(n)} \int_s \underline{\nabla}_t T_{(m)} \bullet \underline{\nabla}_t T_{(n)} dS - \sum_{[n]} i_{[n]} \int_s \underline{\nabla}_t T_{(m)} \bullet \underline{\nabla}_t T_{[n]} \otimes \underline{z}_0 dS \right) \tag{A2.13.18}$$

$$+ - \frac{\sum_{(n)} k_{t(n)}^2 i_{(n)} \int_s \underline{\nabla}_t T_{(m)} \bullet \underline{\nabla}_t T_{(n)} dS}{j\omega\varepsilon_c}$$

* The partial derivatives can now be substituted with absolute derivatives, since "v" and "i" are only dependent to "z."
** The subscript "(m)" represents a generic "TM" mode.

Using the relationships given in Section A2.8 it is simple to recognize that the only nonzero elements are:

a. The first term at first member, if $m \equiv n$
b. The first and third term at second member, if $m \equiv n$.

To simplify the notation, we can set:

$$\int_s \underline{\nabla}_t T \cdot \underline{\nabla}_t T dS \stackrel{+}{=} 1 \qquad (A2.13.19)$$

which does not limit the generality of the solution, since the previous equation set to "1" the constant that appears in "T," but we can insert the constants generality to the constant that appears in "L(z)." So, using A2.13.19, the orthogonality properties given in Section A2.8, A2.13.18 becomes:

$$\frac{dv_{(m)}}{dz} = -\left[j\omega\mu - \frac{k^2_{t(m)}}{j\omega\varepsilon_c} \right] i_{(m)} \qquad (A2.13.20)$$

To obtain an equivalent relationship for the current "$i_{(m)}$" we have to insert Equations A2.13.15 and A2.13.16 into A2.13.8. Then we do a scalar product with "$\underline{\nabla}_t T_{(m)} \otimes \underline{z}_0$" and integrate on a transverse section "S" of our guiding structure. We have:

$$\frac{di_{(m)}}{dz} = -j\omega\varepsilon_c v_{(m)} \qquad (A2.13.21)$$

From Equations A2.13.20 and A2.13.21 we can recognize that the associated network for a "TM" mode inside our guiding structure with curvilinear orthogonal coordinates reference system is as indicated in Figure A2.13.1. The element's values are:

$$L = \mu, \quad C_1 = \varepsilon/-k_t^2, \quad R_1 = -k_t^2/g, \quad C_2 = \varepsilon, \quad R_2 = 1/g \qquad (A2.13.22)$$

It is interesting to observe that if we insert into the definition of propagation constant given in Chapter 1, i.e..

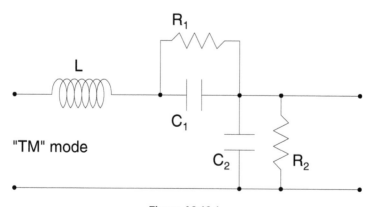

Figure A2.13.1

$$k = \left(Z_s Y_p \right)^{0.5} \tag{1.3.4}$$

the expression of "Z_s" and "Y_p" given by A2.13.20 and A2.13.21, i.e.:

$$Z_s = j\omega\mu - \frac{k_{t(m)}^2}{j\omega\varepsilon_c} \quad \text{and} \quad Y_p = j\omega\varepsilon_c \tag{A2.13.23}$$

we have:

$$k^2 = -\omega^2\mu\varepsilon_c - k_{t(m)}^2 \tag{A2.13.24}$$

So, the value of "k" for the network indicated in Figure A2.13.1 coincides with the value of "k_z" for the "TM" mode inside the guiding structure, obtained by A2.8.6 when evaluated for $\ell_0 \equiv z_0$. Similarly, inserting A2.13.23 into the definition of transmission line impedance we gave in Chapter 1, we have:

$$\zeta \stackrel{\perp}{=} \left(Z_s / Y_p \right)^{0.5} = k_z / j\omega\varepsilon_c \tag{A2.13.25}$$

which is equal to the "TM" mode impedance "ζ_{TM}" we gave in Section A2.8.

2. Associated Network for a "TE" Mode

For the "TE" mode in the previous point 1 we have set:

$$L(z) \stackrel{\perp}{=} v(z) \quad \text{and} \quad \frac{1}{j\omega\mu}\frac{dL}{dz} \stackrel{\perp}{=} -i(z) \tag{A2.13.12}$$

For A2.13.17 let us do a scalar product with "$\nabla_t T_{|m|} \otimes z_0$"* and integrate on a transverse section "S" of our guiding structure. Then, using A2.13.19, the orthogonality properties given in Section A2.8, m = n, we have:**

$$\frac{dv_{[m]}}{dz} = -j\omega\mu i_{[m]} \tag{A2.13.26}$$

To obtain an equivalent relationship for the current "$i_{[m]}$" we have to insert Equations A2.13.15 and A2.13.16 into A2.13.8. Then we do a scalar product with "$\nabla_t T_{[m]}$" and integrate on a transverse section "S" of our guiding structure. We have:

$$\frac{di_{[m]}}{dz} = -\left[j\omega\varepsilon_c - \frac{k_{t[m]}^2}{j\omega\mu} \right] v_{[m]} \tag{A2.13.27}$$

* The subscript "[m]" represents a generic "TE" mode.
** The procedure is similar to what we did in previous point 1 for "TM" mode.

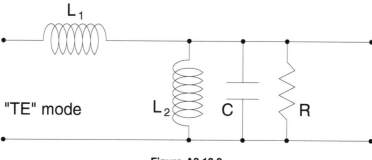

Figure A2.13.2

From Equations A2.13.26 and A2.13.27 we can recognize that the associated network for a "TE" mode inside our guiding structure with curvilinear orthogonal coordinates reference system is as indicated in Figure A2.13.2. The element's values are:

$$L_1 = \mu, \quad L_2 = \mu/-k_t^2, \quad C = \varepsilon, \quad R = 1/g \tag{A2.13.28}$$

Similar to the case of "TM" mode, it is possible to verify that also in this case using the values of "Z_s" and "Y_p" obtained from A2.13.26 and A2.13.27, the values of "k" and "ζ" for the network indicated in Figure A2.13.2 are coincident with "k_z" and "ζ_{TE}," when evaluated for $\ell_0 \equiv z_0$, we gave in Section A2.8.

3. Associated Network for a "TEM" Mode

For a "TEM" mode we set:

$$L(z) \overset{\perp}{=} i(z) \quad \text{and} \quad \frac{1}{j\omega\varepsilon_c}\frac{dL}{dz} \overset{\perp}{=} -v(z) \tag{A2.13.29}$$

In this case "i(z)" and "v(z)" satisfy the transmission line equations, and so it is very simple to associate an equivalent network. Inserting the first equation into the second equation of A2.13.26 we have:

$$\frac{di}{dz} = -j\omega\varepsilon_c v \tag{A2.13.30}$$

Deriving in "z" assuming this equation and supposing to have only the progressive wave we have:

$$\frac{d^2i}{dz^2} = k_z^2 i = -\omega^2\mu\varepsilon_c i = -j\omega\varepsilon_c \frac{dv}{dz} \tag{A2.13.31}$$

from which:

$$\frac{dv}{dz} = -j\omega\mu i \tag{A2.13.32}$$

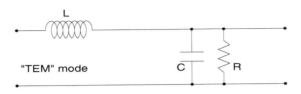

Figure A2.13.3

From Equations A2.13.30 and A2.13.32 we can recognize that the associated network for a "TEM" mode inside our guiding structure with curvilinear orthogonal coordinates reference system is as indicated in Figure A2.13.3. The element's values are:

$$L = \mu, \quad C = \varepsilon, \quad R = 1/g \tag{A2.13.33}$$

For the "TEM" case it is simple to verify that using the values of "Y_p" and "Z_s" obtained from A2.13.30 and A2.13.32, the values of "k" and "ζ" for the network indicated in Figure A2.13.3 are:

$$k = j\omega \left(\mu\varepsilon_c\right)^{0.5} \tag{A2.13.34}$$

$$\zeta = \left(\mu/\varepsilon_c\right)^{0.5} \tag{A2.13.35}$$

coincident with the "k_z" and "ζ_{TEM}" values we gave in A2.11.22 and A2.11.21 for the "TEM" mode inside a coaxial cable.

4. Associated Network for a "UPW"

For a "UPW" it is easy to extract the associated network. From Equations A2.11.29 setting:

$$e_x(z) = v(z), \quad Z_s = j\omega\mu, \quad Y_p = j\omega\varepsilon_c, \quad h_y(z) = i(z) \tag{A2.13.36}$$

we have:

$$\frac{dv}{dz} = -Z_s i \quad \text{and} \quad \frac{di}{dz} = -Y_p v \tag{A2.13.37}$$

From Equation A2.13.37 we can recognize that the associated network for a "UPW" inside a medium with permeability "μ" and permittivity "ε_c" with a Cartesian coordinate system is as indicated in Figure A2.13.4. The element's values are:

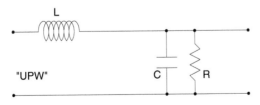

Figure A2.13.4

$$L = \mu, \quad C = \varepsilon, \quad R = 1/g \tag{A2.13.38}$$

Confronting Figure A2.13.3 and A2.13.4 we recognize how a "TEM" mode and an "UPW" have the same type of associated network. Also in this case it is simple to verify that using the values of "Y_p" and "Z_s" obtained from A2.13.37, the values of "k" and "ζ" for the network indicated in Figure A2.13.4 are again given by A2.13.34 and A2.13.35, i.e., coincident with the "k_z" and "ζ_{UPW}" values we gave in A2.11.32 and A2.11.36 for an "UPW."

A2.14 FIELD PENETRATION INSIDE NONIDEAL CONDUCTORS

In all the previous paragraphs we have assumed that all the conductors guiding the e.m. fields are perfect, i.e., they have infinite conductivity, i.e., $g = \infty$. In this section we will study how the e.m. fields penetrate inside a real conductor that is characterized by a finite value of "g." To do that, let us write the Maxwell homogeneous equations:

$$\underline{\nabla} \otimes \underline{e} = -j\omega\mu\underline{h} \quad \text{and} \quad \underline{\nabla} \otimes \underline{h} = -j\omega\varepsilon\underline{e} + g\underline{e} \tag{A2.14.1}$$

Doing the "curl" to both members and applying the identity:

$$\underline{\nabla} \otimes \underline{\nabla} \otimes \underline{e} = \underline{\nabla}\underline{\nabla} \cdot \underline{e} - \nabla^2\underline{e} \tag{A2.14.2}$$

A2.14.1 becomes:

$$\underline{\nabla}\underline{\nabla} \cdot \underline{e} - \nabla^2\underline{e} = -j\omega\mu\underline{\nabla} \otimes \underline{h} \tag{A2.14.3}$$

Since also inside a nonideal conductor there is no volumetric free charge* then $\underline{\nabla} \cdot \underline{e} = 0$, In addition, in a good conductor $g \gg \omega\varepsilon$, and so we can neglect the first term in the second equation of A2.14.1. With these introductions, the previous equation becomes:

$$\nabla^2\underline{e} = j\omega\mu\,\underline{\nabla} \otimes \underline{h} \tag{A2.14.4}$$

Let us resolve A2.14.4 assuming the electric field is dependent only on a coordinate, and to be in the practical situation reported in Figure A2.14.1. The Cartesian coordinate system is as indicated, and the electric field is only dependent on "x." In this case, Equation A2.14.4 becomes:

$$\frac{d^2e_z}{dx^2} = j\omega\mu g e_z \underline{\pm} k^2 e_z, \text{with } k^2 = j\omega\mu g \tag{A2.14.5}$$

Note that the previous equation is similar to what we have resolved in Section A2.3. Then, the general solution is:

$$e_z = C_1 e^{-kx} + C_2 e^{kx} \tag{A2.14.6}$$

The constants "C_1" and "C_2" can be determined observing that for:

---

* Of course, also if we are considering a nonperfect conductor we would always treat with a good conductor, for which the volumetric free charge may be neglected.

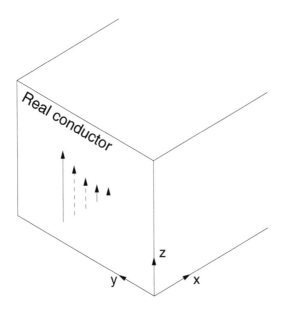

Figure A2.14.1

a. $x = \infty$ the field must be zero. So: $C_2 =! 0$.
b. $x = 0$ the field must be equal to the incident value, that we will indicate as "e_0." So: $C_1 = e_0$.

Consequently, A2.14.6 becomes:

$$e_z = e_0 e^{-kx} \qquad (A2.14.7)$$

Applying the identity $\sqrt{j} = (1 + j)/\sqrt{2}$ to the definition of "k" given in A2.14.5 we have:

$$k = k_r + jk_j = (1 + j)/p \qquad (A2.14.8)$$

where the quantity "p," called "penetration depth," is given by:

$$p \stackrel{\perp}{=} (\pi f \mu g)^{-0.5} \qquad (A2.14.9)$$

with dimensions [meter]. Note that the quantity "p" gives the amplitude decay rate of the electric field inside the good conductor. Note that the smaller the "p" value is, the smaller is the penetration depth, and the higher the "k" value is, the higher is the decay rate.

It is interesting to note that for the magnetic field "h" and the surface current "J" an equation similar A2.14.4 holds. In fact, for the magnetic field we have:

$$\underline{\nabla} \otimes \underline{\nabla} \otimes \underline{h} = \underline{\nabla}\underline{\nabla} \cdot \underline{h} - \nabla^2 \underline{h} = j\omega\varepsilon \underline{\nabla} \otimes \underline{e} + g\underline{\nabla} \otimes \underline{e} \qquad (A2.14.10)$$

Since:

1. from the Maxwell's equations $\underline{\nabla} \cdot \underline{h} = 0$
2. in a good conductor $g \gg \omega\varepsilon$

using the first of A2.14.1 the previous equation becomes:

$$\nabla^2 \underline{h} = j\omega\mu g \underline{h} \tag{A2.14.11}$$

For the surface current density "J" using the first of A2.14.1 we have:

$$\underline{J} = g\underline{e} \rightarrow \underline{\nabla} \otimes \underline{J} = g\underline{\nabla} \otimes \underline{e} = -j\omega\mu g \underline{h} \tag{A2.14.12}$$

Then

$$\underline{\nabla} \otimes \underline{\nabla} \otimes \underline{J} = \underline{\nabla}\underline{\nabla} \cdot \underline{J} - \nabla^2 \underline{J} = -j\omega\mu g \underline{\nabla} \otimes \underline{h} \tag{A2.14.13}$$

Since in a good conductor:

3. $\underline{\nabla} \bullet \underline{J} = 0$
4. $g \gg \omega\varepsilon$

using the second of A2.14.1 the previous equation becomes:

$$\nabla^2 \underline{J} = j\omega\mu g \underline{J} \tag{A2.14.14}$$

So, solutions of the form A2.14.6 and A2.14.7 are also applicable for magnetic field and surface current density.

In our case of nonideal conductors, but still "good" ones, an internal impedance "ζ_i" is defined as the ratio between the electric field on the surface and the total current in the conductor. With reference to Figure A2.14.1, the electric field on the surface, i.e., $x = 0$, is $e_{z0} = J_0/g$. Consequently, the total current "I_z" per unit width of the conductor is:

$$I_z = \int_0^\infty J_z(x)\,dx = \int_0^\infty J_0 e^{\frac{1+j}{p}x}\,dx = J_0 \frac{p}{1+j} \tag{A2.14.15}$$

with dimensions [A/meter]. So, the internal impedance "ζ_i" is.

$$\zeta_i \doteq e_{z0}/I_z \doteq R_s + j\omega L_i = (1+j)/pg \quad [\Omega] \tag{A2.14.16}$$

From this equation we note "ζ_i" has the real part is equal to the imaginary part, which means that:

$$R_s \equiv \omega L_i = 1/pg \qquad\qquad [\Omega] \tag{A2.14.17}$$

The quantity "R_s" is called the "sheet resistance" of the conductor. In literature it is possible to find "R_s" called "surface resistance." We think that this name can generate confusion since "R_s" has [Ω] as physical dimension and not [Ω/unit surface]. In practice, "R_s" is measured as [Ω/square]. The word "square" means that the value of "R_s" is a function of the ratio between the longitudinal "ℓ" and transverse "t" absolute length of the conductor. For example, if one conductor has $\ell = 2t$ its "R_s" along the direction "ℓ_0" is twice that of the same conductor with $\ell = t$ along the direction "ℓ_0."

Inserting A2.14.9 into A2.14.17 we have:

$$R_s = \left(\pi\mu f/g\right)^{0.5} \qquad\qquad [\Omega/\text{square}] \qquad\qquad (A2.14.18)$$

Alternatively, if we extract from A2.14.9 the value of "g" and we insert it into A2.14.17 we have:

$$R_s = \pi\mu f p \qquad\qquad [\Omega/\text{square}] \qquad\qquad (A2.14.19)$$

In Table A2.14.1 we report the conductivity, penetration depth, and "R_s" for typical conductors used in transmission lines:

Table A2.14.1

Metal	g(Sie/m)	p (m)	R_s (W)
Silver	$6.17*10^7$	$0.0642/\sqrt{f}$	$2.52*10^{-7}\sqrt{f}$
Copper	$5.8*10^7$	$0.066/\sqrt{f}$	$2.61*10^{-7}\sqrt{f}$
Gold	$4.35*10^7$	$0.0763/\sqrt{f}$	$3.01*10^{-7}\sqrt{f}$
Aluminum	$3.72*10^7$	$0.0826/\sqrt{f}$	$3.26*10^{-7}\sqrt{f}$
Brass	$1.57*10^7$	$0.127/\sqrt{f}$	$5.01*10^{-7}\sqrt{f}$

It is interesting to observe how copper, which is readily available, has a conductivity second only to silver. However copper oxidizes, and for this reason it is covered with gold when good transmission lines are needed.

Figures A2.14.2 and A2.14.3 report the penetration depth "p" in meters as a function of frequency in MHz, for Silver (Ag), Copper (Cu), and Gold (Au). Note that starting from the frequency of 1 GHz, the penetration depth inside such conductors are only some micrometers and decrease with frequency.

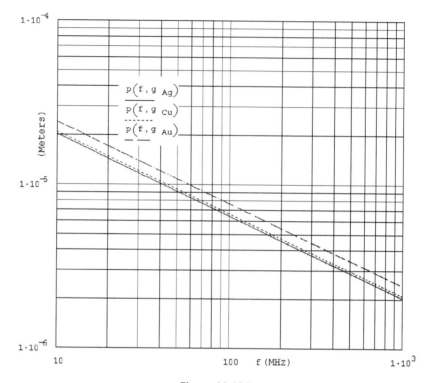

Figure A2.14.2

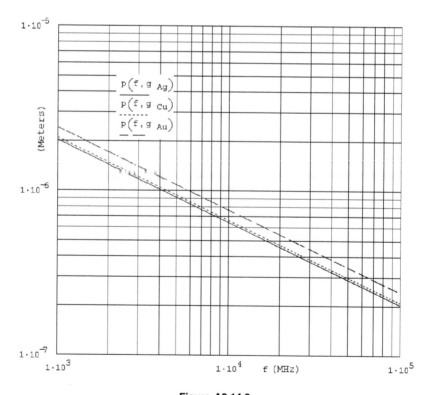

Figure A2.14.3

REFERENCES

1. S. Ramo, J. R. Whinnery, T. Van Duzer, *Fields and Waves in Communication Electronics*, John Wiley & Sons, New York, 568, 1980.
2. G. Barzilai, Fondamenti di elettromagnetismo, Siderea editore, Roma, 125, 1975.
3. R. E. Collin, *Foundations for Microwave Engineering*, McGraw Hill, New York, 1992.
4. G. Gerosa, Appunti di microonde, Università degli studi di Roma La Sapienza, 1980.
5. G. Franceschett, Campi elettromagnetici, Boringhieri editore, 1982.
6. S. Ramo, J. R. Whinnery, T. Van Duzer, *Fields and Waves in Communication Electronics*, John Wiley & Sons, New York, 568, 1980.
7. G. Barzilai, Fondamenti di elettromagnetismo, Siderea editore, Roma, 125, 1975.
8. R. E. Collin, *Foundations for Microwave Engineering*, McGraw Hill, New York, 1992.
9. G. Gerosa, Appunti di microonde, Università degli studi di Roma "La Sapienza," 1980.
10. G. Franceschetti, Campi elettromagnetici, Boringhieri editore, 1982.
11. R. E. Collin, *Foundations for Microwave Engineering*, McGraw Hill, New York, 1992.
12. G. Gerosa, Appunti di microonde, Università degli studi di Roma "La Sapienza," 1980.
13. G. N. Watson, *A Treatise on the Theory of Bessel Functions*, Cambridge University Press, London, 1966.
14. A. Ghizetti, L. Marchetti, A. Ossicini, Lezioni di complementi di matematica, Editrice Veschi, 373, 1976.
15. M. Abramowitz, I. A. Stegun, *Handbook of Mathematical Functions*, Dover, New York, 358, 1970.

Diffusion Parameters and Multiport Devices

A3.1 SIMPLE ANALYTICAL NETWORK REPRESENTATIONS

A generic "n" port electrical network* can be analytically represented using many parameters. The choice of which parameter to use is strongly dependent on the signal frequency that passes in the network. Here we want to introduce the most used representations, two of which are mainly dedicated to the low frequency range, while in the next section we will introduce a representation universally employed at high frequency. For simplicity, we will refer mainly to two port networks.

a. [Z] Matrix

The [Z] matrix is the most intuitive representation of a network since it uses the concepts of impedances, currents, and voltages. In this case, the representation for a two port network is:

$$V_1 = Z_{11}I_1 + Z_{12}I_2 \tag{A3.1.1}$$

$$V_2 = Z_{21}I_1 + Z_{22}I_2 \tag{A3.1.2}$$

where the parameters** "Z," "I," and "V" are, respectively, impedances, currents, and voltages. Note that for the measure of any parameter "Z_{ij}" an open circuit needs to be made at a port. For example, the evaluation of "Z_{11}" needs an open circuit at port "2." For this reason, the "Z" parameters are also called "open circuit parameters." Using matrix notation, the previous system of equations can be rewritten as:***

$$[V] = [Z][I] \tag{A3.1.3}$$

b. [Y] Matrix

In this case, the representation for a two port network is:

$$I_1 = Y_{11}V_1 + Y_{12}V_2 \tag{A3.1.4}$$

$$I_2 = Y_{21}V_1 + Y_{22}V_2 \tag{A3.1.5}$$

---

* In this Appendix, and in this text unless otherwise stated, "network" means an "electrical network."
** For simplicity we omit the subscripts. We will do it every time so no confusion will arise.
*** We assume the reader knows how to transform a system of equations in matrix notation.

where the parameters "I" and "V" are defined as in the previous case a, while the parameters "Y_{ij}" are the admittances between ports "ij" of the network. Note that for the measure of any parameter "Y_{ij}" a short circuit needs to be made at a port. For example, the evaluation of "Y_{11}" needs a short circuit at port "2." For this reason, the "Y" parameters are also called "short circuit parameters." Using matrix notation, the previous system of equation can be rewritten as:

$$[I] = [Y][V] \tag{A3.1.6}$$

c. ABCD or "Chain" Matrix

The "chain" matrix is of greatest interest when it is needed to evaluate the transfer function of a network that is composed of many single two ports connected in series. We have already used the chain matrix in this text, for example in Chapter 5 to study directional couplers.

This representation relates the voltage and current at a port to voltage and current to the other port, according to the following equations:

$$V_1 = AV_2 + BI_2 \tag{A3.1.7}$$

$$I_1 = CV_2 - DI_2 \tag{A3.1.8}$$

First of all, we want to draw the reader's attention to the sign of the two previous equations and the general convention of voltages and currents on a two port network, indicated in Figure A3.1.1. We recognize that the "ABCD" matrix relates the input current at a port with the output current to the other. From this point of view, the chain matrix is the lumped network equivalent of the "reverse transmission matrix" used in transmission line problems.* Another important observation is that the parameters A through C do not have the same dimensions. Note in fact that "A" and "D" are dimensionless while "B" is an impedance and "C" is an admittance.

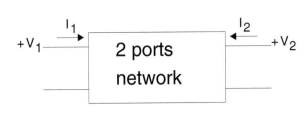

Figuro A3.1.1

It is simple to show how the chain matrix of a network composed of the series connection of two subnetworks, each individuated with a chain matrix, is given by the product of the two chain matrices of the subnetworks. Indicating with a subscript "a" and "b" the two subnetworks and with reference to Figure A3.1.2 we can write, directly from the definition of chain matrix:

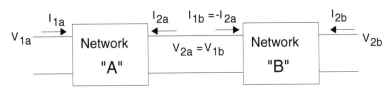

Figure A3.1.2

---

* See Chapter 1 for transmission line matrices.

$$\begin{pmatrix} V_{1a} \\ I_{1a} \end{pmatrix} = \begin{pmatrix} A_a & B_a \\ C_a & D_a \end{pmatrix} \begin{pmatrix} V_{2a} \\ -I_{2a} \end{pmatrix} \qquad (A3.1.9)$$

But $V_{2a} \equiv V_{1b}$ and $-I_{2a} \equiv I_{2b}$, and so:

$$\begin{pmatrix} V_{1a} \\ I_{1a} \end{pmatrix} = \begin{pmatrix} A_a & B_a \\ C_a & D_a \end{pmatrix} \begin{pmatrix} V_{1b} \\ I_{1b} \end{pmatrix} \qquad (A3.1.10)$$

However, among "V_{1b}," "I_{1b}," and "V_{2b}," "I_{2b}" a relationship similar to that in A3.1.9 exists, just by changing the subscripts "a" with "b," to yield:

$$\begin{pmatrix} V_{1a} \\ I_{1a} \end{pmatrix} = \begin{pmatrix} A_a & B_a \\ C_a & D_a \end{pmatrix} \begin{pmatrix} A_b & B_b \\ C_b & D_b \end{pmatrix} \begin{pmatrix} V_{2b} \\ I_{2b} \end{pmatrix} \qquad (A3.1.11)$$

which is what we should verify.

Of course, this result can be generalized to the tandem connection of "n" two ports; the complete chain matrix will be the multiplication of the single "n" chain matrices. Just for an application of the chain matrix definition, the reader can simply verify that the chain matrices "[M1]" and "[M2]" of the two networks indicated in Figure A3.1.3 parts a and b are given respectively by:

$$M_1 = \begin{pmatrix} 1 & Z \\ 0 & 1 \end{pmatrix} \qquad (A3.1.12)$$

$$M_2 = \begin{pmatrix} 1 & 0 \\ Y & 1 \end{pmatrix} \qquad (A3.1.13)$$

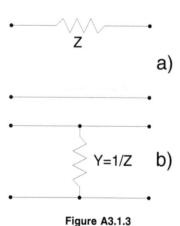

Figure A3.1.3

A3.2 SCATTERING PARAMETERS AND CONVERSION FORMULAS

The parameters we have indicated in the previous section are not well suited for high frequency networks, especially if they are wide bandwidth. This is because it is quite difficult to build a real open or short circuit at these frequencies, and it is practically impossible if the frequency range is

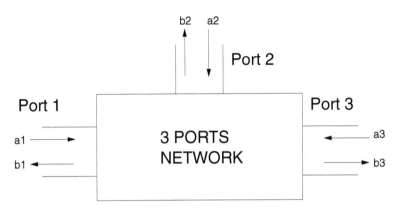

Figure A3.2.1

more than one octave. In practice, "[Z]," "[Y]," and chain matrices are only employed for narrow band operation, while for wide band representation new parameters are introduced, called "scattering parameters" and indicated with "s_{ij}." They are in general complex numbers, and are represented indifferently with modulus and phase or real and imaginary parts. The associated matrix is of course indicated with "scattering matrix." The input and output port quantities, which are related to "s" parameters, are the incident and reflected waves, respectively indicated with "a" and "b." A representation for a three port device is indicated in Figure A3.2.1. We know that the definition of voltage and current is not unique when high frequency guided signals are involved.* For this reason, the incident "a_i" and reflected "b_i" wave at port "i" are defined as:

$$a_i \overset{\perp}{=} v_i^+ / \sqrt{Z_i} \qquad b_i \overset{\perp}{=} v_i^- / \sqrt{Z_i} \tag{A3.2.1}$$

where "v_i^+" and "v_i^-" are the incident and reflected voltages at port "i," and "Z_i" is a reference impedance. Note that "v_i^+" and "v_i^-" are in general complex quantities. Usually, all the "Z_i" are set equal among them and normalized to the measurement system impedance "Z_s," usually 50Ω. Note that with the positions A3.2.1, the incident "W_i^+" and reflected "W_i^-" powers are:

$$W_i^+ = |a_i|^2 / 2 \tag{A3.2.2}$$

$$W_i^- = |b_i|^2 / 2 \tag{A3.2.3}$$

The "a_i" and "b_i" are defined to a reference plane where the field can be assumed to be unimodal,** i.e., the eventual interferences among modes can be neglected. At this reference plane we can write:

$$v_i \overset{\perp}{=} v_i^+ + v_i^- \equiv (a_i + b_i) \sqrt{Z_i} \tag{A3.2.4}$$

where the last coincidence comes from the use of A3.2.1. At the same reference plane, using the transmission line equations we can write:

$$i_i \overset{\perp}{=} (v_i^+ - v_i^-) / Z_i \equiv (a_i - b_i) \sqrt{Z_i} \tag{A3.2.5}$$

---

* See Appendix A2 for definitions on voltages and currents associated to high frequency signals.
** See Appendix A2 for propagation mode definitions.

where the last coincidence comes from the use of A3.2.1. Normalizing the "Z_i" to "Z_s," i.e., setting $Z_i = 1$, we have the following relationships:

$$v_i = a_i + b_i \qquad \text{(A3.2.6)}$$

$$i_i = a_i - b_i \qquad \text{(A3.2.7)}$$

After these definitions, the scattering representation of a two port network is:

$$b_i = s_{11}a_1 + s_{12}a_2 \qquad \text{(A3.2.8)}$$

$$b_2 = s_{21}a_1 + s_{22}a_2 \qquad \text{(A3.2.9)}$$

from which we can say that the scattering parameters relate the output waves at each port with all the input waves. Note that the measurement of the "s" parameters require a matching termination of the network under test. With the convention to set all the "Z_i" equal to "Z_s," the perfect matching is achieved when the ports are terminated with a load of impedance equal to the system reference impedance. For example, the parameter "s_{21}" is given by:

$$s_{21} = b_2/a_1 \quad \text{when} \quad a_2 = 0 \qquad \text{(A3.2.10)}$$

and $a_2 = 0$ if port "2" is terminated with a load of impedance equal to the system reference impedance "Z_s." We can recognize that "s_{11}" and "s_{22}" are the reflection coefficients* at ports "1" and "2," with the other ports terminated in "Z_s." With our convention to use "Z_s" as the reference impedance, the measurements to evaluate the "s_{ii}" parameters give values that represent how far from "Z_s" the port impedance is.

Parameters "s_{21}" and "$|s_{21}|^2$" are known as "forward gain" and "power gain," while "s_{12}" is known as "reverse gain." In general, parameters "s_{21}" and "s_{12}" are also known as "transmission coefficients."

It is simple to verify that the scattering representation doesn't permit the simple multiplication rule to obtain the scattering matrix of a bigger network, like the chain matrix permits. To avoid this difficulty, a new representation related the scattering one is introduced, called "transmission representation." In the case of a two port networks we write:

$$a_1 = t_{11}b_2 + t_{12}a_2 \qquad \text{(A3.2.11)}$$

$$b_1 = t_{21}b_2 + t_{22}a_2 \qquad \text{(A3.2.12)}$$

from which we can say that the transmission parameters relate the input waves at one port with the output waves at the other port. Proceeding as we did for the "[ABCD]" matrix, it is simple to recognize that the "[T]" matrix of a network composed with the series connection "n" subnetworks, each individuated with a "[T]" matrix, is given by the product of the "n" "[T]" matrices of the subnetworks. A lot of conversion relationships are possible between the most used analytical network representation.[1] Here we will relate "s," "t," "z," and chain parameters. Note that since "s" and "t" parameters are dimensionless, then all the chain parameters must be normalized when related to these parameters. The normalization is performed with respect to the system reference impedance, according to the following relationship:

---

* See Chapter 1 for reflection coefficient definition.

$$A_n \equiv A, \quad B_n \stackrel{\perp}{=} B/Z_s, \quad C_n = CZ_s, \quad D_n \equiv D \tag{A3.2.13}$$

where the subscript "n" indicates the normalization.

a. Conversion from "[ABCD]" to "[s]"

$$s_{11} = \frac{A_n + B_n - C_n - D_n}{A_n + B_n + C_n + D_n}, \quad s_{12} = \frac{2(A_n D_n - B_n C_n)}{A_n + B_n + C_n + D_n} \tag{A3.2.14}$$

$$s_{21} = \frac{2}{A_n + B_n + C_n + D_n}, \quad s_{22} = \frac{-A_n + B_n - C_n + D_n}{A_n + B_n + C_n + D_n} \tag{A3.2.15}$$

b. Conversion from "[s]" to "[ABCD]"

$$A_n = \frac{(1+s_{11})(1-s_{22}) + s_{12}s_{21}}{2s_{21}}, \quad B_n = \frac{(1+s_{11})(1+s_{22}) - s_{12}s_{21}}{2s_{21}} \tag{A3.2.16}$$

$$C_n = \frac{(1-s_{11})(1-s_{22}) - s_{12}s_{21}}{2s_{21}}, \quad D_n = \frac{(1-s_{11})(1+s_{22}) + s_{12}s_{21}}{2s_{21}} \tag{A3.2.17}$$

In the above conversions we assume the same load impedance at both ports. If this condition is not satisfied in A3.2.14 and A3.2.15 we have to insert the following "a_n," "b_n," "c_n," and "d_n" instead of "A_n," "B_n," "C_n," and "D_n":

$$a_n = A(R_2/R_1)^{0.5}, \quad b_n = B/(R_1R_2)^{0.5}, \quad c_n = C(R_1R_2)^{0.5}, \quad b_n = D/(R_1/R_2)^{0.5} \tag{A3.2.18}$$

where "R_1" and "R_2" are the terminating resistances at ports "1" and "2." Of course, also for A3.2.16 and A3.2.17, Equation A3.2.18 holds, and it is enough to do the following notational substitutions:

$$A_n - A(R_2/R_1)^{0.5}, \quad B_n - B/(R_1R_2)^{0.5}, \quad C_n - C(R_1R_2)^{0.5}, \quad D_n - D/(R_1/R_2)^{0.5} \tag{A3.2.19}$$

c. Conversion from "[ABCD]" to "[Z]"

$$Z_{11} = A/C, \quad Z_{12} = AD - BC/C \tag{A3.2.20}$$

$$Z_{21} = 1/C, \quad Z_{22} = D/C \tag{A3.2.21}$$

d. Conversion from "[Z]" to "[ABCD]"

$$A = Z_{11}/Z_{21}, \quad B = (Z_{11}Z_{22} - Z_{12}Z_{21})/Z_{21} \tag{A3.2.22}$$

$$C = 1/Z_{21}, \quad D = Z_{22}/Z_{21} \tag{A3.2.23}$$

e. Conversion from "[t]" to "[s]"

$$s_{11} = t_{21}/t_{11}, \quad s_{12} = t_{22} - t_{21}t_{12}/t_{11} \tag{A3.2.24}$$

$$s_{21} = 1/t_{11}, \quad s_{22} = -t_{12}/t_{11} \tag{A3.2.25}$$

f. Conversion from "[s]" to "[t]"

$$t_{11} = 1/s_{21}, \quad t_{12} = -s_{22}/s_{21} \tag{A3.2.26}$$

$$t_{21} = s_{11}/s_{21}, \quad t_{22} = s_{12} - s_{11}s_{22}/s_{21} \tag{A3.2.27}$$

A3.3 CONDITIONS ON SCATTERING MATRIX FOR RECIPROCAL AND LOSSLESS NETWORKS

The most general definition of reciprocal network has been given by Lorentz.* It states that a reciprocal region is where the following relationship is satisfied:

$$\underline{\nabla} \bullet \left(\underline{E}_a \otimes \underline{H}_b - \underline{E}_b \otimes \underline{H}_a \right) = 0 \tag{A3.3.1}$$

where subscripts "a" and "b" specify the fields produced by two sources at same frequency. The operator "$\underline{\nabla}$" performs the "divergence," and it is defined in Appendix A8. Applying the previous relationship to a network, it is possible to show that it is reciprocal if its matrix "[M]" is symmetrical, i.e., the following relationship holds:

$$[M]^T \equiv [M] \tag{A3.3.2}$$

where the superscript "T" specifies the transpose operation.

If the network is lossless, it is possible to verify that the following relationship holds:

$$[s][s]^{T^*} \equiv [I] \tag{A3.3.3}$$

where the superscript "*" specifies the complex conjugate operation and "[I]" is the unitary, or identity matrix, i.e., the matrix with the only nonzero elements on the principal diagonal. Operatively, the previous equation is developed as follows:

1. For every matrix row: the sum of the square of the modulus for any element must be equal to one
2. For any two matrix rows: the product of one element for the complex conjugate of the correspondent element of another row must be equal to zero.

   The condition of reciprocity and/or lossless reduces the number of the independent elements of "[s]," i.e., the number of tests required to evaluate all the elements of the matrix. In general, an $n \times n$ scattering matrix requires n^2 complex elements or $2n^2$ real elements to be determined. In fact, if the network is reciprocal, then a reduction factor "r" for complex quantities can be introduced, given by:

$$r = n(n-1)/2 \tag{A3.3.4}$$

---

* Hendrik Antoon Lorentz, Dutch physicist, born in Arnhem in 1853 and died in Haarlem in 1928.

that is the number of possible combinations of two different elements, the subscripts "i" and "j," in a group of "n" elements. The resulting independent complex elements in this case will be:

$$n^2 - r = n(n+1)/2 \tag{A3.3.5}$$

If the network is lossless, we have two reduction factors "ℓ_1" and "ℓ_2," and exactly:

3. $\ell_1 = n$ real equations, given by point 1 above:
4. $\ell_2 = n(n-1)/2$ complex equations given by point 2 above.

In conclusion, the reduction factor "ℓ" for real quantities is:

$$\ell = \ell_1 + 2\ell_2 \equiv n^2 \tag{A3.3.6}$$

and the independent real elements in this case will be:

$$2n^2 - \ell = n^2 \tag{A3.3.7}$$

Of course, if the network is lossless and reciprocal, then we have the reduction factors "r" and "ℓ" defined above, resulting in a reduction factor "t" for real quantities given by:

$$t = 2r + \ell = 2n^2 - n \tag{A3.3.8}$$

The number of real quantities for the scattering matrix in this case is:

$$2n^2 - t = n \tag{A3.3.9}$$

We see how a great reduction of measurements is possible if the network under test can be approximated as lossless and reciprocal.

A3.4 THREE PORT NETWORKS

We want to verify that a lossless three port network cannot isolate two ports without one of these ports also being isolated with the third. Using the condition A3.3.2 into A3.3.3 we have:

$$|s_{11}|^2 + |s_{12}|^2 + |s_{13}|^2 =! 1 \tag{A3.4.1}$$

$$|s_{12}|^2 + |s_{22}|^2 + |s_{23}|^2 =! 1 \tag{A3.4.2}$$

$$|s_{13}|^2 + |s_{23}|^2 + |s_{33}|^2 =! 1 \tag{A3.4.3}$$

$$s_{11}s_{12}{}^* + s_{12}s_{22}{}^* + s_{13}s_{23}{}^* =! 0 \tag{A3.4.4}$$

$$s_{11}s_{13}{}^* + s_{12}s_{23}{}^* + s_{13}s_{33}{}^* =! 0 \tag{A3.4.5}$$

$$s_{12}s_{13}{}^* + s_{22}s_{23}{}^* + s_{23}s_{33}{}^* =! 0 \tag{A3.4.6}$$

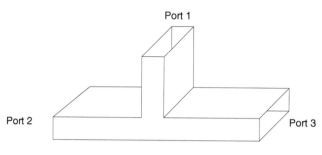

Figure A3.4.1

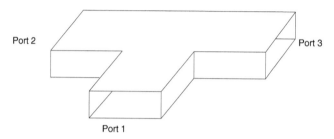

Figure A3.4.2

Now let us assume ports "1" and "2" are isolated, i.e., $s_{12} = 0$. Then from A3.4.4 we have $s_{13}s_{23}{}^* = ! 0$, which implies that $s_{13} = ! 0$ or $s_{23}{}^* = ! 0$, i.e., at least one of ports "1" and "2" is also isolated with port "3." This result is of course in contrast with the starting hypothesis.

The most well-known three port devices are called "T," due to the "T" shape of such devices in waveguide technology. Two types of "T" are possible in rectangular waveguides called "T in E-plane" and "T in H-plane" and indicated respectively in Figures A3.4.1 and A3.4.2. These names comes from the fact that the "T" is built in the plane where the "E" or "H" field of the "TE_{10}" mode lies.* Ports "2" and "3" are called "longitudinal ports" while port "1" is called a "transverse port." These devices, evaluated as reciprocal and lossless, are characterized by a scattering matrix of the form:

$$(s) = \begin{pmatrix} 0 & s_{12} & s_{13} \\ s_{21} \equiv s_{12} & s_{22} & s_{23} \\ s_{31} \equiv s_{13} & s_{32} \equiv s_{23} & s_{33} \equiv s_{22} \end{pmatrix} \tag{A3.4.7}$$

Using the lossless condition for the modulus we have:

$$|s_{12}|^2 + |s_{13}|^2 = ! 1 \tag{A3.4.8}$$

$$|s_{12}|^2 + |s_{22}|^2 + |s_{23}|^2 = ! 1 \tag{A3.4.9}$$

$$|s_{13}|^2 + |s_{23}|^2 + |s_{22}|^2 = ! 1 \tag{A3.4.10}$$

* See Appendix A2 for rectangular waveguide modes.

Note that from A3.4.9 and A3.4.10 we have that $|s_{12}|^2 =! |s_{13}|^2$, and inserting this condition into A3.4.8 we have:

$$|s_{12}| =! |s_{13}| =! 1/\sqrt{2} \qquad (A3.4.11)$$

i.e., the "T" is a power divider, from the port where $s_{11} = 0$.

Using the lossless condition for the phases we have:

$$s_{12}s_{22}{}^* + s_{13}s_{23}{}^* =! 0 \qquad (A3.4.12)$$

$$s_{12}s_{23}{}^* + s_{13}s_{22}{}^* =! 0 \qquad (A3.4.13)$$

$$s_{12}s_{13}{}^* + s_{22}s_{23}{}^* + s_{23}s_{22}{}^* =! 0 \qquad (A3.4.14)$$

Conditions A3.4.12 and A3.4.13 can be rewritten as:

$$\left(s_{12} + s_{13}\right)\left(s_{23}{}^* + s_{22}{}^*\right) =! 0 \qquad (A3.4.15)$$

This condition can be verified if:

$$s_{12} = -s_{13} \qquad (A3.4.16)$$

i.e., the signals at the output ports have 180° phase shift, or if $s_{22}{}^* = -s_{23}{}^*$ which, inserted into A3.4.12, results in

$$s_{12} = s_{13} \qquad (A3.4.17)$$

i.e., the signals at the output ports have zero phase shift.

Observing Figures A3.4.1 and A3.4.2 we can say that a "T in E-plane" respects the condition A3.4.16 entering in port "1," while a "T in H plane" respects the condition A3.4.17 entering in port "1." So, we have the result that a three port passive, reciprocal, and lossless device is a power divider whose reflection coefficient is zero only at one port, and signals at the output ports can be in phase or 180° out of phase, depending on the physical construction.

Another important three port network is the circulator, already encountered in our text. This device has the characteristic that connection between ports is directive, in a circular manner. The schematic symbol is indicated in Figure A3.4.3, while in Figure A3.4.4 we have represented this device in rectangular waveguide technology. It is built with three rectangular waveguides, joined together at 120°, with a magnetized cylinder post at the center of the common junction. The general circulation principle is studied in detail in Appendix A7, and for this reason we will not enter deeply into this operating principle at this time. We will instead show that a lossless circulator has a scattering matrix that can be obtained by:

$$(s) = \begin{pmatrix} 0 & s_{12} & s_{13} \\ s_{21} & 0 & s_{23} \\ s_{31} & s_{32} & 0 \end{pmatrix} \qquad (A3.4.18)$$

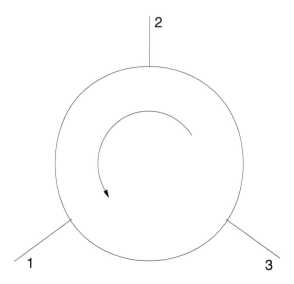

Figure A3.4.3

Since we assume the circulator to be lossless, we have the following conditions:

$$\left|s_{12}\right|^2 + \left|s_{13}\right|^2 =! \, 1 \qquad\qquad (A3.4.19)$$

$$\left|s_{21}\right|^2 + \left|s_{23}\right|^2 =! \, 1 \qquad\qquad (A3.4.20)$$

$$\left|s_{31}\right|^2 + \left|s_{32}\right|^2 =! \, 1 \qquad\qquad (A3.4.21)$$

$$s_{13}s_{23}{}^* = s_{12}s_{32}{}^* = s_{21}s_{31}{}^* =! \, 0 \qquad\qquad (A3.4.22)$$

Since we have a system of five equations* in six unknowns, we can resolve this system only if we define some value *a priori*. So, setting $s_{21} \neq 0$ we have:

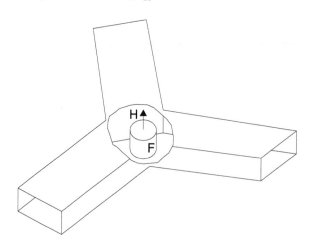

Figure A3.4.4

* Equation A3.4.22 is complex, and corresponds to two real equations.

$$s_{31} =! \, 0 \qquad \text{from A3.4.22} \tag{A3.4.23}$$

$$\left| s_{32} \right| =! \, 1 \qquad \text{from A3.4.21} \tag{A3.4.24}$$

$$s_{12} =! \, 0 \qquad \text{from A3.4.22} \tag{A3.4.25}$$

$$\left| s_{13} \right| =! \, 1 \qquad \text{from A3.4.19} \tag{A3.4.26}$$

$$s_{23} =! \, 0 \qquad \text{from A3.4.22} \tag{A3.4.27}$$

$$\left| s_{21} \right| =! \, 1 \qquad \text{from A3.4.20} \tag{A3.4.28}$$

From Equations A3.4.23 through A3.4.28 we can recognize how the connection between an input and output port is circular, in the direction $1 \rightarrow 3 \rightarrow 2 \rightarrow 1$, as indicated in Figure A3.4.3.

A3.5 FOUR PORT NETWORKS

Among the many four port networks, we are interested in the class of directional couplers. We will assume that these devices are correctly terminated on their own output impedance.

The most general characteristic is that they are always four port and reciprocal devices, whose ports can be divided into two couples, with the following characteristics:

a. Each port is perfectly matched
b. Two ports of any couple are perfectly isolated and
c. Transmission is only possible between ports of different couples.

To do an example, in Figure A3.5.1 we have reported a Lange coupler, microstrip devices studied in Chapter 7. If port "1" is the input, then port "2" is isolated, port "3" is the DC connected, or direct port, and port "4" is the coupled port. So, a directional coupler with such port numbering has a scattering matrix in general given by:

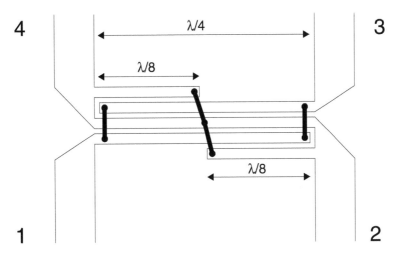

Figure A3.5.1

$$(s) = \begin{pmatrix} 0 & 0 & s_{13} & s_{14} \\ 0 & 0 & s_{23} & s_{24} \\ s_{13} & s_{23} & 0 & 0 \\ s_{14} & s_{24} & 0 & 0 \end{pmatrix} \tag{A3.5.1}$$

Applying the lossless condition, we have:

$$|s_{13}|^2 + |s_{14}|^2 =! 1 \tag{A3.5.2}$$

$$|s_{23}|^2 + |s_{24}|^2 =! 1 \tag{A3.5.3}$$

$$|s_{13}|^2 + |s_{23}|^2 =! 1 \tag{A3.5.4}$$

$$|s_{14}|^2 + |s_{24}|^2 =! 1 \tag{A3.5.5}$$

$$s_{13}s_{23}{}^* + s_{14}s_{24}{}^* =! 0 \tag{A3.5.6}$$

$$s_{13}s_{14}{}^* + s_{23}s_{24}{}^* =! 0 \tag{A3.5.7}$$

From A3.5.2 and A3.5.4 we have:

$$|s_{14}| =! |s_{23}| \tag{A3.5.8}$$

which, inserted into A3.5.2 and compared with A3.5.3, gives:

$$|s_{13}| =! |s_{24}| \tag{A3.5.9}$$

The previous two equations together with A3.5.2 form a system of three equations in four unknowns, which can be resolved only if we define some modulus value *a priori*. In addition, transforming Equations A3.5.6 and A3.5.7 into modulus and phase and using A3.5.8, it is possible to show that Equations A3.5.6 and A3.5.7 result in an equation with four unknown phases. This system can be resolved only if we define three phase values *a priori*. Depending on these arbitrary values, a directional coupler can be classified according to the following cases where we will assume the port numbering and physical meaning as defined above.

a. Hybrid — in this case, connecting the source at port "1" the signals at the output ports "c" and "d" have 180° phase offset. Connecting the source at port "b" the signals at the output ports are in phase.
b. "Magic T" — in this case, together with the previous condition, a, the output signals have equal output power.
c. Symmetric — in this case, connecting the source at port "a" the signals at the output ports "c" and "d" have 90° phase offset.
d. Perfectly symmetric — in this case, together with the previous condition, c, the output signals have equal output power.

Among these directional couplers, the "magic T" is widely used, particularly in mixer circuits. A typical construction of this device in rectangular waveguide technology is indicated in Figure A3.5.2. We can observe that it can be considered as the union of a "T in E" with a "T in

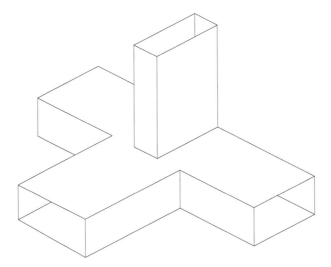

Figure A3.5.2

H," studied in the previous section. Of course, matching circuits like iris and septs are employed, which for simplicity are not indicated in the figure. Now, let us assume application of a signal at the fundamental mode at the transverse port of the "T in E" as indicated in Figure A3.5.3. We see that signals of equal amplitude but 180° out of phase are available at the longitudinal ports of the "T in E",* while ideally no signal is available in the transverse port of the "T in H." Now, let us assume application of the signal at the transverse port of the "T in H," as indicated in Figure A3.5.4. We see that signals of equal amplitude and phase are available at the longitudinal ports of the "T in H,"** while ideally no signal is available in the transverse port of the "T in E." So, the scattering matrix of an ideal "magic T" is given by:

$$(s) = \frac{1}{\sqrt{2}} \begin{pmatrix} 0 & 0 & 1 & 1 \\ 0 & 0 & -1 & 1 \\ 1 & -1 & 0 & 0 \\ 1 & 1 & 0 & 0 \end{pmatrix} \qquad (A3.5.10)$$

Of course, since this directional coupler is reciprocal, it can also be used as a signal adder. So, if in Figure A3.5.3 two generators at the same frequency but 180° out of phase are applied at the longitudinal ports, then a signal with power equal to the sum of the input power appears at the transverse port of "T in E." Of course, a similar discourse can be applied to the device in Figure A3.5.4; in this case, the input signals must be equiphase, and the sum signal comes out from the transverse port of the "T in H."

A3.6 QUALITY PARAMETERS FOR DIRECTIONAL COUPLERS

From the previous section we know that a directional coupler is a four port device, with particular characteristics. Operatively, to measure if the device under test is directional, we apply the source

---

* These ports are also the longitudinal ports of the "T in H."
** These ports are also the longitudinal ports of the "T in E."

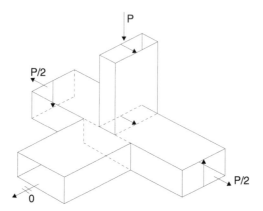

Figure A3.5.3

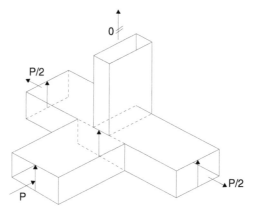

Figure A3.5.4

generator to a port and measure the signal at the other ports. Let us number the port evaluated as input with "1." The following points must be verified:

1. One port exists where the signal has the maximum value between the other ports, at least from the point of view of widest bandwidth. We mean that if we find two output ports with the same signal amplitude,* then expanding the bandwidth only, one of these out ports still continues to have the maximum signal. Let us number this port as "3" and name it as "direct port."
2. One port exists where the signal has an amplitude lower than the amplitude at port "3" at least from the point of view of widest bandwidth. Let us number this port as "4" and name it the "coupled port."
3. The remaining port "2" is isolated.

As an example of port numbering of a directional coupler, Figure A3.6.1 indicates a single step, quarter wavelength, microstrip directional coupler, studied in Chapter 5. Of course, since the device is reciprocal, the previous points 1 to 3 must be verified for any input port, with proper numbering. However, the individuation of a directional coupler may not be so simple, since points 1 to 3 above are frequency dependent. In the following points we will use the port numbering as indicated in Figure A3.6.1.

* This is the case for 3 dB directional couplers, for example.

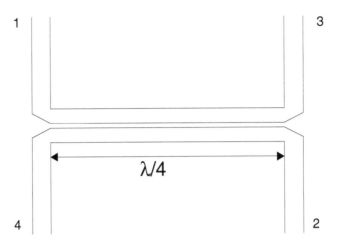

Figure A3.6.1

Now, let us define the quality parameters of these important devices.

a. Coupling. Power coupling "c_{wdB}" in "dB" is defined as:

$$c_{wdB} \triangleq 10 \log \frac{W_{in}}{W_{cou}} \equiv 10 \log \frac{W_1}{W_4} \equiv 10 \log \frac{1}{\left|S_{14}\right|^2} \triangleq 10 \log\left(c_w\right) \qquad (A3.6.1)$$

where the letter "W" individuates powers and "log" is the logarithm in base 10. Coupling can also be defined using voltages "v" or electric fields "E," and in this case we have:

$$c_{vdB} \triangleq 20 \log \frac{\left|E_{in}\right|}{\left|E_{cou}\right|} \equiv 20 \log \frac{\left|E_1\right|}{\left|E_4\right|} \equiv 20 \log \frac{1}{\left|S_{14}\right|} \equiv 20 \log\left(c_v\right) \qquad (A3.6.2)$$

Note "c_w" and "c_v" are related by $c_v^2 = c_w$. Note that with the previous definitions A3.6.1 and A3.6.2 the coupling is a positive number. However, if for the coupling definition the ratios "W_4/W_1" and "E_4/E_1" are used, then the result is a negative number. We think that the use of a negative number, in logarithmic scale, for the coupling is more intuitive, since a coupling on a passive circuit can never generate a signal stronger than the source signal. However, positive values for the coupling are more referenced than the negative ones.

b. Isolation. The isolation "i_{dB}" in "dB" is defined as:

$$i_{dB} \triangleq 10 \log \frac{W_{in}}{W_{iso}} \equiv 10 \log \frac{W_1}{W_2} \equiv 10 \log \frac{1}{\left|S_{21}\right|^2} \triangleq 10 \log\left(i_w\right) \qquad (A3.6.3)$$

For an ideal directional coupler the isolation should be infinite. In practice, the isolation is a function of the coupler technology and mechanical tolerances. This means that the isolation value is also a function of frequency. In waveguide, mean isolation values can be near 30 dB, in stripline near 25 dB, and in microstrip near 20 dB. Note that with the previous definition, the isolation is a positive number.

c. Directivity.   The directivity "d_{dB}" in "dB" is defined as:

$$d_{dB} \overset{\perp}{=} 10 \log \frac{W_{cou}}{W_{iso}} \equiv 10 \log \frac{W_4}{W_2} \equiv 10 \log \frac{|s_{41}|^2}{|s_{21}|^2} \overset{\perp}{=} 10 \log \left(d_w \right) \qquad (A3.6.4)$$

Comparing expressions A3.6.1, A3.6.3, and A3.6.4 it is simple to observe that the following relationship holds:

$$\left| i_{dB} \right| - \left| c_{dB} \right| \equiv d_{dB} \qquad (A3.6.5)$$

For an ideal directional coupler, the directivity should be infinite. In practice, similar to the isolation case, the reachable value is a function of the coupler technology and mechanical tolerances, but also of the designed coupling value. In fact, the higher the coupling in modulus, the lower the directivity. For example, a microstrip 10 dB of coupling directional coupler has a directivity near 18 dB at 10 GHz; if we increase the coupling modulus to 20 dB then the directivity decreases to near 10 dB.

A3.7 SCATTERING PARAMETERS IN UNMATCHED CASE

In Section A3.2 we have defined the scattering parameters, observing that they can be simply defined in the case of matched terminations at the ports, i.e., using the reference system impedance "Z_s." Let us use Figure A3.7.1 as a general "[s]" representation of a two port network.

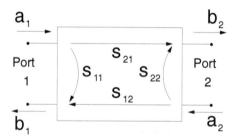

Figure 3.7.1

If the termination load "Z_2" at port "2" is not equal to "Z_s," then the "$s_{11,u}$"* measured value is related to the defined "s_{ii}" by the following relationship:

$$s_{11,u} = s_{11} - \frac{s_{12}s_{21}\Gamma_2}{1 - s_{22}\Gamma_2} \qquad (A3.7.1)$$

where the port "2" load reflection coefficient is defined as:**

$$\Gamma_2 = \frac{Z_2 - Z_s}{Z_2 + Z_s} \qquad (A3.7.2)$$

---

* The subscript "u" will remind us that the measurement is made in the case of unmatched termination, i.e., a termination with different impedance from "Z_s."
** See Chapter 1 for reflection coefficient definition.

Note that if the network is unilateral, i.e., with only the transmission coefficient $s_{21} \neq 0$, then independently from the presence of a "Γ_2" we have $s_{11,u} \equiv s_{11}$. Of course, a similar result can be obtained if we want to evaluate the "s_{22}" and we connect a load "$Z_1 \neq Z_s$" at port "1." In this case we have:

$$s_{22,u} = s_{22} - \frac{s_{12}s_{21}\Gamma_1}{1 - s_{11}\Gamma_1} \qquad (A3.7.3)$$

where the port "1" load reflection coefficient is defined as:

$$\Gamma_1 = \frac{Z_1 - Z_s}{Z_1 + Z_s} \qquad (A3.7.4)$$

So, caution needs to be exercised if a load different from "Z_s" is used to measure the reflection coefficients of two port devices.

These results can of course be generalized to the case of an "n" port device.

REFERENCES

1. D. A. Frickey, Conversions between s, z, y, h, ABCD and t parameters which are valid for complex source and load impedances, *IEEE Trans. on MTT*, 205, Feb. 1994.

Resonant Elements, "Q," Losses

A4.1 THE INTRINSIC LOSSES OF REAL ELEMENTS

Every reactive element possesses a resistive component that is responsible for energy loss inside this element. The most evident effect of this resistance is the warming up of the component when an RF signal passes through it. But also in DC we can observe the presence of a loss. For example, a charged capacitor that ideally should conserve such voltage forever, in practice discharges in a finite time, due to a lossy intrinsic parallel resistor that represents the nonzero conduciveness of the capacitor dielectric. Another example is the inevitable DC resistance of an air conductor coil, due to the noninfinite conductiveness of the real conductors. So, for DC signals, any real capacitor and inductor can be represented as indicated in Figure A4.1.1, parts a and b.

Such simple equivalent circuits become more complicated when the signal frequency increases. In this case, mainly parasitic reactive components arise, which in some cases can transform the component behavior. For example, an inductor can behave as a capacitor and a resistance as a capacitor, or any other combination depending on the frequency of the signal. For example, Figure A4.1.2 represents the equivalent circuit of a capacitor at high frequency. For such a circuit we have:

1. the inductors "L'" and "L'''" represent the wire connection to the concentrated* capacitor
2. "R'" and "R'''" represent the resistances of "L'" and "L''"**
3. "R" represents the DC lossy dielectric
4. "C'" represents the stray capacitance due to capacitor housing or any other coupling.

A frequency response for the $|s_{21}|$*** parameter of the network indicated on the right side of Figure A4.1.2 and for an ideal capacitor are represented in Figure A4.1.3. Here, we have the following values for the elements of Figure A4.1.2: $C = 0.5$ pF, $R = 10$ KΩ, $L' \equiv L'' = 1$ nH, $R' \equiv R'' = 1\,\Omega$, $C' = 1$ pF. Of course, some parasitic elements can be neglected or not, depending on the signal frequency, the capacitor technology, and its housing.

Similar to the equivalent circuit for a capacitor, we can introduce the equivalent circuit for an inductor, as indicated in Figure A4.1.4. The capacitor "C" represents the interwinding stray capacitance of the coil, while the other elements are already defined for Figure A4.1.2. In Figure A4.1.5 we have represented the frequency response for "$|s_{21}|$" parameter of the network in Figure A4.1.4

---

* The definition of a pure capacitor is dependent on the signal frequency. For example, in the GHz region every element is evaluated as a transmission line network. In this case, with "concentrated" element we mean that element without electrical lengths but with only electrical properties.
** At high frequency, "R'" and "R''" also are frequency dependent, due to the skin effect. See Appendix A2 for skin effect discussions.
*** See Appendix A3 for "s" parameter definition.

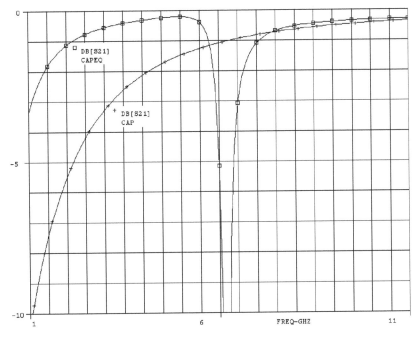

Figure A4.1.3

simple, especially when the losses of the system are not so simple to evaluate. For other systems the procedure is instead simple. To do an example, let us evaluate the "Q" of the series "RLC" circuit indicated in Figure A4.2.1. The resistor "R" can represent the coil's losses. Applying the Kirchhoff* current loop we can write:**

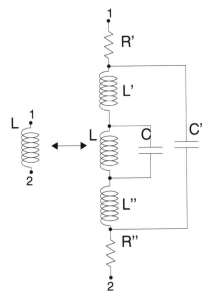

Figure A4.1.4

* Gustav Robert Kirchhoff, German physicist, born in Koenigsberg in 1824 and died in Berlin in 1887.
** For simplicity, we will indicate indifferently with "R," "L" and "C" a physical resistor, inductor, and capacitor or a resistance, inductance, and capacitance, since in what we are going to study we think there should not be confusion.

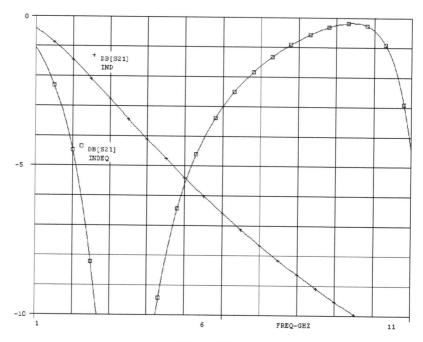

Figure A4.1.5

$$L\frac{di(t)}{dt} + \frac{q(t)}{C} + Ri(t) = 0 \qquad (A4.2.2)$$

which, again derived with respect to the time and omitting time dependence for simplicity:

$$L\frac{d^2i}{dt^2} + \frac{i}{C} + R\frac{di}{dt} = 0 \qquad (A4.2.3)$$

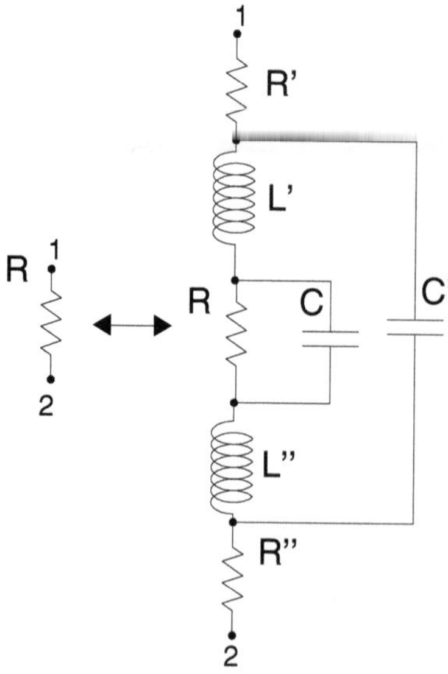

Figure A4.1.6

Theoretically, a "Q" can be defined for any reactive element, just relating its reactance with its losses. For example, a real inductor can be simply thought of as a series between an ideal inductor "L" plus a resistance "R_s," and its "Q" can be defined as:

$$Q_\ell = \omega L/R_s \qquad\qquad (A4.2.11)$$

If this inductor is loaded with a parallel resistance "R_p" and its "R_s" can be neglected, another "Q" can be defined, called loaded "Q," defined as:

$$Q_{\ell\ell} = R_p/\omega L \qquad\qquad (A4.2.12)$$

Similarly, for a capacitor with in series a conductance "G_s" its "Q" is:

$$Q_c = G_s/\omega C \qquad\qquad (A4.2.13)$$

while if the capacitor is loaded with a parallel conductance "G_p" its "Q" is:

$$Q_{\ell c} = \omega C/G_p \qquad\qquad (4.2.14)$$

In particular circuits, it is simple to relate some network function with the "Q" of the circuit. To do an example, let us study the parallel resonant circuit indicated in Figure A4.2.2, evaluating the ratio of the current "I_r" in the resistor with the current "I_g" of the generator. We have:

$$\frac{I_r}{I_g} = \frac{G}{G + j\omega C + 1/j\omega L} \qquad\qquad (A4.2.15)$$

that with the definition of $\omega_0^2 \triangleq 1/LC$ becomes:

$$\frac{I_r}{I_g} = \frac{1}{1 + jQ\left(\omega/\omega_0 - \omega_0/\omega\right)} \qquad\qquad (A4.2.16)$$

where the quality factor "Q" of the parallel resonant circuit indicated in Figure A4.2.2 is defined now

$$Q \triangleq G/\omega_0 C \equiv R/\omega_0 L \qquad\qquad (A4.2.17)$$

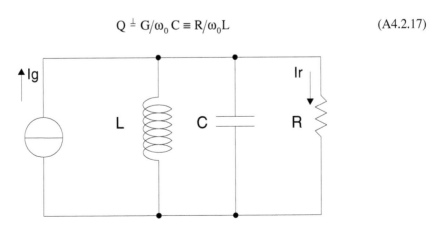

Figure A4.2.2

Note that the "Q" coincides with the loaded "Q" for the inductor or capacitor, reported in Equations A4.2.12 and A4.2.14. In other words, the resistor "R" of the parallel resonant circuit can be associated to a loaded capacitor, with an ideal inductor in parallel, or a loaded inductor, with an ideal capacitor in parallel.

For the case of the series resonant circuit indicated in Figure A4.2.1 we can obtain for the ratio "V_r/V_g" an expression similar to A4.2.16, with "Q" defined as:

$$Q \doteq G/\omega_0 C \equiv \omega_0 L/R \qquad \text{(A4.2.18)}$$

In this case, the "Q" of the circuit can be evaluated from the "Q" of an inductor with a series resistance, with an ideal capacitor in series, or from the "Q" of a capacitor with a series resistance, with an ideal inductor in series.

When the "Q" of a circuit is at least greater than ten, it can be easily evaluated measuring a frequency response of some proper circuit function, like "I_r/I_g" or "V_r/V_g" respectively, for a parallel or series resonant circuit. In this case, it is possible to verify that the "Q" is approximately given by:

$$Q = f\left(f_{+3dB} - f_{-3dB}\right)\big/f_0 \qquad \text{(A4.2.19)}$$

where "f_{+3dB}" and "f_{-3dB}" are respectively the two frequencies, one above and one below the center frequency "f_0" where the response magnitude is below 3 dB from the amplitude at "f_0." As center frequency "f_0," the frequency at which there is the maximum peak of the circuit function is assumed. To do an example, in Figure A4.2.3 we have represented the modulus of A4.2.16, for "Q" values of 2, 5, and 10. Here we can see how if the "Q" is not higher than 10, then the modulus is not

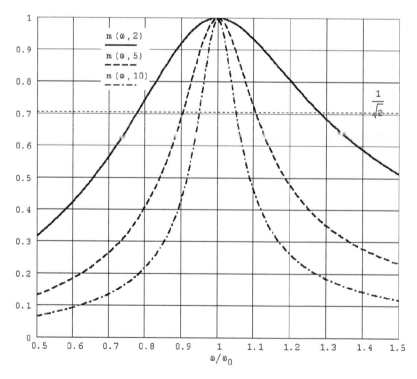

Figure A4.2.3

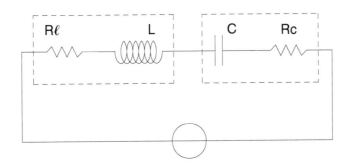

Figure A4.2.4

symmetric with respect to "ω_0" and the value extracted from A4.2.19 is not accurate. The 3 dB points are individuated by the crossing of the line at "$1/\sqrt{2}$" ordinate with each modulus. So, from a measure like that indicated in Figure A4.2.3 we can extract the "Q" using A4.2.19.

For some circuits it is simple to relate the "Q" of the single component with the "Q" of the whole circuit. To show this, let us examine the series circuits indicated in Figure A4.2.4, where each reactive element has its loss. Using definitions A4.2.11 and A4.2.13 for the "Q" of the elements, we can write:

$$R_\ell = \omega_0 L / Q_\ell \tag{A4.2.20}$$

$$R_c = 1/\omega_0 C Q_c \tag{A4.2.21}$$

The whole resistance of the circuit is:

$$R = R_\ell + R_c = \omega_0 L / Q_\ell + 1/\omega_0 C Q_c \tag{A4.2.22}$$

Associating such "R" only to "L" we have a "Q" given by:

$$Q = \omega_0 L / R \rightarrow R = \omega_0 L / Q \tag{A4.2.23}$$

Then, inserting A4.2.20 and A4.2.21 into A4.2.22 we have:

$$R = \omega_0 L / Q_\ell + 1/\omega_0 C Q_c \tag{A4.2.24}$$

and equating the last equation in A4.2.23 with A4.2.24 we obtain:

$$1/Q = 1/Q_\ell + 1/Q_c \tag{A4.2.25}$$

This expression is true for all forms of resonant circuit, where any component has its own "Q."

A4.3 ELEMENTS OF FILTER THEORY

For every filtering structure we have indicated in the text, if it is built using distributed devices, its transfer function must also belong to the general filter theory.

In this section we will review the most general definitions of functions and their relationships for filter theory. Since our text is not oriented to filter theory and synthesis, and also because it is impossible to cover in just a few pages the large amount of filter theory, we assume the reader knows these concepts, here reported for convenience. Texts entirely dedicated to general filter theory and synthesis are indicated in the bibliography.[1,2,3,4,5]

A general passive, reciprocal network, studied from a filter theory point of view, possesses an input and an output port. It can be electrically individuated by a transfer function, which relates the power available "W_a" from the generator connected to the network input port and the power "W_2" given by the network to a load connected at its output port.

So, a function "H(s)" called the "transmission function" is defined as:

$$|H(s)|^2 \doteq W_a/W_2 \qquad (A4.3.1)$$

where "s" is the Laplace variable.* Since $W_2 \le W_a$, then $|H(s)|^2 \ge 1$, and for this reason the characteristic function is set as:

$$|H(s)|^2 \doteq 1 + |K(s)|^2 \qquad (A4.3.2)$$

where $|K(s)|^2 \ge 0$ and "K(s)" is called the "characteristic function" of the filter. In this case, the attenuation "A(s)" in dB inserted by the filter is:

$$A(s) \doteq 10 \lg t\left(1 + |K(s)|^2\right) \qquad (A4.3.3)$$

The determination of "K(s)" is the most important, and in general quite difficult, action for filter synthesis.

The transmission function of the filter terminated on reference impedances is just the reciprocal of the "$s_{21}(s)$"** of the filter, so that we can write:

$$|H(s)|^2 \doteq C/|s_{21}(s)|^2 \qquad (A4.3.4)$$

where "C" is a generic constant. From A4.3.2 we have:

$$|K(s)|^2/|H(s)|^2 = 1 - 1/|H(s)|^2 = 1 - W_2/W_a \qquad (A4.3.5)$$

From the general transmission line theory described in Chapter 1, the last member of A4.3.5 is just equal to the square of the reflection coefficient, and so we can write:

$$|s_{11}(s)|^2 = 1 - W_2/W_a = |K(s)|^2/|H(s)|^2 \qquad (A4.3.6)$$

Using A4.3.4 and the previous equation we have:

$$K(s) = Ds_{11}(s)/s_{21}(s) \qquad (A4.3.7)$$

where "D" is a generic constant. So we have the result that the characteristic function is related to the ratio of two important "s" parameters of any network, i.e., the reflection and transmission coefficients.

---

* Pier Simon de Laplace, French scientist, born in Beaumont an Auge in 1749 and died in Paris in 1827.
** See Appendix A3 for "s" parameter definitions.

Is is possible to show that in the most general case the functions "H(s)" and "K(s)" are rational functions and, since from A4.3.2 it follows that these functions have the same denominator, they can be written as:

$$H(s) \doteq e(s)/p(s) \quad K(s) \doteq f(s)/p(s) \tag{A4.3.8}$$

and from A4.3.6:

$$s_{11}(s) = f(s)/e(s) \tag{A4.3.9}$$

with "e(s)," "f(s)," and "p(s)" polinomials with real coefficients. Using A4.3.8 into A4.3.2 we have:

$$|e(s)|^2 \doteq |p(s)|^2 + |f(s)|^2 \tag{A4.3.10}$$

Relationships A4.3.8 through A4.3.10 are very important in filter theory and can be used to simplify the synthesis procedure, also using the following conditions, that we give without proof,[6] that "e(s)" and "p(s)" must satisfy:

1. the zeros of "e(s)" must have negative real parts, i.e., "e(s)" is said to be an Hurwitz* polynomial.
2. the degree of "e(s)" must be at least equal to that of "f(s)" or "p(s)"
3. "p(s)" must be an even or odd polynomial

In the next section we will describe how the synthesis procedure can be greatly simplified if the desired filter belongs to one of three general groups, for which many tables are available.[7,8]

A4.4 BUTTERWORTH, CHEBYSHEV, AND CAUER LOW PASS FILTERS

To simplify the filter synthesis procedure, in the literature tables are available[9,10] where poles and zeros of attenuation functions are defined, according to three families of functions called the Butterworth, Chebyshev,** and Cauer functions. These attenuation functions are all relative to low pass shapes, since with opportune variable transformation, any other filter shape can be obtained*** from a low pass prototype. These families produce different attenuation shapes, that can be chosen according to the system requirements where the filter is employed.

Before we describe the frequency characteristics of any family, we need to define three terms which are common among them.

1. *Band Pass Region* — It is defined as that frequency range where the attenuation is lower than a specified value. Since we are dealing with low pass filters, the lower frequency of the pass band is the DC. The highest acceptable frequency of the band pass region is called the "cut-off" frequency.

2. *Stop Band Region* — It is defined as that frequency range where the attenuation is higher than a specified value. Since we are dealing with low pass filters, the highest frequency of the stop band is at infinity.

---

* Adolf Hurwitz, German mathematician, born in Hannover in 1859 and died in Zurig 1919.
** Pafnutij Chebyshev, Russian mathematician, born in Okatovo in 1821 and died in Petersburg in 1894.
*** This will be discussed in the next section.

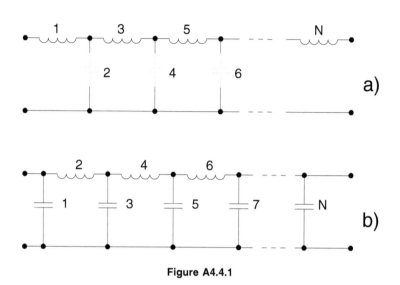

Figure A4.4.1

3. Transition Band Region — It is defined as that frequency range between the band pass region and stop band regions. The increase of attenuation in the transition band is strongly dependent on the filter topology, as will be discussed later.

Now, we will define the attenuation characteristics of any family, specializing in the case of low pass filter. We will normalize the cut-off frequency to a unitary value.

a. Butterworth filters.

These filters are characterized as having the attenuation in the pass band region without any ripple, i.e., "maximally flat" as it is normally said. Any filter of this family introduces an attenuation of 3 dB at the cut-off frequency. These filters are characterized by an attenuation function "$A_B(f)$" in dB given by:

$$A_B(f) = 10 \lg t \left[1 + (f)^n\right]^2 \qquad (A4.4.1)$$

The number "n" is called the "order" of the filter. It can be an even or odd number and gives the total number of reactive elements employed in the filter. In Figure A4.4.1 parts a and b we have represented the two possible general structures of a low pass filter. We assume an odd number of elements, as in the great number of cases. However, an even number of sections is also possible, if this is not a preferred solution. Note in the part b topology, the minimum number of inductors are needed for a fixed order "n." The filter structures indicated in Figure A4.4.1 can be used to synthesize both Butterworth and Chebychev type, described in the next point. The attenuation shape in dB for Butterworth low pass filters for order "2" to "5" is indicated in Figure A4.4.2.

b. Chebyshev filters

These filters have the same topology as the Butterworth type, indicated in Figure A4.4.1, parts a and b. The characteristic of this family is that in the pass band the attenuation has an equi-ripple shape, and the cut-off frequency is defined as the highest frequency for the equi-ripple. Above the cut-off frequency, the attenuation increases indefinitely, with a higher rate than the Butterworth counterpart. It can be shown that the Chebyshev filters have the widest band pass bandwidth of any other filter, assuming that:

b1. the attenuation has to increase monotonically above the pass band

b2. a fixed maximum attenuation, or ripple, in the pass band is defined.

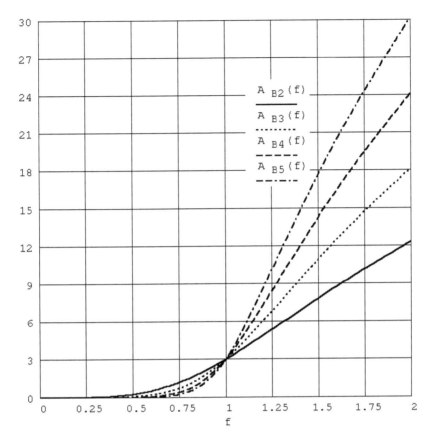

Figure A4.4.2

If the previous point b1 is removed, then another family of filters exists, which has a still sharper increase of attenuation in the transition band. These are the "Cauer filters" discussed in the next point.

The attenuation introduced by these filters depends on the Chebychev polynomials "Tn(x)" defined as:

$$T_n(f) \doteq \cos\left[n * a\cos(f)\right] \qquad (A4.4.2)$$

where "n" is the filter order and gives the total number of reactive elements employed in the filter. The attenuation function "$A_T(f)$" in dB is given by:

$$A_T(f) = 10 \lg t\left\{1 + \left[\varepsilon T_n(f)\right]^2\right\} \qquad (A4.4.3)$$

where the quantity "ε" is related to the band pass ripple through the relationship:

$$\varepsilon = \left[10^{(0.1r)} - 1\right]^{0.5} \qquad (A4.4.4)$$

The positive real number "r" plays an important role in Chebyschev filters, and represents the peak undulation in dB in the pass band. These filters also have the characteristic that once the filter order "n" is defined, if the ripple in the pass band is increased, then the rate of the attenuation in

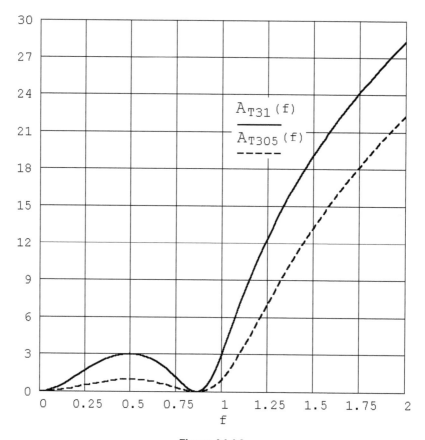

Figure A4.4.3

the transition band also increases. To show it, in Figure A4.4.3 we have reported the attenuation for a third order filter, once with $\varepsilon = 1$ and indicated with "$A_{T31}(f)$," and the other with $\varepsilon = 0.5$ and indicated with "$A_{T305}(f)$." In Figure A4.4.4 we have instead represented the attenuation for filters from second to fifth order. Note Chebyschev filters are individuated by two parameters, i.e., the filter order and the ripple.

c. Cauer filters

These filters, also named "elliptic filters"* are characterized by:

c1. a pass band shape that is similar to the Chebyschev case

c2. a minimum attenuation in the stop band.

To achieve such performances, the low pass filter topology is different from the types shown in Figure A4.4.1. In the present case, resonant elements are used, and each one of these counts as "1" in the final filter order "n." Cauer filter topologies are shown in Figure A4.4.5, where in parts a1, a2 and b1, b2 are respectively represented in the cases for "n" odd or even. Note that independently from the filter order, the topologies with parallel resonant circuits need the minimum number of inductors.

In Figure A4.4.6 we have reported the attenuation graphs for a Chebyshev and Cauer third order low pass filter, respectively indicated with "$A_{T3}(f)$" and "$A_{C3}(f)$." Ripple has been defined near 0.3 dB, and for the Cauer filter a minimum attenuation value of 42 dB has been used. Note as in the bandwidth, and also until 10% above the cut-off frequency, the two filters give practically the same shape. In Figure A4.4.7 we have reported the same attenuation functions, but in a wider

---

* This name comes from the position in the complex plane of the zeros of the rational attenuation function.

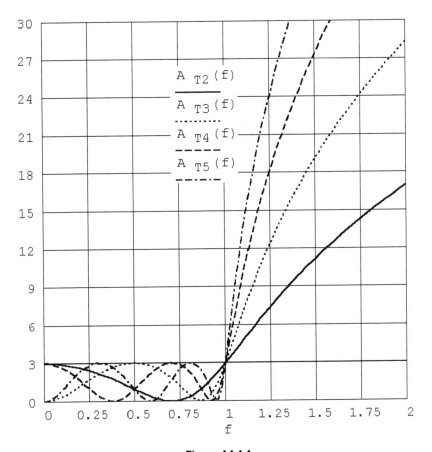

Figure A4.4.4

frequency range. Note how the two attenuations are different. In fact, the Chebyshev filter increases attenuation indefinitely, while to the Cauer filter can be associated a stop band with attenuation value never lower than a defined quantity. Also note that the Cauer filter gives a higher attenuation slope in the transition band, which makes this filter family the preferred choice of all the electronic equipment where a multicarrier communication system is employed.

Every Cauer filter attenuation function presents at least a biquadratic term "$B(s)$" of the form:

$$B(s) = C_i \frac{s^2 - 2\alpha_i s + \tau_i}{s^2 + o_i^2} \tag{A4.4.5}$$

Odd order filters together with the previous elementary biquadratic expression also have a single multiplying linear factor "$L(s)$" given by:

$$L(s) = s - \alpha_1 \tag{A4.4.6}$$

For example, a third order Cauer filter is represented by the following characteristic function:

$$C_3(s) = C \frac{(s - \alpha_1)(s^2 - 2\alpha s + \tau)}{s^2 + o^2} \tag{A4.4.7}$$

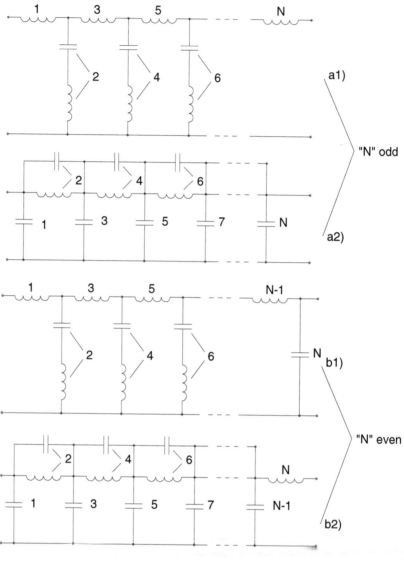

Figure A4.4.5

where factors "α_1," "α," "τ," "o," and "C" are tabulated in many of the filter design texts indicated in the bibliography, depending on the desired attenuation shape. Note also that Cauer filters are individuated by two parameters, i.e., the filter order and the pass band ripple. In Figures A4.4.8 and A4.4.9 we have drawn the attenuations for a Cauer filter of 3, 4, 5, and 6 order, respectively, for a frequency sweep up to the stop band and in the pass band only. To fix some reference parameter, we have chosen all the filters for 0.3 dB of peak ripple and 60 dB of minimum attenuation value. According to the use of the elementary biquadratic and linear functions reported in Equations A4.4.5 and A4.4.6, the attenuations for these filters are:

$$C_4(s) = C\frac{\left(s^2 - 2\alpha_1 s + \tau_1\right)\left(s^2 - 2\alpha_2 s + \tau_2\right)}{s^2 + o^2} \qquad (A4.4.8)$$

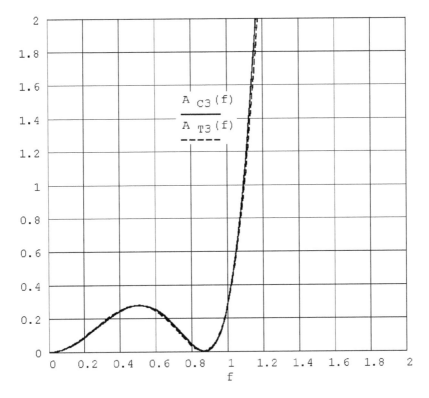

Figure A4.4.6

$$C_5(s) = C \frac{(s - \alpha_1)(s^2 - 2\alpha_2 s + \tau_2)(s^2 - 2\alpha_3 s + \tau_3)}{(s^2 + o_1^2)(s^2 + o_2^2)} \qquad (A4.4.9)$$

$$C_6(s) = C \frac{(s^2 - 2\alpha_1 s + \tau_1)(s^2 - 2\alpha_2 s + \tau_2)(s^2 - 2\alpha_3 s + \tau_3)}{(s^2 + o_1^2)(s^2 + o_2^2)} \qquad (A4.4.10)$$

Of course, the relationship between the characteristic function and the attenuation in

$$A_{ci}(s) \doteq 10 \lg t \left| C_i(s) \right| \qquad (A4.4.11)$$

All the coefficients that appear in the previous equations are tabulated, and are different among expressions, and for notational simplicity we have used the same subscripts for different attenuation functions.

We conclude this overview on filter families noting that other filters are possible, which respond to different analytical elementary functions. These are, for example, Gaussian or Bessel filters. However, the families we have reported here are surely the most used in filtering devices.

A4.5 FILTER GENERATION FROM A NORMALIZED LOW PASS

The extraction of elements value for any filter whose shape belongs to one of the three previous families can be simply performed using tables, indicated in the bibliography, for normalized low pass filter values.

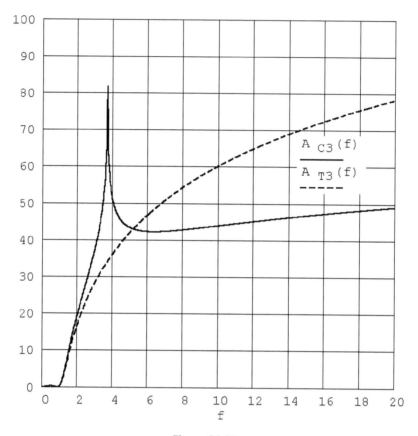

Figure A4.4.7

The normalization has proved to be very useful for synthesizing any kind of filter, be it a low pass, high pass, band pass, or band stop type. The normalization procedure starts in finding a reference frequency "f_r" and a reference resistor "R_r." "f_r" is assumed to be the cut-off frequency of the high or low pass filter, or the center frequency of the band pass or band stop filter. As "R_r" it is assumed to be the generator impedance. With these choices of reference quantities, a reference inductor "L_r" and reference capacitor "C_r" are defined according to the following positions:

$$L_r \overset{\perp}{=} R_r / 2\pi f_r \qquad\qquad (A4.5.1)$$

$$C_r \overset{\perp}{=} 1/2\pi f_r R_r \qquad\qquad (A4.5.2)$$

Once such reference quantities are found, the normalization procedure is simply done according to the following positions:

$$L_n \overset{\perp}{=} L/L_r, \quad C_n \overset{\perp}{=} C/C_r, \quad R_n \overset{\perp}{=} R/R_r, \quad f_n \overset{\perp}{=} f/f_r \qquad\qquad (A4.5.3)$$

Since the tabulated attenuation functions are relative to normalized low pass, any desired low pass filter can be simply designed by choosing the desired response and denormalizing the elements values. The other filter topologies require a change in the frequency variable of the respective attenuation function, which we are going to discuss.

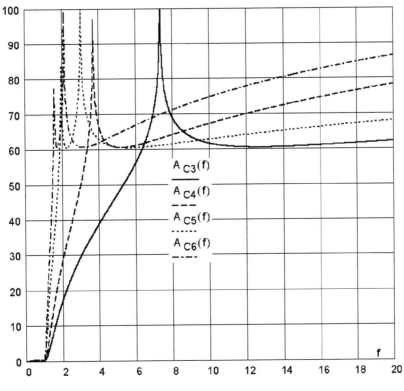

Figure A4.4.8

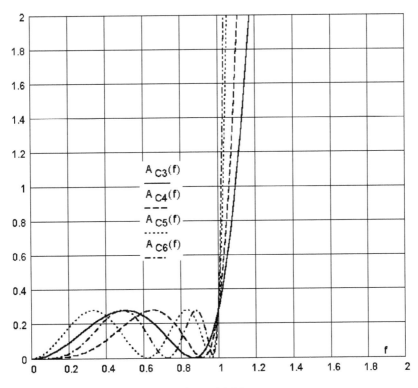

Figure A4.4.9

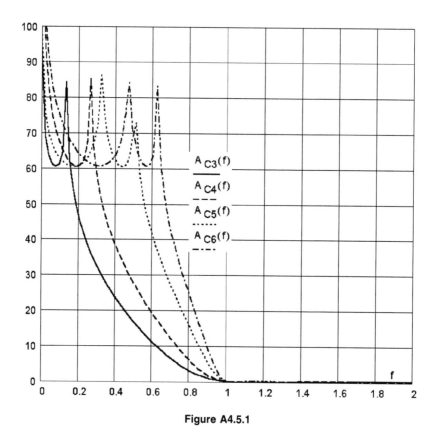

$$A_{C3}(f)$$
$$A_{C4}(f)$$
$$A_{C5}(f)$$
$$A_{C6}(f)$$

Figure A4.5.1

a. High Pass Filters

These filters can be designed using the following variable transformation:

$$s = 1/s_{hp} \qquad (A4.5.4)$$

where "s_{hp}" is the variable of the high pass attenuation function. For example, applying the previous equation to the Cauer low pass functions A4.4.7 through A4.4.10 we have the attenuation shapes reported in Figure A4.5.1. The electric circuit for high pass filters can be easily obtained from the simple low pass circuits indicated in Figure A4.4.1 exchanging in it the positions of inductors with those of capacitors. For example, an elliptical high pass filter is indicated in Figure A4.5.2. Of course, element valuse cannot be exchanged so easily, and opportune transformation is needed.

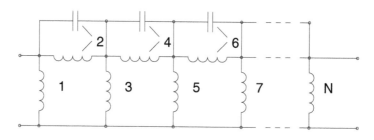

Figure A4.5.2

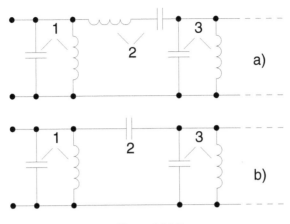

Figure A4.5.3

b. Band Pass Filters

Band pass filters can be obtained by applying the following variable transformation:

$$s = (1/\Delta f)\left(s_{bp} + 1/s_{bp}\right) \tag{A4.5.5}$$

where "s_{bp}" is the variable of the band pass attenuation function and "Δf" is the pass band width, given by:

$$\Delta f = f_h - f_\ell \tag{A4.5.6}$$

where "f_h" and "f_ℓ" are, respectively, the upper and lower frequencies of the pass band. Filters synthesized with this procedure have "f_h" and "f_ℓ" related in geometric symmetry, i.e.:

$$f_h f_\ell \stackrel{\perp}{=} 1 \tag{A4.5.7}$$

Consequently, the left side of the band pass attenuation response will have a higher attenuation slope with respect to the right side.

The electric circuit for Chebyshev or Butterworth band pass filters can be easily obtained from the simple low pass circuits indicated in Figure A4.4.1, substituting every inductor with a series resonant circuit and every capacitor with a parallel resonant circuit. An example is indicated in Figure A4.5.3a. However, such direct transformation can produce unpractical element values. To avoid such difficulty, a useful circuit variation is indicated in Figure A4.5.3b, called a "coupled parallel resonator band pass filter." These kinds of filters are synthesized in a very simple way, using dedicated tables of normalized values, called "K and Q values."

Elliptical band pass filters can also be obtained with the element transformations indicated above, but since the direct resulting circuit has a great number of inductors, it is modified to reduce this number to a minimum. The resulting unit cell can be represented as indicated in Figure A4.5.4, and for its extraction, we send the reader to the specified filter texts.

c. Band Stop Filters

These filters can be obtained applying the following variable transformation:

$$s = (1/\Delta f)\left(s_{bs} + 1/s_{bs}\right)^{-1} \tag{A4.5.8}$$

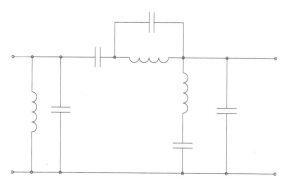

Figure A4.5.4

where "s_{bs}" is defined as in A4.5.5. In this case "f_h" and "f_ℓ" are respectively the lower frequency of the upper pass band and higher frequency of the lower pass band. Also band stop filters synthesized with this procedure have "f_h" and "f_ℓ" related in geometric symmetry. Consequently, the left side of the stop band response will have a higher attenuation slope with respect to the right side.

Using A4.5.8, the resulting band stop filter topology can be obtained from the low pass prototype substituting every inductor with a parallel resonant circuit and every capacitor with a series resonant circuit. Also in this case, elliptical band stop filter circuits can be obtained with this element transformation, but some modification is needed to reduce the number of coils.[11,12]

We want to conclude these brief notes on filter theory and practice noting that the simple variable transformation in the low pass attenuation functions always leads to manipulation of the low pass prototype tabulated values for poles and zeros, from which the desired filter can be designed.

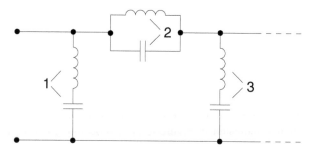

Figure A4.5.5

A4.6 FILTERS WITH LOSSY ELEMENTS

Every real reactive component has losses, as already discussed at the beginning of this appendix. These losses cause a lot of problems in filters. In fact, the attenuation shapes of typical filters given in the previous section are obtained in the ideal case, i.e., with no losses in the reactive elements.

In practice, inductors have higher losses with respect to capacitors. For example, inductors in the 10 to 100 MHz range have typical "Q" inside 100 to 50, while capacitors have "Q" values inside 1000 to 500. This means that losses are quite often due only to the inductors. For this reason, it is preferable to use filters with a minimum number of coils. In addition, coils also have higher temperature variation than capacitors. To do an example, in Figure A4.6.1 we have represented the attenuation vs. frequency in MHz of a third order Chebyshev low pass filter of the type shown in Figure A4.4.1 b, i.e., "CLC" sequence.* The solid and dashed lines, respectively, indicate the ideal case of no losses

* We mean "capacitor, inductor, capacitor" as first, second, and third element of the low pass.

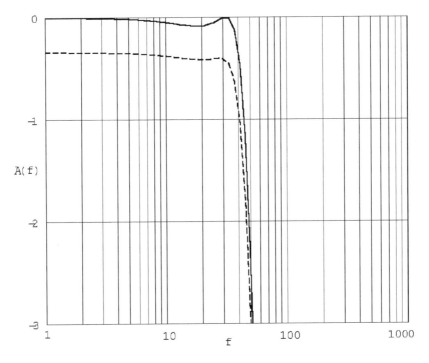

Figure A4.6.1

and the practical case of $Q_c = 100$ and $Q_\ell = 50$. Figure A4.6.2 instead represents the attenuation vs. frequency in MHz of a third order Chebyshev low pass filter of the type shown in Figure A4.4.1 a, i.e., "LCL" sequence. With solid and dashed lines we have respectively indicated the ideal case of no losses and the practical case of $Q_c = 100$ and $Q_\ell = 50$. Note the case of higher disequalization of attenuation in the pass band of this last case with respect to the "CLC" sequence of Figure A4.6.1.

In general, losses give the following effects for a filter:

1. increase attenuation in the pass band
2. rounded transition points between pass band and transition band
3. finite values at the attenuation peaks*

Points 1 and 2 above are more and more evident when decreasing the pass band and/or increasing the filter order.

The effect of losses are even more evident in band pass filters. For example, Figure A4.6.3 reports the attenuation vs. frequency in MHz of a third order Chebyshev coupled resonator band pass filter. Note the effect of $Q_c = 100$ and $Q_\ell = 50$ indicated with the dashed line with respect to the ideal case indicated with the solid line.

In practice it has been verified that the minimum "Q" that reactive elements need to have depends on the center frequency "fc" and bandwidth "Δf" of the band pass or band stop filter, through the following expression:

$$Q \geq 20 f_c / \Delta f \qquad\qquad (A4.6.1)$$

For any practical filter the previous expression is easily satisfied by capacitors, while for inductors Equation A4.6.1 is seldom verified.

---

* For example, the values of attenuation peaks for Cauer filters are theoretically infinite.

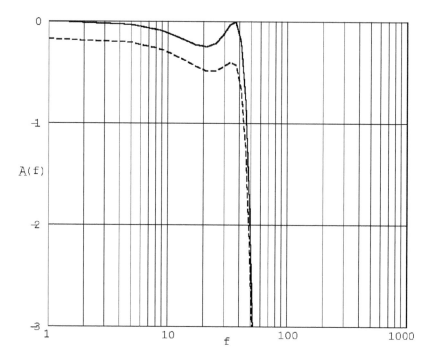

Figure A4.6.2

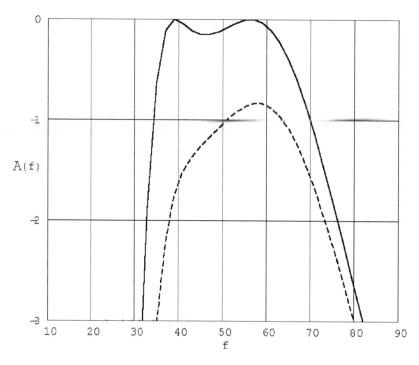

Figure A4.6.3

REFERENCES

1. *The Circuits and Filters Handbook*, IEEE and CRC Press, Boca Raton, 1995.
2. E. A. Guillemin, *Synthesis of Passive Networks*, John Wiley & Sons, New York, 1964.
3. L. Weinberg, *Network Analysis and Synthesis*, McGraw Hill, New York, 1962.
4. G. Martinelli, Sintesi delle reti elettriche. Bipoli, Siderea, 1986.
5. G. C. Themes, S. K. Mittra, *Modern Filter Theory and Design*, John Wiley & Sons, New York, 1973.
6. A. I. Zverev, *Handbook of Filter Synthesis*, John Wiley & Sons, New York, 1967.
7. A. I. Zverev, *Handbook of Filter Synthesis*, John Wiley & Sons, New York, 1967.
8. R. Saal, *Handbook of Filter Design*, AEG Telefunken, 1978.
9. A. I. Zverev, *Handbook of Filter Synthesis*, John Wiley & Sons, New York, 1967.
10. R. Saal, *Handbook of Filter Design*, AEG Telefunken, 1978.
11. A. I. Zverev, *Handbook of Filter Synthesis*, John Wiley & Sons, New York, 1967.
12. R. Saal, *Handbook of Filter Design*, AEG Telefunken, 1978.

Charges, Currents, Magnetic Fields, and Forces

A5.1 INTRODUCTION

In many chapters of this book we discuss ferrimagnetic devices. Appendix A7 covers the behavior of ferrite inside magnetic fields. For a proper reading of that appendix, it is important to know some fundamental concepts of magnetism and atomic physics. This appendix will discuss the fundamental elements of atomic physics, required for the reader to properly understand Appendix A7. Unless otherwise stated, the MKSA reference unit system will be used.

It is assumed the reader knows the topics which we are going to discuss, so that it is only necessary to remind the reader of the most important relationships. In addition, since our text is not a book on physics, many expressions will be given without proof. Deeper insight into this branch of physics can be obtained from the books indicated in references.[1,2,3]

In Appendix A6 we will study the foundations of magnetism.

A5.2 SOME IMPORTANT RELATIONSHIPS OF CLASSIC MECHANICS

Classic mechanics is that branch of physics that deals with the motion and forces on bodies. The formulas obtained with these mechanics are continuous and deterministic, i.e., at any time it is possible to determine the speed, position, and mass of a body. In the following discussion, the reference system is considered "inertial," i.e., completely fixed. In addition, the dimensions of the quantity we will define are assumed to be known by the reader and, in general, they will not be given.

We will now discuss concepts and formulas that will be useful in this context. Let us begin with the three principles of dynamics, according to I. Newton.*

a. First Principle of Dynamics

Every body not subjected to forces stays in its status, be it quiet or rectilinear, uniform motion.

b. Second Principle of Dynamics

The acceleration "$\underline{a}$" of a body in motion is due to a force "F" so that $\underline{F} = m\underline{a}$, where "m" is the mass of the body. "a" is measured in m/sec² "F" in newtons, and "m" in kilos.

c. Third Principle of Dynamics

The mutual forces between two bodies have equal intensities but opposite directions. This principle is also known as the "action and reaction principle."

Now we will give some general definitions that will be successively useful.

* Isaac Newton, English physicist, born in Woolsthorpe in 1642 and died in Kensington in 1727.

1. Work of a Force

Given a force "F," the work "L_{12}" to move a body from position "P_1" to position "P_2" is given by:

$$L_{12} = \int_1^2 \underline{F} \cdot d\underline{\ell} \tag{A5.2.1}$$

2. Momentum of Inertia of a Body with Respect to an Axis

Given a body of mass "m," homogeneous with constant density, its "momentum of inertia $\Im$" with respect to an axis passing through the body is:

$$\Im = \int_1 r^2 dm \tag{A5.2.2}$$

where "r" is the distance from "a" of the infinitesimal mass "dm." Of course, it is required that the integral in A5.2.2 be finite. In the case of a sphere with radius "r" and mass "m," its momentum of inertia with respect to an axis passing through its center is

$$\Im = 2mr^2/5 \tag{A5.2.3}$$

In the case of a torus of mass "m" and mean radius "r," whose section has negligible dimension compared to "r," its "$\Im$" with respect to an axis orthogonal to the torus plane and passing for its center is:

$$\Im = mr^2 \tag{A5.2.4}$$

The previous expression is also valid for a mass "m" moving in a circular orbit of radius "r" when the mass has negligible dimensions compared to "r." This is the case of an electron moving around its atomic nucleus.

3. "Vector Momentum" of a Vector

Define a point "o" in the space, the origin of a reference system. Also define a point "p" where a vector "$\underline{f}$" presents its action. The "vector momentum $\underline{M}$" of the vector "$\underline{f}$" with respect to "o" the quantity is defined as:

$$\underline{M} \stackrel{\perp}{=} \underline{r} \otimes \underline{f} \tag{A5.2.5}$$

where "r" is the distance between "o" and "p." By extension, if the point "o" is on an axis the previous result is called, "axial vector momentum."

4. Couple and Momentum of a Couple

A "Couple" is defined as the set of two parallel vectors "$\underline{f_1}$" and "$\underline{f_2}$," with same modulus but oriented in different directions, working on two points "P_1" and "P_2" so that the segment "P_1P_2" is not parallel to "$\underline{f_1}$."* Figure A5.2.1 represents the situation.

---

* Since "$\underline{f_1}$" is parallel to "$\underline{f_2}$" it is enough to refer to one vector.

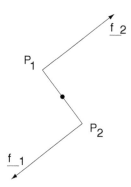

Figure 5.2.1

The momentum of a couple is also defined, and since "$\underline{f}_1$" and "$\underline{f}_2$" have the same modulus only a vector can be considered. To this purpose, we define an orientation for the segment "P_1P_2," i.e., the segment "P_1P_2" becomes a vector "$\underline{r}$," then the vector momentum "$\underline{M}$" of the couple is defined as:

$$\underline{M} \overset{\perp}{=} \underline{r} \otimes \underline{f}_1 \equiv -\underline{r} \otimes \underline{f}_2$$

5. Vector "Quantity of Motion" and Vector "Angular Orbital Momentum of the Quantity of Motion"

A body of mass "m" with speed "$\underline{v}$" possesses a vector "quantity of motion p" given by:

$$\underline{p} = m\underline{v} \tag{A5.2.6}$$

If we derive the time "t" with respect to the previous equation we have:

$$d\underline{p}/dt = \underline{v}dm/dt + md\underline{v}/dt \tag{A5.2.7}$$

Then, not considering the variation of mass with time* we have:

$$\underline{F} = d\underline{p}/dt \tag{A5.2.8}$$

which is a different representation of the second principle of dynamics. With point "o" defined as the origin of the reference system, this body possesses a momentum "L" of quantity of motion given by:

$$\underline{L} = \underline{r} \otimes \underline{p} \tag{A5.2.9}$$

Note that "$\underline{L}$" is orthogonal to the plane formed by "$\underline{r}$" and "$\underline{p}$." If the body moves with a circular uniform motion with tangential speed "v" from the previous equation we have the modulus:

$$L = mvr \quad \left[Kg.m^2/sec\right] \tag{A5.2.10}$$

---

* This text does not aim to be a text on physics. Consequently, the relativistic effects concerning "dm/dt" are not evaluated. In general, in this appendix, and in the following A6, only the relativistic effects strictly necessary to Appendix A7 will be treated.

In this case "L" is called the "angular orbital momentum of the quantity of motion," and it is directed orthogonally and at the center of the plane of the body orbit. In the study of circular motion in orbits with radius "r," the angular speed "ω" is used, defined as:

$$v = \omega r \tag{A5.2.11}$$

whose vector "$\underline{\omega}$" is orthogonal and at the center of the plane of the body orbit and directed where the body appears to move counterclockwise. The acceleration "a" is:

$$a = v^2/r = \omega^2 r \tag{A5.2.12}$$

Using the definition of "ω," A5.2.9 can be written as:

$$\underline{L} = mr^2\underline{\omega} \tag{A5.2.13}$$

Note as "mr^2" is the momentum of inertia of the mass "m" in circular uniform motion, so A5.2.13 can be rewritten as:

$$\underline{L} = \Im\underline{\omega} \tag{A5.2.14}$$

This expression is quite general, and can be applied to a body rotating on axis passing through it. In this case we can define a new vector, as defined in the next point.

6. Vector "Angular Intrinsic Momentum of the Quantity of Motion"

A body of mass "m" rotating on an axis passing through its center of mass possesses an "angular intrinsic momentum of the quantity of motion,"* with respect to this axis, called "spin" and indicated with "S."** For the case of a sphere of radius "r" rotating with angular speed "ω" we have:

$$\underline{S} = 2mr^2\,\underline{\omega}/5 \tag{A5.2.15}$$

7. The Theorem of the Quantity of Motion

Let us consider an origin "o" of a fixed reference system. The vector "$\underline{r}$" with origin in "o," points to a body of mass "m" with speed "$\underline{v}$." Using A5.2.8, multiplied at left vectorally by "r," we have:

$$\underline{r} \otimes \underline{F} \overset{\perp}{=} \underline{M} \equiv \underline{r} \otimes d\underline{p}/dt \tag{A5.2.16}$$

Deriving the time with respect to A5.2.9 we have:

$$d\underline{L}/dt = (d\underline{r}/dt) \otimes \underline{p} + \underline{r} \otimes d\underline{p}/dt \tag{A5.2.17}$$

Since $\underline{v} = d\underline{r}/dt$, the first term in A5.2.17 is zero, and we can write:

$$\underline{M} = d\underline{L}/dt \tag{A5.2.18}$$

---

* Note how with respect to the case studied in point 5, now the word "orbital" does not appear.
** "Spin" is used only when discussing the quantity of motion; otherwise "S" is also called "spin quantity of motion."

$$d\underline{B} = \frac{\mu_0 i}{4\pi r^3} d\underline{\ell} \otimes \underline{r}$$

(A5.5.1)

Note that "$d\underline{B}$" is always orthogonal to the plane determined by "$d\underline{\ell}$" and "$\underline{r}$." This means that if "r" moves on a circle, "B" does the same resulting in the lines of "B" being continuous, closed lines, orthogonal to "$d\ell$." As a consequence, the flux of "B" in a closed surface "S" is zero, since every induction line of "B" that enters "S" must also exit from "S." Integrating the A5.5.1 on the whole closed conductor, we can evaluate the whole magnetic induction in a point "P" located by "r," provided that r >> d.

2. Biot* e Savart** Expression

These two French physicists experimentally found one expression for "B" in the case of a very long straight wire supporting a current "i." Given a point "P" whose orthogonal distance from the wire is "r," then in "P" the magnetic induction "B" is:

$$B = \mu i / 2\pi r$$

(A5.5.2)

where "μ" is the absolute permeability of the medium surrounding the wire.

It is possible to show that applying the Laplace Equation A5.5.1 to this particular case of straight conductor leads to the Biot e Savart expression.

3. Ampere's*** Expression

Let us assume a region exists where "n" wires are carrying currents, and consider a closed loop "c" surrounding these wires. Ampere's law states that:

$$\oint_c \underline{B} \cdot d\underline{\ell} = \mu \sum_{k=1}^{k=n} i_k$$

(A5.5.3)

The currents "i_k" need to be evaluated according to a defined direction assumed as positive for the closed loop "c." In other words, if a current generates a magnetic induction whose line forces are concordant to the defined positive direction, then it must be evaluated with the positive sign, otherwise it will be considered with a negative sign. Of course, this expression can also be obtained by the application of the Laplace equation to our particular case, but the application of A5.5.3 can sometimes simplify the calculations. An example is the evaluation of "B" inside a solenoid.

Ampere's law can be transformed in a differential expression using the Stokes theorem.**** Let us consider the surface "S" whose perimeter is the closed line "c" used at the first member of Equation A5.5.3. Associating with each current "i_k" its current density "J_k," with $[J_k]$ = Amp/m^2, we can write:

$$\sum_{k=1}^{k=n} i_k = \int_S \underline{J} \cdot \underline{n} dS$$

(A5.5.4)

---

* Jean Baptiste Biot, French physicist, born in Paris in 1797 and died there in 1862.
** Felix Savart, French scientist, born in Mezieres an Auge in 1791 and died in Paris in 1841.
*** Andrè Marie Ampere, French scientist, born in Lyon in 1775 and died in Marseilles in 1836.
**** See Appendix A8 for the Stokes theorem.

since J ≠ 0 only on the surface intersection between wires and "S." "$\underline{n}$" is the normal to the surface "S" directed where the path on "c" is seen counterclockwise. Then A5.5.3 is written as:

$$\oint_c \underline{B} \bullet d\underline{\ell} = \mu \int_S \underline{J} \bullet \underline{n} dS \tag{A5.5.5}$$

Applying the Stokes* theorem to Equation A5.5.5 we have:

$$\int_S \underline{\nabla} \otimes \underline{B} \bullet \underline{n} dS = \mu \int_S \underline{J} \bullet \underline{n} dS \rightarrow \int_S \left(\underline{\nabla} \otimes \underline{B} - \mu \underline{J} \right) \bullet \underline{n} dS = 0 \tag{A5.5.6}$$

from which it must be:

$$\nabla \otimes \underline{B} = \mu \underline{J} \tag{A5.5.7}$$

The previous equation is also called the "differential Ampere's law."

A5.6 TWO IMPORTANT RELATIONSHIPS OF QUANTUM MECHANICS

Quantum mechanics states that many of the classical mechanics equations given before cannot be evaluated as continuous when applied to the atomic structure. In fact the values that they can attain are a multiple of a minimum quantity, called "quanta" and are given by:

$$\hbar \overset{\perp}{=} h/2\pi \tag{A5.6.1}$$

where "h" is Planck's** constant, equal to $6.62*10^{-34}$ Joule.sec.

We will now give some important relationships of quantum mechanics that will be used later and in the following appendices.

1. Indetermination Principle

This principle was formulated by W. Heisenberg,*** and states that for a body the indeterminations "Δr" in the exact location "r" and "Δp" for its quantity of motion cannot be reduced to zero, but instead:

$$\Delta p \Delta r \geq h \tag{A5.6.2}$$

Due to the small value of the Planck's constant, for the bodies, speeds, and distances that one uses, the previous equation has little utility, while for the atomic structure the Heisenberg's indetermination principle is of extraordinary importance.

2. The Energy-Frequency Relationship

This principle was formulated by A. Einstein,**** and states that the energy quantum "$\mathcal{E}$" of a radiation with frequency "f" is given by:

* George Gabriel Stokes, English mathematician, born in 1819, died in Cambridge in 1903.
** Max Planck, German physicist, born in Kiel in 1858 and died in Gottinga in 1947.
*** Werner Heisenberg, German physicist, born in Wuerzburg in 1901 and died in Monaco in 1976.
**** Albert Einstein, German physicist, born in Ulm in 1879 and died in Princeton in 1955.

A new quantic number was then introduced, just to quantize "L" when the atom is inside an induction magnetic field "B." In this case "L" is written as "L_B" and set:

$$L_B = m_L \hbar \tag{A5.7.11}$$

where "m_L" is called the "magnetic quantic number," given by:

$$m_L = 0, \pm 1, \pm 2, \ldots \pm \ell_e \tag{A5.7.12}$$

Due to other experiments, the electron was supposed to rotate around an axis passing through its center with angular speed "ω_S" and a "spin quantity of motion momentum, S" and a "spin magnetic momentum, μ_S" were introduced. "S" was then quantized according to a number "m_S" called the "spin magnetic quantic number," of values:

$$m_S = \pm 1/2 \tag{A5.7.13}$$

The ratio between the spin magnetic momentum and the spin quantity of motion momentum "S" is called "spin gyromagnetic ratio, $\gamma_{S,}$"* i.e.:

$$\gamma_S \overset{\perp}{=} \mu_S/S = -e/m \tag{A5.7.14}$$

and its value was determined by the physicist P. Dirac.** If for the determination of magnetic moments "H" is used instead of "B," in the previous equation there would appear "μ_0" as multiplier at last member.

So, when the atom is inside an induction magnetic field "B," then "S" is written as "S_B," and set:

$$S_B = m_S \hbar \tag{A5.7.15}$$

It is interesting to observe that "$\underline{L}$," "$\underline{\omega_e}$" and "$\underline{\mu_L}$" are oriented in the same direction and pass through the orbit center, while "$\underline{S}$" and "$\underline{\mu_S}$" are oriented in opposite directions and pass through the electron center. Due to the principle of minimum energy of natural systems, also according to A5.4.7, "$\underline{\mu_L}$" and "$\underline{\mu_s}$" have equal direction.

When dealing with the hydrogen atom, which has only one electron, a total magnetic momentum "$\underline{\mu_T}$" and a total quantity of motion momentum "T" are defined as:

$$\underline{\mu_T} \overset{\perp}{=} \underline{\mu_L} + \underline{\mu_S} \tag{A5.7.16}$$

$$\underline{T} \overset{\perp}{=} \underline{L} + \underline{S} \tag{A5.7.17}$$

Caution needs to be used when applying the previous two relationships with atoms with more than one electron, since it is not certain that "$\underline{\mu_L}$" and "$\underline{\mu_s}$" are parallel, and consequently "$\underline{\mu_T}$" cannot be parallel to "$\underline{T}$."

We conclude this brief introduction to atom theory giving values of mass and charge for the quiet electron and some values for the hydrogen atom. Note that $1\text{Å} = 10^{-7}$ mm:

---

* The "spin gyromagnetic ratio" is also simply called "gyromagnetic ratio."
** Paul Dirac, English physicist, born in Bristol in 1902 and died in Tallahassee, FL, U.S., in 1984.

Electron

Mass:	$9.1066*10^{-28}$ grams
Charge:	$1.6*10^{-19}$ Coulomb

Hydrogen Atom

Quantic number "n"	Mean orbit radius (Å)	$\mathcal{E}_t/10^{-23}$ erg
1	0.53	−217.3
2	2.12	−54.3
3	4.77	−24.2

A5.8 THE ATOM STRUCTURE IN QUANTUM MECHANICS

Some of the first quantic expressions developed by Bohr and Sommerfeld have been successively modified to improve the accuracy of the results of experiments on excited atoms. The quantity of motion momentum "$\underline{L}$" associated with the electron elliptical orbit is quantized according to:

$$L = \left(\ell^2 + \ell\right)^{0.5} \hbar \qquad \text{with } \ell = 0,1,2... \qquad (A5.8.1)$$

Note it is different from the original expression of Bohr given in A5.7.1. The number "ℓ" is called the "orbital quantic number."

Using Equation A5.8.1 and the orbital gyromagnetic ratio "γ_L" given by A5.7.10 we obtain the new expression of the orbital magnetic momentum "μ_L:"

$$\mu_L \doteq \gamma_L \left(\ell^2 + \ell\right)^{0.5} \hbar \qquad (A5.8.2)$$

The factor "$|\gamma_L|\hbar$" is called "Bohr's magneton" and is indicated with "μ_b," i.e.:

$$\left|\gamma_L\right| \hbar = e\hbar/2m \doteq \mu_b \qquad (A5.8.3)$$

with $\mu_b = 0.927*10^{-23}$ and dimensions [A*m² ≡ Joule*m²/Weber].*

With the introduction of the orbital quantic number given in A5.8.1 the magnetic quantic number "m_L" given by A5.7.11 now assumes another set of values, given by:

$$m_L = 0, \pm1, \pm2... \pm \ell \qquad (A5.8.4)$$

and it is interpreted as an index that represents the projections of "L" along the direction of the applied magnetic field. These projections "$\underline{L}_B$" are given by:

$$\underline{L}_B = m_L \hbar \underline{b}_0 \qquad (A5.8.5)$$

where "$\underline{b}_0$" is the versor of "$\underline{B}$." A case for $\ell = 3$ is represented in Figure A5.8.1.

The spin quantity of motion momentum is still quantized along the direction of the applied magnetic field according to A5.7.14, and to "S" is associated a modulus given by:

---

* Wilhelm Eduard Weber, German physicist, born in Wittenberg 1804 and died in Gottingen in 1891.

It is important at this point refer to an important principle verified in practice and formulated by the physicist W. Pauli,* which states that two electrons with the same spin direction cannot exist in an orbit. Due to this principle, for a complete atom** there is no net contribution of spin magnetic momentum, and of course no net contribution of spin quantity of motion momentum.

Concerning the total magnetic momentum "μ_T" defined in A5.7.16, for atoms with more than one electron, it is possible to define a projection "μ_{TT}" of "μ_T" along "$\underline{T}$" since it is not assured that "μ_T" is parallel to "$\underline{T}$." This projection is:

$$\underline{\mu}_{TT} = g\gamma_L \underline{T} \tag{A5.8.12}$$

where "g" is the Lande*** factor given by:

$$g = 1 + \frac{j(j+1) + s(s+1) - \ell(\ell+1)}{2j(j+1)} \tag{A5.8.13}$$

In some texts "μ_{TT}" is simply indicated with "μ_T," but we prefer to use the double subscript to emphasize the concept of projection. The quantity "total gyromagnetic ratio, γ_T" is defined as:

$$\gamma_T = g\gamma_L \tag{A5.8.14}$$

Using A5.8.3 and A5.8.10 the modulus of A5.8.12 becomes:

$$\mu_{TT} = -g\mu_b\left(j^2 + j\right)^{0.5} \hbar \tag{A5.8.15}$$

Similar to "L" and "S," "T" also assumes discrete projections along the direction of the applied magnetic field, with a picture similar to that shown in Figure A5.8.1. So, a new quantic number "m_T" is introduced, called the "total magnetic quantum number" given by:

$$m_T = 0, \pm 1, \pm 2 \ldots \pm j \tag{A5.8.16}$$

and it is interpreted as an index that represents the components "$\underline{T}_B$" of "T" along the direction of the applied magnetic field, given by:

$$\underline{T}_B = m_T \hbar \underline{b}_0 \tag{A5.8.17}$$

Using the total gyromagnetic ratio "γ_T," the projections "μ_{TB}" of "μ_T" along "$\underline{B}$," are given by:

$$\underline{\mu}_{TB} = \gamma_T \underline{T}_B \overset{\perp}{=} -m_T g\mu_b \underline{b}_0 \tag{A5.8.18}$$

In general, in an atom "μ_T" can be due to both "μ_L" and "μ_S" or only one of them. In the cases when "μ_T" is due to "μ_S" it is interesting to show that the previous equation evaluated for the case of a single electron gives the same value of Equation A5.8.9. In fact, in this case we have $\ell = 0$ and from A5.8.11 it follows that $j \equiv s$ and from A5.8.13 we have $g = 2$. Since we only have one electron, then $m_T \equiv m_s$, and the previous equation is coincident with A5.8.9.

* Wolfgang Pauli, Swiss physicist, born in Vienna in 1900 and died in Zurich in 1958.
** With "complete atom" it is intended an atom with every orbit occupied with two electrons.
*** Alfred Lande, German physicist, born in Elberfeld in 1888 and died in Columbus in 1976.

Since the function "Φ" is representative of the probability of finding the electron in the space, to the integral of "Φ²" extended to an infinite volume "V" is associated the certainty of finding the electron, i.e.:

$$\int_V \Phi^2 \, dV = 1 \qquad (A5.10.6)$$

The solutions of the Schroedinger equation, which also satisfy the previous equation, are chosen as the solutions of the wave mechanics applied to atomic structure. These solutions are called "orbitals" and are individualized by four parameters, as in the case of quantum mechanics. For this reason the four parameters are associated with the same four letters used in quantum mechanics, i.e., "n," "ℓ," "m_ℓ" and "m_s," with values:

$$n = 1, 2, 3 \ldots \qquad (A5.10.7)$$

$$\ell = 0, 1, 2 \ldots (n-1) \qquad (A5.10.8)$$

$$m_\ell = 0, \pm 1, \pm 2, \ldots \pm \ell \qquad (A5.10.9)$$

$$m_s = \pm 1/2 \qquad (A5.10.10)$$

Since the fourth number is known *a priori* in its value, it is usually not employed in the notation of the orbital, unless two electrons are on the same orbital. So, each orbital is generally indicated with "$\Phi_{n,\ell,m\ell}$," with a specified value for the three subscripts. It is possible to show that the index "n" quantizes the electron energy, as Bohr did with Equation A5.7.6, while the index "ℓ" strongly affects the orbital shape. Depending on the values of "ℓ," the orbitals are named according to the following Table A5.10.1:

Table A5.10.1

ℓ	Orbital name
0	s (**sharp**)
1	p (**principal**)
2	d (**diffuse**)
3	f (**fundamental**)

and from $\ell \geq 4$ the orbitals are called as the letters of the alphabet after "f," i.e., "g," "h,".... Orbitals that have the same values of "n" and "ℓ," i.e., they only differ for "m_ℓ" and/or "m_s," are called "degenerate," since they have practically the same energy. For any value of "n" a number of "n^2" orbitals exists, and since a maximum of two electrons are possible for each orbital, then the maximum number "Z" of electrons is $Z = 2n^2$. A list of some orbitals are indicated in the following Table A5.10.2 where we have indicated the degenerate orbitals with underlined symbols.

When some possible degenerate orbitals have no electrons, the electrons that enter first occupy all these orbitals with same direction magnetic spin, of course one electron for one orbital according to Pauli's principle discussed before. When other electrons enter in these orbitals, they complete each orbital with their spin magnetic momentum in the opposite direction to the spin of the already present electron. This principle is called the "maximum multiplicity principle," according to the physicist F. Hund.* For example, the oxygen atom has 8 electrons: two are in "1s" orbital, two in

---

* Friedrich Hund, German physicist, born in Karlsruhe in 1896 and died in Gottingen in 1997.

Table A5.10.2

n	ℓ	m_ℓ	Notation	Name
0	0	0	Not possible	
1	0	0	Φ_{100}	1s
2	0	0	Φ_{200}	2s
	1	−1	Φ_{21-1}	2p
	1	0	Φ_{210}	2p
	1	1	Φ_{211}	2p
3	0	0	Φ_{300}	3s
	1	−1	Φ_{31-1}	3p
	1	0	Φ_{310}	3p
	1	1	Φ_{311}	3p
	2	−2	Φ_{32-2}	3d
	2	−1	Φ_{32-1}	3d
	2	0	Φ_{320}	3d
	2	1	Φ_{321}	3d
	2	2	Φ_{322}	3d
4	0	0	Φ_{400}	4s
	1	−1	Φ_{41-1}	4p
	1	0	Φ_{410}	4p
	1	1	Φ_{411}	4p
	2	−2	Φ_{42-2}	4d
	2	−1	Φ_{42-1}	4d
	2	0	Φ_{420}	4d
	2	1	Φ_{421}	4d
	2	2	Φ_{422}	4d
	3	−3	Φ_{43-3}	4f
	3	−2	Φ_{43-2}	4f
	3	−1	Φ_{43-1}	4f
	3	0	Φ_{430}	4f
	3	1	Φ_{431}	4f
	3	2	Φ_{432}	4f
	3	3	Φ_{433}	4f

"2s," the remaining 4 electrons are 3 in each "2p" degenerate orbitals, and the remaining completes the first degenerate orbital.

Different from classical or quantum mechanics, in wave mechanics electron orbits are not only spherical or elliptical. In Figures A5.10.2 through A5.10.4 we have represented some orbitals. The areas with more concentrated dots indicate where the probability of finding electrons is greatest.

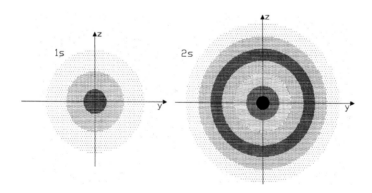

Figure A5.10.2

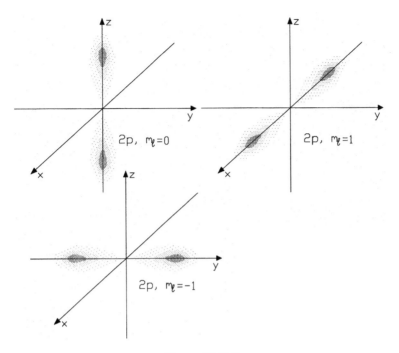

Figure A5.10.3

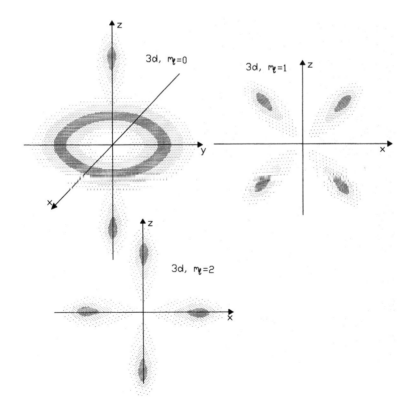

Figure A5.10.4

Orbitals "s" always have spherical symmetry, and in Figure A5.10.2 we have represented for simplicity a cross-sectional view.

In contrast, "p" orbitals can be approximated with a ring shape, each one lying on a plane, and some transverse sections are represented in Figure A5.10.3.

"d" orbitals have more complicated shapes, in general combinations of rings oriented in more planes.

Of course, all the shapes depend on the values of the letters "n," "ℓ," and "m_ℓ."

REFERENCES

1. F. Bitter, H. A. Medicus, *Fields and Particles*, Elsevier, Amsterdam, 1973.
2. R. P. Feynman, R. B. Leighton, M. Sands, *The Feynman Lectures on Physics*, Addison-Wesley, New York, 1963.
3. D. Sette, *Lezioni di fisica*, Veschi, 1963.
4. L. Landau, E. Lifshitz: "Teorija kondensirovannogo sostojanija"(in Russian). This book can be found translated in many languages. For Italian language see: Fisica statistica: teoria dello stato condensato, Editori riuniti, edizioni MIR, p. 345, 1981.

The Magnetic Properties of Materials

A6.1 INTRODUCTION

The study of the magnetic properties of materials and interactions of magnetic fields with materials requires a lot of knowledge of physics and chemistry. It is not possible to concentrate all these science branches in only a few pages, and this is not a physics or chemistry text. Nevertheless, in many chapters of our book we discuss ferrimagnetic devices, particularly in Appendix A7 where we study the behavior of ferrite inside magnetic fields. For a proper reading of that appendix it is important to know some fundamental concepts of magnetism and atomic physics. This last topic has been discussed in Appendix A5, while in this appendix we will discuss the fundamental elements concerning the macroscopic physical properties of materials inside magnetic fields. We will report only the relationships required to introduce the reader to properly understanding Appendix A7, just to make a bridge between Appendix A5 and A7. It is assumed that the reader knows the topics that we are going to discuss, so that it is only necessary to remind him or her of the relationships here indicated. Deeper insight into this branch of physics can be obtained from the books indicated in the references.[1,2,3]

We will concentrate our study on ferromagnetic and ferrimagnetic devices, and we will show how they are closely related. The other materials classifications and characteristics from a magnetic point of view will be briefly discussed, while deeper insights can be found in specific texts.[4]

A6.2 FUNDAMENTAL RELATIONSHIPS FOR STATIC MAGNETIC FIELDS AND MATERIALS

Every time in a free space pervaded with a static induction magnetic field "B_0" a material is inserted, a new magnetic situation arises. In the free space, the magnetic situation is completely defined with the two equations:

$$\underline{\nabla} \bullet \underline{B}_0 = 0 \qquad \text{(A6.2.1)}$$

$$\underline{\nabla} \otimes \underline{B}_0 = \mu_0 \underline{J} \qquad \text{(A6.2.2)}$$

i.e., one of the static Maxwell's equations* and the Ampere's law.** When a homogeneous body is inserted in this space, we also have to evaluate the current density "J_a" representing the whole atomic current density that generates the body magnetization "$\mathcal{M}$."*** So, Equation A6.2.2 becomes:

---

* See Appendix A2 for Maxwell's equations definitions.
** See Appendix A5 for Ampere's law.
*** See Appendix A5 for magnetization definition.

called respectively "relative" and "absolute" permeability. Inserting Equation A6.2.10 into A6.2.8 and assuming the medium is also isotropic we can write:

$$\underline{\nabla} \otimes \underline{B} = \mu \underline{J} \qquad (A6.2.12)$$

and since the induction magnetic field always has null divergence:

$$\underline{\nabla} \bullet \underline{B} = 0 \qquad (A6.2.13)$$

The two previous equations are completely similar to A6.2.1 and A6.2.2 since the characteristics of the material inserted there, "B_0", are included inside the definition of "B" or "μ."

A6.3 THE DEFINITIONS OF MATERIALS IN MAGNETISM

Materials are likely to be defined according their electrical properties, also from the magnetic point of view they can be divided mainly into four groups, i.e.: diamagnetic, paramagnetic, ferromagnetic, or ferrimagnetic materials. The characterization is done according to the behavior of "χ," as we will now define.

a. Diamagnetic Materials

Diamagnetic materials can be considered as the magnetic equivalent of the dielectric counterparts. These materials present a small value of "χ," typically inside the range -10^{-4} to -10^{-3}, and consequently their relative permeability, using Equation A6.2.11, is nearly closed to one. Using Equation A6.2.9, the magnetization "$\mathcal{M}$" is also -10^{-4} to -10^{-3} times smaller than "H." Note that "χ" is negative, meaning that "$\mathcal{M}$" is in the opposite direction to the applied field. Diamagnetic materials also have the characteristic that "χ" is nearly independent of temperature, which is a unique characteristic, dependent on the fact that they are quite insensitive to magnetic fields. Some materials belonging to the diamagnetic family are copper (Cu), silver (Ag), water (H_2O), bismuth (Bi), lead (Pb).

b. Paramagnetic Materials

These materials have positive values of susceptibility that are nearly ten times those of the paramagnetic counterparts, i.e., in the range 10^{-3} to 10^{-2}. Paramagnetic materials have "χ" moving in temperature nearly according to Curie's* law:

$$\chi = C/T \qquad (A6.3.1)$$

The constant "C," called "Curie's constant," is measured experimentally for each material, so as to match experimental values to Equation A6.3.1. For temperatures closer to T = 0 Curie's law doesn't accurately represent the behavior of "χ" since the susceptibility tends to a constant value. Typical paramagnetic materials are platinum (Pt), aluminium (Al), magnesium (Mg), and air.

c. Ferromagnetic Materials

For these materials the susceptibility reaches the highest value, typically some units of 10^4. Together with the ferrimagnetic materials that we will discuss later, ferromagnetic materials have

---

* Pierre Curie, French physicist, born in Paris in 1859 and died there in 1906.

the ability to retain memory of previous magnetization with respect to the magnetic experiment we make, but only if the antecedent applied magnetic field has been stronger than a value.* These high values of "χ" are obtained for operating temperatures lower than a precise one "T_{cf}," called the "ferromagnetic Curie's temperature." For temperatures higher than "T_{cf}" ferromagnetic materials become paramagnetic. A lot of ferromagnetic materials are good conductors, different from their ferrimagnetic counterparts, which are good insulators. Typical ferromagnetic materials are iron (Fe) with $T_{cf} = 775$ °C, nickel (Ni) with $T_{cf} = 360$ °C, cobalt (Co) with $T_{cf} = 1100$ °C.

d. Antiferromagetic Materials

To this family belong all those materials that have equal macroscopic characteristics of ferromagnetic materials, like molecular structure, but have no ferromagnetic properties, and can be evaluated as paramagnetic. Typical antiferromagnetic materials are manganese (Mn) and chrome (Cr), also if bonded with other materials, the resulting one is ferromagnetic.

e. Ferrimagnetic Materials

Ferrimagnetic materials take their name from the chemical compound of iron (Fe) with other substances like oxygen (O), magnesium (Mg), manganese (Mn), and nickel (Ni). The first discovered natural ferrimagnetic material was called "magnetite" whose chemical composition is Fe_3O_4. A generic ferrimagnetic material is usually called "ferrite," also if ferrite is only a particular ferrimagnetic material. Unless otherwise stated, we also use this convention. One characteristic common to all ferrites is the fact that they are good insulators, typically 10^{-6} times the conductibility of ferromagnetic materials. But in any case, their susceptibility is very high, typically near some thousand.

Due to their wide use in high frequency signal conditioning, the following Appendix A7 is completely dedicated to theory and applications of ferrite, while in this appendix we will try to define the physical reasons that permit the material to be ferro-ferrimagnetic.

A6.4 STATISTICS FUNCTIONS FOR PARTICLE DISTRIBUTION IN ENERGY LEVELS

The oldest magnetic materials to be studied are the ferromagnetic ones. Of course, in RF devices these materials have no comparable use with that of ferrites, but the study of ferromagnetic materials has guided researchers to find a theory for the ferrimagnetic ones. In nature all ferromagnetic materials are metals, i.e., good conductors. Since in this appendix we study the macroscopic characteristics of materials inside a magnetic field, in this case we have to introduce what is a metal.

A metal is composed of a lot of small geometric cells, like cubes,** where atoms are placed at the vertices. A structure of this kind is in general said to be a crystal. From an electrical point of view, one of the first interpretations of metals was made by the physicist P. Drude.*** He supposed the metal to be composed of an electron's gas confined in the metal, where atoms share electrons between them. Until today Drude's interpretation of metals is still employed. For most metals used in electrical circuits like copper (Cu), silver (Ag), gold (Au), aluminium (Al),**** every atom typically shares 2 electrons with the other atoms. The electrical and magnetic macroscopic properties of such a system, composed of many particles of the same kind, have to be studied

---

* We will return later to this topic.
** Not all metals have such a simple, unitary geometric cell.
*** Paul Drude, German physicist, born in Brunswick in 1863 and died in Berlin in 1906.
**** Such metals are also used to make networks and devices using transmission lines.

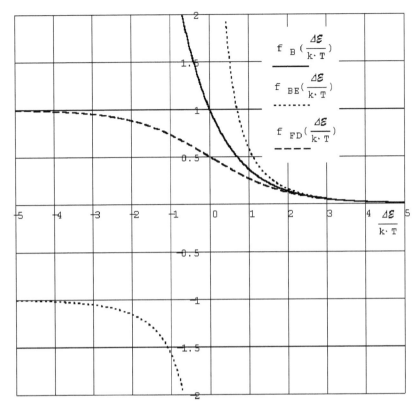

Figure A6.4.1

levels is lower than the probability for low energy levels. So the principle that at high energy levels we have a low probability of finding particles is respected for all statistics. Conversely, at very low temperature, typically some Kelvin degrees,* some gases liquefy and A6.4.10 is not valid. Consequently, B-function is not accurate to predict particles' distribution, while it has been proved that "$f_{BE}(\mathcal{E})$" is the best choice. Particles belonging to a system that obeys the "B-E" function are called "bosons." At temperatures higher than some tens of Kelvin degrees** every real gas** can be approximately represented with the B-function.

From Figure A6.4.1 we can also observe how the "$f_{BE}(\mathcal{E})$" is infinite for $\mathcal{E} = \mathcal{E}_{BE}$. It means that this statistic doesn't assume energy values lower than "$\mathcal{E}_{BE}$," which is also the energy level to have the highest probability of being occupied. Conversely, neither "$f_{FD}(\mathcal{E})$" nor "$f_B(\mathcal{E})$" has a low energy limit.

The "F-D" distribution has the characteristic that energy level $\mathcal{E} \equiv \mathcal{E}_F$ has equal probability to be free or occupied by a particle since $f_{FD}(\mathcal{E}_F) = 0.5$. Particles belonging to a system that obeys the "B-E" function are called "fermions." The best example of a gas represented by the "F-D" distribution is the electron gas inside a metal. This system is equivalent to a real gas only from the point of view of the particle motion while the particles' quantity and particle mass are now respectively extremely higher and lower than any particle of a real gas, also at ambient temperature. From this other point of view, the electron gas is a high density system at ambient temperature, and in this case the other two functions fall in error. Due to the high importance of the "F-D" function in the systems with which we are dealing,*** we think it is important to evaluate the "$\mathcal{E}_F$" expression, since it assumes a large importance in the application of the "F-D" function. In Figure A6.4.2 we have represented this function with $\mathcal{E}_F = 5eV$**** at three temperatures, i.e., 1,

* William Thomson Kelvin, English physicist, born in Belfast in 1824 and died in Netherhall in 1907.
** With "real gas" we mean not an electron gas.
*** Do not forget that all natural ferromagnetic materials are good conductors which are represented as metals.

The susceptibility obtained by Equation A6.5.13 has been verified to be accurate for paramagnetic materials, also if the Boltzmann distribution statistic for the material's atoms has been used. In general, the total macroscopic magnetic moment obtained from A6.5.7 gives values that are practically coincident with those measured for paramagnetic materials. Also for ferromagnetic materials a good coincidence is verified, but it is necessary to use the spin magnetic moment "μ_S" instead of "μ_T," i.e., $\mu_T \equiv \mu_S$. This fact has been explained assuming that in metals the chemical bond is mainly performed by the most external orbitals. Since these orbitals share electrons among them, these also create an orbital magnetic momentum "μ_L." However, these orbitals are involved just in the chemical binding, and they cannot move the plane of their orbitals to align "μ_L" with "$\underline{B}$." The spin momentum can instead do that and this causes $\mu_T \equiv \mu_S$. The phenomenon of the external orbital blocking is called "quenching."

The extension to the ferromagnetic materials of the Brillouin result to explain the paramagnetic effect of these materials brings high errors if another internal field is not taken into account. This internal field is called the "molecular magnetic field" and will be studied in a next section.

A6.6 ANISOTROPY, MAGNETOSTRICTION, DEMAGNETIZATION IN FERROMAGNETIC MATERIALS

In this section we will discuss some important properties of ferromagnetic materials, which cause these materials to be unique in the area of magnetism.

a. Magnetization Anisotropy

It has been verified in practice that for any ferromagnetic material the energy "$\mathcal{E}$" required to magnetize it in a direction can be different from that required energy to magnetize it in another direction. The directions for which the minimum "$\mathcal{E}_m$" and maximum "$\mathcal{E}_M$" are necessary are called the "direction of easy magnetization" and "direction of difficult magnetization." The energy difference "$\mathcal{E}_a$" between "$\mathcal{E}_M$" and "$\mathcal{E}_m$" is called "anysotropy energy," and it has been verified that this value decreases if the material is composed of hypersymmetric* geometric microscopic structure. For example, iron, which possesses a cubic crystalline structure has lower "$\mathcal{E}_a$" than cobalt, which possesses an axial crystalline structure.

The explanation of this behavior has been done assuming that the "quenching" is not uniform with the applied direction of the external magnetic field.

b. Magnetostriction

This effect could be thought of as the magnetic counterpart of the dielectric effect. When a magnetic field is applied to a ferromagnetic material, it modifies its geometric structure, typically inside a variation range $\Delta \ell = 10$ to 30 ppm. These movements are not the same for any ferromagnetic materials, since some lengthen and others shorten. In addition, these behaviors are also dependent on the direction and strength of the applied magnetic field. For example, in the direction of easy and difficult magnetization, iron possesses respectively a positive and negative "$\Delta \ell$." Conversely, nickel always possesses a negative "$\Delta \ell$." The explanation of this behavior is still under investigation, but it is assumed that an internal magnetic energy exists, which the ferromagnetic material tries to release when placed inside a magnetic field. The magnetostriction process ends when a balance is reached between the reduction of the internal magnetic energy and the increase of the potential energy due to the material deformation.

---

* By hypersymmetrism we mean a symmetry in many directions.

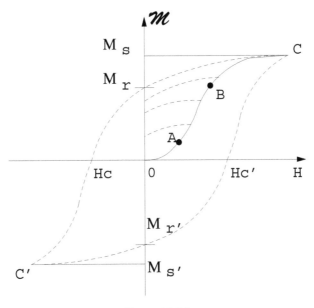

Figure A6.8.2

c. First Magnetization Curve and Hysteresis Loop

If we take a ferromagnetic material never magnetized and we begin to apply a static magnetic field "H," then a magnetization begins to appear, following the line indicated with "OC" in Figure A6.8.2.

If we remove the magnetic field until "H" magnitude is below that relative to point "A" the material returns to its starting status, i.e., practically without a net magnetization. This means that below a defined value of magnetic field, the movements of Bloch walls are reversible. Instead, if we decrease the external magnetic field when it has a magnitude greater than that corresponding to point "A," the magnetization follows the dashed curves indicated in the figure, until intersecting the ordinate axis at a point with a value called the "residual magnetization." The ferromagnetic material has become permanently magnetized, i.e., it is a magnet. This is what happens for any removal point inside the path "AC" as indicated by the dashed curves in the segment "AC," and the relative residual magnetization is usually indicated with "$\mathcal{m}_r$." This behavior can be understood observing Figure A6.8.3, where we have represented a typical graph of total energy "$\mathcal{E}_t$" of the

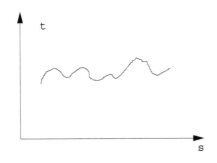

Figure A6.8.3

crystal varying the position "s" of Bloch's wall. We see that some minimums are possible. In fact, when Bloch's wall moves, a potential energy arises due to material compression. This energy produces heat, which is partially lost if no other compression is performed, i.e., if the applied field doesn't increase. Due to this energy loss, there can be a new minimal total energy, which corresponds to overcoming a peak of Figure A6.8.3 and falling into another valley. Of course, other energies are involved in this procedure, like magnetostriction and anysotropy, but the result can always be represented as indicated in Figure A6.8.3.

Increasing the strength of the applied field "H" from point "A" in Figure A6.8.2 we move toward point "B," i.e., near the flex of the first magnetization curve. At this point all of the Bloch's wall is at the final position, i.e., all the crystals have only one domain. At point "C" the magnetization has reached the maximum value for the ferromagnetic material under test, and this value is called the "saturation magnetization" usually indicated with "$\mathcal{M}_s$." At this point, all the domains are aligned in the same direction of "$\underline{H}$." The resulting values of "$\mathcal{M}_s$" are strongly dependent on the material under test. For some compounds, as we will study later, the residual magnetization can be very high, near 5000 Gauss in "CGSA" units system, or in the "MKSA" units system the corresponding value "$\mu_0\mathcal{M}_r$," ranges between 0.5 Weber/m². It is important to note that while the "$\mathcal{M}_s$" value is unique for that material, since it depends on its molecular structure, this does not happen for "$\mathcal{M}_r$" because this value is relative to the value of the external field before its removal. In our figure, with "$\mathcal{M}_r$" we have indicated the residual magnetization relative to "$\mathcal{M}_s$."

If we now reverse the direction of "$\underline{H}$," that is we apply the external field in the opposite direction of "$\mathcal{M}_s$," the magnetization moves on the curve indicated with "M_rC'" until it reaches a value of zero for an external magnetic field of value "H_c," called the "coercive field." At point "C'" the magnetization is that of saturation, and if we remove the external field we still have a residual magnetization "$\mathcal{M}_r'$," following the path "$C'\mathcal{M}_r'$," usually equal in value to "$\mathcal{M}_r$" but obviously in the opposite direction. If we now reverse again the direction of the external magnetic field we may again set the magnetization to zero when the external field has a value "H_c'." If we increase the value of "H_e" we again reach the point "C" but, as is evident from the figure, the path "OC" is never more covered. This is called the "hysteresis loop" the path "$C\mathcal{M}_rC'\mathcal{M}_r'C$," while the path "OC" is called the "first magnetization curve." The aspect of the hysteresis loop changes with the characteristics of ferrimagnetic material, so that higher crystalline purity materials have a more rectangular shape, always symmetrical with respect to the origin of the axis drawn in Figure A6.8.2. It is important to note that the aspect is also dependent on the frequency of "H" since for frequencies above some hundreds Hertz of the external magnetic field, ferromagnetic materials present increasing losses.*

With reference to Figure A6.8.2, two values of permeability are defined. One, called "initial permeability" and indicated with "μ_i" is defined as the slope of the tangent to the initial magnetization curve at the point H = 0. Values of "μ_i" range from some units to some thousands, depending on the material. The other, called "maximum permeability" and indicated with "μ_m," is defined as the slope of the line passing through the origin of the coordinate and tangent to the higher knee of the first magnetization curve. Of course "μ_m" has higher values of "μ_i" also reaching values of some tens of thousands.

A6.9 THE HEISENBERG THEORY FOR THE MOLECULAR FIELD

Weiss' molecular field "B_m," has been extremely important in explaining some ferromagnetic characteristics, but no reason was formulated to permit the existence of such a field. The physicist W. Heisenberg** was the first to formulate a theory that can explain the existence of "B_m" and also other important magnetic characteristics of materials. According to Heisenberg, indicating with

---

* A similar argument for ferrites will be studied in Appendix A7.
** Werner Heisenberg, German physicist, born in Wuerzburg in 1901 and died in Monaco in 1976.

A6.10 FERROMAGNETIC MATERIALS AND THEIR APPLICATIONS

Ferromagnetic materials can be divided in two groups, depending on how easily the material can be magnetized. This net division in two groups is not found in natural ferromagnetic materials, and for this reason these materials are chemical compounds. These two groups are called "soft" and "hard" materials whose characteristics we are going to discuss.

1. "Soft" Materials

These materials require a small intensity external field to be saturated, and maintain low residual magnetization, near some tenths of Weber/m². Their hysteresis loop is quite rectangular, which corresponds to having a very pure crystal as the compound unit cell. The permeability is very high, usually near some tens of thousands. Typical materials of this group are nickel-iron compounds. For example, the so called "permalloy" is made of 80% nickel and 20% iron and its permeability is near 20.000. The "supermalloy" is composed of 80% nickel, 15% iron, and 5% molybdenum, and possesses a permeability near some hundreds of thousands. All these high permeabilities are obtainable up to some KHz of the applied signal. Consequently, the applications of these materials are in low frequency devices, like audio transformers and microphones. When high power is required, for example in low frequency transformers like those used to convert the 220V A.C. in lower voltage, some compound of iron and silica is employed.

a. Hard Materials

These materials can be regarded as the reciprocal of the "soft" counterpart. In fact, these materials require a high intensity external field to be saturated and maintain high residual magnetization. Their hysteresis loop is quite wide. Typical materials of this group are compounds of aluminium-nickel-cobalt, called "Alnico," or platinum-cobalt. The "Alnico" compounds are indicated by a number, which gives the percentage of the single materials. For example, "Alnico 5" is capable of 0.1 Weber/m² of residual magnetization, while some compounds of platinum-cobalt reach 0.5 Weber/m². The use of these materials is in the realization of permanent magnets. These devices are also used in the field of transmission line devices like in ferrimagnetic circulators or phase shifters, as studied in our text. Magnets are also used in high power microwave tubes "TWT" to focus the electron beam.

A6.11 ANTIFERROMAGNETISM

One of the first studies of antiferromagnetism was made by the physicist L. Neel.* He supposed that in antiferromagnetic materials the magnetic atomic spin was antiparallel, resulting in a practically null net magnetization. Later, Heisenberg with his theory on molecular fields gave an important interpretation of this phenomenon. Evaluating the exchange integral for typical antiferromagnetic materials like manganese, chrome, and palladium, its value is negative. This corresponds to a high exchange energy, which results in an antiferromagnetism, for the reasons discussed in Section A6.9.

Neel also observed that the small residual magnetization moves with temperature, increasing its value until we reach a particular temperature "T_{Na}" called "Neel's antiferromagnetic temperature." Typical "T_{Na}" are below 200 °C. For example, the manganese oxide, MnO, possesses a $T_{Na} = 120$ °C. For temperatures higher than "T_{Na}" the antiferromagnetic materials become paramagnetic. Neel has suggested an expression for the susceptibility "χ" above "T_{Na}" given by:

* Louis Neel, French physicist, born in Lyon in 1904.

$$\chi = \frac{C_N}{T + T_{Na}}$$

(A6.11.1)

The previous expression has proved to give accurate results for $T > T_N$, properly choosing the "Neel's constant, C_N," after measurements on the material.

Neel' expression A6.11.1 is empirical, and a complete study to analytically evaluate such expression, together with "T_N," still needs to be developed.

A6.12 FERRIMAGNETISM

It is usual to name any material of this family with the same term "ferrite" and if we have already said that the "ferrite" is a particular ferromagnetic material. Ferrimagnetic materials present the same static properties of their ferromagnetic counterparts, but with less intensity. Two characteristics are really only relative to ferrimagnetic materials:

a. All the ferrites are insulators, with a resistivity 10^4 to 10^{10} times the resistivity of the ferromagnetic materials
b. Considering an applied time varying magnetic field, ferrites maintain a high permeability for a wider bandwidth than the ferromagnetic materials.

The previous point b will be discussed in Appendix A7, which is completely dedicated to ferrites. Here we only say that for this reason ferrites are widely employed in RF and microwave devices, as we have discussed in our text.

Characteristics like domains of spontaneous magnetization, hysteresis loop, magnetostriction, and anisotropy are also pertinent to ferrites. The reason for the low magnetization values in the case of ferrites can be explained with the Neel interpretation of ferrimagnetism. Neel supposed that the total magnetic spin vectors for each atom are disposed in an antiparallel configuration among them, as in the antiferromagnetic materials. However, in this case a net magnetization arises because the intensities of the antiparallel atoms do not have the same value. A qualitative representation of this interpretation is indicated in Figure A6.12.1.

Similar to the ferromagnetic case, a temperature "T_{Nf}" exists over which the ferrite becomes paramagnetic. This "T_{Nf}" is called "Neel's ferrimagnetic temperature." Typical "T_{Nf}" are below 200 °C. The evaluation of "T_{Nf}" follows, in general, the procedure we used for the ferromagnetic materials, but it is more complicated since all ferrites are compounds and the chemical bond theory must be known. This procedure can be found in the specified literature.[5]

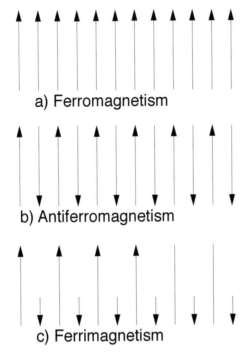

a) Ferromagnetism

b) Antiferromagnetism

c) Ferrimagnetism

Figure A6.12.1

REFERENCES

1. F. Bitter, H. A. Medicus, *Fields and Particles*, Elsevier, Amsterdam, 1973.
2. R. P. Feynman, R. B. Leighton, M. Sands, *The Feynman Lectures on Physics*, Addison-Wesley, New York, 1963.
3. D. Sette, *Lezioni di Fisica*, Veschi, 1963.
4. C. Heck, *Magnetic Materials and Their Applications*, Butterworth Co., U.K., 1974.
5. R. F. Soohoo, *Theory and Applications of Ferrites*, Prentice Hall, New York, 1960.

The Electromagnetic Field and the Ferrite

A7.1 INTRODUCTION

To understand the behavior of ferrite inside a magnetic field, static or time varying, it is important to know some fundamental concepts of magnetism and atomic physics. In this appendix we will assume that the reader knows these fundamental concepts well, but for those who desire to see these concepts again, we refer them to Appendices A5 and A6 where we have treated all the topics that we assume are needed to understand what is going to be covered here.

For a complete analysis of the ferrites, it is necessary to also have fundamental concepts of chemistry since the magnetic properties of these materials change when their components change. A book to introduce the reader to chemistry is indicated in the reference section,[1] together with other books that speak about the general theory of ferrites.[2,3]

In addition, when we cover the problem of propagation of electromagnetic waves inside the ferrite, we have to know the fundamental concepts of electromagnetism, and for this purpose we refer the reader to Appendix A2. In that appendix we have written everything necessary to introduce the electromagnetic wave propagation for the understanding of what is said in this book. If the reader wants to know the whole theory of electromagnetic wave propagation, some books are indicated in the references.[4,5,6]

At this point the reader will recognize how many physics topics are included in the term "ferrite."

This appendix will discuss the first application of ferrite in the field of microwaves, i.e., in devices using waveguides, as well as the modern use of ferrite in planar transmission lines. We will do it in a theoretical approach, since the specific applications together with practical considerations are treated in the chapters of this book.

A7.2 THE CHEMICAL COMPOSITION OF FERRITES

"Ferrite" means a material composed of particular substances so that the final composite has magnetic properties that are similar to that of ferromagnetic materials,* but with a resistivity that can be 10^{10} times the resistivity of the latter, which instead are good conductors. Ferrites are consequently realized as good insulator materials, with a quite complicated molecular geometric aspect. Simplifying, the general chemical formula can be indicated as:

---

* See Appendix A6 for foundations on magnetism.

$$Me\ O\ Fe_2O_3 \qquad\qquad (A7.2.1)$$

where "Me" is a metal, like iron "Fe," nickel "Ni," cobalt "Co," copper "Cu," aluminium "$A\ell$," magnesium "Mg," or their composites, and "O" is the oxygen. From A7.2.1 we can deduce that it is the big quantity of metal oxides that induce in ferrites its high resistivity. If, in the previous equation, iron is substituted for Me, we obtain the only natural ferrimagnetic material, i.e., magnetite, whose chemical formula is Fe_3O_4.

All ferrites possess a grey or grey-black color with a smooth surface, and result in a hard material with no flexibility at all. Consequently, if we try to bend ferrites they break very easily. As told, the geometric aspect of the ferrite molecule is quite complicated, resembling a cubical shape, and to this structure is given the name "spinel." In particular, the oxygen atoms connect together to form a plane and the whole material is constructed with vertical connections of such planes, so that oxygen atoms of each plane fit the spaces between the oxygen atoms of the other layers. The remaining atoms that form the complete molecule lie in the remaining space between the oxygen atoms. A look inside this geometric structure shows us that two crystalline forms are possible, one tetrahedral and one octahedral, both with oxygen atoms at the corners. The magnetic situation is that the atoms interact in an antiferromagnetic way, as is said in Appendix A6, forming a net magnetic moment in microscopic areas, called "domains." As a result, ferrites have a magnetic domain organization like ferromagnetic materials.

A7.3 THE FERRITE INSIDE A STATIC MAGNETIC FIELD

In this Section we want to summarize the behavior of ferrite inside a magnetic field, be it static or time varying. More in-depth explanations on these topics may be found in Appendix A6. We remind the reader that underlined letters refer to vectors.

a. Permanent Magnetization

All ferrites may be permanently magnetized if introduced inside a magnetic field of suitable strength. In this case they become magnets, generating a residual magnetic field of induction vector "$\underline{B}_r$," with a magnitude called "magnetization" which will be indicated with "$\mathcal{M}$." If the "MKSA" unit system is used, "$\mathcal{M}$" has the same dimensions as the magnetic field "H," i.e., Amp.turn/meter, and "$\mathcal{M}$" is usually indicated with "M." If the "CGSA" unit system is used, "$\mathcal{M}$" is measured in "Gauss,"[*] and "$\mathcal{M}$" is usually indicated with "$4\,\pi$ M." The value of "$\mathcal{M}$" depends on the type of ferrite, usually ranging from some hundredth to some tenth of Weber/m.[2,**] In ferrite magnetics the "CGSA" units system is most used, where magnetic fields "H" are measured in Oersted[***] (Oe) and the magnetization "$4\,\pi$ M" is measured in "Gauss." In this case "$\mathcal{M}$" ranges from some hundreds to some thousands of "Gauss." It is important to note that ferrites are not made to be used as permanent magnets, since ferromagnetic magnets produce a stronger field if their use is restricted to power electronics and low frequency applications. For instance, ferromagnetic magnets are used to magnetize ferrites in circulators, as has been seen in Chapter 7 for microstrip devices, or to guide an electron beam in a Travelling Wave Tube, TWT.

If ferrites never come inside a magnetic field, they do not present a macroscopic net magnetic field, and if inside a "domain" a magnetic field exists due to the natural presence of the molecular field, as stated in Appendix A6.

---

* Carl Friedrich Gauss, German physicist, born in Brunswick in 1777 and died in Gottingen in 1855.
** Wilhelm Eduard Weber, German physicist, born in Wittenberg 1804 and died in Gottingen in 1891.
*** Hans Christian Oersted, Danish physicist, born in Rudkoping in 1777 and died in Copenhagen in 1851.

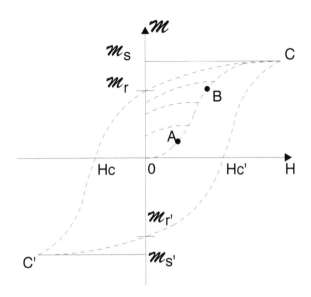

Figure A7.3.1

b. First Magnetization Curve and Hysteresis Loop

The definition and physical aspects of this most important behavior of ferrites are the same as that studied in Appendix A6 for ferromagnetic materials, synthetically indicated in Figure A7.3.1. Only the following differences need to be outlined:

1. Typical values of "$\mathcal{M}_s$" in the "CGSA" unit system range between 1500 to 5000 "Gauss," or in the "MKSA" units system the corresponding value "$\mu_0\mathcal{M}_s$" ranges between 0.15 and 0.5 Weber/m^2.
2. Also for the ferrimagnetic case the hysteresis loop is dependent on the frequency of the applied external magnetic field "H_e." In addition, as will be explained later, in our case for some frequencies of "H_e" ferrites present energy absorption. In particular applications, as in radio frequency circuits, ferrites are constructed to be insensitive to the frequency of the magnetic field applied so that wideband circuits may be realized.
3. Typical values for "initial permeability" range from some units to some thousands, depending on the material, while the "maximum permeability" reaches values of some tens of thousands.

c. Paramagnctiom

The characteristics of the previous two points a and b are verified until the ferrite temperature is below a particular value "T_n," called the "Neel* temperature" while if the temperature is higher than "T_n," ferrite becomes paramagnetic.** The name of this temperature is taken from the name of French physicist Louis Neel, in honor of his work and study in this field. Neel was one of the first scientists to suppose that the origin of ferrimagnetism was in the antiparallel alignment of atomic magnetic moments. Sometimes in ferrimagnetism, Neel's temperature is improperly called "Curie's temperature," in honor of the French physicist Pierre Curie,*** who experimentally discovered the variation with temperature of susceptibility for ferrimagnetic materials. Neel temperature ranges between 100 °C and 500 °C. Usually these values of temperature do not create problems in commercial devices, while attention is required in space or military devices that are subject to severe environmental forces.

---

* Louis Neel, French physicist, born in Lyons in 1904.
** See Appendix A6 for paramagnetic behavior of ferromagnetic materials.
*** Pierre Curie, French physicist, born in Paris in 1859 and died there in 1906.

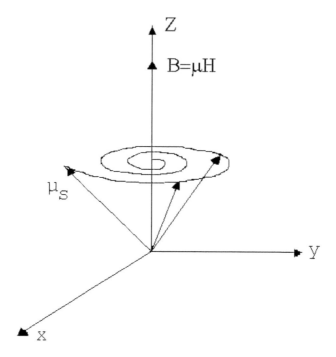

Figure A7.3.2

d. Precession Motion

As known from Appendix A6, the spontaneous magnetization inside ferrite domains is due to an alignment of electron spin magnetic moments "μ_s" of the material. This alignment is formed by the action of the natural induction magnetic field, called the "molecular field," inside the domain. When the ferrite is inside an external magnetic field, the spin magnetic moment vector begins a precession motion around the direction of the external field and consequently the magnetization does the same. This motion is subjected to loss inside the material and, as a consequence, the extreme of the spin magnetic moment vector describes a spiral motion, as indicated in Figure A7.3.2.

It is known that the angular frequency "ω_p" is proportional to the internal magnetic field "H" induced by the externally applied field "H_e." For an electron moving around an atomic nucleus, so generating an orbital magnetic moment, we have:

$$\omega_p = -\mu_0 \gamma_s H \tag{A7.3.1}$$

where "μ_0" is vacuum permeability and "γ_s" is the electron spin gyromagnetic ratio, given by:

$$\gamma_s = e/m \tag{A7.3.2}$$

where "e" is the electron charge and "m" its mass. Note that "γ_s" is a negative number. Both "e" and "m" are known, so the gyromagnetic ratio in the "MKSA" units system has a value of $-1.75696 \times (10)^{11}$ Coulomb/Kg.

It is interesting to obtain from A7.3.1 the value of "ω_p" for a typical induction magnetic field, let us say 0.1 Weber/m². Multiplying this value by the gyromagnetic ratio we have $\omega_p = 1.75696 \times (10)^{10}$ radians/sec, which is 2.7977 GHz. We see how we can simply reach precession frequencies in the microwave regions. We know* that the precession motion is damped by the

---

* Precession motion is studied in more detail in Appendix A5.

$$\mu = \mu_0 (1 + \chi) \tag{A7.4.26}$$

As we said, in ferrimagnetic materials the susceptibility is a matrix, that in our case of isodirectional propagation becomes "$[\chi_z]$." So, transforming the previous equation in matrices, we have:

$$[\mu_z] = \mu_0 \left([i] + [\chi_z] \right) \tag{A7.4.27}$$

where "$[i]$" is the unit matrix, i.e., the matrix that has nonzero elements, the elements that lie on the main diagonal, which are all equal to one. Doing calculations involved in the previous equation, we have:

$$[\mu_z] = \mu_0 \begin{bmatrix} 1 + x_p & jx_\ell & 0 \\ -jx_\ell & 1 + x_p & 0 \\ 0 & 0 & 1 \end{bmatrix} \doteq \mu_0 \begin{bmatrix} \mu_{xx} & j\mu_\ell & 0 \\ -j\mu_\ell & \mu_{yy} & 0 \\ 0 & 0 & \mu_{zz} \end{bmatrix} \tag{A7.4.28}$$

where obviously:

$$\mu_{xx} = 1 + x_p \equiv \mu_{yy} \doteq \mu_p \tag{A7.4.29}$$

$$\mu_\ell \equiv x_\ell \tag{A7.4.30}$$

$$\mu_{zz} = 1 \tag{A7.4.31}$$

The last matrix of A7.4.28, with the notations A7.4.29 to A7.4.31, is usually indicated as "$[\mu_{fz}]$," that is:

$$[\mu_{fz}] \doteq \begin{bmatrix} \mu_p & j\mu_\ell & 0 \\ -j\mu_\ell & \mu_p & 0 \\ 0 & 0 & 1 \end{bmatrix} \tag{A7.4.32}$$

It is very important to observe that the elements of "$[\mu_{fz}]$," as the elements of "$[\chi_z]$," implicitly contain all the quantities of our problem, that is "ω," "H_s," and "$\mathcal{M}_s$." It is now the case to note as with our hypothesis of statically saturated ferrite with small time varying magnetic field "$\underline{h}$," the ferrite does not respond to any variation of "$\underline{h}$" along the direction, "$\underline{z}_0$" in our study, of static saturating magnetic field "$\underline{H}_s$." This fact corresponds to having $x_{zz} = 0$ and $\mu_{zz} = 1$ inside the matrix. The hypothesis of statically saturated ferrite reflects some practical useful behavior of this material in such magnetic conditions when used in electromagnetic devices, as we will see later. However, note that in the other directions, that is "$\underline{x}_0$" and "$\underline{y}_0$" in our coordinate system, the ferrite is sensitive to "h" since the matrix elements for these axes are nonzero.

We now want to see what the denominator "D," which appears in A7.4.21 and A7.4.22, becomes when "α" is negligible with respect to one. In this case, A7.4.17 becomes:

$$D \equiv -\omega^2 + \omega_p^2 \tag{A7.4.33}$$

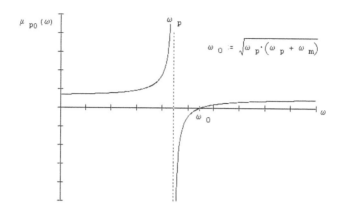

Figure A7.4.1a

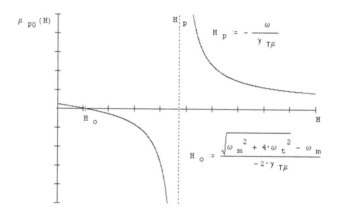

Figure A7.4.1b

$$\mu_{p0} \overset{\perp}{=} \mu_p(\alpha = 0) \overset{\perp}{=} 1 + \omega_m \frac{\omega_p}{\omega_p^2 - \omega^2} \qquad (A7.4.41)$$

$$\mu_{\ell0} \overset{\shortmid}{=} \mu_\ell(\alpha = 0) \overset{\perp}{=} \omega_m \frac{\omega}{\omega_p^2 - \omega^2} \qquad (A7.4.42)$$

The graphs of the two previous equations, as a function of "ω" or function of "H_s," are represented in the following Figures A7.4.1 and A7.4.2. Concerning Figures A7.4.2a and b, it is very important to remember that the range of variability of "H" must have a minimum value that is greater than the minimum value required to saturate the ferrite.

A situation that often happens in applications of ferromagnetic devices is where "$\underline{H}_{sz}$" and "$\underline{h}$" are still parallel, but each point in the opposite direction. We will show that in this case the susceptibility matrix changes form from the one "$[\chi_z]$" seen in A7.4.23. Then, if "$\underline{H}_{sz}$" and "$\underline{h}$" are parallel but point in opposite directions, A7.4.1 and A7.4.2 become:

$$\underline{Hh} = h_x \underline{x}_0 + h_y \underline{y}_0 + (h_z - H_{sz}) \underline{z}_0 \qquad (A7.4.43)$$

$$[\chi_x] = \begin{bmatrix} 0 & 0 & 0 \\ 0 & x_p & jx_\ell \\ 0 & -jx_\ell & x_p \end{bmatrix} \tag{A7.4.67}$$

and consequently:

$$[\mu_x] = \mu_0 \begin{bmatrix} 1 & 0 & 0 \\ 0 & \mu_p & j\mu_\ell \\ 0 & -j\mu_\ell & \mu_p \end{bmatrix} \tag{A7.4.68}$$

From A7.4.28, A7.4.64, and A7.4.68 we recognize how in all the permeability matrices the same terms always appear but changed in position. Also note that to extract these matrices we have not mentioned the reciprocal orientation between "$\underline{H}$" and "$\underline{h}$" or fixed a particular polarization for "$\underline{h}$." The choice of which of these matrices we have to use is dependent on how we name the axis of magnetization in our electromagnetic problem geometry. Later we will use some of these different forms of permeability matrices.

We now want to conclude this section doing an important consideration. Here we have studied a material whose magnetization is saturated under a static magnetic field "H_s," and in which a time varying magnetic field "h" has been impressed. Note also that we have not specified a value for "ω." From these points of view, the ferrite is not the only one possible material to employ in our experiments and studies, since the ferromagnetic materials have the same properties. But if we give to "ω" a value greater than a few kHz times 2π, the ferrite becomes the only one material to be successfully employed. This is because, as known, ferrimagnetic materials are good insulators, while the ferromagnetic ones are good conductors, so that electromagnetic energy at high frequency can propagate through ferrites while it cannot propagate inside ferromagnetic materials. This is the reason why ferrites are so widely employed in RF and microwave devices as we will see later. Of course, other types of losses exist inside ferrites, which will be studied in Section A7.11.

A7.5 "TEM" WAVE INSIDE AN ISODIRECTIONAL MAGNETIZED FERRITE

In the previous section we extracted the expressions of dynamic susceptibility and permeability of ferrites, while they are statically magnetized at saturation from a magnetic field "H_s."

Here we want to study the case when a "TEM" wave* with magnetic field "h" propagates inside a ferrite, saturated with internal static magnetic field "H_s." Let us also assume that the direction of propagation of the wave is coincident with the direction of the saturating internal field "$\underline{H}_s$." We will refer to this situation as the case of an electromagnetic wave inside an "isodirectional magnetized" ferrite. Since ferrites are sensitive to magnetic fields, we concentrate our study on the magnetic field of the "TEM" wave, which we will write as:

$$\underline{h}(z) \stackrel{\perp}{=} \underline{h}_t e^{(-j\beta z)} \tag{A7.5.1}$$

where:

$$\underline{h}_t \stackrel{\perp}{=} h_x \underline{x}_0 + h_y \underline{y}_0 \tag{A7.5.2}$$

---

* See Appendix A2 for wave definitions.

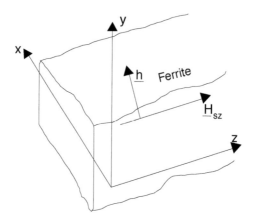

Figure A7.5.1

This situation is geometrically indicated in Figure A7.5.1. As A7.5.1 shows, we introduce a Cartesian coordinate system, where the "TEM" wave propagates along the "z" axis and the static magnetic field "$\underline{H}_{sz}$" is also directed along "z." Of course, we will suppose that the time variability of "h(z)" will be sinusoidal so that when we want to obtain the time dependence, it is enough to multiply the quantity times "$e^{(j\omega t)}$" where "ω" is the angular frequency. Another expression for "h(z)" often used may be obtained by inserting A7.5.2 in A7.5.1 and doing multiplication. We will have:

$$\underline{h}(z) = h_x(z)\underline{x}_0 + h_y(z)\underline{y}_0 \tag{A7.5.3}$$

where:

$$h_x(z) \equiv h_x \, e^{(-j\beta z)}$$

$$h_y(z) \equiv h_y \, e^{(-j\beta z)}$$

If we also want to obtain the expression that contains the time dependence we may apply the procedure seen in Appendix A2 in Section A2.3. Since this aspect is not important here, we will not do it.

We now want to remember that in matrix form the vector "$[\underline{h}]$" is written as:

$$\underline{h} \equiv \left[\underline{x}_0 \underline{y}_0 \underline{z}_0\right] \begin{bmatrix} h_x \\ h_y \\ 0 \end{bmatrix} \tag{A7.5.4}$$

and we will define "versor matrix $[\underline{u}]$" with the following notation:

$$[\underline{u}] \stackrel{\perp}{=} \left[\underline{x}_0 \underline{y}_0 \underline{z}_0\right] \tag{A7.5.5}$$

After these discussions, the Maxwell* equations to be resolved in our system are:

---

* James Clark Maxwell, English physicist, born in Edinburgh in 1831, died in Cambridge in 1879.

$$\underline{\nabla} \otimes (\underline{e}) = -j\omega[\underline{u}][\mu_z][h] \tag{A7.5.6}$$

$$\underline{\nabla} \otimes (\underline{h}) = j\omega\varepsilon\underline{e} \tag{A7.5.7}$$

In A7.5.6 the "component matrix [h]" is defined as:

$$[h] \overset{\perp}{=} \begin{bmatrix} h_x \\ h_y \\ 0 \end{bmatrix} \tag{A7.5.8}$$

while in A7.5.7 "ε" is the dielectric constant of ferrite. The operators we have used above are defined in Appendix A8. To resolve the Maxwell's equations we may proceed similarly to what we have done in Appendix A2 when discussing wave equations. In our case we have matrix equations. So, applying the curl operator to A7.5.7 and inserting in the expression A7.5.6 we have:

$$\underline{\nabla} \otimes [\underline{\nabla} \otimes (\underline{h})] = \omega^2\varepsilon[\underline{u}][\mu_z][h] \tag{A7.5.9}$$

It is known that the following relationship holds:

$$\underline{\nabla} \otimes [\underline{\nabla} \otimes (\underline{h})] = \underline{\nabla}(\underline{\nabla} \bullet \underline{h}) - \nabla^2(\underline{h}) \tag{A7.5.10}$$

Remembering our hypothesis of "TEM" wave with transversal components independent to "x" and "y" it is simple to recognize that $\underline{\nabla} \bullet \underline{h} = 0$, and A7.5.9, A7.5.10 become:

$$-\nabla^2(\underline{h}) = \omega^2\varepsilon[\underline{u}][\mu_z][h] \tag{A7.5.11}$$

Using the "Laplacian" expression in Cartesian coordinates given in Appendix A8, inserting A7.4.28 and equating the components along the same axis, A7.5.11 becomes:

$$\beta^2 h_x = \omega^2\mu_0\varepsilon\left(\mu_n h_x + j\mu_\ell h_y\right) \tag{A7.5.12}$$

$$\beta^2 h_y = \omega^2\mu_0\varepsilon\left(-j\mu_\ell h_x + \mu_p h_y\right) \tag{A7.5.13}$$

These last two equations form a homogeneous system, with "h_x" and "h_y" as unknowns. Setting to zero the determinant of the coefficients, we have:

$$\beta = \pm\omega\left(\mu_0\varepsilon\right)^{0.5}\left(\mu_0 \mp \mu_\ell\right)^{0.5} \tag{A7.5.14}$$

The first double sign "$\pm$" is relative to the two possible directions of propagations of the wave along the "z" axis. The second double sign "$\mp$" means instead that two possible waves are generated inside the ferrite. In other words, a fixed direction of propagation inside the ferrite, for instance the one corresponding to the "+" sign in the "$\pm$" of A7.5.14; when a "TEM" wave is launched in the isodirectional magnetically saturated material this wave divides itself into two waves,* one with phase constant

* This is called "birefringence."

$$\beta_d \overset{\perp}{=} \omega \left(\mu_0 \varepsilon\right)^{0.5} \left(\mu_p - \mu_\ell\right)^{0.5} \tag{A7.5.15}$$

and the other with phase constant

$$\beta_c \overset{\perp}{=} \omega \left(\mu_0 \varepsilon\right)^{0.5} \left(\mu_p + \mu_\ell\right)^{0.5} \tag{A7.5.16}$$

We will return shortly to the subscripts "d" and "c." It is very important at this point to say that since "μ_p" and "μ_ℓ" are in general complex quantities, as seen in the previous section, the same has to be said regarding "β_d" and "β_c." It follows that, since the propagation factor is "$e^{(-j\beta z)}$" when "β" (that is "β_d" and "β_c") becomes imaginary negative, the exponent in the propagation factor becomes real negative, which means that the wave is attenuated exponentially through propagation inside the ferrite. Applications of this phenomenon will be studied later in this appendix.

If we now insert A7.5.13 in A7.5.12 we obtain:

$$h_y = \pm j h_x \tag{A7.5.17}$$

Since a minus "–" means a phase shift of 180° and "j" means a phase shift of 90°, the previous equation shows us that inside the ferrite the two waves are both spatially circularly polarized, but one clockwise and the other counterclockwise. These directions of rotation are relative when looking in the same direction of propagation of the wave. We will call the counterclockwise rotation, which is relative to "+" sign in A7.5.17, the "discordant wave" while the remaining one will be called the "concordant wave." These names are relative to the concordance or discordance of the direction of the circular polarization with the direction of the precession motion of the total magnetic moment of the ferrite, when looking in the same direction of propagation of the wave. The discordant wave has phase constant "β_d" given in A7.5.15 while the concordant wave has phase constant "β_c" given in A7.5.16. It is now interesting to evaluate "$\mu_p \pm \mu_\ell$" in the case of negligible losses. Using A7.4.41 and A7.4.42 we have:

$$\left(\mu_p - \mu_\ell\right)\Big|_{(\alpha=0)} \equiv \mu_{p0} - \mu_{\ell0} = 1 + \frac{\omega_m}{\omega_p + \omega} \overset{\perp}{=} \mu_{d0} \tag{A7.5.18}$$

$$\left(\mu_p + \mu_\ell\right)\Big|_{(\alpha=0)} \equiv \mu_{p0} + \mu_{\ell0} = 1 + \frac{\omega_m}{\omega_p - \omega} \overset{\perp}{=} \mu_{c0} \tag{A7.5.19}$$

and the graphs of these functions vs. "ω" are represented in Figure A7.5.2. From this figure we may see that the permeability presented to the concordant wave becomes infinite when the frequency of the "TEM" wave is equal to the precession frequency created by "H_{sz}." Also note that since "μ_d" and "μ_c" are functions of frequency, from A7.5.15 and A7.5.16 it follows that the diagram "ω/β" is not constant with the frequency, which means that magnetized ferrite is a dispersive medium. To have the graphs of A7.5.18 and A7.5.19 as functions of "H" we have to remember what we said about Figure A7.4.2, i.e., the range of variability of "H" must have a minimum value that is greater than the minimum value required to saturate the ferrite. If we also want to examine the behavior of the previous equations, we must consider the effect of "H" and magnetization. In first analysis, we may use the hysteresis diagram seen in A7.3 point b, to remember the relationship between "$\mathcal{M}$" and "H." If we denote with "$\mathcal{M}(H)$" the function representing the curve "OC" of Figure A7.3.1, the two previous equations may be synthetically written as:

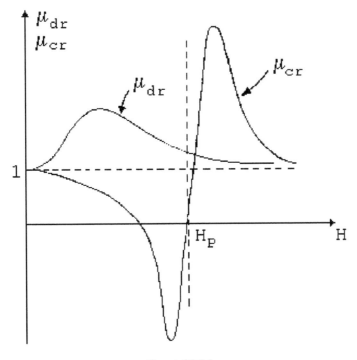

Figure A7.5.5

analogy to the terminology used for waveguide as seen in Appendix A2, and will be useful to synthetically explain the field displacement effect of ferrites as will be seen later in Section A7.14.

Another interesting point of view may be obtained by drawing the phase velocity of the concordant and discordant waves. From A7.5.15 and A7.5.16, and remembering from Appendix A2 that the phase velocity "v_f" is given by:

$$V_f \overset{\perp}{=} \omega/\beta$$

we may calculate the ratio "R_f" between the phase velocity and the speed of light in the ferrite when this is evaluated like a medium with "ε," and $\mu_i = 1$. We obtain:

$$R_{fd} \equiv 1/\sqrt{\mu_d} \tag{A7.5.22}$$

$$R_{fc} \equiv 1/\sqrt{\mu_c} \tag{A7.5.23}$$

The graphs of these two equations, in the case of zero losses, are represented in Figure A7.5.6.

Since the imaginary part of "R_{fd0}" is always zero, it has not indicated in the figure. As it is clear from the figure, when the frequency of the wave is equal to precession one created by "H," then $v_{fd} = 0$. This means that the concordant wave is locked inside the ferrite. In addition, another frequency "$\omega_p + \omega_m$" exists where $v_{fd} = \infty$. This last result, of course, does not mean that energy propagates at infinite speed, since it should be clear from Appendix A2 that electromagnetic energy propagates at group speed. It is also interesting to note that the phase velocities tend to reach the speed of light in the medium for higher and higher frequencies.

It is important at this point to say that all these concepts, which are relative to circular polarized concordant and discordant waves generated from the "TEM" wave, are also valid if a circular

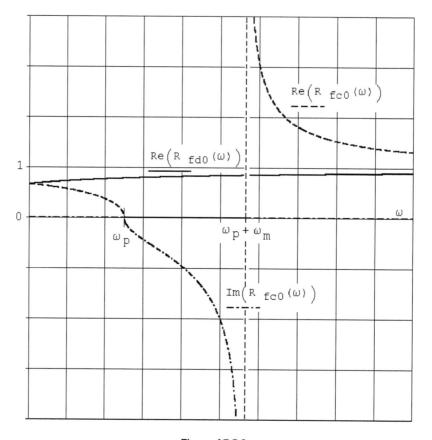

Figure A7.5.6

polarized wave is traveling inside the ferrite. That is, the effects we have studied are true independently of how a circular polarized electromagnetic wave is generated inside an isodirectional magnetized ferrite.

We now want to show that the division in circularly polarized waves only depends on the existence of both the two transverse components of "h." In fact, supposing $\underline{h}_t \equiv h_x \underline{x}_0$ from A7.5.12 and A7.5.13 we have:

$$\beta^2 h_x = \omega^2 \mu_0 \varepsilon \mu_p h_x \qquad (A7.5.24)$$

$$0 = -j\omega^2 \mu_0 \varepsilon \mu_\ell h_x \qquad (A7.5.25)$$

From A7.5.24, which is the only physical solution, we can say that if our "TEM" wave is linear polarized inside the ferrite it remains linear polarized along its travel, and only one phase constant exists. However, we will show in the next section that in this case another important physical phenomena occurs, the so-called "Faraday rotation," i.e., the rotation of the plane of linear polarization.

To conclude this section we have to say that in some literature on the topic it is possible to find the ferrite described by a 2×2 matrix of the form:

$$\left[\mu_{fTEM}\right] \stackrel{\perp}{=} \begin{bmatrix} \mu_p & j\mu_\ell \\ -j\mu_\ell & \mu_p \end{bmatrix} \qquad (A7.5.26)$$

This happens because the $h_z = 0$ for the "TEM" wave in our coordinate system, so that all the fields may be represented by two element vectors; for instance, the "$\underline{h}$" field may be written as:

$$[\underline{h}] = \begin{bmatrix} \underline{x}_0 & \underline{y}_0 \end{bmatrix} \begin{bmatrix} h_x \\ h_y \end{bmatrix} \tag{A7.5.27}$$

Anyway, we prefer to use the 3×3 matrices defined previously since the 2×2 can confuse the reader in other studies where non- "TEM" waves or nonisodirectional propagation are involved.

A7.6 LINEAR POLARIZED UNIFORM PLANE WAVE INSIDE AN ISODIRECTIONAL MAGNETIZED FERRITE: THE FARADAY ROTATION

In this section we will study an important phenomenon of ferrite called the "Faraday rotation."

Let us consider a plane wave with linear polarization, which after its propagation in the vacuum, starts to travel inside an isodirectional magnetized ferrite. The static magnetic field is, of course, of enough strength to saturate the ferrite, as we intend when we speak about "magnetized ferrite" unless otherwise noted. Let us introduce a Cartesian coordinate system, with the static magnetic field "$\underline{H}_{sz}$" oriented in the "z" direction and with the transverse magnetic field "$\underline{h}_t$" of the wave oriented along "x" so that we may write:

$$\underline{h}_t = h_x \underline{x}_0 \tag{A7.6.1}$$

Note that "$\underline{h}_t$" is orthogonal to "$\underline{H}_z$."

The spatial dependence is defined by the usual factor "$e^{(-j\beta z)}$," writing:

$$\underline{h}(z) = \underline{h}_t \, e^{(-j\beta z)} \tag{A7.6.2}$$

"h_x" is the only magnetic component of our wave since by hypothesis we are examining a linear polarized plane wave. It is clear that we do not make an error if we introduce a null "h_y" field so written:

$$\underline{h}_y \overset{\perp}{=} \left(j\frac{h_x}{2} - j\frac{h_x}{2} \right) \underline{y}_0 \tag{A7.6.3}$$

Summing this field to A7.6.1 we may write:

$$\underline{h}_t \equiv \left(\frac{h_x}{2} + \frac{h_x}{2} \right) \underline{x}_0 + \left(j\frac{h_x}{2} - j\frac{h_x}{2} \right) \underline{y}_0 \equiv h_x \underline{x}_0 \tag{A7.6.4}$$

We may recognize in this equation two circular contro-polarized waves, that when propagating inside the ferrite are given by:

$$\underline{h}_d = \left(\frac{h_x}{2} \underline{x}_0 + j\frac{h_x}{2} \underline{y}_0 \right) e^{(-j\beta_d z)} \tag{A7.6.5}$$

$$\underline{h}_c = \left(\frac{h_x}{2} \underline{x}_0 - j\frac{h_x}{2} \underline{y}_0 \right) e^{(-j\beta_c z)} \tag{A7.6.6}$$

where "β_c" and "β_d" are given by A7.5.15 and A7.5.16.

It is important to observe that a difference exists between these circular polarized waves and the two ones we have studied in the previous section. There, from a "TEM" wave, two circular polarized waves are effectively generated inside the ferrite. Here, we have rewritten a linear polarized wave as the sum of two circular contro-polarized waves; this notation, mathematically exact, will help us to understand the Faraday rotation encountered by our linear polarized plane wave inside an isodirectional magnetized ferrite.

After this note, let us rewrite the two previous equations joining together the components along the same axis. We have:

$$\underline{h}_x = \left[\frac{h_x}{2} e^{(-j\beta_d z)} + \frac{h_x}{2} e^{(-j\beta_c z)} \right] \underline{x}_0 \tag{A7.6.7}$$

$$\underline{h}_y = \left[j\frac{h_x}{2} e^{(-j\beta_d z)} - j\frac{h_x}{2} e^{(-j\beta_c z)} \right] \underline{y} \tag{A7.6.8}$$

If now we rewrite "β_d" and "β_c" according to:

$$\beta_d \overset{\perp}{=} \frac{\beta_c + \beta_d}{2} - \frac{\beta_c - \beta_d}{2} \overset{\perp}{=} \beta_+ - \beta_- \tag{A7.6.9}$$

$$\beta_c \overset{\perp}{=} \frac{\beta_c + \beta_d}{2} + \frac{\beta_c - \beta_d}{2} \overset{\perp}{=} \beta_+ + \beta_- \tag{A7.6.10}$$

where, of course, we have:

$$\beta_+ \overset{\perp}{=} \frac{\beta_c + \beta_d}{2} \tag{A7.6.11}$$

$$\beta_- \overset{\perp}{=} \frac{\beta_c - \beta_d}{2} \tag{A7.6.12}$$

and insert them in A7.6.7 and A7.6.8, after doing simplifications we obtain:

$$\underline{h}_x = h_x e^{(-j\beta_+ z)} \cos(\beta_- z) \underline{x}_0 \tag{A7.6.13}$$

$$\underline{h}_y = -h_x e^{(-j\beta_+ z)} \sin(\beta_- z) \underline{y}_0 \tag{A7.6.14}$$

It is now important to note that the spatial phase terms of "$\underline{h}_x$" and "$\underline{h}_y$" are the same for both these vectors, since they are "$\beta_+ z$" and "$\beta_- z$." Note that since "$\underline{h}_x$" moves as cosinus while "$\underline{h}_y$" moves as sinus the resultant vector rotates clockwise moving along the direction of propagation "z," as shown in Figure A7.6.1.

This is the phenomenon called the "Faraday rotation," in honor of the English physicist and chemist Michael Faraday, (born at Newington in 1791 and died at Hampton Court in 1867), who

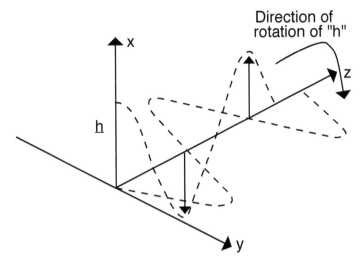

Figure A7.6.1

was the first to prove the movement of electrical circuits when inside magnetic fields. Do not confuse this movement of the propagation plane with a circular polarization of the wave. In fact, temporarily we have that the resultant vector "$\underline{h}$" inside the ferrite always moves on a line, i.e., it is always linearly polarized. In fact we have:

$$\underline{h}(z) = h_x \left[\cos\left(\beta_- z\right) \underline{x}_0 - \sin\left(\beta_- z\right) \underline{y}_0 \right] e^{(-j\beta_+ z)} \qquad (A7.6.15)$$

and, as seen in Appendix A2:

$$\underline{h}(z,t) \equiv \text{Im}\left[\underline{h}(z) e^{(-j\omega t)} \right] \qquad (A7.6.16)$$

Executing the calculations at the right member of the previous equation we have:

$$\underline{h}(z,t) = h_x \left[\cos\left(\beta_- z\right) \underline{x}_0 - \sin\left(\beta_- z\right) \underline{y}_0 \right] \sin\left(\omega t - j\beta_+ z\right) \qquad (A7.6.17)$$

from which it is clear that when a point "z" is fixed, then "$\underline{h}(z,t)$" has only one sinusoidal time dependence, i.e., it moves linearly polarized.

Two important points have now to be studied, which show us that the direction of rotation is only dependent on the direction of the static saturating magnetic field "H_{zs}" and not on the direction of propagation of the wave "z_0." Let us start to note from A7.4.41 and A7.4.42 that "μ_{p0}" does not change sign if we reverse the direction of "H_{sz}" since both "ω_p" and "ω_m" change their sign, while "$\mu_{\ell 0}$" changes its sign. From A7.5.15 and A7.5.16 it follows that "β_{d0}" and "β_{c0}" interchange, and consequently from A7.6.11 and A7.6.12 "β_+" does not change its sign while "β_-" does. Remembering now that the cosinus function does not change its sign if we change the sign to its argument, while the sinus function does, from A7.6.17 we may recognize that if we change the direction of "H_{sz}" then the "h" field inside the ferrite becomes:

$$\underline{h}(z,t) = h_x \left[\cos\left(\beta_- z\right) \underline{x}_0 + \sin\left(\beta_- z\right) \underline{y}_0 \right] \sin\left(\omega t - j\beta_+ z\right) \qquad (A7.6.18)$$

which is a field that rotates counterclockwise along the direction of propagation "$\underline{z}_0$." If now we remember the definition of tangent, for the angle "θ" of "$\underline{h}$" with respect to "$\underline{y}_0$" we may write:

$$\tan(\theta) = \frac{h_y(z)}{h_x(z)} \qquad (A7.6.19)$$

Using A7.6.13 and A7.6.14 in the previous equation we have:

$$\theta = \frac{\beta_d - \beta_c}{2} \qquad (A7.6.20)$$

Let us define as the positive direction of the angle the clockwise one, when looking in the same direction of propagation. In this case we have that if the wave propagates toward "$-\underline{z}_0$" but we are looking toward "$\underline{z}_0$" then from our point of view "θ" changes its sign. In addition, when we introduced A7.5.14 we said that both "β_d" and "β_c" change their signs when the waves propagates toward "$-\underline{z}_0$." The result is that the double change of sign in A7.6.20 when the wave changes the direction of propagation makes the angle "θ" independent of this direction. This important nonreciprocal effect of the Faraday rotation has been widely used to realize ferrite isolators, as we will show later.

We want to conclude this section indicating to the reader that the entity of rotation by unit length is dependent on the intensity of the applied static magnetic field. The nonreciprocal effect is maximum when the ferrite is statically magnetized to saturation. A typical graph of "θ" as function of "H" is given in Figure A7.6.2.

We see that as the applied static magnetic field reaches a value "H_p" which creates a precession frequency equal to the frequency of the electromagnetic wave, then the rotation becomes infinite, of course theoretically, and in the case of zero losses.

When the Farady rotation is used for a particular device, the applied magnetic field must range in the values required for saturation of ferrite, but with a frequency intensity equal to that of the electromagnetic wave, otherwise the device will not work.

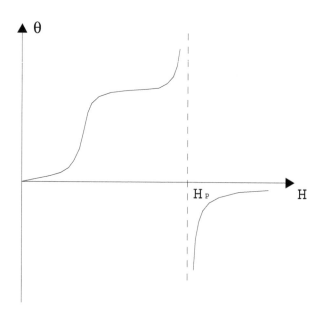

Figure A7.6.2

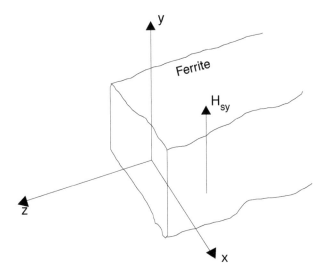

Figure A7.7.1

A7.7 ELECTROMAGNETIC WAVE INSIDE A TRANSVERSE MAGNETIZED FERRITE

In this section we will show that some physical properties of ferrite are only dependent on the reciprocal orientation between the time varying magnetic field "$\underline{h}$" and the static magnetic field "$\underline{H}_s$."

Let us assume that an electromagnetic wave propagates inside a saturated ferrite with internal static magnetic field "$\underline{H}_s$." Let us use a Cartesian coordinate system and write:

for the static magnetic field:

$$\underline{H}_s = H_{sy}\underline{y}_0 \qquad (A7.7.1)$$

for the magnetic field of the wave:

$$\underline{h} = h_x\underline{x}_0 + h_y\underline{y}_0 + h_z\underline{z}_0 \qquad (A7.7.2)$$

$$\underline{h}(x,z) = \underline{h}e^{(-k_x x \, -k_z z)} \qquad (A7.7.3)$$

Of course, the electric field of the wave will have equations similar to the two previous ones. Note the result of A7.7.3 is that a wave is propagating orthogonally to the direction "$\underline{y}_0$" of the static saturating internal magnetic field "$\underline{H}_s$," and the wave has no spatial dependence along "$\underline{y}_0$." We will call this situation the case of an electromagnetic wave inside a "transverse magnetized" ferrite. This scenario may be graphically represented as that in Figure A7.7.1. Since we have decided to mark as "y" the axis where "H" is directed, we have to use "$[\mu_y]$" in our problem. So, the Maxwell's equations to be resolved in our system are:

$$\underline{\nabla} \otimes (\underline{e}) = -j\omega[\underline{u}]\left[\mu_y\right][h] \qquad (A7.7.4)$$

$$\underline{\nabla} \otimes (\underline{h}) = j\omega\varepsilon\underline{e} \qquad (A7.7.5)$$

where "$[\mu_y]$" is the ferrite permeability matrix given by A7.4.64. So executing calculations in A7.7.4 and equating terms with same components along the coordinate axis, we have:

$$k_z e_y \underline{x}_0 = ! -j\omega\mu_0 \left(h_x \mu_p - j\mu_\ell h_z \right) \underline{x}_0 \qquad (A7.7.6)$$

$$\left(k_x e_z - k_z e_x \right) \underline{y}_0 = ! -j\omega\mu_0 h_y \underline{y}_0 \qquad (A7.7.7)$$

$$-k_x e_y \underline{z}_0 = ! -j\omega\mu_0 \left(h_z \mu_p + j h_x \mu_\ell \right) \underline{z}_0 \qquad (A7.7.8)$$

Doing similar calculations for A7.7.5 we have:

$$k_z h_y \underline{x}_0 = ! -j\omega\varepsilon e_x \underline{x}_0 \qquad (A7.7.9)$$

$$\left(k_x h_z - k_z h_x \right) \underline{y}_0 = ! j\omega\varepsilon e_y \underline{y}_0 \qquad (A7.7.10)$$

$$-k_x h_y \underline{z}_0 = ! j\omega\varepsilon e_z \underline{z}_0 \qquad (A7.7.11)$$

In the previous six equations we may recognize two groups, one that only contains "h_x," "h_z," and "e_y" and another that only contains "e_x," "e_z," and "h_y." The wave that contains "h_x," "h_z," and "e_y" is a "TE"* wave since the only transverse component to the direction of propagation is the electric field "e_y." This situation is represented in Figure A7.7.2. Similarly, the wave that contains "e_x," "e_z," and "h_y" is a "TM" wave, since the only transversal component to the direction of propagation is the electric field "h_y." This situation is indicated in Figure A7.7.3. So, the first result we have is that our wave when propagating inside the ferrite, with spatial independence only orthogonal to the direction of the applied static field, has divided itself in two waves, a result similar to that deduced in Section A7.5. So, we may expect that these two waves are subjected to different actions by the saturated ferrite. To examine what happens, let us begin to resolve the system of equations for the "TE" wave, that is the group of Equations A7.7.6, A7.7.8, and A7.7.10. This is an homogeneous system of equations, which has a nonzero solution when the determinant of the coefficient is zero. Doing such calculations we obtain the condition that has to satisfy "k_x" and "k_z"** to assure the solution of Maxwell's equation, that is:

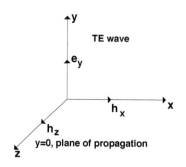

Figure A7.7.2

* See Appendix A2 for definition.
** Remember we have assumed that the e.m. fields have zero dependence with the "y" axis, i.e., $k_y = 0$, as indicated by A7.7.3.

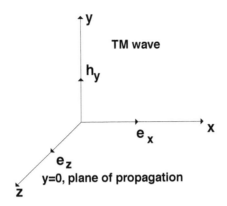

Figure A7.7.3

$$k_x^2 + k_z^2 = ! -\omega^2\mu_0\varepsilon\frac{\mu_p^2 - \mu_\ell^2}{\mu_p} \overset{\perp}{=} k_{teq\perp}^2 \qquad (A7.7.12)$$

The term that indicates the permeabilities ratio in A7.7.12 is called the "equivalent permeability" and is indicated with "$\mu_{eq\perp}$," i.e.:

$$\mu_{eq\perp} \overset{\perp}{=} \frac{\mu_p^2 - \mu_\ell^2}{\mu_p} \qquad A7.7.13)$$

The subscript "$\perp$" we have used in the two previous equations will remind us that these expressions are obtained for the case of "$\underline{h}$" which is orthogonal to "$\underline{H}_s$." Also note from A7.7.12 that "$k_{teq\perp}^2$" is a negative number. Observing Figure A7.7.1 and Figure A7.7.2 we note that the "TE" mode of our electromagnetic wave has the magnetic field that is orthogonal to the direction of application of the static magnetic field "H_s."

Let us now resolve the equation system for the "TM" mode, that is the group of Equations A7.7.7, A7.7.9 and A7.7.11. Proceeding in a similar way to what we did for the "TE" mode, we obtain:

$$k_x^2 + k_z^2 = ! - \omega^2\mu_0\varepsilon \overset{\perp}{=} k_t^2 \qquad (A7.7.14)$$

Note that in this equation any term of the permeability matrix "$[\mu_y]$,"does not appear, which means that this mode propagates inside a ferrite as if it was a nonferrimagnetic medium, with relative permeability equal to one. From A7.7.14 we see that also "k_t^2" is a negative number, as "$k_{teq\perp}^2$." Observing Figure A7.7.1 and Figure A7.7.3 we note that the "TM" mode of our electromagnetic wave has the magnetic field that is parallel to the direction of application of the static magnetic field "H_s."

The graph of "$\mu_{eq\perp}$" as a function of frequency, and with no losses in the ferrite, is given in Figure A7.7.4. We have indicated this quantity as "$\mu_{eq\perp0}$." Note that at the precession frequency "$\mu_{eq\perp0}$" is not infinite, while this happens at the frequency "ω_∞" which is higher than "ω_p."

The graph of "$\mu_{eq\perp0}$" as a function of "H" is indicated in Figure A7.7.5, where frequency is assumed to be constant. We must repeat here what was said at the time of Figure A7.5.3, that is, the graph at the left of the point "$N_d\mathcal{M}_s$" is relative to values of "H" too low to saturate the ferrite.

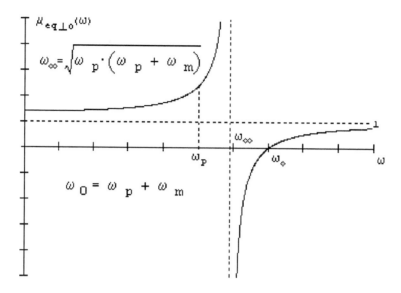

Figure A7.7.4

In analogy to Figure A7.7.4, we have that at the precession field "H_p" the permeability "$\mu_{eq\perp0}$" is not infinite while this happens at the value "H_∞."

It is very important to note the equivalence between the graph of "μ_{c0}" in Figure A7.5.2 and that of "$\mu_{eq\perp0}$" in Figure A7.7.4. This is not a coincidence since the circular polarized concordant wave inside an isodirectional magnetized ferrite has the magnetic field that is orthogonal to the directions of propagation and that of "$\underline{H}_s$." But this also is the case for our "TE" wave in a transversal magnetized ferrite, as may be recognized after viewing Figures A7.7.1 and A7.7.2.

We now want to show that "$\mu_{eq\perp}$" is only dependent on the reciprocal orientation among "$\underline{H}_s$," the direction of propagation of the wave, and the directions of the nonzero spatial dependence of the wave. This situation is contrary to what happens for "$[\chi]$" and "$[\mu]$," which are dependent on the coordinate system. Let us suppose the following expression for the magnetic field of the wave:

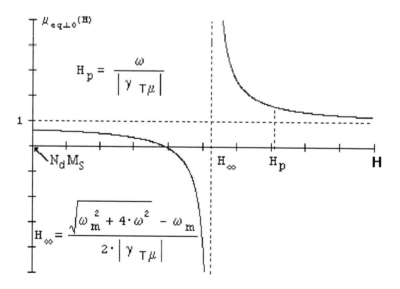

Figure A7.7.5

$$\underline{h}(y, z) = \underline{he}^{(-k_y Y - k_z Z)} \qquad (A7.7.15)$$

i.e., the wave is propagating orthogonally to the direction "$\underline{x}_0$." The vector "$\underline{h}$" is still given by A7.7.2. In the same direction "$\underline{x}_0$" we apply the static saturating internal magnetic field "$\underline{H}_s$." Now:

1. Employing "$[\mu_x]$," given in A7.4.68, and inserting A7.7.15 in the Maxwell's equations A7.7.4)
2. Proceeding as we did in obtaining A7.7.6 through A7.7.11,

we have two groups of equations, one that only contains "h_y," "h_z," and "e_x" relative to "TE" mode and another that only contains "e_y," "e_z," and "h_x" relative to "TM" mode. The conditions that have to satisfy "k_y" and "k_z" to assure the solution of Maxwell's equations are:

$$k_y^2 + k_z^2 = ! - \omega^2 \mu_0 \varepsilon \frac{\mu_p^2 - \mu_\ell^2}{\mu_p} \pm k_{teq\perp}^2 \qquad (A7.7.16)$$

$$k_y^2 + k_z^2 = ! - \omega^2 \mu_0 \varepsilon \pm k_t^2 \qquad (A7.7.17)$$

respectively for "TE" and "TM" modes. These two equations are equal respectively to Equations A7.7.12 and A7.7.14, from which we may see that "$\mu_{eq\perp}$" is independent of the coordinate system, while it only depends on the reciprocal orientation between "$\underline{H}_s$," and the magnetic field of the wave.

Of course, equations for "$k_{teq\perp}^2$" and "k_t^2" similar to A7.7.16 and A7.7.17 can be obtained using "$[\mu_z]$" and a wave whose magnetic field is given by $\underline{h}(x,y) = \underline{he}^{(-k_x X - k_y Y)}$, resulting in:

$$k_x^2 + k_y^2 = ! - \omega^2 \mu_0 \varepsilon \frac{\mu_p^2 - \mu_\ell^2}{\mu_p} \pm k_{teq\perp}^2 \qquad (A7.7.18)$$

$$k_x^2 + k_y^2 = ! - \omega^2 \mu_0 \varepsilon \pm k_t^2 \qquad (A7.7.19)$$

It is important to show that the existence of "$[\mu_{eq\perp}]$" is in general a function of the number of "h" components in the coordinate and the transverse spatial dependence of the original wave. To show it, let us assume the electromagnetic case governed by Equations A7.7.6 through A7.7.11, and write $\underline{h} = h_x \underline{x}_0$. A7.7.6 through A7.7.11 become:

$$k_z e_y \underline{x}_0 = ! - j\omega\mu_0 h_x \mu_p \underline{x}_0 \qquad (A7.7.20)$$

$$-k_x e_y \underline{z}_0 = ! \omega\mu_0 h_x \mu_\ell \underline{z}_0 \qquad (A7.7.21)$$

$$-k_z h_x \underline{y}_0 = ! j\omega\varepsilon e_y \underline{y}_0 \qquad (A7.7.22)$$

which represent a "TE" wave for our reference system shown in Figure A7.7.1. Note that in this case there is no wave splitting. Simply manipulating these equations we have:

$$k_z^2 = -\omega^2 \mu_0 \mu_p \varepsilon \qquad (A7.7.23)$$

$$k_x^2 = \omega^2 \mu_0 \varepsilon \frac{\mu_\ell^2}{\mu_p} \qquad (A7.7.24)$$

or:

$$k_z^2 + k_x^2 = -\omega^2 \mu_0 \varepsilon \frac{\mu_p^2 - \mu_\ell^2}{\mu_p} \pm k_{teq\perp}^2 \qquad (A7.7.25)$$

that is A7.7.12, also if the original wave has only one magnetic component "h_x." Similarly, for our reference system shown in Figure A7.7.1 let us write $\underline{h} = h_y \underline{y}_0$. The Equations A7.7.6 through A7.7.11 become:

$$\left(k_x e_z - k_z e_x\right) \underline{y}_0 =! -j\omega\mu_0 h_y \underline{y}_0 \qquad (A7.7.26)$$

$$k_z h_y \underline{x}_0 =! j\omega\varepsilon e_x \underline{x}_0 \qquad (A7.7.27)$$

$$-k_x h_y \underline{z}_0 =! j\omega\varepsilon e_z \underline{z}_0 \qquad (A7.7.28)$$

which represent a "TM" wave. Note that also in this case there is no wave splitting. Simply manipulating these equations we have:

$$k_x^2 + k_z^2 =! -\omega^2\mu_0\varepsilon \pm k_t^2 \qquad (A7.7.29)$$

that is the same result of A7.7.14.

We can conclude these notes recognizing that in all ferrimagnetic problems it is very important to know the direction of the applied static magnetic field, the number of the electromagnetic wave components, and the wave spatial dependence. It is important to give a table where we summarize all the change of signs for the ferrite permeability elements and for the wave elements, when "$\underline{H}$" or the propagation change direction, since it will be very useful in our studies. So, we have the following Table A7.7.1:

Table A7.7.1

Change of sign when $\underline{H}$ change dir.		Change of sign when prop. change dir.	
"μ_p"	NO	"μ_p"	NO
"μ_ℓ"	YES	"μ_ℓ"	NO
"$\mu_{eq\perp}$"	NO	"$\mu_{eq\perp}$"	NO
"β"	NO	"β_z"	YES

Also note it is not difficult to realize the reciprocal directions between all these vectors, as was shown in the figures of this section. Think of the fundamental mode "qTEM"* in a microstrip realized on a substrate of ferrite. If we magnetize with "H_s" in a direction orthogonal to the ground plane, then the magnetic field of the wave is "$\mu_{eq\perp}$." In this way we may create interesting nonreciprocal devices, as seen in Chapter 7, called "field displacement devices." Of course, similar components may be realized in stripline, as shown in Chapter 8, since the fundamental mode in a stripline is the "TEM." In this appendix in Section A7.14 we will analyze field displacement devices in waveguide technology, the first microwave components using this phenomenon.

A7.8 CONSIDERATIONS ON DEMAGNETIZATION AND ANISOTROPY

In Section A6.6 of Appendix A6 we discussed the demagnetization field, anisotropy, and magnetostriction inside ferromagnetic materials. These concepts equally apply to ferrimagnetic

---

* "qTEM" means "quasi TEM," where quasi means almost.

materials, like ferrite, since mainly the difference between ferro- and ferrimagnetic materials is that the last are good insulators.

We also have to say that in the study we made concerning ferrite we assumed that the internal fields are uniform. In principle this is not always true, since demagnetization and anisotropy fields are strongly dependent on the geometric aspect of our ferrimagnetic material and usually they are not uniform inside the specimen. The relationship inside the ferrite among demagnetization "H_d," anisotropy "H_a," and external "H_e" magnetic fields is as follow:

$$\underline{H} = \underline{H}_e + \underline{H}_d + \underline{H}_a \tag{A7.8.1}$$

where "$\underline{H}$" is the total internal magnetic field. Usually "H_a" is lower than "H_d" and may be neglected. From this equation we recognize that if "$\underline{H}_d$" and "H_a" are nonuniform, the same happens for "$\underline{H}$," also if "$\underline{H}_e$" is uniform inside the ferrite.

The most important consequence of not considering the fields "H_d" and "H_a" is that the precession frequency has a slightly different value from what we have from the known relationship:

$$\omega_p = -\gamma_{T\mu} H \tag{A7.4.7}$$

If we take into account high crystalline ferrites so that anisotropy effects may be neglected, it is not very complicated conceptually to obtain the new precession frequency. Then, let us consider the two fundamental relationships between magnetic induction "b," magnetic field "h," and magnetization "m," which in our case of ferrite, are:

$$[b] = [\mu][h] \tag{A7.8.2}$$

$$[b] = \mu_0([h] + [m]) \tag{A7.8.3}$$

where the matrices "[b]," "[h]," and "[m]" are column 1×3 and "[μ]" is squared 3×3. Let us assume in the physical situation of A7.4.1, i.e., with "$\underline{H} = H_{sz} \underline{z}_0$." Then, using for "[$\mu$]" the expression of "[$\mu_z$]" given in A7.4.28, the previous equations become:

$$b_x = \mu_0\left(1 + x_p\right)h_x + j\mu_0 x_\ell h_y \tag{A7.8.4}$$

$$b_y = -j\mu_0 x_\ell h_x + \mu_0\left(1 + x_p\right)h_y \tag{A7.8.5}$$

$$b_z = \mu_0 h_z \tag{A7.8.6}$$

$$b_x = \mu_0\left(h_x + m_x\right) \tag{A7.8.7}$$

$$b_y = \mu_0\left(h_y + m_y\right) \tag{A7.8.8}$$

$$b_z = \mu_0 h_z \tag{A7.8.9}$$

Inserting the second members of A7.8.7 through A7.8.9 in the first members of A7.8.4 through A7.8.6 we have:

$$h_x + m_x = \left(1 + x_p\right)h_x + jx_\ell h_y \tag{A7.8.10}$$

A7.9 THE BEHAVIOR OF NOT STATICALLY SATURATED FERRITE

It is very important to discuss the elements of "$[\chi]$" given in A7.4.23 and the starting hypothesis used. The most important condition we assumed was the existence of a static saturating magnetic field "H_s" applied in a direction of a Cartesian coordinate system used inside the ferrite, in a direction that we assumed to be "z." If we remember the fundamental concepts of magnetism given in Appendix A6, when a static saturating field is applied to a ferrite, all the Bloch boundaries* are moved and distorted in order to realize a whole macro domain inside the ferrite. In addition, the molecular field "$\underline{B}_m$" of the macro domain becomes aligned with the direction of "$\underline{H}_s$." Ferrites used in microwave devices, which usually require a static saturation, are known to require an easily reachable value of "$\underline{H}_s$" for saturating, typically ranging from some tenths to some hundreds of A.t/m (Ampere.turn/meter).

We now want to ask ourselves what happens when "$\underline{H}_s$" is not strong enough to saturate, or it is just null.

In the case of $H_s = 0$, observing A7.4.21 and A7.4.22 we could conclude that "x_p" and "x_ℓ" are both equal to zero since $\mathcal{M}_s = 0$ if $H_s = 0$, because we have defined "$\mathcal{M}_s$" as the saturation magnetization reached by the application of "H_s." A similar discussion may be done for "$[\mu_z]$" given in A7.4.28, from A7.4.29 to A7.4.31 obtaining that $\mu_p \equiv 1$ and $\mu_\ell \equiv 0$. But the conclusions $x_p = 0$, $x_\ell = 0$, $\mu_\ell \equiv 0$, and $\mu_p \equiv 1$ would be generated by a conceptual error, since the formulas were created with the assumption of an existing "H_s" not equal to zero.

Important studies have been made[9] for the behavior of ferrites when $H_s = 0$ which show that "μ_ℓ" is really equal to zero, while "μ_p" assumes the value:

$$\mu_{pH0} = 1/3\left[1 + 2\left(1 - \omega_{\pi m}^2/\omega^2\right)^{0.5}\right] \qquad (A7.9.1)$$

where "$\omega_{\pi m}$" is the value of "ω_m" in the "CGSA" system, i.e.:

$$\omega_{\pi m} = -\gamma_{T\mu}\left(4\pi M_s\right) \qquad (A7.9.2)$$

The value "M_s" in the previous equation, and which implicitly appears through "$\omega_{\pi m}$" in A7.9.1, cannot be the magnetization created by a static magnetic field, since now $H_s = 0$, but it is the reachable magnetization of the ferrite under test. All the quantities in A7.9.1 and A7.9.2 must be measured with the "CGSA" unit system, so in A7.9.2 "M_s" must be measured in "Oersted." We want to remember again that caution must be used when using the "CGSA" or "MKSA" unit systems. For instance, the relationship between Oersted and Amp.t/m is:

$$1 \text{ Oersted} = 10^3/4\pi \text{ Amp.t/m} \qquad (A7.9.3)$$

In addition, in "CGSA" the quantity "$4\pi M_s$" is called "magnetization," and its measuring dimension is the "Gauss." The relationship between Gauss and Weber/m² of "MKSA" is:

$$1 \text{ Gauss} = 10^{-4} \text{ Web}/\text{m}^2 \qquad (A7.9.4)$$

Remember that in "MKSA" the dimension of magnetization is Amp.t/m, i.e., like the magnetic fields.

---

* See Appendix A5.

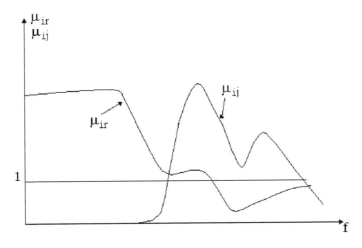

Figure A7.9.1

One important aspect of A7.9.1 is that for an angular frequency "ω" below "$\omega_{\pi m}$" the quantity "μ_{pH0}" is complex. This means that an electromagnetic wave with $\omega < \omega_{\pi m}$ that propagates spatially inside this ferrite according to "$e^{-j\beta z}$" with $\beta = \omega(\mu_{pH0}\varepsilon)^{0.5}$, will have attenuation due to the fact that when "μ_{pH0}" is complex "β" also becomes complex, and the term "$e^{-j\beta z}$" will become the product of two exponentials, one real and one imaginary, with the real one responsible for attenuation. These magnetic losses must not be confused with the losses due to the resistivity of the material.

Another characteristic of ferrites when no static magnetic field has been applied may be deduced from its hysteresis loop. As has been seen in A7.3 point b, when $H = 0$ the work point of ferrites is in the region where the permeability is defined as "initial." This operating region is mainly used in RF and UHF applications, as Section A7.15 will show. The initial permeability "μ_i" is a complex quantity that can be written as:

$$\mu_i \stackrel{\perp}{=} \mu_{ir} + j\mu_{ij} \qquad (A7.9.5)$$

with "μ_{ir}" and "μ_{ij}," respectively, the real and imaginary parts of "μ_i." Exact formulation of "μ_i" does not exist due to the complex connections among physics, chemistry, and magnetism involved in ferrites. Typical behavior of "μ_i" as a function of frequency may be deduced after experiments, giving graphs as indicated in Figure A7.9.1. The expression A7.9.1 represents in some way the graph of Figure A7.9.1 in that range of frequency where "μ_{ir}" grows to reach the value "1." Today, the maximum frequency where "μ_{ij}" starts to rise is near 2.5 GHz, also depending on the type of ferrite. For instance, some ferrites optimized for audio frequency applications have a rise point of "μ_{ij}" near some MHz. From Figure A7.9.1 it is interesting to note that a region exists for "μ_{ir}" where it does not change so much with frequency. This means, remembering what we said in Appendix A2, that ferrite is not dispersive. This characteristic is very important when ferrite is used for wideband circuits in the UHF region, as Section A7.15 will show.

Let us discuss the case when the static magnetic field is not zero, but not enough to saturate the ferrite. If we remember the topics of magnetism treated in Appendix A6, it is evident that we cannot treat the ferrite as only one macro domain. In this case we have to define a magnetization vector "$\mathcal{M}_{nsz}$" which is an average along the direction "z" of application of "H." In this situation "x_p" is negligible with respect to "x_ℓ" which assumes the expression:

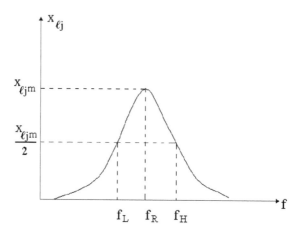

Figure A7.10.1

From the previous equation we may see that the lower the loss factor, "α," the thinner the line width with the characteristic that when $\alpha = 0$ the "$x_{\ell j}$" is null everywhere with the exception of resonance, where it becomes infinite.

A characteristic of all ferrites is that "α" increases as their crystal purity decreases, as in polycrystalline ferrites, with the consequent increase of linewidth. In contrast, monocrystalline ferrites have very narrow linewidth due to their high crystal purity and material homogeneity. Remember from Section A7.3 and A6.8c that low loss ferrites also have very narrow and sharp hysteresis diagrams.

Studies concerning the origins of losses and their relationships with the chemical composition of ferrites are still under development due to confluence in these subjects of many scientific branches like chemistry and magnetism. To a first approximation, losses are due to energy exchange between atomic spins, and between atoms and the crystal that composes the ferrite. In the next section we will define the most important losses inside ferrite.

A7.11 LOSSES IN FERRITES

In the most general case, the total loss "L_t" in a ferro-ferri-magnetic material is due to the actions of three terms[12] i e :

$$L_t = L_c + L_i + L_r \tag{A7.11.1}$$

where "L_c" are losses due to induced conduction current, "L_i" are losses inside the hysteresis loop and "L_r" are losses due to movements and resonances of Bloch boundaries.* Sometimes "L_r" are called "residual losses." Let us start to define in more detail the single components of A7.11.1.

a. The Conduction Losses "L

These losses are directly proportional to the frequency "f" of the electromagnetic wave propagating inside the material and inversely proportional to its resistivity "ρ" according to the relationship:

$$L_c = k_c \, f^2 / \rho \tag{A7.11.2}$$

* See Appendix A5 for Bloch's wall theory.

where "k_c" is a factor that is dependent on the geometric aspect of the material. This means that different material geometries (also if they are the same chemical material and are posed in the same position) may cause different attenuation to the electromagnetic wave propagating in the space. From the previous equation, remembering that the ratio between ferrite resistivity to ferromagnetic material may approach 10^{10} for some ferrites used in the microwave region, it is evident how "L_c" of ferrite is negligible with respect to the "L_c" for ferromagnetic materials. In any case, conduction losses are also negligible in ferrimagnetic materials, when compared to the other origin of losses.

b. The Hysteresis Loop Losses "L

Refer to the hysteresis loop indicated in Figure A7.3.1. In Appendix A6 we stated that the external static field "H" must work against the forces inside the material when creating the hysteresis loop. A similar case is when "H" moves the Bloch boundaries of the magnetic material. Remember that for a diamagnetic material, the field "H" does not do any work since diamagnetic materials have no hysteresis loop.

When we go along the hysteresis loop, the dissipated energy "E_d" inside the volume "V" is proportional to the area of this path, according to:

$$E_d = V \oint H dB \qquad (A7.11.3)$$

where the circular integral must be done on the hysteresis loop. From the previous equation it is evident that to lower the energy dissipated inside the material (energy that is taken from "H") it is desirable to have a material with no hysteresis. This is not possible, due to the existence of the Bloch boundaries between magnetic domains. So, what we can do is to use material with a hysteresis loop as sharp as possible, reducing more and more the impurity particles that do not belong to the material crystal. By controlling the manufacturing process, we can have low loss ferrites with a sharp and quite rectangular shape of the hysteresis. This kind of material is mainly used for RF transformers that require a high linearity between input and output, which cannot be reached if the hysteresis loop possesses the curved shape indicated in Figure A7.3.1. For a sinusoidal time varying relationship of "H" in the primary winding, a same time varying relationship of "B" inside the ferrite does not correspond just because the graph of Figure A7.3.1 is not linear. So, at the secondary winding the induced voltage is not of an exact sinusoidal time varying relationship. It is also the case that when the material is very linear, if too much signal is applied to the winding of the transformer the ferrite will saturate, distorting the output signal.

In any case the losses "L_i" are proportional to the frequency "f" of the signal, according to the following relationship:

$$L_i = k_i f \qquad (A7.11.4)$$

where "k_i" is a factor that is dependent on the geometric aspect of the material.

c. The Residual Losses "L

These losses are relative to movements and resonances of the Bloch boundaries. Experiments on nonmagnetized ferrite has shown that the imaginary part "μ_{ij}" of the initial permeability versus the frequency of the applied "RF" signal presents two peaks; one in the range of 10 to 100 Mhz and the other between 1 GHz and 2 GHz. This situation has been drawn in Figure A7.9.1. The position of these peaks may be changed, inside the specified range, by proper chemical composition of ferrite. Experimentally it has been shown that the higher the purity of the material, the lower the first peak. This result has led to the idea that the highest peak is due to the rotations of Bloch boundaries, since the behavior in the peak is as if the ferrite were composed of a big macrodomain, a situation that is nearer to the concept of the rotation of all the boundaries rather than the resonances of them. So, the first peak is attributed to the resonances of the Bloch boundaries.

Similar to the case of "L_i," the losses "L_r" are proportional to the frequency "f" of the signal, according to the following relationship:

$$L_i = k_r f \tag{A7.11.5}$$

where "k_r" is a factor that is dependent on the geometric aspect of the material.

The determination of the coefficients "k_c" "k_i" and "k_r" is quite complicated, and the procedures are still under study.

A7.12 ISOLATORS, PHASE SHIFTERS, CIRCULATORS IN WAVEGUIDE WITH ISODIRECTIONAL MAGNETIZATION

The first transmission lines used in the microwave region were the waveguides. As shown in Appendix A2, waveguides are really nonplanar transmission lines so, from this point of view, their study is outside the object of this book. It is not possible to avoid their study because in our case the analysis of waveguide loaded with ferrite, can help us to better understand the behavior of planar transmission lines loaded with ferrite which are also studied in our book. Note that waveguides are not dead since in the case of high power applications, let us say for power ratings higher than 100 watts, and where low loss microwave transmission lines are required, waveguides are still the better solution.

So, let us start with the first historically used device in waveguides loaded with ferrite, which uses the Faraday effect.

To have a good understanding of what we are going to explain, it is important to know the concepts of electromagnetic propagation inside a waveguide. To this purpose, the reader has the Appendix A2 where all the necessary information is given.

a. Nonreciprocal Isolators

A "nonreciprocal isolator," or only "isolator," is defined as a device that is able to permit the propagation of the e.m. energy in one direction while presenting the highest attenuation in the opposite direction. The nonreciprocal nature of the device is evident.

In the previous sections we have seen how ferrite presents a nonreciprocal effect, i.e., the Faraday rotation, when a linear polarized wave propagates inside it. The waveguide isolator just uses this phenomenon in a classical geometrical configuration indicated in Figure A7.12.1. A circular waveguide is connected at its extremes to two rectangular waveguides, so that these two form an angle of 45 degrees between their major axes. At the center of the circular waveguide is inserted a cylinder of ferrite, supported in this position by low loss material, of length near one half of that of the circular waveguide. The ferrite is magnetized, not at saturation, using a permanent magnet surrounding the circular waveguide or using a solenoid, wound on this waveguide. The use of solenoid permits more flexibility since by reversing the direction of current flowing in the solenoid, the direction of the magnetic field may be reversed. To explain the behavior of the isolator it is usual to refer to the electrical field propagating in the waveguides. This is not a problem since the electrical and magnetic fields propagating the energy are orthogonal, and to obtain the direction of the magnetic field it is enough to rotate 90 degrees.

Remembering the explanation of the Faraday effect seen in Section A7.6 it is simple to understand how the device works. Suppose that the static magnetic field magnetizing the ferrite cylinder is in the direction indicated by the arrow in Figure A7.12.1a and that the fundamental mode "TE_{10}" in

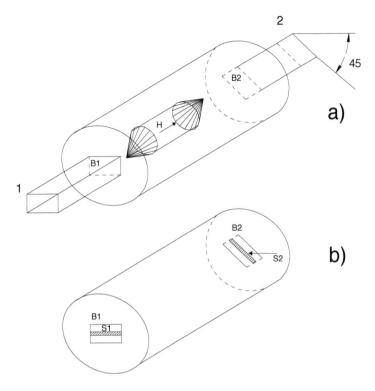

Figure A7.12.1

the rectangular waveguide is in isodirectional propagation. From Appendix A2 we remember that the magnetic field is linearly polarized in the center of the circular waveguide, and the same happens for the magnetic field in the middle of the rectangular waveguide. In this case, the wave, when passing from the rectangular waveguide to the circular one, will mainly generate inside it the fundamental mode "TE_{11}" which, as seen in Appendix A2, has the magnetic field linearly polarized in the center of the circular waveguide. When propagating inside this waveguide, this linear polarized magnetic field also propagates inside the ferrite where it starts to rotate. With the proper choice of "H" and/or length of the circular waveguide, the magnetic field will arrive to the aperture "B2" of the rectangular waveguide "2" rotated clockwise 45 degrees, i.e., in the proper direction to generate the fundamental mode "TE_{10}." As a result, when the signal propagates in the direction of the applied field "H," in this case in the direction "1–2," the e.m. wave will encounter low attenuation, typically in the range 1 to 2 dB depending on the frequency of the signal. When the signal propagates in the direction "2–1," the magnetic field still rotates clockwise 45 degrees, for the reasons seen in Section A7.6, and when it arrives at the aperture "B1" it is under cut-off and cannot generate propagation inside the rectangular waveguide "1." So, the reflected signal starts to travel in the direction "1–2," passes another time in the ferrite, rotates 45 degrees and arrives at "B2" under cut-off. It is still reflected at "B2," starts to travel in the direction "2–1," passes another time in the ferrite, rotates 45 degrees and arrives at "B1" in the proper direction to generate the fundamental mode "TE_{10}" but in opposition of phase to the "TE_{10}" applied in "1." If we could not apply a remedy to this phenomenon, some distortion on the passing signal would be generated by the isolator, and its isolation in the direction "2–1" would be low. To this purpose, one sept for each transition between rectangular to circular waveguides is applied, as indicated in Figure A7.12.1b. We can see how the septs are in the proper positions* to attenuate the electric

---

* See Appendix A2 for "TE_{10}" mode field components.

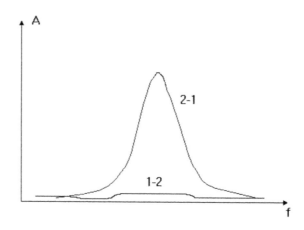

Figure A7.12.2

field, which tries to enter below cut-off in the rectangular waveguides and is consequently reflected back. In this explanation we have referred to magnetic field rotation, but in the literature the explanation can be found using the electric field as the reference field. Of course, this is only another way to explain the working principle, since to every rotation of magnetic field is associated a rotation of the electric field, due to the local orthogonality of these fields.

To give some number, a Faraday isolator working at 10 GHz has a loss near 1.5 dB in the direction 1–2 and 30 dB in the direction 2–1, with a bandwidth of at least 20 dB of attenuation near 10%, as indicated in Figure A7.12.2. Note that this graph is similar to that of Figure A7.5.4, which is relative to "μ_{dj}" and "μ_{cj}." In Section A7.5 we said the imaginary parts of "μ" insert loss in the propagation of the e.m. wave. We do not confuse the peak in Figure A7.5.4 with that in Figure A7.12.2 since in the former figure we use resonance inside the ferrite, while now the peak is due to exact 45° rotation, two reflections for cut-off and two attenuations of the septs "S1" and "S2" indicated in Figure A7.12.1b. In other words, in the Faraday isolator we do not use resonance phenomena.

Typical applications of isolators are to separate an oscillator from the load, since quite often a microwave oscillator has a frequency that is dependent on the value of load, if no particular expedient is applied.

h Nonreciprocal Phase Shifters

To introduce this device let us start with a consideration of the Farady isolator studied in the previous point. Note that the magnetic field rotates passing through the ferrite. So, if for instance we regulate the intensity of "H" and the length of the ferrite, we may rotate 180° the field from port "B1" to port "B2," that is we have created a 180° phase shifting between input and output. In this case, to use the shifted signal the output rectangular waveguide must be in the same position as the input one. Only this value of phase shifting can be done with such a device since to propagate the fundamental mode, we need the electric field to be aligned with the shorter walls of the rectangular waveguide. As a result, the Faraday rotation may be used to insert phase shifting between signals, but the device indicated in Figure A7.12.1 is not well suited and it is only used as an isolator.

A more commonly used waveguide phase shifter based on the Faraday effect is indicated in Figure A7.12.3 b. This device uses the circular polarization of the magnetic field inside the rectangular waveguide, as seen in Appendix A2. The signal in the rectangular waveguide is coupled to the circular waveguide through a hole at the center of the end surface of the circular waveguide,

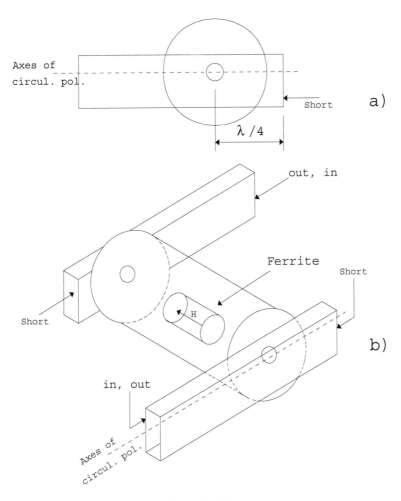

Figure A7.12.3

so that the hole has its center on the axis of magnetic circular polarization for the rectangular waveguide. In this way, the wave propagating in the rectangular waveguide is able to launch a circularly polarized magnetic field inside the circular waveguide. So, we can expect that two phase constants will exist, depending on the direction of propagation or direction of "$\underline{H}$" as studied in Section A7.5. This is why phase shifters are called "differential phase shifters" since the measure that is done with such devices is the difference between the phases of the same wave passing in the two directions. The measurement unit for a differential phase shifter is usually degrees per unit length of the device.

As indicated in detail in Figure A7.12.3a, each rectangular waveguide is short circuited at one quarter of wavelength after the coupling hole, thus assuring an electric field maximum at the coupling hole. With such an arrangement, energy traveling in the rectangular waveguide passes in the circular waveguide, travels inside the ferrite cylinder and comes out to the other rectangular waveguide. The cylinder is held with low loss material, as in the isolators, and the magnetization can be obtained in the same way, that is using a permanent magnet or a solenoid wound on the circular waveguide. The structure indicated in Figure A7.12.3 is not the only possible mounting,[13] even if it is one of the most used. Note that in this arrangement the devices do not suffer from the limitation of the Faraday isolator thought of as a phase shifter. In this case, whichever the phase shifting is, the wave is always circular polarized in the transition points, and so the energy exchange between the waveguides is always at its maximum.

If we indicate with "β_c" and "β_d" the phase constants respectively in the same or opposite direction of "$\underline{H}$," the differential phase shift "$\Delta\theta$" will be:

$$\Delta\theta \propto \left(\beta_c - \beta_d\right)\ell \qquad (A7.12.1)$$

where "ℓ" is the length of the ferrite cylinder. We may arrive at the same value of differential phase shifting if we leave the wave to pass in a fixed direction but do two measurements of phase, each for a fixed direction of "$\underline{H}$." This kind of nonreciprocal phase shifter, in spite of its big size, is a research topic to increase the operating bandwidth and decrease its dimensions.[14]

c. Circulators

By "circulator" we mean a transmission line device with the characteristic that there is propagation between the lines only in one circular direction, while in the opposite circular direction there is more attenuation than possible. A typical system block diagram is indicated in Figure A7.12.4a. In waveguide components, the circulator is realized by adding a third rectangular waveguide to the isolator, as indicated in Figure A7.12.4b, which is orthogonal to port "1." A signal "TE_{10}" coming from "1" rotates its electric field clockwise and exits from port "2" for the reason explained in the previous point a. A signal coming from port "2" is reflected to port "1" and when coming back, it is in the proper orientation to exit from port "3." A signal coming from port "3" rotates still clockwise toward port "1" and arrives there under cut-off. So it is reflected back, arrives at port "2" rotating clockwise 45° and still under cut-off. It is reflected to port "2" and finally it arrives at port "1" in the proper orientation to exit from this port. So, in the connection between port "3-1" we may attend a little more attenuation in with respect to the other possible connections "1–2" and "2–3."

Typical applications of circulators are to connect one antenna to a receiver and a transmitter. For instance, the antenna is connected to port "2," the transmitter to port "1," and the receiver to port "3." Sometimes they are also used as isolators, connecting a load to one port. The value of the isolation is a little less than a pure isolator, mainly due to a little increase of mismatch to the ports.

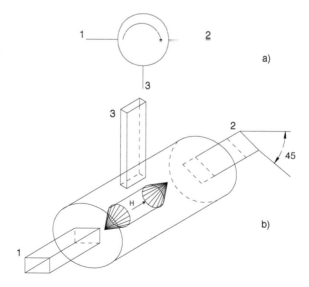

Figure A7.12.4

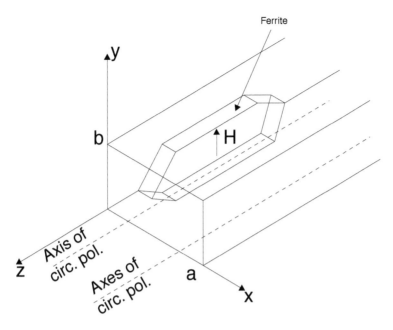

Figure A7.13.1

Appendix A2 showed that in rectangular waveguides two longitudinal axes exist where the "TE$_{10}$" has its magnetic field circular polarized. We have used these points to realize the phase shifter in Figure A7.12.3 which also employs isodirectional propagation inside the circular waveguide.

This section will show how these points of circular polarization may be used to realize devices in transversal propagation.

a. Nonreciprocal Isolators

An isolator that uses propagation in transversal magnetized ferrite is indicated in Figure A7.13.1. A slab of ferrite is located along one axis of circular polarization, and it is magnetized along "$\underline{y}_0$," that is in transversal direction to "$\underline{z}_0$," which is the direction of propagation. The intensity of "H" is to create a precession frequency equal to that of the passing signal. For this reason, this kind of isolator is often called a "resonance isolator." So, this static magnetic situation is completely different from that employed for the isodirectional isolator studied in the previous section where the static magnetic field does not create resonance phenomena, that is the coincidence of the precession frequency with the frequency of the signal. If this were the case, in Figure A7.12.1 the signal would be attenuated for both the directions of propagations. In our case instead, the circular polarized magnetic field, which lies in a plane orthogonal to "$\underline{H}_y$," has right and left rotation in the two axes of circular polarization, which interchange when changing the direction of propagation. So, in the direction of propagation where the magnetic field has the same direction of the precession motion, the e.m. energy is passed to the ferrite, while in the opposite direction the signal flows with minimal attenuation. Note that when the ferrite absorbs the e.m. energy, it starts to increase its temperature, so care must be taken when this kind of isolator is used in high power applications. The heating of ferrite in this device is a phenomenon not present in the isolator with isodirectional propagation seen in the previous section. If we want to change the attenuated direction of propa-

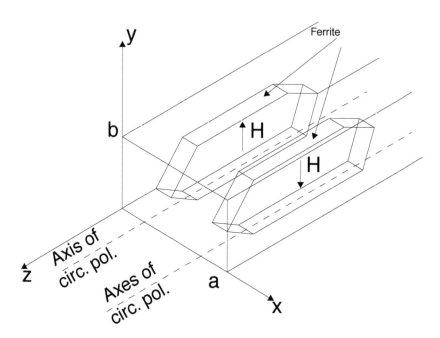

Figure A7.13.2

gation, we may move the ferrite slab to the other axis of circular polarization or, alternatively, reverse the direction of "H."

A characteristic of this resonance isolator is that the ratio between attenuated and nonatttenuated value is greater than the Faraday isolator. The length of the Faraday isolator is smaller compared to the type under study since here the value of the attenuation introduced on the signal depends on the length of the slab. Typically, the normalized value of attenuation for the resonance isolator is 6 or 7 dB per centimeter.

b. Phase Shifters

Using transverse magnetization it is possible to realize both nonreciprocal and reciprocal type phase shifters, resulting in a greater versatility with respect to the Faraday phase shifters.

A typical nonreciprocal phase shifter is indicated in Figure A7.13.2. Note it is very similar to the resonance isolator with the difference that in this case two ferrite slabs are used and the intensity of "H" must not create resonance with the frequency of the signal to be phase shifted. It is important to see that the static magnetic field in one slab is in the opposite direction in comparison to the static field in the other slab. This is done just to phase shift the wave passing in only one direction since for each axis of circular polarization "h" changes the direction of rotation. So, with a fixed direction of propagation where we insert phase shifting, we have to change the direction of "H" inside each slab if we want to change the direction of propagation where we insert the higher phase shifting.

To understand how this device works, study Figure A7.13.3. Remember from the Appendix A2 that the magnetic field of the fundamental mode "TE_{10}" lies in a plane orthogonal to the "y" axis. This means we are in the magnetic situation studied in Section A7.5.

Part a of Figure A7.13.3 represents the situation where the wave propagates along "z_0." In this case, in the left slab the magnetic field of the wave is clockwise circular polarized, while in the right slab it is counterclockwise polarized. Then the wave is subjected to "β_d," since in each slab the direction of circular polarization is discordant to the direction of the precession motion.

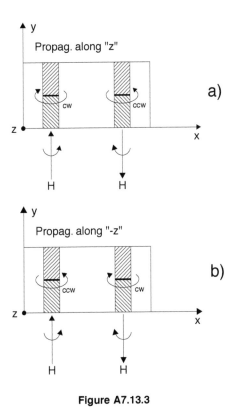

Figure A7.13.3

Part b of Figure A7.13.3 represents the situation where the wave propagates along "$-z_0$." It is simple to recognize that the wave is subjected to "β_c" since in each slab the direction of circular polarization is concordant to the direction of the precession motion.

So, the differential phase shift "$\Delta\varphi$" for the device indicated in Figure A7.13.2 will be:

$$\Delta\varphi = \left(\beta_d - \beta_c\right)\ell \tag{A7.13.1}$$

If we would realize a reciprocal phase shifter it is enough to magnetize the two ferrite slabs with a magnetic static field in the same direction for each slab. In this way the wave will have the same phase shifting for each direction of propagation, and the differential phase shift "$\Delta\varphi$" will be zero. From a practical point of view, such devices are seldom used since it is the differential phase shift that is important more than the absolute phase shifting.

While the Faraday isolator is usually preferred to the resonance one, in the case of phase shifters the contrary holds and the transverse magnetization type is preferred due to the smaller size of the device under study. Waveguide phase shifters due to their importance in high quality telecom systems are a research topic.[15]

c. Circulators

A circulator with transversal magnetization is bigger than the Faraday counterpart, and usually the last is preferred. Anyway, we will show how in our case a circulator may be built and how it works.

Such a device can be built using two 90 degree 3 dB splitter/adders, of the type seen in Appendix A3, and two phase shifters studied in the previous point b. The device is schematically indicated

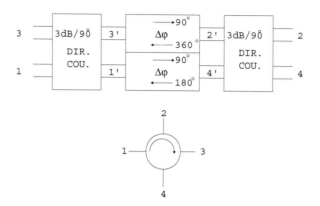

Figure A7.13.4

in Figure A7.13.4. With the blocks indicated as "3dB/90° Dir. Cou." we intend a directional coupler so that for any port considered as input, the signal outgoing from the direct opposite port is in phase with the input signal while the signal outgoing from the other opposite port is 90° delayed with the input signal. If the phase shifters are tuned to realize the phase shifting indicated in the figure, not considering any multiple of 2π, it is simple to recognize that the device is a circulator in the direction "1 → 4." Suppose that a signal is entering port "1." It is equally divided at ports "1'" and "3'," with the phase at this port 90° delayed with respect to port "1'." A signal will arrive at port "2'" 180° delayed, while the other will arrive at port "4'" 90° delayed. This last signal will arrive at port "2" with a total 180° of delay, so in phase with the signal that from port "2'" arrive at port "2." In addition, the signal in "2'" will also arrive at port "4" 280° delayed while the signal from "4'" will also arrive at port "4" 90° delayed, that is in phase opposition with the signal coming from "2'." As a result, the signals in "2'" and "4'" at port "2" will enter while at port "4" no signal ideally will exit.

Similarly proceeding for the other ports valuated as inputs, it is possible to verify the circulation in the direction "1 → 4."

This kind of isolator is seldom used, mainly due to the wide dimensions. The literature reports other constructions to reduce the dimensions, which partially avoid the use of ferrite and so the use of magnets.[16]

A7.14 FIELD DISPLACEMENT ISOLATORS AND PHASE SHIFTERS

The field displacement effect of ferrites is a nonreciprocal behavior of these elements that is usually employed so that in a direction of propagation of the e.m. energy, the maximum fields of the wave are pushed on one extreme of ferrite, while in the opposite direction of propagation in the same position there is now the minimum fields.

The first device to employ the field displacement effect was built in a rectangular waveguide and this appendix will cover such a device, while the chapters of this book have covered the planar counterparts of the waveguide ones. Before studying the field displacement effect we prefer first to show in Figure A7.14.1, how this phenomenon acts on the electric field of the fundamental mode "TE_{10}" inside the rectangular waveguide. We have represented a transversal section of a rectangular waveguide. A slab of ferrite is positioned inside it in a precise position that is not necessarily the position of circular polarization of "h," as it is instead used for the devices studied in the previous section. The dashed and solid curved lines represent the intensity along "x" of the component "e_y" of the electric field of the wave.* The ferrite is opportunely magnetized, without creating resonance

---

* See Appendix A2 for the representation of the fields inside a rectangular waveguide.

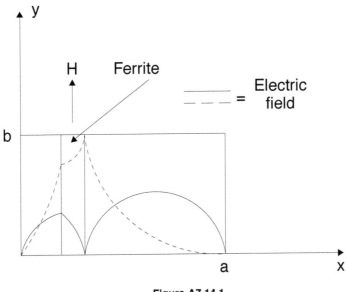

Figure A7.14.1

with the signal passing in the waveguide. What happens is that for a direction of propagation the electric field has its maximum on a side of the slab, as indicated by the dashed lines, while in the opposite direction in this position, there is a minimum of the field, as indicated by the solid lines. Note that an analogy with the devices of the previous section exists, since also now we are in the case of propagation inside a transversal magnetized ferrite.

Theoretically, this effect may be explained with the cut-off phenomenon of ferrite, as stated in Section A7.5. Here we will fix the exact formulation of the problem, but we have to say that for a complete understanding of what we are going to say it is important to clearly understand the basis of the wave propagation and the representation of the modes inside a rectangular waveguide. The reader can read Appendix A2 for the necessary background.

So, let us examine the situation represented in Figure A7.14.2. Part a indicates a three dimensional representation of the device we use to study the field displacement effect, while part b indicates a cross-sectional view. Referring to part b of Figure A7.14.2, the saturating static magnetic field "H" is uniform and internal to the ferrite slab. In the waveguide part where the slab is present we define three regions where we have to find the e.m. fields — two air regions "A1" and "A2," and one region "F1" filled by the slab, that is the inside of the slab. We assume inside the waveguide, and far away from the slab, only the fundamental mode "TE$_{10}$ exists." This mode has the only nonzero field components "h$_x$," "h$_z$," and "e$_y$" with coordinate dependence along only the "x" and "z" axes. Since the propagation of the wave is inside a transversal magnetized ferrite, we may recognize that we are in the same situation studied in Section A7.7, where "$\mu_{eq\perp}$" was defined. The fields of the wave must so respect the "TE" modes equations defined in Section A7.7, that is:

$$k_z e_y \underline{x}_0 = ! - j\omega\mu_0\left(h_x\mu_p - j\mu_\ell h_z\right)\underline{x}_0 \tag{A7.7.6}$$

$$-k_x e_y \underline{z}_0 = ! - j\omega\mu_0\left(h_z\mu_p + jh_x\mu_\ell\right)\underline{z}_0 \tag{A7.7.8}$$

$$\left(k_x h_z - k_z h_x\right)\underline{y}_0 = ! j\omega\varepsilon e_y\underline{y}_0 \tag{A7.7.10}$$

From A7.7.6 and A7.7.8 we may obtain:

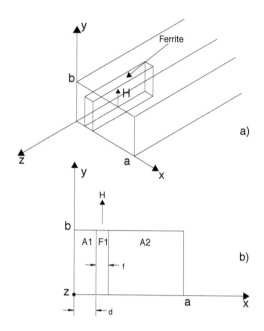

Figure A7.14.2

$$h_x = \frac{-\mu_\ell k_x - j\mu_p k_z}{\omega\mu_0\left(\mu_\ell^2 - \mu_p^2\right)} e_y \qquad (A7.14.1)$$

$$h_z = \frac{j\mu_p k_x - \mu_\ell k_z}{\omega\mu_0\left(\mu_\ell^2 - \mu_p^2\right)} e_y \qquad (A7.14.2)$$

Now we have to remember that A7.7.6, A7.7.8, and A7.7.10 have been determined assuming that only the progressive* wave exists because we assume a homogeneous and uniform media filled with ferrite. In our case, we want to evaluate the most general case where both the progressive and the regressive wave exists along the "x" axis. In this case, the two previous equations become.

$$h_x(x,z) = \left[P\frac{-\mu_\ell k_x - j\mu_p k_z}{\omega\mu_0\left(\mu_\ell^2 - \mu_p^2\right)} e^{(-k_x X)} + R\frac{\mu_\ell k_x - j\mu_p k_z}{\omega\mu_0\left(\mu_\ell^2 - \mu_p^2\right)} e^{(k_x X)} \right] e^{(-k_z Z)} \qquad (A7.14.3)$$

$$h_z(x,z) = \left[P'\frac{j\mu_p k_x - \mu_\ell k_z}{\omega\mu_0\left(\mu_\ell^2 - \mu_p^2\right)} e^{(-k_x X)} + R'\frac{-j\mu_p k_x - \mu_\ell k_z}{\omega\mu_0\left(\mu_\ell^2 - \mu_p^2\right)} e^{(k_x X)} \right] e^{(-k_z Z)} \qquad (A7.14.4)$$

where "P," "P'," "R," and "R'" are generic constants, with the dimensions of Volt/m. Refer to Appendix A2 to see how these formulas come from A7.14.1 and A7.14.2. The explanation lies on how the operation $\underline{\nabla} \otimes (\underline{V})$ changes when the generic vector "$\underline{V}$" has the coordinate dependence $e^{(\pm\Sigma_i K_i i)}$ with $i = x, y, z$, as we now assume to be the case along the "x" axis.

To evaluate "e_y" we may proceed with the variable separation method, i.e., we will suppose that the dependence of this field on "x" and "z" may be written as:

* See Chapter 1 for wave definitions.

$$e_y(x, y) = e_y(x)e_y(y) \qquad (A7.14.5)$$

Proceeding as in Appendix A2, we may say that the generic function $e_y(i)$ with $i = x, z$ will respect the equation:

$$\frac{\partial^2 e_y(i)}{\partial i^2} = k_i^2 \, e_y(i) \qquad (A7.14.6)$$

with the condition

$$k_{xf}^2 + k_z^2 = ! - \omega^2 \mu_0 \mu_{eq\perp} \varepsilon \qquad (A7.14.7)$$

inside the ferrite region "F1" and

$$k_{x0}^2 + k_z^2 = ! - \omega^2 \mu_0 \varepsilon_0 \qquad (A7.14.8)$$

inside the air regions "A1" and "A2." As seen in Chapter 1 and Appendix A2, two equivalent forms of solutions exist for A7.14.6. We will give both forms, i.e.:

$$e_y(x, z) = \left[e^+ e^{(-k_x x)} + e^- e^{(-k_x x)} \right] e^{(-k_z z)} \qquad (A7.14.9)$$

$$e_y(x, z) = \left[E^+ \operatorname{senh}(k_x x) + E^- \cosh(k_x x) \right] e^{(-k_z z)} \qquad (A7.14.10)$$

where the constants "e^+," "e^-," "E^+," "E^-," "k_x," and "k_z" will be determined applying the conditions that the e.m. field must satisfy on the border of the structure. Of course, "e^+," "e^-," "E^+," and "E^-" have dimensions of Volt/m. Note that in the two previous equations we have assumed only the progressive wave inside the waveguide while, as we said, we are considering both the wave along the "x" axis. It is more useful to use A7.14.9 inside the ferrite slab "F1" and to use the A7.14.10 on the air regions "A1" and "A2."

So, in "F1" and along "x" we will write:

$$e_{yf}(x) = Ae^{(-k_{xf} x)} + Be^{(k_{xf} x)} \qquad (A7.14.11)$$

The constant "k_{xf}" is in general a complex number that is also a function of "H" since "$\mu_{eq\perp}$" is just a function of "H" as we saw in Section A7.7.

In the regions "A1" and "A2" and along "x" we may write, from A7.14.10:"

$$e_{y1}(x) = C\operatorname{senh}(k_{x0} x) + D\cosh(k_{x0} x) \qquad (A7.14.12)$$

$$e_{y2}(x) = E\operatorname{senh}(k_{x0} x) + F\cosh(k_{x0} x) \qquad (A7.14.13)$$

where the constants "C," "D," "E," and "F" will be determined applying the conditions that the e.m. field must satisfy on the border of the structure while "k_{x0}" is known from A7.14.8, which assuming the absence of loss, can be rewritten as:

$$k_{x0}{}^2 = !\,\beta_z{}^2 - \omega^2\mu_0\varepsilon_0 \qquad (A7.14.14)$$

since for zero losses

$$k_z \equiv j\beta_z \qquad (A7.14.15)$$

Considering the rectangular waveguide as a perfect conductor we know that "e_y" must be zero for x = 0 and x = a. So, from A7.14.12 for x = 0 it follows that D = !0, and we have:

$$e_{y1}(x) \equiv C\,\text{senh}\left(k_{x0}x\right) \qquad (A7.14.16)$$

From A7.14.13 for x = a it follows that F = !0, and we have:

$$e_{y2}(x) \equiv E\,\text{senh}\left[k_{x0}\left(a - x\right)\right] \qquad (A7.14.17)$$

The other border conditions the e.m. field must satisfy are that the tangential magnetic and electric field at the air-ferrite separation must be continuous since the ferrite is valuated as an insulator. The tangential electric field at this boundary is "e_y," which is known from A7.14.11, A7.14.16, and A7.14.17. The tangential magnetic field is "h_{zf}" which we can obtain from A7.14.4 after inserting "e_{yf}" from A7.14.11, resulting:

$$h_{zf}(x) = \left(M - Q\right)Ae^{\left(-k_{xf}\,x\right)} - \left(M + Q\right)Be^{\left(-k_{xf}x\right)} \qquad (A7.14.18)$$

where:

$$M \triangleq \frac{jk_x}{\omega^2\mu_0\mu_{eq\perp}} \qquad (A7.14.19)$$

$$Q \triangleq \frac{j\beta_z}{\omega^2\mu_0\mu_{eq\perp}}\frac{\mu_\ell}{\mu_p} \qquad (A7.14.20)$$

The last field we have to evaluate is "h_z" inside the regions "A1" and "A2." We can get help from the general Maxwell's equation:

$$\underline{\nabla} \otimes \left(\underline{e}\right) = -j\omega\mu\underline{h}$$

which, is applied to A7.14.16 and A7.14.17, gives:

$$h_{z1}(x) = \frac{k_{x0}C}{j\omega\mu}\cosh\left(k_{x0}x\right) \qquad (A7.14.21)$$

$$h_{z2}(x) = \frac{k_{x0}E}{j\omega\mu}\cosh\left[k_{x0}(a - x)\right] \qquad (A7.14.22)$$

$$\text{senh}(j\alpha) \equiv j\text{sen}(\alpha) \qquad\qquad (A7.14.27)$$

from the expressions A7.14.16 and A7.14.17 we may recognize that when "k_{x0}" is real, then the electric field starts to have exponential shape along the "x" axis, as shown by the dotted line in Figure A7.14.1. Conversely, when "k_{x0}" is imaginary from A7.14.27 it follows that the electric field starts to have a sinusoidal shape along the "x" axis, as shown by the solid line in Figure A7.14.1. As a consequence, by carefully tuning the device it is possible to have on a side of the ferrite, a maximum of the electric field in a direction of propagation, and a minimum of this field for the opposite direction of propagation.

After such a brief theoretical explanation of the field displacement effect, let us start to explain what kind of device we can build using this effect.

a. Nonreciprocal Isolators

An isolator using the field displacement effect is realized, as the first thing, finding the position "d" where the maximum ratio exists between the strength of the electric field in the two directions of propagation. As a first approximation, this value of "d" may be found setting x = d + f, i.e.:

$$e_{y2}(d + f) = 0$$

From A7.14.17 it follows that:

$$k_{x0}(a - d - f) =! 0$$

Inserting this condition in the transcendental equation we may obtain the coordinate "d" (where to place the ferrite slab). The exact position to maximize the ratio of the electric field in the two directions of propagation has to be found experimentally since for a perfect operation, too much tolerance on the value of "d." K.J. Button has shown[17] that for a waveguide with a = 4 cm, the maximum tolerance it is not allowed on "d" is near one millimeter. In any case, the position "d" is always near one side of the waveguide, typically near the 10% or less of the dimension "a." The thickness "f" of the ferrite slab is inversely proportional to the difficulty in finding the exact position "d" since the thinner the slab, the more critical is the exact position "d."

Once the proper value "d" has been found, to build an isolator it is enough to place a sheet of resistive material on the side where the ratio of the electric field is maximized, as shown in Figure A7.14.4a. In such a case, in the direction of propagation where the electric field is maximum the signal will be absorbed by the resistive sheet, while in the opposite direction the signal will pass with small attenuation since it has been displaced from the slab, as shown in Figure A7.14.1. To give some value of the field displacement isolater shown in Figure A7.14.4a, experiments report a maximum attenuation in the pass direction near 1 dB for a device working at 10 GHz, a minimum attenuation in the stop direction near 40 dB, and a bandwidth near 10%.

To increase the ratio of attenuations between the two directions of propagation it is possible to use two slabs of ferrite, doubling the field displacement effect. This situation is indicated in Figure A7.14.4 part b. We see that the two ferrite slabs "F1" and "F2" are magnetized opposite each other. S. Weisbaum and H. Boyet have realized[18] such a device working at 11 GHz reaching a maximum attenuation in the pass direction of 1.2 dB, a minimum attenuation in the stop direction of 64 dB, and a bandwidth of 10%.

The field displacement isolator can handle more power than the Faraday counterpart, with power handling capability similar to that of the resonance isolator. Due to the difficulty in finding the exact value of "d," the field displacement isolator is seldom used, especially in low power appli-

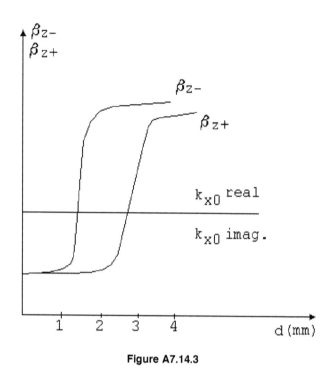

Figure A7.14.3

cations, and the Faraday isolator is preferred. Nevertheless, experiments on f.d. isolators are continuing[19] to improve the performances of this device.

Caution must be used to attribute a physical mechanism for a device depicted as in Figure A7.14.4. Note how this construction is very similar to the resonance isolator indicated in Figure A7.13.1. So, at first sight the two devices may be confused in the physical mechanism involved for the nonreciprocal isolation, but note that the resonance and the field displacement effects are completely different. Usually, the field displacement isolator may be recognized due to the little distance "d" from the side wall of the waveguide and the slab, typically of some millimeters or maximum a 10% of the dimension "a."

b. Phase Shifters

It is simple to understand that nonreciprocal phase shifters may be realized using the field displacement effect. Note from Figure A7.14.3 how some values of "d" exist where it is possible to have quite different values of "β_{z+}" and "β_{z-}." Placing the magnetized ferrite slab in such positions, the differential phase shift "$\Delta\phi$" for a given slab length "ℓ" will be:

$$\Delta\phi = \left(\beta_{z+} - \beta_{z-}\right)\ell \qquad\qquad (A7.14.28)$$

It is not necessary to place the slab in the position for the maximum ratio between the electric fields in the two directions of propagation. If this position is used, sometimes on the surface of the slab where the field ratio is maximum, a little sheet of high dielectric constant material is added to increase the phase constant of the wave in the direction of propagation where the electric field is maximum. In such a way, the differential phase shift is increased for a given slab length "ℓ."

While the Faraday isolator is usually preferred to the field displacement one, in the case of phase shifters, the contrary holds and the field displacement type is preferred due to the smaller size of the device under study.

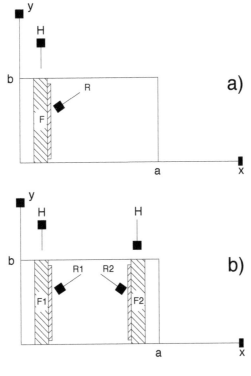

Figure A7.14.4

A7.15 THE FERRITE IN PLANAR TRANSMISSION LINES

As seen in the chapters of this book, ferrite is also widely used in planar transmission lines. Circulators, phase shifters, and isolators are typical devices used in a lot of transmission equipment and their use is growing. For instance, many cellular phones use circulators or isolators in microstrip technology.

In the chapters of this book we have described how these devices are built and where they are used. Since in this Appendix we have described the behavior of ferrite under static and time varying magnetic fields, in this section we will study these devices from a more theoretical point of view, using all the concepts we have introduced up to now. The complete vision of all the possible ferrimagnetic devices can be seen in the chapters of this book.

Let us start with the case of the ferrite used as substrate in microstrip transmission lines, in presence or less of static magnetic field, since microstrips are the most used planar transmission lines. This subject was not treated in Chapter 7, which is completely dedicated to microstrip devices since the case of zero or moderate static magnetic field is not often used in practice. But this case is very important to join together all the concepts we have studied in this Appendix with practical microstrip ferrite devices.

a. The Ferrite As Microstrip Substrate

A study of the case of microstrips using ferrite as a substrate has been done by D. J. Massè and R. A. Pucel[20] for the case of nonmagnetized or at residual magnetization ferrites. In the case of residual magnetization, the ferrite has been magnetized with a static magnetic field with direction parallel to the substrate and in the direction of the hot conductor.* To understand their results we

* See Chapter 2 for mechanic construction of microstrip transmission lines.

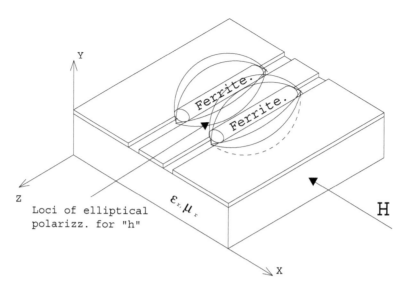

Figure A7.15.2

For the graphs at the left of the line at "f_m" the behavior of ferrite has no meaning since the impedance of the microstrip becomes complex, i.e., it is reactive to the propagation of the signal.

The studies made by D. J. Massè and R. A. Pucel may be considered as the fundamental ones to understand the behavior of partially magnetized ferrite or under residual magnetization. We have seen in Chapter 7 for microstrips and Chapter 8 for striplines how many devices are used that employ residual magnetization.

b. Nonreciprocal Isolators

A clear example of how we can get directional behavior using a planar transmission line may be obtained with reference to a coplanar waveguide, CPW, that we have studied in Chapter 10. Let us examine the situation indicated in Figure A7.15.2. A coplanar waveguide has a cylinder of ferrite posed in a slot. A static magnetic field "H" magnetizes the ferrite at resonance. In this case we use the elliptical polarization of the magnetic field "h," which lies on planes passing along the slots. The loci of elliptical polarization for "h" are represented in Figure A7.15.2.* The reader may recognize that the magnetic situations, that is the vectorial relationships between "h," "H" and the direction of propagation, are as we studied for the waveguide resonance isolator. That is, we are in the case of propagation in transversal magnetized ferrite. In fact, in both situations the magnetic field "h" is polarized on a plane orthogonal to "H" and the wave propagates orthogonally to "H."

In our case, the right and left directions of rotation of the elliptically polarized magnetic field interchange when changing the direction of propagation. So, in the direction of propagation where the magnetic field has the same rotation as the precession motion the e.m. energy is passed to the ferrite, while in the opposite direction the signal flows with minimum attenuation. The attenuation in the two directions of propagation will have a graph of the type indicated in Figure A7.12.2 for the resonance isolator in waveguide.

The fact that an elliptical polarization of "h" may be employed must not surprise the reader. In fact, what is important about having energy absorption in the ferrite is that the direction of the rotation of the precession motion of the total atomic moment of the ferrite is concordant with the

---

* See Chapter 10 for "CPW" characteristics.

direction of rotation of the magnetic field of the wave and, of course, the frequency of these motions is the same. In addition, in coplanar waveguides there exists some point where the eccentricity of the elliptical polarization is very low and the polarization of "h" may be valuated as circular.

Many experiments on "CPW" using ferrite have been done by C. P. Wen,[21] from which it appears that the "CPW" is a planar transmission line that is very suitable to building nonreciprocal devices.

In Chapter 10 we saw other constructions of "CPW" isolators and other devices using ferrite. The reader may refer to that chapter to have a global view on devices using this kind of transmission line.

c. Phase Shifters

Using all the concepts we have introduced concerning ferrites under magnetic fields, the reader may recognize that the circular polarization of the RF magnetic field is the phenomenon that is best suited to realize phase shifters. Some planar transmission lines do not have this polarization, such as microstrips, for instance. We have seen that it is possible, however, to create such a polarization in microstrips, using meander lines. As we have seen in Chapter 7 with microstrip meander lines on ferrite substrate we can build phase shifters. We know that other planar transmission lines have elliptical polarization of the signal magnetic field, which is also suitable to interact in some way with precession motion of magnetized ferrite. These are coplanar waveguides and slotlines.* We will refer to "CPW," in analogy to the previous point b.

C. P. Wen[22] has made experiments on a "CPW" realized with ferrite as the substrate and transversely magnetized, as indicated in Figure A7.15.3a, measuring the differential phase shift "$\Delta\varphi$" in the frequency range 5 to 7 GHz. The results are indicated in Figure A7.15.3b. Note also in this case that we are in transversal magnetized ferrite, like in the case of the waveguide phase shifters studied in Section A7.13 or like in the previous point b, and the differential phase shift will be a function of "β_d" and "β_c" according to:

$$\Delta\varphi = \left(\beta_d - \beta_c\right)\ell$$

where "ℓ" is the length of the ferrite cylinders.

d. Three Port Circulators

A typical stripline three port circulator is indicated in Figure A7.15.4. In part a we have drawn a transversal view of the device. Two ferrite cylinders "F1" and "F2" are posed on each side of a circular conductor of maximum diameter not greater than the diameters of the ferrites. This circular conductor has attached on its border three striplines "L_1," "L_2," and "L_3," each one forming an angle of 120° with the nearest stripline. On the other side of the ferrite cylinders, a ground plane "M1" and "M2" is attached, each for ferrite. These ground planes realize the stripline technology of the device. At the external of the ground planes, two magnets generate the proper static magnetic field that biased the ferrites. Sometimes it is used as a magnet only.

A top view of the device is drawn in Figure A7.15.4b with the upper magnet, the top ground plane "M1," and the top ferrite "F1" removed. The dashed line represents the ferrite "F2" under the disk conductor. We have indicated the origin of the angles "θ" at the angular center of port in "L_1."

After these mechanical considerations, we want to explain how the circulator works.

Let us start with the analysis of the structure indicated in Figure A7.15.4 with the ferrite never magnetized, that is valuated with its "μ_r" end "ε_r." It is possible to demonstrate that the signal

---

* See Chapter 9 on slotline characteristics.

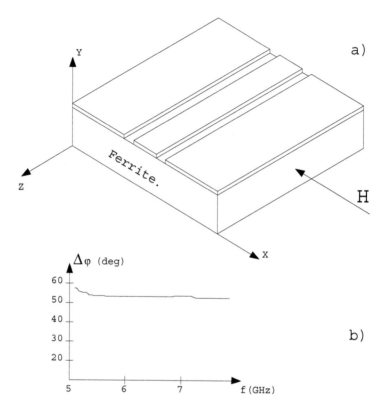

Figure A7.15.3

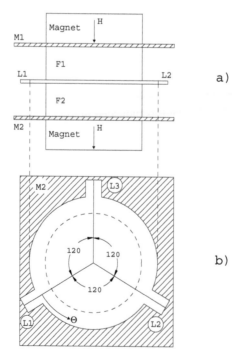

Figure A7.15.4

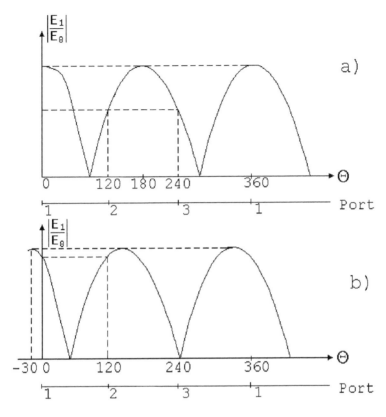

Figure A7.15.5

coming from any of the ports "L_i," with i = 1 to 3, generates a pattern of stationary wave, with angular shape as indicated in Figure A7.15.5. Here we have assumed to connect the signal source to port "L_1." On the ordinate axis we have indicated the ratio "E_1/E_θ" between the field intensity "E_1" at port "L_1" and the field intensity "E_θ" at a generic angle "θ." We see in Figure A7.15.5a how the signal at ports "L_2" and "L_3" have the same amplitude. Since this is true for any port valuated as input, we conclude that in this situation a signal coming into any port will exit from the other two ports with the same amplitude.

If we apply a suitable intensity of "H" it is possible to offset the pattern of standing wave of 30°, as indicated in Figure A7.15.5b. In this case we see how at port "L_2" a signal exists, while at port "L_3" we have a minimum of signal. In this case, a signal incoming at port "L_1" will exit at port "L_2" but will be strongly attenuated at port "L_3." This is true for any input port so, for instance, if the signal is injected in port "L_3" it will exit at port "L_1" and will be strongly attenuated at port "L_2." The device is clearly a circulator. Of course, the direction of rotation depends on the direction of "<u>H</u>."

The theoretical explanation of this phenomenon may be obtained using the concepts of propagation in striplines, studied in Chapter 3, and the theory of ferrites we have studied in this appendix. In particular, the magnetic field "h" of the wave is parallel to the conductors and the wave propagates in a transversely magnetized ferrite. So, according to the theory treated in Section A7.7, this wave will be under the effect of "$\mu_{eq\perp}$" and will be composed of a "TE" and a "TM" wave. In addition, from Chapter 3 we know that the fundamental propagation mode of the stripline is the "TEM" mode. So, remembering the theory treated in Section A7.5 we know that the two waves in which this "TEM" mode divides, that is the "TE" and the "TM" modes, have two phase constants as a function of "H," that is "β_c" and "β_d." In the sensitivity to "H" of these two phase constants is the explanation of the rotation of the standing wave pattern when "H" is applied to the structure. There is more than a theory to explain the operation principle of the circulator, usually different for the technology of the device[23,24,25,26,27] as these aspects are treated in the chapters of this book for the particular transmission line involved in the circulator.

A7.16 OTHER USES OF FERRITE IN THE MICROWAVE REGION

In this section we will indicate other devices that still use ferrites but they do not have nonreciprocal behavior in contrast to the devices studied in the previous section. These devices are not studied in the chapters because they are not strictly connected to the planar transmission lines. We think that the study of these subjects in the appendix will help in a complete understanding of the ferrite world.

a. Variable Frequency Oscillators "VFO"

The most used type of ferrite for these devices is a particular crystal configuration called "garnet" composed of yttrium and iron with the formula $Y_3Fe_5O_{12}$. The final composite is called "YIG," an abbreviation of "Yttrium Iron Garnet." This ferrite has very low losses which makes it very suitable for microwave applications. VFO using YIG are commonly called "YIG oscillators." These oscillators use the very low line width* of YIG to stabilize the oscillation frequency. The most employed form of the ferrite used is the sphere, for two reasons: first, the sphere reduces the criticalness of positioning it inside the electronic circuit because the demagnetizing factors are independent of the direction of the applied static magnetic field, as was stated in Section A7.8; second, a sphere is not as difficult to realize with good precision. Care must be taken to pose the sphere in an area of uniform static magnetic field, otherwise spurious oscillation can take place. The static magnetic field is realized with a solenoid, sometimes wound on a ferromagnetic material to increase and concentrate the intensity on the sphere. The electronic circuit is usually coupled to the YIG sphere through a magnetic feedback loop, as in the classical theory of oscillators.

Since it is known from this appendix that the precession frequency may be changed by varying the intensity of the applied static magnetic field "H," to change the oscillation frequency (which is stabilized from the sphere) it is enough to change the intensity of "H."

Typical data of a YIG oscillator are:

Bandwidth	2...18 GHz
Current variation for tunability	800 mA
Output power	15 dBm, on 50 Ω
Phase noise, a 10 kHz from carr.	−90 dBc/Hz
Spurious and harmonics	−15 dBc

Of course other YIG oscillators are available, more or less optimized for the required bandwidth.

To conclude this point, we have to say that when a tunability of more than two octaves is required, YIG oscillators have no competitors.

b. Tunable Filters

Good band filters, i.e., stop-band or pass-band, may be realized with ferrites. Also in this case, the most commonly used ferrite is the YIG type, and these filters are called "YIG filters." The same care of the previous point must be taken for ferrite, i.e., to use a sphere of this material and to place it in a uniform "H."

---

* See Section A7.10 for definition of line width.

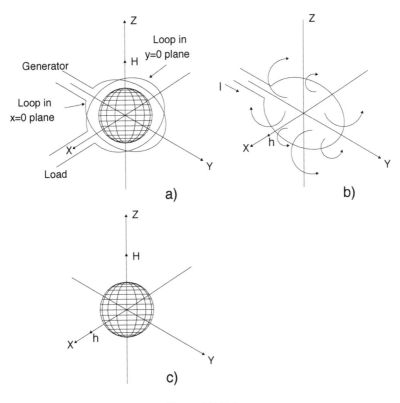

Figure A7.16.1

To understand the operating principle of the YIG filters, let us examine the structure indicated in Figure A7.16.1a. At the loop in the plane, x = 0 is connected to a signal generator; we will call this loop the "x-loop." At the loop in the plane, y = 0 is connected a load; we will call this loop the "y-loop." As is indicated, the two loops are orthogonal. The magnetic field produced by the x-loop has the lines of strength as indicated in Figure A7.16.1b, and it is orthogonal to the y-loop. Consequently, as it is known from Appendix A6, no induced current will go in the load. Now, let us insert at the center of the Cartesian coordinate system a sphere of ferrite of such dimensions so that the generated field "h" is uniform inside the sphere. From the situation indicated in Figure A7.16.1c we may recognize that we are in a situation where the linear polarized magnetic field "h" is orthogonal to "H," that is a situation similar to the one studied in Section A7.6. The difference is that in this case we do not have propagation inside the ferrite sphere. From Section A7.6 we know that the magnetic field "h" begins to rotate along the "z" axis, and when the frequency of "h" is equal to the precession frequency induced in the sphere by "H," the frequency of rotation becomes theoretically infinite as indicated in Figure A7.6.2. Associated to "h," the ferrite creates a magnetization "m" and, in particular, components along the "y" axis appear. This situation is indicated in Figure A7.16.2a. It is this component that is able to create an induced current into the "y"-loop, current that is proportional to the input current, and consequently to transmit the information of the input signal to the load. A concentrated constants equivalent circuit to this phenomenon is indicated in Figure A7.16.2b. The two transformers "T1" and "T2" represent the energy exchange due to ferrimagnetic coupling while the resonance phenomena is represented by the resonant circuit "RLC."

A simplified stripline YIG band pass filter is indicated in Figure A7.16.2c. The two center conductors of the striplines are set at 90° to minimize the coupling when not desired. In the intermediate ground plane "M2" a small aperture is created where a ferrite sphere "S" is located,

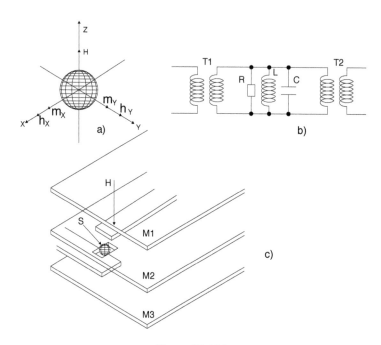

Figure A7.16.2

suspended by low loss material. When a band pass function at a desired frequency "f" is desired between the two strip lines, a static magnetic field "H" is applied, so as to create in the ferrite a precession frequency "f_p" of the total atomic moment equal to "f."

The typical tunability of YIG filters are near two octaves, with "Q" around some hundreds, with maximum central frequencies near 20 GHz.

A7.17 USE OF FERRITE UNTIL UHF

What we have studied up to now is only a particular use of ferrite, mainly in the transmission line field and in microwave region. In this section we will discuss the use of ferrite below the microwave region, that is starting from some hundreds of kHz until the maximum frequency of the UHF band, i.e., 3 GHz. The number of applications of ferrite in this region of frequency exceeds that in the microwave region.

In this band of frequency it is usual to employ ferrite without any static magnetic field, and the most used characteristics of these elements are the value of the initial permeability and the shape of the hysteresis loop. These magnetic characteristics are in common with the ferromagnetic materials, but what makes ferrimagnetic materials the preferred ones is their low loss compared to losses in the ferromagnetic ones, especially the induced current losses as we have studied in Section A7.11. As a result, above some kHz the ferrites are the only used magnetic materials.

One of the first applications of ferrite was as a holder for wire wound antennas in pocket radios. As studied in Appendix A5, the high permeability of ferrite concentrates the magnetic induction inside it, and consequently the induced voltage on the loops around it will be higher compared to that induced in the same loops without the ferrite as a holder.

Another application of ferrites employed as material of high permeability is in transformers. For this application a high constancy of "μ_i" with frequency is required, at least in the band of the signal to be applied to the transformer, and also small losses or at least a constancy of losses with frequency. From Section A7.9, we know that losses are created by the imaginary part "μ_{ij}" of "μ_i." Some typical shapes of initial permeabilities are drawn in Figure A7.9.1.

Also even though these results are good, this device has not been used much since with the improvement in technology today it is possible to directly generate such output frequencies and output powers without the need to duplicate any frequency.

Another device that can be realized using A7.18.5 is a detector. If "h_x" and "h_y" have different frequencies, for instance one of these two components is modulated, in "m_z" will be present sum and difference frequencies. Some researchers[29] have built test devices to prove this theory and effectively they have obtained a ferrite detector, both in circular and rectangular waveguide. Also even if these assemblies are very creative from a mechanical point of view, they have not gotten much practical success, due the advent of semiconductor detectors in the period (1958) of the experiments.

A7.19 MAIN RESONANCE REDUCTION AND SECONDARY RESONANCE IN FERRITE

Some researchers[30] noted that when high power signals were applied to ferrites, some effects not directly correlated to small signal theory appeared in the ferrite behavior. First of all, the peak value of the signal absorption at resonance was lower and wider; in addition a new small absorption peak appeared for a value "H_2" of "H" lower than the value "H_p" for the main resonance. Graphically, the situation of these effects is represented in Figure A7.19.1, where the dashed line graph is relative to the small RF signal case and the solid line one is relative to the high power of RF signal. The new absorption peak is called the "secondary resonance" or "auxiliary resonance." This phenomena begins quite abruptly when the RF signal exceeds a critical value "h_c" which is dependent on the shape of the ferrite specimen.

The theory that can explain these phenomena is known as the "theory of spin waves," developed by the two physicists C. Herring and C. Kittel[31] in approximately 1951. This formidable theory requires, for a deep understanding, a high familiarity with physical relationships and mathematical analysis. Appendix A5 can help the reader with a review of the fundamental concepts involved in

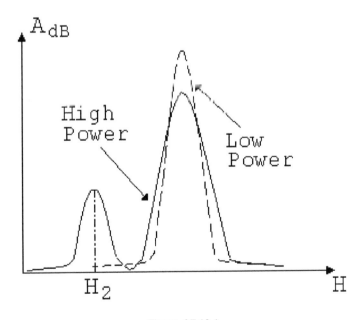

Figure A7.19.1

Experimentally, the researcher H. Suhl[35] has proved that the value of "H_2" can also be half of "H," but in any case a good agreement exists between the theoretical value of A7.19.7 and the measured value.

Different from the previous point a, the secondary resonance is a phenomenon that can be advantageously used to build power limiters. If we apply the static magnetic field "H_2" to ferrite, no attenuation takes place until the input power creates a magnetic field equal to or greater than "h_{c2}" given in A7.19.8, above which the attenuation will increase sharply. Such a device has been built[36] originally to perform some experiments to verify the theory discussed here. This device appears as a Faraday isolator that is as a cylindrical waveguide with a pole of ferrite held at the center. A power attenuation of 5 dB has been obtained for an input power near 9 kW at a frequency of 9.375 GHz. The ferrite cylinder was 5.08 cm long with a diameter of 1.27 cm, and longitudinally magnetized with 450 Oe. From this experiment the characteristic appeared that the value of the attenuation increases with input signal level, so that the output power does not increase when the input power increases. Other experiments made by Soohoo[37] in a rectangular waveguide have shown that values of 10 dB of attenuation are possible. This experiment has been made with a signal at 9.6 GHz, and the minimum input power to reach at least 5 dB is 1.3 kW.

Anyway the power levels indicated in these experiments are destructive for modern solid state circuits, and for these circuits power limiters near some watts maximum are required. We may note that the values of "h_c" and "h_{cs}" are the internal fields, and in all these expressions the demagnetizing factors appear. So, it is logical to think that the type and shape of the ferrite and its place in the device can affect the performances of the power limiter. Soohoo has made an experiment in a rectangular waveguide in which the limiting action starts near one watt for a signal frequency in "x" band, i.e., 8.2 to 12.4 GHz.

Other experiments have been made trying the realization of power limiters in microstrip technology. The researchers Roome and Hair[38] have made experiments of microstrips with ferrite as substrate. They have found the critical signal magnetic field "h_{cs}" for secondary resonance is given by:

$$h_{cs} \overset{+}{=} 4\omega\Delta H_k \Big/ \left\{ \omega_m \left[3 - \left(1 + 2\omega^2 / \omega_m^2 \right) \right]^{0.5} \right\} \tag{A7.19.9}$$

and the associated critical power "W_{cs}" is:

$$W_{cs} \approx 6.414 \; w^2 h_{cs}^2 \zeta \tag{A7.19.10}$$

In the previous equation "w" is the width of microstrip in centimeters, "h_{cs}" must be measured in Oersted, and "ζ" is the microstrip impedance. Their experiments have also proved a good agreement between the values of the theoretical and practical fields. In particular, for a frequency of 2.2 GHz they have obtained a value of "h_{cs}" corresponding to a power near 2 watts.

REFERENCES

1. P. Silvestroni, Fondamenti di chimica, Veschi, Editor, Roma, 1976.
2. R. F. Soohoo, *Theory and Applications of Ferrites*. Prentice Hall, New York, 1960.
3. R. A. Waldron, *Ferrites. An Introduction for Microwave Engineers*, D. Van Nostrand Co. Ltd, 1972.
4. S. Ramo, J. R. Whinnery, T. Van Duzer, *Fields and Waves in Communication Electronics*, John Wiley & Sons, New York, 1965.
5. R. E. Collin, *Foundations for Microwave Engineering*, McGraw Hill, New York, 1992.
6. G. Gerosa, Lezioni di Microonde, University La Sapienza of Roma, Rome, 1980.
7. C. Kittel, On the theory of ferromagnetic resonance absorption, *Physical Rev.*, Rome, 73(73), 15, 1948.

8. C. Kittel, On the theory of ferromagnetic resonance absorption, *Physical Rev.*, Rome, 73(73), 15, 1948.

9. J. Green, E. Shloemann, F. Sandy, Characterization of the microwave tensor permeability of partially magnetized materials, Rome Air Development Center, 69(93), Feb. 1969.

10. J. Green, F. Sandy, C. Patton, Microwave properties of partially magnetized ferrites, Rome Air Development Center, 68(312), Aug., 1968.

11. J. Green, F. Sandy, Microwave characterization of partially magnetized ferrites, *IEEE Trans. on MTT*, 22(6), June 1974.

12. R. F. Soohoo, *Theory and Applications of Ferrites*. Prentice Hall, New York, 1960.

13. G. Matthaei, L. Young, E. M. T. Jones, *Microwave Filters, Impedance Matching Networks and Coupling Structures*, Artech House, Norwood, MA, 1980.

14. C. R. Boyd, Jr., A latching ferrite rotary field phase shifter, *Microwave Symp.*, 1, 103, 1995.

15. W. Junding, Y. Z. Xiong, M. J. Shi, G. F. Chen, M. De Yu, Analysis of twin ferrite toroidal phase shifter in grooved waveguide, *IEEE Trans. on MTT*, 616, Apr. 1994.

16. S. K. Koul, B. Bhat, *Microwave and Millimeter Wave Phase Shifters*, Vol. 1, Artech House, Norwood, MA, 1991.

17. K. J. Button, Theoretical analysis of the operation of the field displacement ferrite isolator, *IRE Trans. on MTT*, July 1958.

18. S. Weisbaum, H. Boyet, A double slab ferrite field displacement isolator at 11kmc, *Proc. of the IRE*, 554, Apr. 1956.

19. W. Junding, W. Che, Y. Xiong, Y. Wen, Operation of new type field displacement isolator in ridged waveguide, *IEEE Trans. on MTT*, 698, May 1997.

20. D. J. Massè, R. A. Pucel, Microstrip propagation on magnetic substrates. *IEEE Trans. on MTT*, MTT.20, (5), May 1972.

21. C. P. Wen, Coplanar waveguide, a surface strip transmission line suitable for non reciprocal gyromagnetic device applications, *IEEE Trans. on MTT*, MTT.17, (12), Dec. 1969.

22. C. P. Wen, Coplanar waveguide, a surface strip transmission line suitable for non reciprocal gyromagnetic device applications, *IEEE Trans. on MTT*, MTT.17, (12), Dec. 1969.

23. C. E. Fay R. L. Comstock, Operation of the ferrite junction circulator, *IEEE Trans. on MTT*, 15, Jan. 1965.

24. Y. S.Wu, F. J. Rosenbaum, Wide band operation of microstrip circulators, *IEEE Trans. on MTT*, 22(10), Oct. 1974.

25. B. A. Auld, The synthesis of symmetrical waveguide circulators, *IRE Trans. on MTT*, 238, Apr. 1959.

26. U. Milano, J. H. Saunders, L. Davis, Jr., A Y junction stripline circulator, *IRE Trans. on MTT*, 346, May 1960.

27. E. K. N. Yung, R. S. Chen, K. Wu, D. X. Wang, Analysis and development of millimeter wave waveguide junction circulator with a ferrite sphere, *IEEE Trans. on MTT*, 1721, Nov. 1998.

28. J. L. Melchor, W. P. Ayres, P. H. Vartanian, Microwave frequency doubling from 9 to 18 GHz in ferrites, *Proc. of the IRE*, 643, May 1957.

29. D. Jaffe, J. C. Cacheris, N. Karayianis, Ferrite microwave detector, *Proc. of the IRE*, March 1958.

30. N. G. Sakiotis, H. N. Chait, M. L. Kales, Non linearity of propagation in ferrite media, *Proc. of the IRE*, 1011, Aug. 1955.

31. C. Herring, C. Kittel, On the theory of spin waves in ferromagnetic media, *Physical Rev.*, 81(5), 869, 1951.

32. H. Suhl, The non linear behavior of ferrites at high microwave signal levels, *Proc. of the IRE*, 1270, Oct. 1956.

33. H. Suhl, The non linear behavior of ferrites at high microwave signal levels, *Proc. of the IRE*, 1270, Oct. 1956.

34. H. Suhl, The non linear behavior of ferrites at high microwave signal levels, *Proc. of the IRE*, 1270, Oct. 1956.

35. H. Suhl, The non linear behavior of ferrites at high microwave signal levels, *Proc. of the IRE*, 1270, Oct. 1956.

36. N. G. Sakiotis, H. N. Chait, M. L. Kales, Non linearity of propagation in ferrite media, *Proc. of the IRE*, 1011, Aug. 1955.

37. R. F. Soohoo, *Theory and Applications of Ferrites*. Prentice Hall, New York, 1960.

38. G. T. Roome, H. A. Hair, Thin ferrite devices for microwave integrated circuits, *IEEE Trans. on MTT*, 16(7), 411, 1968.

Symbols, Operators Definitions and Analytical Expressions

A8.1 INTRODUCTION

In many parts of the book we use some symbols which we think are generally known to the reader. Other symbols are defined by the author of this book, because he thinks that they can help the reader to remember or to understand the analytical relations where they appear.

In addition, in this book we use some mathematical operator that should also be known by the reader.

In any case, in this Appendix we will collect all these symbols and mathematical operators, thinking to help the reader to read this book.

Of course, this book is neither a mathematical nor a physical one, and consequently we will only indicate the operators we use in the text, leaving to other books the rigorous verifications and related physical aspects.

A8.2 DEFINITIONS OF SYMBOLS AND ABBREVIATIONS

Here symbol character identification follows, with its name and its definition. This will be done dividing the objects by argument.

A8.2.1 Associated to Vectors

_ (underline): vector. Every time a symbol is underlined it is represented as a vector.

underline + subscript 0: versor. The versor is a particular vector which only gives a direction. Consequently, it has no dimensions and its modulus is the unity. For instance, if the vector "$\underline{e}$" is directed along the "x" axis, then we can write $\underline{e} = e_x \underline{x}_0$, with "$\underline{x}_0$" the versor of the "x" axis.

$\perp$: orthogonality between vectors. So, if "$\underline{v}_1$" and "$\underline{v}_2$" are two vectors, then "$\underline{v}_1 \perp \underline{v}_2$" means that "$\underline{v}_1$" and "$\underline{v}_2$" are orthogonal.

$\otimes$: vector product. If "$\underline{v}_1$" and "$\underline{v}_2$" are two vectors, then "$\underline{v}_1 \perp \underline{v}_2$" represents the vector product between "$\underline{v}_1$" and "$\underline{v}_2$." The result is of course a vector.

"$\bullet$": scalar product. If "$\underline{v}_1$" and "$\underline{v}_2$" are two vectors, then "$\underline{v}_1 \bullet \underline{v}_2$" represents the scalar product between "$\underline{v}_1$" and "$\underline{v}_2$." The result is of course a scalar. Sometimes the symbol "$\bullet$" is applied at the point of an arrow representing a vector. In this case we mean that the vector is directed with the arrow toward us; it is arriving toward us.

x : Sometimes the letter "x" is applied at the end of an arrow representing a vector. In this case we mean that the vector is going away from us.

A8.2.2 Mathematical

$\equiv$: coincidence.   This symbol is inserted in any equation which results as a particular case of a more general expression. For example, if we have a general equation $y = 2x + 1$ if $x = 2$ then $y \equiv 5$.

$\doteq$: equality by definition.   This symbol is used every time that an expression is defined simply with another symbol or name. For example $\cosh(z) \doteq (e^z + e^{-z})/2$.

$= !$: forced equality.   This symbol is employed in any equation that must be assured to permit a consequence. For example, if we want Equation A2.8.1 to be equal to zero, then $x = !\ -1/2$.

$\neq !$: forced inequality.   By extension of the forced equality.

tg, tan: tangent function.   If an "h" is appended, the hyperbolic counterparts are involved.

atg, atan: arc tangent function.   If an "h" is appended, the hyperbolic counterparts are involved.

ctg, cotg: cotangent function.   If an "h" is appended, the hyperbolic counterparts are involved.

actg, acot: arc cotangent function.   If an "h" is appended, the hyperbolic counterparts are involved.

quantity in square brackets: dimensions.   For example, if "R" is a resistance, then $[R]$ = Ohm.

ln, ℓn: natural logarithm

log, lgt: base 10 logarithm

superscript *: complex conjugate.   Every time a star "*" is used as superscript to a letter, we refer to its complex conjugate, i.e., "n*" is the complex conjugate of "n."

$\rightarrow$: becomes.   The expression on the left of the arrows becomes the expression on the right of the arrow.

$'$ (prime): derivative.   For example, given the function "f(x)" then $df(x)/dx \doteq f'(x)$. The symbol " $'$ " as derivative is obviously used where we think no misunderstanding can arise.

$\wedge$: raise to a power.   For example a^2 means "a" raised to the square.

exp( ): natural number "e" raised to ( ).

A8.2.3 General

ul, u.l.: unit length

e.m.: electromagnetic

c or v_0: vacuum light speed.   Numerically $c \doteq 1/(\mu_0\varepsilon_0)^{0.5} = v_0$

t.l.: transmission line.

P.C.B.: printed circuit board.

ω: angular frequency, measured in radians/sec

f: frequency.   It is known that $\omega \doteq 2\pi f$.

λ: wavelength

ζ_v: vacuum characteristic impedance $\zeta_v = 120\pi\ \Omega$

ε_r: relative dielectric constant*

---

* The dielectric constant is also called "permittivity."

A8.3.2 ∇^2: Laplacian or Square Delta

The Laplacian is simply defined as:

$$\underline{\nabla} \bullet \underline{\nabla} \doteq \nabla^2 [\qquad\qquad (A8.3.2)$$

Also this operator is strongly dependent on the reference system employed in the region under study. Note as ∇^2 is a scalar operator and for this reason when it operates on vectors, it does not need a scalar or vector product. For this reason, when ∇^2 operates on scalar the result is a scalar, when it operates on vector the result is a vector. The expressions of ∇^2 will be given later.

A8.3.3 Operator identities.

The most used identities applicable to "$\underline{\nabla}$" are:

$$\underline{\nabla} \otimes \underline{\nabla} \otimes \underline{v} = \underline{\nabla}\underline{\nabla} \bullet \underline{v} - \nabla^2 \underline{v} \qquad\qquad (A8.3.3)$$

$$\underline{\nabla} \bullet \underline{\nabla} \otimes \underline{v} = 0 \qquad\qquad (A8.3.4)$$

$$\underline{\nabla} \otimes \underline{\nabla}s = 0 \qquad\qquad (A8.3.5)$$

$$\underline{\nabla} \otimes s\underline{v} = \underline{\nabla}s \otimes \underline{v} + s\underline{\nabla} \otimes \underline{v} \qquad\qquad (A8.3.6)$$

$$\underline{\nabla} \bullet s\underline{v} = \underline{v} \bullet \underline{\nabla}s + s\underline{\nabla} \bullet \underline{v} \qquad\qquad (A8.3.7)$$

$$\underline{\nabla} \bullet \underline{A} \otimes \underline{B} = \underline{B} \bullet \underline{\nabla} \otimes \underline{A} - \underline{A} \bullet \underline{\nabla} \otimes \underline{B} \qquad\qquad (A8.3.8)$$

$$\underline{A} \otimes \underline{B} \otimes \underline{C} = \underline{B}(\underline{A} \bullet \underline{C}) - \underline{C}(\underline{A} \bullet \underline{B}) \qquad\qquad (A8.3.9)$$

$$\underline{A} \bullet \underline{B} \otimes \underline{C} = \underline{C} \bullet \underline{A} \otimes \underline{B} = \underline{B} \bullet \underline{C} \otimes \underline{A} \qquad\qquad (A8.3.10)$$

A8.4 DELTA OPERATOR FUNCTIONS IN CARTESIAN ORTHOGONAL COORDINATE SYSTEM

In this reference system a vector "v" is indicated as:

$$\underline{v} = v_x \underline{x}_0 + v_y \underline{y}_0 + v_z \underline{z}_0 \qquad\qquad (A8.4.1)$$

where:

$$v_x = v_x(x, y, z), \quad v_y = v_y(x, y, z), \quad v_z = v_z(x, y, z)$$

and a scalar is represented in the most general way as $f = f(x, y, z)$.

The "delta" is given by:

$$\underline{\nabla} \doteq \frac{\partial}{\partial_x}\underline{x}_0 + \frac{\partial}{\partial_y}\underline{y}_0 + \frac{\partial}{\partial_z}\underline{z}_0 \qquad\qquad (A8.4.2)$$

The "Laplacian" is:

$$\nabla^2 \doteq \frac{\partial^2}{\partial x^2} + \frac{\partial^2}{\partial y^2} + \frac{\partial^2}{\partial z^2} \qquad \text{(A8.4.3)}$$

As was stated previously, the Laplacian may be applied to scalar, returning a scalar, or to a vector, returning a vector. The application of this operator to scalars is immediate, resulting in:

$$\nabla^2(f) \doteq \frac{\partial^2 f}{\partial x^2} + \frac{\partial^2 f}{\partial y^2} + \frac{\partial^2 f}{\partial z^2} \qquad \text{(A8.4.4)}$$

and the application to vector "$\underline{v}$" is defined as:

$$\nabla^2(\underline{v}) \doteq \nabla^2(v_x)\underline{x}_0 + \nabla^2(v_y)\underline{y}_0 + \nabla^2(v_z)\underline{z}_0 \qquad \text{(A8.4.5)}$$

In a lot of e.m. problems the geometric structure has a cylindrical symmetry. In this case it is always possible to evaluate "$\underline{\nabla}$" and "∇^2" as the sum of a transverse operator plus a longitudinal one. For example, if "x" and "y" are the transverse coordinates we can write:

$$\underline{\nabla} \doteq \underline{\nabla}_t + \frac{\partial}{\partial z}\underline{z}_0 \qquad \text{(A8.4.6)}$$

$$\nabla^2 \doteq \nabla_t^2 + \frac{\partial^2}{\partial z^2} \qquad \text{(A8.4.7)}$$

where:

$$\underline{\nabla}_t \doteq \frac{\partial}{\partial x}\underline{x}_0 + \frac{\partial}{\partial y}\underline{y}_0 \qquad \text{(A8.4.8)}$$

$$\underline{\nabla}_t^2 \doteq \frac{\partial^2}{\partial x^2} + \frac{\partial^2}{\partial y^2} \qquad \text{(A8.4.9)}$$

An example where we have used such composition is the case of rectangular waveguides, as indicated in Figure A8.4.1. Quite often, when such a composition is employed the function where delta operates has the longitudinal component only dependent on the longitudinal coordinate "z." When it is true the partial derivative in A8.4.6 and A8.4.7 can be replaced with the absolute derivative.

The "gradient" is of course the direct application to a scalar of the "delta" operator indicated in A8.4.2, i.e.:

$$\underline{\nabla}(f) \doteq \frac{\partial f}{\partial x}\underline{x}_0 + \frac{\partial f}{\partial y}\underline{y}_0 + \frac{\partial f}{\partial z}\underline{z}_0 \qquad \text{(A8.4.10)}$$

The "divergence" is:

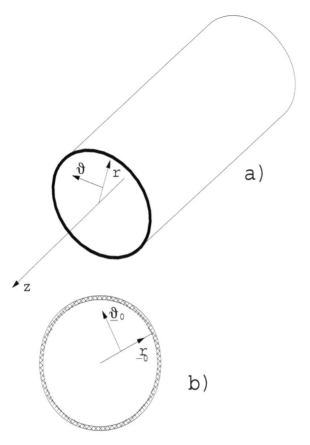

Figure A8.5.1

A8.6 DELTA OPERATOR FUNCTIONS IN A SPHERICAL COORDINATE SYSTEM

In this reference system, represented in Figure A8.6.1, a vector "V" is indicated as:

$$\underline{v} = v_r \underline{r}_0 + v_\theta \underline{\theta}_0 + v_\varphi \underline{\varphi}_0 \tag{A8.6.1}$$

where:

$$v_r = v_r(r,\theta,\varphi), \quad v_\theta = v_\theta(r,\theta,\varphi), \quad v_\varphi = v_\varphi(r,\theta,\varphi)$$

and a scalar is represented in the most general way as $f = f(r,\theta,\varphi)$.
The "delta" is given by:

$$\underline{\nabla} \doteq \frac{\partial}{\partial r}\underline{r}_0 + \frac{\partial}{r\partial\theta}\underline{\theta}_0 + \frac{\partial}{r\,sen\theta\partial\varphi}\underline{\varphi}_0 \tag{A8.6.2}$$

Concerning Laplacian, it is necessary to define if it is applied to a scalar or to a vector. So we have:

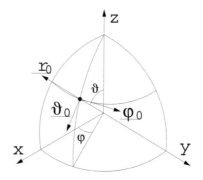

Fiure A8.6.1

Laplacian applied to a scalar:

$$\nabla^2 f \doteq \frac{\partial}{r^2 \partial r}\left(r^2 \frac{\partial f}{\partial r}\right) + \frac{\partial}{r^2 sen\theta \partial\theta}\left(sen\theta \frac{\partial f}{\partial\theta}\right) + \frac{\partial^2 f}{(rsen\theta)^2 \partial\varphi^2} \qquad (A8.6.3)$$

Laplacian applied to a vector:
in this case, it is convenient to apply the vector identity given in A8.3.3, from which we have:

$$\nabla^2 \underline{v} = \underline{\nabla}\underline{\nabla} \bullet \underline{v} - \underline{\nabla} \otimes \underline{\nabla} \otimes \underline{v} \qquad (A8.3.3)$$

where the divergence and curl will be given later.

Also in this case it is always possible to evaluate "$\underline{\nabla}$" and "∇^2" as the sum of a transverse operator plus a longitudinal one. For example, if "φ" and "θ" are the transverse coordinates we can write:

$$\underline{\nabla} \doteq \underline{\nabla}_t + \frac{\partial}{\partial r}\underline{r}_0 \qquad (A8.6.4)$$

$$\underline{\nabla}^2 \doteq \underline{\nabla}_t^2 + \frac{\partial^2}{\partial r^2} \qquad (A8.6.5)$$

where:

$$\underline{\nabla}_t \doteq \frac{\partial}{r\partial\theta}\underline{\theta}_0 + \frac{\partial}{rsen\theta\partial\varphi}\underline{\varphi}_0 \qquad (A8.6.6)$$

$$\nabla_t^2 \doteq \frac{\partial}{r^2 sen\theta\partial\theta}\left(sen\theta \frac{\partial}{\partial\theta}\right) + \frac{\partial^2}{r^2(sen\theta)^2 \partial\varphi^2} \qquad (A8.6.7)$$

The gradient is the direct application to a scalar of the delta operator indicated in A8.6.2, i.e.:

$$\underline{\nabla}f \doteq \frac{\partial f}{\partial r}\underline{r}_0 + \frac{\partial f}{r\partial\theta}\underline{\theta}_0 + \frac{\partial f}{rsen\theta\partial\varphi}\underline{\varphi}_0 \qquad (A8.6.8)$$

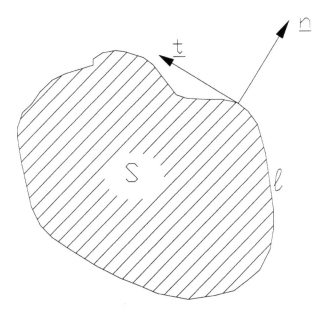

Figure A8.7.1

c. Green First Identity

$$\oint_S g\underline{\nabla}f \bullet \underline{n}dS = \int_Q \left(\underline{\nabla}g \bullet \underline{\nabla}f + g\nabla^2 f\right)dQ \tag{A8.7.3}$$

d. Green Second Identity

$$\oint_S \left(g\underline{\nabla}f - f\underline{\nabla}g\right) \bullet \underline{n}dS = \int_Q \left(g\underline{\nabla}^2 f - f\nabla^2 g\right)dQ \tag{A8.7.4}$$

For both points c and d the definition of "S," "Q," and "n" is the same we have used for the Gauss theorem.

e. Green Two Dimensional First Identity

$$\oint_\ell g\underline{\nabla}_t f \bullet \underline{n}d\ell = \int_S \left(\underline{\nabla}_t g \bullet \underline{\nabla}f + g\nabla_t^2 f\right)dS \tag{A8.7.5}$$

f. Green Two Dimensional Second Identity

$$\oint_\ell \left(g\underline{\nabla}_t f - f\underline{\nabla}_t g\right) \bullet \underline{n}d\ell = \int_S \left(g\nabla_t^2 f - f\nabla_t^2 g\right)dS \tag{A8.7.6}$$

For both points e and f the definition of "ℓ," "S," and "n" is the same we have used for the Stokes theorem.

A8.8 ELLIPTIC INTEGRALS AND THEIR APPROXIMATIONS

In many chapters of this text we have encountered the complete elliptic integral of first kind, in particular in expressions where the ratio of these integrals appears. This integral is a function of a parameter "p," and one definition* is:

$$K(p) = \int_0^1 \frac{dt}{\sqrt{(1-t^2)(1-p^2t^2)}} \tag{A8.8.1}$$

This integral appears in solution of electromagnetic problems through the conformal transformation method, studied in Appendix A3. Depending on the particular e.m. problem, the parameter "p" can assume different expressions. Associated to "p" there is the complementary parameter "p'," defined as:**

$$p' \overset{\perp}{=} (1-p^2)^{0.5} \tag{A8.8.2}$$

The ratio $K(p)/K(p')$ can be calculated using the tabulated values of the elliptic integrals,[1] but they have been approximated by the researcher W. Hilberg[2] with closed form expressions as follows:

$$0 \le p \le 1/\sqrt{2} \quad \text{then} \quad \frac{K(p')}{K(p)} = \frac{1}{\pi} \ln\left[2 \frac{1+(p')^{0.5}}{1-(p')^{0.5}} \right] \tag{A8.8.3}$$

$$1/\sqrt{2} < p \le 1 \quad \text{then} \quad \frac{K(p')}{K(p)} = \pi \left\{ \ln\left[2 \frac{1+(p)^{0.5}}{1-(p)^{0.5}} \right] \right\}^{-1} \tag{A8.8.4}$$

These expressions are widely used in our text. Sometimes, the range limit $1/\sqrt{2}$ is varied to 0.5, for a simple notation. This does not affect the approximation much.

The complete elliptic integral of first kind is a particular case of the "elliptic integral of first kind,"[3] indicated with:

$$F(\xi,p) = \int_0^\xi \left[(1-y^2)(1-p^2y^2) \right]^{-0.5} dy \tag{A8.8.5}$$

and the following relationships hold:

$$K(p) = F(1,p) \tag{A8.8.6}$$

$$K(p) = F(1/p, p) = K(p) + jK(p') \tag{A8.8.7}$$

---

* Our text is not a book of mathematical analysis. So we will define the elliptic integral in one of the most used expressions. Rigorous definitions of elliptic integrals can be found in mathematical analysis books or mathematical handbooks. See for example: M. Abramowitz, I. A. Stegun, Handbook of Mathematical Functions, Dover, New York, 1970.
** Sometimes in literature "K(p')" is indicated with "K'(p)." This is only a different symbology since operatively the integral is evaluated for $p' = (1-p^2)^{0.5}$.

REFERENCES

1. M. Abramowitz, I. A. Stegun, Handbook of Mathematical Functions. Dover, New York, 1970.
2. W. Hilberg, From approximations to exact relations for characteristic impedances, *IEEE Trans. on MTT*, May 1969.
3. M. Abramovitz, I. A. Stegun, Handbook of Mathematical Functions, Dover, New York, 1970.

INDEX